CALCULUS
Theory and Applications
Volume 2

CALCULUS
Theory and Applications

Volume 2

Kenneth Kuttler
Brigham Young University, USA

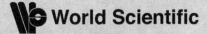
World Scientific

NEW JERSEY · LONDON · SINGAPORE · BEIJING · SHANGHAI · HONG KONG · TAIPEI · CHENNAI

Published by

World Scientific Publishing Co. Pte. Ltd.

5 Toh Tuck Link, Singapore 596224

USA office: 27 Warren Street, Suite 401-402, Hackensack, NJ 07601

UK office: 57 Shelton Street, Covent Garden, London WC2H 9HE

British Library Cataloguing-in-Publication Data
A catalogue record for this book is available from the British Library.

CALCULUS — Volume 2
Theory and Applications

ISBN-13 978-981-4324-27-4
ISBN-10 981-4324-27-2
ISBN-13 978-981-4329-70-5 (pbk)
ISBN-10 981-4329-70-3 (pbk)

Printed in Singapore.

Preface

This volume is mainly devoted to multivariable calculus. It is generally easier to consider linear functions than nonlinear ones. Therefore, the linear functions are presented first in terms of basic linear algebra. Then this is used to unify the presentation of nonlinear functions. All theorems are proved although the most difficult ones are in the appendices along with some other applications such as curvilinear coordinates.

Like Volume I, really difficult theoretical sections have a dragon at the beginning. These sections are optional and are there for anyone who is interested. It seems to me that a math book should provide explanations of the theorems, but these explanations can be skipped if there is no interest.

Supplementary material for this text including routine exercise sets is available at http://www.math.byu.edu/~klkuttle/CalculusMaterials.

I am grateful to World Scientific for publishing this volume and also to Kate Phillips, and Maple for drawing the pictures.

Contents

Chapter 1

Matrices And Linear Transformations

1.1 Matrix Arithmetic

1.1.1 *Addition And Scalar Multiplication Of Matrices*

Numbers are also called **scalars**. In this book, scalars will be real numbers or complex numbers.

A **matrix** is a rectangular array of numbers. Several of them are referred to as **matrices**. For example, here is a matrix.

$$\begin{pmatrix} 1 & 2 & 3 & 4 \\ 5 & 2 & 8 & 7 \\ 6 & -9 & 1 & 2 \end{pmatrix}$$

The size or dimension of a matrix is defined as $m \times n$ where m is the number of rows and n is the number of columns. The above matrix is a 3×4 matrix because there are three rows and four columns. The first row is (1 2 3 4), the second row is (5 2 8 7) and so forth. The first column is $\begin{pmatrix} 1 \\ 5 \\ 6 \end{pmatrix}$. When specifying the size of a matrix, you always list the number of rows before the number of columns. Also, you can remember the columns are like columns in a Greek temple. They stand upright while the rows just lay there like rows made by a tractor in a plowed field. Entries of the matrix are identified according to position in the matrix. For example, 8 is in position 2, 3 because it is in the second row and the third column. You might remember that you always list the rows before the columns by using the phrase **Row**man Catholic.

The symbol (a_{ij}) refers to a matrix. The entry in the ith row and the jth column of this matrix is denoted by a_{ij}. Using this notation on the above matrix, $a_{23} = 8, a_{32} = -9, a_{12} = 2$, etc.

There are various operations which are done on matrices. Matrices can be added, multiplied by a scalar, and multiplied by other matrices. To illustrate scalar multiplication, consider the following example in which a matrix is being multiplied by the scalar 3.

$$3\begin{pmatrix} 1 & 2 & 3 & 4 \\ 5 & 2 & 8 & 7 \\ 6 & -9 & 1 & 2 \end{pmatrix} = \begin{pmatrix} 3 & 6 & 9 & 12 \\ 15 & 6 & 24 & 21 \\ 18 & -27 & 3 & 6 \end{pmatrix}.$$

The new matrix is obtained by multiplying every entry of the original matrix by the given scalar. If A is an $m \times n$ matrix, $-A$ equals $(-1)A$ because $A + (-1)A = 0$.

Two matrices must be the same size to be added. The sum of two matrices is a matrix which is obtained by adding the corresponding entries. Thus

$$\begin{pmatrix} 1 & 2 \\ 3 & 4 \\ 5 & 2 \end{pmatrix} + \begin{pmatrix} -1 & 4 \\ 2 & 8 \\ 6 & -4 \end{pmatrix} = \begin{pmatrix} 0 & 6 \\ 5 & 12 \\ 11 & -2 \end{pmatrix}.$$

Two matrices are equal exactly when they are the same size and the corresponding entries are identical. Thus

$$\begin{pmatrix} 0 & 0 \\ 0 & 0 \\ 0 & 0 \end{pmatrix} \neq \begin{pmatrix} 0 & 0 \\ 0 & 0 \end{pmatrix}$$

because they are different sizes. As noted above, you write (c_{ij}) for the matrix C whose ij^{th} entry is c_{ij}. In doing arithmetic with matrices you must define what happens in terms of the c_{ij} sometimes called the **entries** of the matrix or the **components** of the matrix.

The above discussion stated for general matrices is given in the following definition.

Definition 1.1. (Scalar Multiplication) If $A = (a_{ij})$ and k is a scalar, then $kA = (ka_{ij})$.

Example 1.1. $7\begin{pmatrix} 2 & 0 \\ 1 & -4 \end{pmatrix} = \begin{pmatrix} 14 & 0 \\ 7 & -28 \end{pmatrix}.$

Definition 1.2. (Addition) If $A = (a_{ij})$ and $B = (b_{ij})$ are two $m \times n$ matrices. Then $A + B = C$ where

$$C = (c_{ij})$$

for $c_{ij} = a_{ij} + b_{ij}$.

Example 1.2.

$$\begin{pmatrix} 1 & 2 & 3 \\ 1 & 0 & 4 \end{pmatrix} + \begin{pmatrix} 5 & 2 & 3 \\ -6 & 2 & 1 \end{pmatrix} = \begin{pmatrix} 6 & 4 & 6 \\ -5 & 2 & 5 \end{pmatrix}.$$

To save on notation, we will often use A_{ij} to refer to the ij^{th} entry of the matrix A.

Definition 1.3. (The **zero matrix**) The $m \times n$ zero matrix is the $m \times n$ matrix having every entry equal to zero. It is denoted by 0.

Example 1.3. The 2×3 zero matrix is $\begin{pmatrix} 0 & 0 & 0 \\ 0 & 0 & 0 \end{pmatrix}$.

Note there are 2×3 zero matrices, 3×4 zero matrices, etc. In fact there is a zero matrix for every size.

Definition 1.4. (Equality of matrices) Let A and B be two matrices. Then $A = B$ means that the two matrices are of the same size and for $A = (a_{ij})$ and $B = (b_{ij})$, $a_{ij} = b_{ij}$ for all $1 \leq i \leq m$ and $1 \leq j \leq n$.

The following properties of matrices can be easily verified. You should do so.

- Commutative law of addition.

$$A + B = B + A, \tag{1.1}$$

- Associative law for addition.

$$(A + B) + C = A + (B + C), \tag{1.2}$$

- Existence of an additive identity.

$$A + 0 = A, \tag{1.3}$$

- Existence of an additive inverse.

$$A + (-A) = 0. \tag{1.4}$$

Also for α, β scalars, the following additional properties hold.

- Distributive law over matrix addition.

$$\alpha (A + B) = \alpha A + \alpha B, \tag{1.5}$$

- Distributive law over scalar addition.

$$(\alpha + \beta) A = \alpha A + \beta A, \tag{1.6}$$

- Associative law for scalar multiplication.

$$\alpha (\beta A) = \alpha \beta (A), \tag{1.7}$$

- Rule for multiplication by 1.

$$1A = A. \tag{1.8}$$

As an example, consider the commutative law of addition. Let $A + B = C$ and $B + A = D$. Why is $D = C$?

$$C_{ij} = A_{ij} + B_{ij} = B_{ij} + A_{ij} = D_{ij}.$$

Therefore, $C = D$ because the ij^{th} entries are the same. Note that the conclusion follows from the commutative law of addition of numbers.

1.1.2 *Multiplication Of Matrices*

Definition 1.5. Matrices which are $n \times 1$ or $1 \times n$ are called **vectors** and are often denoted by a bold letter. Thus the $n \times 1$ matrix

$$\mathbf{x} = \begin{pmatrix} x_1 \\ \vdots \\ x_n \end{pmatrix}$$

is also called a **column vector.** The $1 \times n$ matrix

$$(x_1 \cdots x_n)$$

is called a **row vector**.

Although the following description of matrix multiplication may seem strange, it is in fact the most important and useful of the matrix operations. To begin with, consider the case where a matrix is multiplied by a column vector. We will illustrate the general definition by first considering a special case.

$$\begin{pmatrix} 1 & 2 & 3 \\ 4 & 5 & 6 \end{pmatrix} \begin{pmatrix} 7 \\ 8 \\ 9 \end{pmatrix} = ?$$

This equals

$$7 \begin{pmatrix} 1 \\ 4 \end{pmatrix} + 8 \begin{pmatrix} 2 \\ 5 \end{pmatrix} + 9 \begin{pmatrix} 3 \\ 6 \end{pmatrix} = \begin{pmatrix} 50 \\ 122 \end{pmatrix}$$

In general, here is the definition of how to multiply an $(m \times n)$ matrix times a $(n \times 1)$ matrix.

Definition 1.6. Let $A = (A_{ij})$ be an $m \times n$ matrix

$$A = \begin{pmatrix} \mathbf{a}_1 & \mathbf{a}_2 & \cdots & \mathbf{a}_n \end{pmatrix}$$

where the ith column of A is denoted by $\mathbf{a}_i$, and let $\mathbf{v}$ be an $n \times 1$ matrix

$$\mathbf{v} = \begin{pmatrix} v_1 \\ \vdots \\ v_n \end{pmatrix}$$

Then $A\mathbf{v}$ is an $m \times 1$ matrix equal to

$$v_1 \mathbf{a}_1 + v_2 \mathbf{a}_2 + \cdots + v_n \mathbf{a}_n = \sum_{k=1}^{n} v_k \mathbf{a}_k$$

It follows from the observation that the jth column of A is

$$\begin{pmatrix} A_{1j} \\ A_{2j} \\ \vdots \\ A_{mj} \end{pmatrix},$$

that the above sum is of the form

$$v_1 \begin{pmatrix} A_{11} \\ A_{21} \\ \vdots \\ A_{m1} \end{pmatrix} + v_2 \begin{pmatrix} A_{12} \\ A_{22} \\ \vdots \\ A_{m2} \end{pmatrix} + \cdots + v_k \begin{pmatrix} A_{1n} \\ A_{2n} \\ \vdots \\ A_{mn} \end{pmatrix}$$

It follows that the ith entry of the $m \times 1$ matrix or column vector which results is

$$\sum_{j=1}^{n} A_{ij} v_j$$

Here is another example.

Example 1.4. Compute

$$\begin{pmatrix} 1 & 2 & 1 & 3 \\ 0 & 2 & 1 & -2 \\ 2 & 1 & 4 & 1 \end{pmatrix} \begin{pmatrix} 1 \\ 2 \\ 0 \\ 1 \end{pmatrix}.$$

First of all, this is of the form $(3 \times 4)(4 \times 1)$ and so the result should be a (3×1). Note how the inside numbers cancel. Then this equals

$$1 \begin{pmatrix} 1 \\ 0 \\ 2 \end{pmatrix} + 2 \begin{pmatrix} 2 \\ 2 \\ 1 \end{pmatrix} + 0 \begin{pmatrix} 1 \\ 1 \\ 4 \end{pmatrix} + 1 \begin{pmatrix} 3 \\ -2 \\ 1 \end{pmatrix} = \begin{pmatrix} 8 \\ 2 \\ 5 \end{pmatrix}$$

The next task is to multiply an $m \times n$ matrix times an $n \times p$ matrix. Before doing so, the following may be helpful.

For A and B matrices, in order to form the product AB the number of columns of A must equal the number of rows of B.

$$\overbrace{(m \times \quad n)}^{\text{these must match!}} (n \times p \quad) = m \times p$$

Note the two outside numbers give the size of the product. Remember:

> **The two middle numbers MUST match**

Definition 1.7. When the number of columns of A equals the number of rows of B the two matrices are said to be **conformable** and the product AB is obtained as follows. Let A be an $m \times n$ matrix and let B be an $n \times p$ matrix, B of the form

$$B = \begin{pmatrix} \mathbf{b}_1 & \cdots & \mathbf{b}_p \end{pmatrix}$$

where $\mathbf{b}_k$ is an $n \times 1$ matrix or column vector. Then the $m \times p$ matrix AB is defined as follows:

$$AB \equiv \begin{pmatrix} A\mathbf{b}_1 & \cdots & A\mathbf{b}_p \end{pmatrix} \tag{1.9}$$

where $A\mathbf{b}_k$ is an $m \times 1$ matrix or column vector which gives the kth column of AB.

Example 1.5. Multiply the following.

$$\begin{pmatrix} 1 & 2 & 1 \\ 0 & 2 & 1 \end{pmatrix} \begin{pmatrix} 1 & 2 & 0 \\ 0 & 3 & 1 \\ -2 & 1 & 1 \end{pmatrix}$$

The first thing you need to check before doing anything else is whether it is possible to do the multiplication. The first matrix is a 2×3 and the second matrix is a 3×3. Therefore, is it possible to multiply these matrices. According to the above discussion it should be a 2×3 matrix of the form

$$\left(\overbrace{\begin{pmatrix} 1 & 2 & 1 \\ 0 & 2 & 1 \end{pmatrix} \begin{pmatrix} 1 \\ 0 \\ -2 \end{pmatrix}}^{\text{First column}}, \overbrace{\begin{pmatrix} 1 & 2 & 1 \\ 0 & 2 & 1 \end{pmatrix} \begin{pmatrix} 2 \\ 3 \\ 1 \end{pmatrix}}^{\text{Second column}}, \overbrace{\begin{pmatrix} 1 & 2 & 1 \\ 0 & 2 & 1 \end{pmatrix} \begin{pmatrix} 0 \\ 1 \\ 1 \end{pmatrix}}^{\text{Third column}} \right)$$

You know how to multiply a matrix times a vector and so you do so to obtain each of the three columns. Thus

$$\begin{pmatrix} 1 & 2 & 1 \\ 0 & 2 & 1 \end{pmatrix} \begin{pmatrix} 1 & 2 & 0 \\ 0 & 3 & 1 \\ -2 & 1 & 1 \end{pmatrix} = \begin{pmatrix} -1 & 9 & 3 \\ -2 & 7 & 3 \end{pmatrix}.$$

Example 1.6. Multiply the following.

$$\begin{pmatrix} 1 & 2 & 0 \\ 0 & 3 & 1 \\ -2 & 1 & 1 \end{pmatrix} \begin{pmatrix} 1 & 2 & 1 \\ 0 & 2 & 1 \end{pmatrix}$$

First check if it is possible. This is of the form $(3 \times 3)\,(2 \times 3)$. Aren't these the same two matrices considered in the previous example? Yes they are. It is just that here they are in a different order. This shows something you must always remember about matrix multiplication.

$$\boxed{\textbf{Order Matters!}}$$

This is very different than multiplication of numbers!

1.1.3 The ij^{th} Entry Of A Product

It is important to describe matrix multiplication in terms of entries of the matrices. What is the ij^{th} entry of AB? It would be the ith entry of the jth column of AB. Thus it would be the ith entry of $A\mathbf{b}_j$. Now

$$\mathbf{b}_j = \begin{pmatrix} B_{1j} \\ \vdots \\ B_{nj} \end{pmatrix}$$

and from the above definition, the ith entry is

$$\sum_{k=1}^{n} A_{ik} B_{kj}. \tag{1.10}$$

In terms of pictures of the matrix, you are doing

$$\begin{pmatrix} A_{11} & A_{12} & \cdots & A_{1n} \\ A_{21} & A_{22} & \cdots & A_{2n} \\ \vdots & \vdots & & \vdots \\ A_{m1} & A_{m2} & \cdots & A_{mn} \end{pmatrix} \begin{pmatrix} B_{11} & B_{12} & \cdots & B_{1p} \\ B_{21} & B_{22} & \cdots & B_{2p} \\ \vdots & \vdots & & \vdots \\ B_{n1} & B_{n2} & \cdots & B_{np} \end{pmatrix}$$

Then as explained above, the jth column is of the form

$$\begin{pmatrix} A_{11} & A_{12} & \cdots & A_{1n} \\ A_{21} & A_{22} & \cdots & A_{2n} \\ \vdots & \vdots & & \vdots \\ A_{m1} & A_{m2} & \cdots & A_{mn} \end{pmatrix} \begin{pmatrix} B_{1j} \\ B_{2j} \\ \vdots \\ B_{nj} \end{pmatrix}$$

which is an $m \times 1$ matrix or column vector which equals

$$\begin{pmatrix} A_{11} \\ A_{21} \\ \vdots \\ A_{m1} \end{pmatrix} B_{1j} + \begin{pmatrix} A_{12} \\ A_{22} \\ \vdots \\ A_{m2} \end{pmatrix} B_{2j} + \cdots + \begin{pmatrix} A_{1n} \\ A_{2n} \\ \vdots \\ A_{mn} \end{pmatrix} B_{nj}.$$

The second entry of this $m \times 1$ matrix is

$$A_{21} B_{ij} + A_{22} B_{2j} + \cdots + A_{2n} B_{nj} = \sum_{k=1}^{m} A_{2k} B_{kj}.$$

Similarly, the ith entry of this $m \times 1$ matrix is

$$A_{i1} B_{ij} + A_{i2} B_{2j} + \cdots + A_{in} B_{nj} = \sum_{k=1}^{m} A_{ik} B_{kj}.$$

This shows the following definition for matrix multiplication in terms of the ij^{th} entries of the product coincides with Definition 1.7.

Definition 1.8. Let $A = (A_{ij})$ be an $m \times n$ matrix and let $B = (B_{ij})$ be an $n \times p$ matrix. Then AB is an $m \times p$ matrix and

$$(AB)_{ij} = \sum_{k=1}^{n} A_{ik} B_{kj}. \tag{1.11}$$

A useful description of this is as follows. The ij^{th} entry of AB equals the real dot product of the ith row of A with the jth column of B.

$$(AB)_{ij} = \left(i^{th} \text{ row of } A \right) \cdot \left(j^{th} \text{ column of } B \right)$$

Example 1.7. Multiply if possible $\begin{pmatrix} 1 & 2 \\ 3 & 1 \\ 2 & 6 \end{pmatrix} \begin{pmatrix} 2 & 3 & 1 \\ 7 & 6 & 2 \end{pmatrix}$.

First check to see if this is possible. It is of the form $(3 \times 2)(2 \times 3)$ and since the inside numbers match, the two matrices are conformable and it is possible to do the multiplication. The result should be a 3×3 matrix. The answer is of the form

$$\left(\begin{pmatrix} 1 & 2 \\ 3 & 1 \\ 2 & 6 \end{pmatrix} \begin{pmatrix} 2 \\ 7 \end{pmatrix}, \begin{pmatrix} 1 & 2 \\ 3 & 1 \\ 2 & 6 \end{pmatrix} \begin{pmatrix} 3 \\ 6 \end{pmatrix}, \begin{pmatrix} 1 & 2 \\ 3 & 1 \\ 2 & 6 \end{pmatrix} \begin{pmatrix} 1 \\ 2 \end{pmatrix} \right)$$

where the commas separate the columns in the resulting product. Thus the above product equals

$$\begin{pmatrix} 16 & 15 & 5 \\ 13 & 15 & 5 \\ 46 & 42 & 14 \end{pmatrix},$$

a 3×3 matrix as desired. In terms of the ij^{th} entries and the above definition, the entry in the third row and second column of the product should equal

$$\sum_j a_{3k} b_{k2} = a_{31} b_{12} + a_{32} b_{22} = 2 \times 3 + 6 \times 6 = 42.$$

You should try a few more such examples to verify the above definition in terms of the ij^{th} entries works for other entries.

Example 1.8. Multiply if possible $\begin{pmatrix} 1 & 2 \\ 3 & 1 \\ 2 & 6 \end{pmatrix} \begin{pmatrix} 2 & 3 & 1 \\ 7 & 6 & 2 \\ 0 & 0 & 0 \end{pmatrix}$.

This is not possible because it is of the form $(3 \times 2)(3 \times 3)$ and the middle numbers don't match. In other words the two matrices are not conformable in the indicated order.

Example 1.9. Multiply if possible $\begin{pmatrix} 2 & 3 & 1 \\ 7 & 6 & 2 \\ 0 & 0 & 0 \end{pmatrix} \begin{pmatrix} 1 & 2 \\ 3 & 1 \\ 2 & 6 \end{pmatrix}$.

This is possible because in this case it is of the form $(3 \times 3)(3 \times 2)$ and the middle numbers do match so the matrices are conformable. When the multiplication is done it equals

$$\begin{pmatrix} 13 & 13 \\ 29 & 32 \\ 0 & 0 \end{pmatrix}.$$

Check this and be sure you come up with the same answer.

Example 1.10. Multiply if possible $\begin{pmatrix} 1 \\ 2 \\ 1 \end{pmatrix}$ (1 2 1 0).

In this case you are trying to do $(3 \times 1)(1 \times 4)$. The inside numbers match so you can do it. Verify

$$\begin{pmatrix} 1 \\ 2 \\ 1 \end{pmatrix} (1 \ 2 \ 1 \ 0) = \begin{pmatrix} 1 & 2 & 1 & 0 \\ 2 & 4 & 2 & 0 \\ 1 & 2 & 1 & 0 \end{pmatrix}$$

1.1.4 *Properties Of Matrix Multiplication*

As pointed out above, sometimes it is possible to multiply matrices in one order but not in the other order. What if it makes sense to multiply them in either order? Will the two products be equal then?

Example 1.11. Compare $\begin{pmatrix} 1 & 2 \\ 3 & 4 \end{pmatrix} \begin{pmatrix} 0 & 1 \\ 1 & 0 \end{pmatrix}$ and $\begin{pmatrix} 0 & 1 \\ 1 & 0 \end{pmatrix} \begin{pmatrix} 1 & 2 \\ 3 & 4 \end{pmatrix}$.

The first product is

$$\begin{pmatrix} 1 & 2 \\ 3 & 4 \end{pmatrix} \begin{pmatrix} 0 & 1 \\ 1 & 0 \end{pmatrix} = \begin{pmatrix} 2 & 1 \\ 4 & 3 \end{pmatrix}.$$

The second product is

$$\begin{pmatrix} 0 & 1 \\ 1 & 0 \end{pmatrix} \begin{pmatrix} 1 & 2 \\ 3 & 4 \end{pmatrix} = \begin{pmatrix} 3 & 4 \\ 1 & 2 \end{pmatrix}.$$

You see these are not equal. Again you cannot conclude that $AB = BA$ for matrix multiplication even when multiplication is defined in both orders. However, there are some properties which do hold.

Proposition 1.1. If all multiplications and additions make sense, the following hold for matrices A, B, C and a, b scalars.

$$A(aB + bC) = a(AB) + b(AC) \tag{1.12}$$

$$(B + C)A = BA + CA \tag{1.13}$$

$$A(BC) = (AB)C \tag{1.14}$$

Proof: Using Definition 1.8,

$$(A(aB + bC))_{ij} = \sum_k A_{ik}(aB + bC)_{kj} = \sum_k A_{ik}(aB_{kj} + bC_{kj})$$

$$= a \sum_k A_{ik} B_{kj} + b \sum_k A_{ik} C_{kj} = a \left(AB \right)_{ij} + b \left(AC \right)_{ij} = \left(a \left(AB \right) + b \left(AC \right) \right)_{ij}.$$

Thus $A \left(B + C \right) = AB + AC$ as claimed. Formula (1.13) is entirely similar.
    Formula (1.14) is the associative law of multiplication. Using Definition 1.8,

$$
\begin{aligned}
\left(A \left(BC \right) \right)_{ij} &= \sum_k A_{ik} \left(BC \right)_{kj} = \sum_k A_{ik} \sum_l B_{kl} C_{lj} \\
&= \sum_l \left(AB \right)_{il} C_{lj} = \left(\left(AB \right) C \right)_{ij}.
\end{aligned}
$$

This proves (1.14). ■

1.1.5 *The Transpose*

Another important operation on matrices is that of taking the **transpose**. The
following example shows what is meant by this operation, denoted by placing a T
as an exponent on the matrix.

$$
\begin{pmatrix} 1 & 4 \\ 3 & 1 \\ 2 & 6 \end{pmatrix}^T = \begin{pmatrix} 1 & 3 & 2 \\ 4 & 1 & 6 \end{pmatrix}
$$

What happened? The first column became the first row and the second column
became the second row. Thus the 3×2 matrix became a 2×3 matrix. The number
3 was in the second row and the first column and it ended up in the first row and
second column. Here is the definition.

Definition 1.9. Let A be an $m \times n$ matrix. Then A^T denotes the $n \times m$ matrix
which is defined as follows.

$$
\left(A^T \right)_{ij} = A_{ji}
$$

Example 1.12.

$$
\begin{pmatrix} 1 & 2 & -6 \\ 3 & 5 & 4 \end{pmatrix}^T = \begin{pmatrix} 1 & 3 \\ 2 & 5 \\ -6 & 4 \end{pmatrix}.
$$

The transpose of a matrix has the following important properties.

Lemma 1.1. *Let A be an $m \times n$ matrix and let B be an $n \times p$ matrix. Then*

$$
\left(AB \right)^T = B^T A^T \tag{1.15}
$$

and if α and β are scalars,

$$
\left(\alpha A + \beta B \right)^T = \alpha A^T + \beta B^T \tag{1.16}
$$

Proof: From the definition,

$$\left((AB)^T\right)_{ij} = (AB)_{ji} = \sum_k A_{jk} B_{ki} = \sum_k \left(B^T\right)_{ik} \left(A^T\right)_{kj} = \left(B^T A^T\right)_{ij}$$

The proof of Formula (1.16) is left as an exercise. ∎

Definition 1.10. An $n \times n$ matrix A is said to be **symmetric** if $A = A^T$. It is said to be **skew symmetric** if $A = -A^T$.

Example 1.13. Let

$$A = \begin{pmatrix} 2 & 1 & 3 \\ 1 & 5 & -3 \\ 3 & -3 & 7 \end{pmatrix}.$$

Then A is symmetric.

Example 1.14. Let

$$A = \begin{pmatrix} 0 & 1 & 3 \\ -1 & 0 & 2 \\ -3 & -2 & 0 \end{pmatrix}$$

Then A is skew symmetric.

1.1.6 *The Identity And Inverses*

There is a special matrix called I and referred to as the identity matrix. It is always a square matrix, meaning the number of rows equals the number of columns and it has the property that there are ones down the main diagonal and zeroes elsewhere. Here are some identity matrices of various sizes.

$$(1), \begin{pmatrix} 1 & 0 \\ 0 & 1 \end{pmatrix}, \begin{pmatrix} 1 & 0 & 0 \\ 0 & 1 & 0 \\ 0 & 0 & 1 \end{pmatrix}, \begin{pmatrix} 1 & 0 & 0 & 0 \\ 0 & 1 & 0 & 0 \\ 0 & 0 & 1 & 0 \\ 0 & 0 & 0 & 1 \end{pmatrix}.$$

The first is the 1×1 identity matrix, the second is the 2×2 identity matrix, the third is the 3×3 identity matrix, and the fourth is the 4×4 identity matrix. By extension, you can likely see what the $n \times n$ identity matrix would be. It is so important that there is a special symbol to denote the ij^{th} entry of the identity matrix

$$I_{ij} = \delta_{ij}$$

where δ_{ij} is the **Kroneker symbol** defined by

$$\delta_{ij} = \begin{cases} 1 \text{ if } i = j \\ 0 \text{ if } i \neq j \end{cases}$$

It is called the **identity matrix** because it is a **multiplicative identity** in the following sense.

Lemma 1.2. *Suppose A is an $m \times n$ matrix and I_n is the $n \times n$ identity matrix. Then $AI_n = A$. If I_m is the $m \times m$ identity matrix, it also follows that $I_m A = A$.*

Proof:

$$(AI_n)_{ij} = \sum_k A_{ik}\delta_{kj}$$
$$= A_{ij}$$

and so $AI_n = A$. The other case is left as an exercise for you. ∎

Definition 1.11. An $n \times n$ matrix A has an inverse A^{-1} if and only if $AA^{-1} = A^{-1}A = I$. Such a matrix is called **invertible**.

It is very important to observe that the inverse of a matrix, if it exists, is unique. Another way to think of this is that if it acts like the inverse, then it is the inverse.

Theorem 1.1. *Suppose A^{-1} exists and $AB = BA = I$. Then $B = A^{-1}$.*

Proof:

$$A^{-1} = A^{-1}I = A^{-1}(AB) = (A^{-1}A)B = IB = B. \quad ∎$$

Unlike numbers, it can happen that $A \neq 0$ but A may fail to have an inverse. This is illustrated in the following example.

Example 1.15. Let $A = \begin{pmatrix} 1 & 1 \\ 1 & 1 \end{pmatrix}$. Does A have an inverse?

One might think A would have an inverse because it does not equal zero. However,

$$\begin{pmatrix} 1 & 1 \\ 1 & 1 \end{pmatrix}\begin{pmatrix} -1 \\ 1 \end{pmatrix} = \begin{pmatrix} 0 \\ 0 \end{pmatrix}$$

and if A^{-1} existed, this could not happen because you could write

$$\begin{pmatrix} 0 \\ 0 \end{pmatrix} = A^{-1}\left(\begin{pmatrix} 0 \\ 0 \end{pmatrix}\right) = A^{-1}\left(A\begin{pmatrix} -1 \\ 1 \end{pmatrix}\right)$$

$$= (A^{-1}A)\begin{pmatrix} -1 \\ 1 \end{pmatrix} = I\begin{pmatrix} -1 \\ 1 \end{pmatrix} = \begin{pmatrix} -1 \\ 1 \end{pmatrix},$$

a contradiction. Thus the answer is that A does not have an inverse.

Example 1.16. Let $A = \begin{pmatrix} 1 & 1 \\ 1 & 2 \end{pmatrix}$. Show $\begin{pmatrix} 2 & -1 \\ -1 & 1 \end{pmatrix}$ is the inverse of A.

To check this, multiply

$$\begin{pmatrix} 1 & 1 \\ 1 & 2 \end{pmatrix}\begin{pmatrix} 2 & -1 \\ -1 & 1 \end{pmatrix} = \begin{pmatrix} 1 & 0 \\ 0 & 1 \end{pmatrix}$$

and

$$\begin{pmatrix} 2 & -1 \\ -1 & 1 \end{pmatrix}\begin{pmatrix} 1 & 1 \\ 1 & 2 \end{pmatrix} = \begin{pmatrix} 1 & 0 \\ 0 & 1 \end{pmatrix}$$

showing that this matrix is indeed the inverse of A.

There are various ways of finding the inverse of a matrix. One way will be presented in the discussion on determinants. You can also find them directly from the definition provided they exist.

In the last example, how would you find A^{-1}? You wish to find a matrix $\begin{pmatrix} x & z \\ y & w \end{pmatrix}$ such that

$$\begin{pmatrix} 1 & 1 \\ 1 & 2 \end{pmatrix} \begin{pmatrix} x & z \\ y & w \end{pmatrix} = \begin{pmatrix} 1 & 0 \\ 0 & 1 \end{pmatrix}.$$

This requires the solution of the systems of equations,

$$x + y = 1, \ x + 2y = 0, \ z + w = 0, \ z + 2w = 1.$$

The first pair of equations has the solution $y = -1$ and $x = 2$. The second pair of equations has the solution $w = 1$, $z = -1$. Therefore, from the definition of the inverse,

$$A^{-1} = \begin{pmatrix} 2 & -1 \\ -1 & 1 \end{pmatrix}.$$

To be sure it is the inverse, you should multiply on both sides of the original matrix. It turns out that if it works on one side, it will always work on the other. The consideration of this as well as a more detailed treatment of inverses is a good topic for a linear algebra course. However, see the appendix on determinants which discusses this topic from the point of view of determinants.

1.2 Linear Transformations

An $m \times n$ matrix can be used to transform vectors in $\mathbb{R}^n$ to vectors in $\mathbb{R}^m$ through the use of matrix multiplication.

Example 1.17. Consider the matrix $\begin{pmatrix} 1 & 2 & 0 \\ 2 & 1 & 0 \end{pmatrix}$. Think of it as a function which takes vectors in $\mathbb{R}^3$ and makes them into vectors in $\mathbb{R}^2$ as follows. For $\begin{pmatrix} x \\ y \\ z \end{pmatrix}$ a vector in $\mathbb{R}^3$, multiply on the left by the given matrix to obtain the vector in $\mathbb{R}^2$. Here are some numerical examples.

$$\begin{pmatrix} 1 & 2 & 0 \\ 2 & 1 & 0 \end{pmatrix} \begin{pmatrix} 1 \\ 2 \\ 3 \end{pmatrix} = \begin{pmatrix} 5 \\ 4 \end{pmatrix}, \begin{pmatrix} 1 & 2 & 0 \\ 2 & 1 & 0 \end{pmatrix} \begin{pmatrix} 1 \\ -2 \\ 3 \end{pmatrix} = \begin{pmatrix} -3 \\ 0 \end{pmatrix},$$

$$\begin{pmatrix} 1 & 2 & 0 \\ 2 & 1 & 0 \end{pmatrix} \begin{pmatrix} 10 \\ 5 \\ -3 \end{pmatrix} = \begin{pmatrix} 20 \\ 25 \end{pmatrix}, \begin{pmatrix} 1 & 2 & 0 \\ 2 & 1 & 0 \end{pmatrix} \begin{pmatrix} 0 \\ 7 \\ 3 \end{pmatrix} = \begin{pmatrix} 14 \\ 7 \end{pmatrix}.$$

More generally,

$$\begin{pmatrix} 1 & 2 & 0 \\ 2 & 1 & 0 \end{pmatrix} \begin{pmatrix} x \\ y \\ z \end{pmatrix} = \begin{pmatrix} x + 2y \\ 2x + y \end{pmatrix}$$

The idea is to define a function which takes vectors in $\mathbb{R}^3$ and delivers new vectors in $\mathbb{R}^2$.

This is an example of something called a linear transformation.

Definition 1.12. Let $T : \mathbb{R}^n \to \mathbb{R}^m$ be a function. Thus for each $\mathbf{x} \in \mathbb{R}^n$, $T\mathbf{x} \in \mathbb{R}^m$. Then T is a **linear transformation** if whenever α, β are scalars and $\mathbf{x}_1$ and $\mathbf{x}_2$ are vectors in $\mathbb{R}^n$,

$$T(\alpha \mathbf{x}_1 + \beta \mathbf{x}_2) = \alpha_1 T\mathbf{x}_1 + \beta T\mathbf{x}_2.$$

In words, linear transformations distribute across $+$ and allow you to factor out scalars. At this point, recall the properties of matrix multiplication. The pertinent property is (1.13) on Page 9. Recall it states that for a and b scalars,

$$A(aB + bC) = aAB + bAC$$

In particular, for A an $m \times n$ matrix and B and C $n \times 1$ matrices (column vectors) the above formula holds which is nothing more than the statement that matrix multiplication gives an example of a linear transformation.

Definition 1.13. A linear transformation is called **one to one** (often written as $1 - 1$) if it never takes two different vectors to the same vector. Thus T is one to one if whenever $\mathbf{x} \neq \mathbf{y}$ $T\mathbf{x} \neq T\mathbf{y}$. Equivalently, if $T(\mathbf{x}) = T(\mathbf{y})$, then $\mathbf{x} = \mathbf{y}$.

In the case that a linear transformation comes from matrix multiplication, it is common usage to refer to the matrix as a one to one matrix when the linear transformation it determines is one to one.

Definition 1.14. A linear transformation mapping $\mathbb{R}^n$ to $\mathbb{R}^m$ is called **onto** if whenever $\mathbf{y} \in \mathbb{R}^m$ there exists $\mathbf{x} \in \mathbb{R}^n$ such that $T(\mathbf{x}) = \mathbf{y}$.

Thus T is onto if everything in $\mathbb{R}^m$ gets hit. In the case that a linear transformation comes from matrix multiplication, it is common to refer to the matrix as onto when the linear transformation it determines is onto. Also it is common usage to write $T\mathbb{R}^n$, $T(\mathbb{R}^n)$, or $\mathrm{Im}(T)$ as the set of vectors of $\mathbb{R}^m$ which are of the form $T\mathbf{x}$ for some $\mathbf{x} \in \mathbb{R}^n$. In the case that T is obtained from multiplication by an $m \times n$ matrix A, it is standard to simply write $A(\mathbb{R}^n)$, $A\mathbb{R}^n$, or $\mathrm{Im}(A)$ to denote those vectors in $\mathbb{R}^m$ which are obtained in the form $A\mathbf{x}$ for some $\mathbf{x} \in \mathbb{R}^n$.

1.3 Constructing The Matrix Of A Linear Transformation

It turns out that if T is any linear transformation which maps $\mathbb{R}^n$ to $\mathbb{R}^m$, there is always an $m \times n$ matrix A with the property that

$$A\mathbf{x} = T\mathbf{x} \qquad (1.17)$$

for all $\mathbf{x} \in \mathbb{R}^n$. Here is why. Suppose $T : \mathbb{R}^n \to \mathbb{R}^m$ is a linear transformation and you want to find the matrix defined by this linear transformation as described in (1.17). Then if $\mathbf{x} \in \mathbb{R}^n$ it follows

$$\mathbf{x} = \sum_{i=1}^{n} x_i \mathbf{e}_i = x_1 \begin{pmatrix} 1 \\ 0 \\ \vdots \\ 0 \end{pmatrix} + x_2 \begin{pmatrix} 0 \\ 1 \\ \vdots \\ 0 \end{pmatrix} + \cdots + x_n \begin{pmatrix} 0 \\ 0 \\ \vdots \\ 1 \end{pmatrix}$$

where as implied above, $\mathbf{e}_i$ is the vector which has zeros in every slot but the ith and a 1 in this slot. Then since T is linear,

$$T\mathbf{x} = \sum_{i=1}^{n} x_i T(\mathbf{e}_i) = \begin{pmatrix} | & & | \\ T(\mathbf{e}_1) & \cdots & T(\mathbf{e}_n) \\ | & & | \end{pmatrix} \begin{pmatrix} x_1 \\ \vdots \\ x_n \end{pmatrix} \equiv A \begin{pmatrix} x_1 \\ \vdots \\ x_n \end{pmatrix}$$

and so you see that the matrix desired is obtained from letting the ith column equal $T(\mathbf{e}_i)$. This yields the following theorem.

Theorem 1.2. *Let T be a linear transformation from $\mathbb{R}^n$ to $\mathbb{R}^m$. Then the matrix A satisfying (1.17) is given by*

$$\begin{pmatrix} | & & | \\ T(\mathbf{e}_1) & \cdots & T(\mathbf{e}_n) \\ | & & | \end{pmatrix}$$

where $T\mathbf{e}_i$ is the ith column of A.

Sometimes you need to find a matrix which represents a given linear transformation which is described in geometrical terms. A good example of this is the problem of rotation of vectors.

Why is such a transformation linear? Consider the following picture which illustrates a rotation.

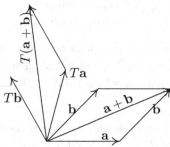

To get $T(\mathbf{a}+\mathbf{b})$, you can add $T\mathbf{a}$ and $T\mathbf{b}$. Here is why. If you add $T\mathbf{a}$ to $T\mathbf{b}$, you get the diagonal of the parallelogram determined by $T\mathbf{a}$ and $T\mathbf{b}$. This diagonal also results from rotating the diagonal of the parallelogram determined by $\mathbf{a}$ and $\mathbf{b}$. This is because the rotation preserves all angles between the vectors as well as their lengths. In particular, it preserves the shape of this parallelogram. Thus both $T\mathbf{a}+T\mathbf{b}$ and $T(\mathbf{a}+\mathbf{b})$ give the same directed line segment. Thus T distributes across $+$ where $+$ refers to vector addition. Similarly, if k is a number $Tk\mathbf{a}=kT\mathbf{a}$ (draw a picture) and so you can factor out scalars also. Thus rotations are an example of a linear transformation as claimed. ∎

Example 1.18. Determine the matrix which represents the linear transformation defined by rotating every vector through an angle of θ.

Let $\mathbf{e}_1 \equiv \begin{pmatrix} 1 \\ 0 \end{pmatrix}$ and $\mathbf{e}_2 \equiv \begin{pmatrix} 0 \\ 1 \end{pmatrix}$. These identify the geometric vectors which point along the positive x axis and positive y axis as shown.

From the above, you only need to find $T\mathbf{e}_1$ and $T\mathbf{e}_2$, the first being the first column of the desired matrix A and the second being the second column.

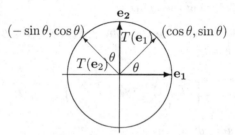

From drawing a picture and doing a little geometry, you see that

$$T\mathbf{e}_1 = \begin{pmatrix} \cos\theta \\ \sin\theta \end{pmatrix}, T\mathbf{e}_2 = \begin{pmatrix} -\sin\theta \\ \cos\theta \end{pmatrix}.$$

Therefore, from Theorem 1.2,

$$A = \begin{pmatrix} \cos\theta & -\sin\theta \\ \sin\theta & \cos\theta \end{pmatrix}$$

Example 1.19. Find the matrix of the linear transformation which is obtained by first rotating all vectors through an angle of ϕ and then through an angle θ. Thus you want the linear transformation which rotates all angles through an angle of $\theta + \phi$.

Let $T_{\theta+\phi}$ denote the linear transformation which rotates every vector through an angle of $\theta + \phi$. Then to get $T_{\theta+\phi}$, you could first do T_ϕ and then do T_θ where T_ϕ

is the linear transformation which rotates through an angle of ϕ and T_θ is the linear transformation which rotates through an angle of θ. Denoting the corresponding matrices by $A_{\theta+\phi}$, A_ϕ, and A_θ, you must have for every $\mathbf{x}$

$$A_{\theta+\phi}\mathbf{x} = T_{\theta+\phi}\mathbf{x} = T_\theta T_\phi \mathbf{x} = A_\theta A_\phi \mathbf{x}.$$

Consequently, you must have

$$A_{\theta+\phi} = \begin{pmatrix} \cos(\theta+\phi) & -\sin(\theta+\phi) \\ \sin(\theta+\phi) & \cos(\theta+\phi) \end{pmatrix} = A_\theta A_\phi$$

$$= \begin{pmatrix} \cos\theta & -\sin\theta \\ \sin\theta & \cos\theta \end{pmatrix} \begin{pmatrix} \cos\phi & -\sin\phi \\ \sin\phi & \cos\phi \end{pmatrix}.$$

You know how to multiply matrices. Do so to the pair on the right. This yields

$$\begin{pmatrix} \cos(\theta+\phi) & -\sin(\theta+\phi) \\ \sin(\theta+\phi) & \cos(\theta+\phi) \end{pmatrix}$$

$$= \begin{pmatrix} \cos\theta\cos\phi - \sin\theta\sin\phi & -\cos\theta\sin\phi - \sin\theta\cos\phi \\ \sin\theta\cos\phi + \cos\theta\sin\phi & \cos\theta\cos\phi - \sin\theta\sin\phi \end{pmatrix}.$$

Don't these look familiar? They are the usual trig. identities for the sum of two angles derived here using linear algebra concepts.

Example 1.20. A linear transformation T is obtained by first rotating every vector through an angle of $\pi/6$ and then reflecting across the x axis. Find $T\begin{pmatrix} 1 \\ 2 \end{pmatrix}$. Consider the same problem by doing the two linear transformations in the opposite order.

First you do $\begin{pmatrix} \cos\left(\frac{\pi}{6}\right) & -\sin\left(\frac{\pi}{6}\right) \\ \sin\left(\frac{\pi}{6}\right) & \cos\left(\frac{\pi}{6}\right) \end{pmatrix}$ to a given vector and then you do $\begin{pmatrix} 1 & 0 \\ 0 & -1 \end{pmatrix}$ to what you got. This second matrix is the matrix of the linear transformation which reflects through the x axis because if you reflect through the x axis, you leave $\mathbf{e}_1$ unchanged but make $\mathbf{e}_2$ into $-\mathbf{e}_2$. Hence the end result of doing the two linear transformations in the order listed is

$$\begin{pmatrix} 1 & 0 \\ 0 & -1 \end{pmatrix} \begin{pmatrix} \cos\left(\frac{\pi}{6}\right) & -\sin\left(\frac{\pi}{6}\right) \\ \sin\left(\frac{\pi}{6}\right) & \cos\left(\frac{\pi}{6}\right) \end{pmatrix} = \begin{pmatrix} \frac{1}{2}\sqrt{3} & -\frac{1}{2} \\ -\frac{1}{2} & -\frac{1}{2}\sqrt{3} \end{pmatrix}$$

Now what does this do to the given vector?

$$\begin{pmatrix} \frac{1}{2}\sqrt{3} & -\frac{1}{2} \\ -\frac{1}{2} & -\frac{1}{2}\sqrt{3} \end{pmatrix} \begin{pmatrix} 1 \\ 2 \end{pmatrix} = \begin{pmatrix} \frac{1}{2}\sqrt{3} - 1 \\ -\frac{1}{2} - \sqrt{3} \end{pmatrix}$$

I think you would have had a hard time figuring this out using only methods of trigonometry.

Does it make a difference if you do the two given linear transformations in the opposite order? If you think about it geometrically, you will see that it does. Now what would you get if you did them in the other order? The matrix would be

$$\begin{pmatrix} \cos\left(\frac{\pi}{6}\right) & -\sin\left(\frac{\pi}{6}\right) \\ \sin\left(\frac{\pi}{6}\right) & \cos\left(\frac{\pi}{6}\right) \end{pmatrix} \begin{pmatrix} 1 & 0 \\ 0 & -1 \end{pmatrix} = \begin{pmatrix} \frac{1}{2}\sqrt{3} & \frac{1}{2} \\ \frac{1}{2} & -\frac{1}{2}\sqrt{3} \end{pmatrix}$$

which is a different matrix than the above. Also

$$\begin{pmatrix} \frac{1}{2}\sqrt{3} & \frac{1}{2} \\ \frac{1}{2} & -\frac{1}{2}\sqrt{3} \end{pmatrix}\begin{pmatrix} 1 \\ 2 \end{pmatrix} = \begin{pmatrix} \frac{1}{2}\sqrt{3}+1 \\ \frac{1}{2}-\sqrt{3} \end{pmatrix}$$

also different than what was obtained in the other order.

You do not have to stop with two dimensions. You can consider rotations and other geometric concepts in any number of dimensions. This is one of the major advantages of linear algebra. You can break down a difficult geometrical procedure into small steps, each corresponding to multiplication by an appropriate matrix. Then by multiplying the matrices, you can obtain a single matrix which can give you numerical information on the results of applying the given sequence of simple procedures. That which you could never visualize can still be understood to the extent of finding exact numerical answers. The following is a more routine example quite typical of what will be important in the calculus of several variables.

Example 1.21. Let $T(x_1, x_2) = \begin{pmatrix} x_1 + 3x_2 \\ x_1 - x_2 \\ x_1 \\ 3x_2 + 5x_1 \end{pmatrix}$. Thus $T : \mathbb{R}^2 \to \mathbb{R}^4$. Explain

why T is a linear transformation and write $T(x_1, x_2)$ in the form $A\begin{pmatrix} x_1 \\ x_2 \end{pmatrix}$ where A is an appropriate matrix.

From the definition of matrix multiplication,

$$T(x_1, x_2) = \begin{pmatrix} 1 & 3 \\ 1 & -1 \\ 1 & 0 \\ 5 & 3 \end{pmatrix}\begin{pmatrix} x_1 \\ x_2 \end{pmatrix}$$

Since $T\mathbf{x}$ is of the form $A\mathbf{x}$ for A a matrix, it follows T is a linear transformation. You could also verify directly that $T(\alpha\mathbf{x} + \beta\mathbf{y}) = \alpha T(\mathbf{x}) + \beta T(\mathbf{y})$.

1.4 Exercises

(1) Here are some matrices:

$$A = \begin{pmatrix} 1 & 2 & 3 \\ 2 & 1 & 7 \end{pmatrix}, B = \begin{pmatrix} 3 & -1 & 2 \\ -3 & 2 & 1 \end{pmatrix},$$

$$C = \begin{pmatrix} 1 & 2 \\ 3 & 1 \end{pmatrix}, D = \begin{pmatrix} -1 & 2 \\ 2 & -3 \end{pmatrix}, E = \begin{pmatrix} 2 \\ 3 \end{pmatrix}.$$

Find if possible $-3A, 3B - A, AC, CB, AE, EA$. If it is not possible explain why.

(2) Here are some matrices:

$$A = \begin{pmatrix} 1 & 2 \\ 3 & 2 \\ 1 & -1 \end{pmatrix}, B = \begin{pmatrix} 2 & -5 & 2 \\ -3 & 2 & 1 \end{pmatrix},$$

$$C = \begin{pmatrix} 1 & 2 \\ 5 & 0 \end{pmatrix}, D = \begin{pmatrix} -1 & 1 \\ 4 & -3 \end{pmatrix}, E = \begin{pmatrix} 1 \\ 3 \end{pmatrix}.$$

Find if possible $-3A, 3B - A, AC, CA, AE, EA, BE, DE$. If it is not possible explain why.

(3) Here are some matrices:

$$A = \begin{pmatrix} 1 & 2 \\ 3 & 2 \\ 1 & -1 \end{pmatrix}, B = \begin{pmatrix} 2 & -5 & 2 \\ -3 & 2 & 1 \end{pmatrix},$$

$$C = \begin{pmatrix} 1 & 2 \\ 5 & 0 \end{pmatrix}, D = \begin{pmatrix} -1 & 1 \\ 4 & -3 \end{pmatrix}, E = \begin{pmatrix} 1 \\ 3 \end{pmatrix}.$$

Find if possible $-3A^T, 3B - A^T, AC, CA, AE, E^TB, BE, DE, EE^T, E^TE$. If it is not possible explain why.

(4) Let $A = \begin{pmatrix} 1 & 1 \\ -2 & -1 \\ 1 & 2 \end{pmatrix}$,

$B = \begin{pmatrix} 1 & -1 & -2 \\ 2 & 1 & -2 \end{pmatrix}$, and $C = \begin{pmatrix} 1 & 1 & -3 \\ -1 & 2 & 0 \\ -3 & -1 & 0 \end{pmatrix}$.

Find if possible.

(a) AB
(b) BA
(c) AC
(d) CA
(e) CB
(f) BC

(5) Let $A = \begin{pmatrix} 1 & 2 \\ 3 & 4 \end{pmatrix}, B = \begin{pmatrix} 1 & 2 \\ 3 & k \end{pmatrix}$. Is it possible to choose k such that $AB = BA$? If so, what should k equal?

(6) Let $A = \begin{pmatrix} 1 & 2 \\ 3 & 4 \end{pmatrix}, B = \begin{pmatrix} 1 & 2 \\ 1 & k \end{pmatrix}$. Is it possible to choose k such that $AB = BA$? If so, what should k equal?

(7) Let $\mathbf{x} = (-1, -1, 1)$ and $\mathbf{y} = (0, 1, 2)$. Find $\mathbf{x}^T\mathbf{y}$ and $\mathbf{x}\mathbf{y}^T$ if possible.

(8) Find the matrix for a linear transformation which rotates every vector in $\mathbb{R}^2$ through an angle of $\pi/4$.

(9) Find the matrix for a linear transformation which rotates every vector in $\mathbb{R}^2$ through an angle of $-\pi/3$.

(10) Find the matrix for a linear transformation which rotates every vector in $\mathbb{R}^2$ through an angle of $2\pi/3$.

(11) Find the matrix for a linear transformation which rotates every vector in $\mathbb{R}^2$ through an angle of $\pi/12$. **Hint:** Note that $\pi/12 = \pi/3 - \pi/4$.

(12) Let $T(x_1, x_2) = \begin{pmatrix} x_1 + 4x_2 \\ x_2 + 2x_1 \end{pmatrix}$. Thus $T : \mathbb{R}^2 \to \mathbb{R}^2$. Explain why T is a linear transformation and write $T(x_1, x_2)$ in the form $A \begin{pmatrix} x_1 \\ x_2 \end{pmatrix}$ where A is an appropriate matrix.

(13) Let $T(x_1, x_2) = \begin{pmatrix} x_1 - x_2 \\ x_1 \\ 3x_2 + x_1 \\ 3x_2 + 5x_1 \end{pmatrix}$. Thus $T : \mathbb{R}^2 \to \mathbb{R}^4$. Explain why T is a linear transformation and write $T(x_1, x_2)$ in the form $A \begin{pmatrix} x_1 \\ x_2 \end{pmatrix}$ where A is an appropriate matrix.

(14) Let
$$T(x_1, x_2, x_3, x_4) = \begin{pmatrix} x_1 - x_2 + 2x_3 \\ 2x_3 + x_1 \\ 3x_3 \\ 3x_4 + 3x_2 + x_1 \end{pmatrix}.$$
Thus $T : \mathbb{R}^4 \to \mathbb{R}^4$. Explain why T is a linear transformation and write $T(x_1, x_2, x_3, x_4)$ in the form $A \begin{pmatrix} x_1 \\ x_2 \\ x_3 \\ x_4 \end{pmatrix}$ where A is an appropriate matrix.

(15) Let $T(x_1, x_2) = \begin{pmatrix} x_1^2 + 4x_2 \\ x_2 + 2x_1 \end{pmatrix}$. Thus $T : \mathbb{R}^2 \to \mathbb{R}^2$. Explain why T cannot possibly be a linear transformation.

(16) Suppose A and B are square matrices of the same size. Which of the following are correct?

(a) $(A - B)^2 = A^2 - 2AB + B^2$

(b) $(AB)^2 = A^2 B^2$

(c) $(A + B)^2 = A^2 + 2AB + B^2$

(d) $(A + B)^2 = A^2 + AB + BA + B^2$

(e) $A^2 B^2 = A(AB)B$

(f) $(A + B)^3 = A^3 + 3A^2 B + 3AB^2 + B^3$

(g) $(A + B)(A - B) = A^2 - B^2$

(17) Let $A = \begin{pmatrix} -1 & -1 \\ 3 & 3 \end{pmatrix}$. Find $\boxed{\text{all}}$ 2×2 matrices B such that $AB = 0$.

(18) In (1.1) - (1.8) describe $-A$ and 0.

(19) Let A be an $n \times n$ matrix. Show A equals the sum of a symmetric and a skew symmetric matrix. **Hint:** Consider the matrix $\frac{1}{2}\left(A + A^T\right)$. Is this matrix symmetric?

(20) If A is a skew symmetric matrix, what can be concluded about A^n where $n = 1, 2, 3, \cdots$?

(21) Show every skew symmetric matrix has all zeros down the main diagonal. The main diagonal consists of every entry of the matrix which is of the form a_{ii}. It runs from the upper left down to the lower right.

(22) Using only the properties (1.1) - (1.8) show $-A$ is unique.

(23) Using only the properties (1.1) - (1.8) show 0 is unique.

(24) Using only the properties (1.1) - (1.8) show $0A = 0$. Here the 0 on the left is the scalar 0 and the 0 on the right is the zero for $m \times n$ matrices.

(25) Using only the properties (1.1) - (1.8) and previous problems show $(-1) A = -A$.

(26) Prove (1.16).

(27) Prove that $I_m A = A$ where A is an $m \times n$ matrix.

(28) Let $A = \begin{pmatrix} 1 & 2 \\ 2 & 1 \end{pmatrix}$. Find A^{-1} if possible. If A^{-1} does not exist, determine why.

(29) Let $A = \begin{pmatrix} 1 & 0 \\ 2 & 3 \end{pmatrix}$. Find A^{-1} if possible. If A^{-1} does not exist, determine why.

(30) Let $A = \begin{pmatrix} 3 & 2 \\ 2 & 1 \end{pmatrix}$. Find A^{-1} if possible. If A^{-1} does not exist, determine why.

(31) Give an example of matrices A, B, C such that $B \neq C$, $A \neq 0$, and yet $AB = AC$.

(32) Suppose $AB = AC$ and A is an invertible $n \times n$ matrix. Does it follow that $B = C$? Explain why or why not. What if A were a non invertible $n \times n$ matrix?

(33) Find your own examples:

    (a) 2×2 matrices A and B such that $A \neq 0, B \neq 0$ with $AB \neq BA$.

    (b) 2×2 matrices A and B such that $A \neq 0, B \neq 0$, but $AB = 0$.

    (c) 2×2 matrices A, D, and C such that $A \neq 0, C \neq D$, but $AC = AD$.

(34) Explain why if $BA = CA$ and A^{-1} exists, then $B = C$.

(35) Give an example of a matrix A such that $A^2 = I$ and yet $A \neq I$ and $A \neq -I$.

(36) Show that if A^{-1} exists for an $n \times n$ matrix, then it is unique. That is, if $BA = I$ and $AB = I$, then $B = A^{-1}$.

(37) Show $(AB)^{-1} = B^{-1}A^{-1}$.

(38) Show that if A is an invertible $n \times n$ matrix, then so is A^T and $\left(A^T\right)^{-1} = \left(A^{-1}\right)^T$.

(39) Show that if A is an $n \times n$ invertible matrix and $\mathbf{x}$ is a $n \times 1$ matrix such that $A\mathbf{x} = \mathbf{b}$ for $\mathbf{b}$ an $n \times 1$ matrix, then $\mathbf{x} = A^{-1}\mathbf{b}$.

(40) Prove that if A^{-1} exists and $A\mathbf{x} = \mathbf{0}$ then $\mathbf{x} = \mathbf{0}$.

(41) Show that $(ABC)^{-1} = C^{-1}B^{-1}A^{-1}$ by verifying that

$$(ABC)\left(C^{-1}B^{-1}A^{-1}\right) = \left(C^{-1}B^{-1}A^{-1}\right)(ABC) = I.$$

(42) Prove that $T\mathbf{v} \equiv \text{proj}_{\mathbf{u}}(\mathbf{v})$ is a linear transformation and find the matrix of $\text{proj}_{\mathbf{u}}(\mathbf{v})$ where $\mathbf{u} = (1,2,3)$. Recall $\text{proj}_{\mathbf{u}}(\mathbf{v}) \equiv \frac{\mathbf{u}\cdot\mathbf{v}}{|\mathbf{u}|^2}\mathbf{u}$. This involves the properties of the dot product.

(43) Let A and be a real $m \times n$ matrix and let $\mathbf{x} \in \mathbb{R}^n$ and $\mathbf{y} \in \mathbb{R}^m$. Show $(A\mathbf{x}, \mathbf{y})_{\mathbb{R}^m} = \left(\mathbf{x}, A^T\mathbf{y}\right)_{\mathbb{R}^n}$ where $(\cdot, \cdot)_{\mathbb{R}^k}$ denotes the dot product in $\mathbb{R}^k$. In the notation above, $A\mathbf{x} \cdot \mathbf{y} = \mathbf{x} \cdot A^T\mathbf{y}$. Use the definition of matrix multiplication to do this.

(44) Use the result of Problem 43 to verify directly that $(AB)^T = B^T A^T$ without making any reference to subscripts.

(45) *Suppose A is a 3×3 matrix and $\det(A) \neq 0$. Show that $\left(A^{-1}\right)_{ks} = \frac{1}{2\det(A)}\varepsilon_{rps}\varepsilon_{ijk}A_{pj}A_{ri}$. Here ε_{ijk} is the permutation symbol discussed in Volume 1.

Chapter 2

Determinants

2.1 Basic Techniques And Properties

2.1.1 Cofactors And 2 × 2 Determinants

Let A be an $n \times n$ matrix. The **determinant** of A, denoted as $\det(A)$, is a number. If the matrix is a 2×2 matrix, this number is very easy to find.

Definition 2.1. Let $A = \begin{pmatrix} a & b \\ c & d \end{pmatrix}$. Then

$$\det(A) \equiv ad - cb.$$

The determinant is also often denoted by enclosing the matrix with two vertical lines. Thus

$$\det \begin{pmatrix} a & b \\ c & d \end{pmatrix} = \begin{vmatrix} a & b \\ c & d \end{vmatrix}.$$

Example 2.1. Find $\det \begin{pmatrix} 2 & 4 \\ -1 & 6 \end{pmatrix}$.

From the definition this is just $(2)(6) - (-1)(4) = 16$.

Having defined what is meant by the determinant of a 2×2 matrix, what about a 3×3 matrix?

Definition 2.2. Suppose A is a 3×3 matrix. The ij^{th} **minor,** denoted as minor$(A)_{ij}$, is the determinant of the 2×2 matrix which results from deleting the ith row and the jth column.

Example 2.2. Consider the matrix

$$\begin{pmatrix} 1 & 2 & 3 \\ 4 & 3 & 2 \\ 3 & 2 & 1 \end{pmatrix}.$$

The $(1, 2)$ minor is the determinant of the 2×2 matrix which results when you delete the first row and the second column. This minor is therefore

$$\det \begin{pmatrix} 4 & 2 \\ 3 & 1 \end{pmatrix} = -2.$$

The $(2,3)$ minor is the determinant of the 2×2 matrix which results when you delete the second row and the third column. This minor is therefore

$$\det \begin{pmatrix} 1 & 2 \\ 3 & 2 \end{pmatrix} = -4.$$

Definition 2.3. Suppose A is a 3×3 matrix. The ij^{th} **cofactor** is defined to be $(-1)^{i+j} \times \left(ij^{th} \text{ minor} \right)$. In words, you multiply $(-1)^{i+j}$ times the ij^{th} minor to get the ij^{th} cofactor. The cofactors of a matrix are so important that special notation is appropriate when referring to them. The ij^{th} cofactor of a matrix A will be denoted by $\text{cof}\,(A)_{ij}$. It is also convenient to refer to the cofactor of an entry of a matrix as follows. For a_{ij} an entry of the matrix, its cofactor is just $\text{cof}\,(A)_{ij}$. Thus the cofactor of the ij^{th} entry is just the ij^{th} cofactor.

Example 2.3. Consider the matrix

$$A = \begin{pmatrix} 1 & 2 & 3 \\ 4 & 3 & 2 \\ 3 & 2 & 1 \end{pmatrix}.$$

The $(1,2)$ minor is the determinant of the 2×2 matrix which results when you delete the first row and the second column. This minor is therefore

$$\det \begin{pmatrix} 4 & 2 \\ 3 & 1 \end{pmatrix} = -2.$$

It follows

$$\text{cof}\,(A)_{12} = (-1)^{1+2} \det \begin{pmatrix} 4 & 2 \\ 3 & 1 \end{pmatrix} = (-1)^{1+2}(-2) = 2$$

The $(2,3)$ minor is the determinant of the 2×2 matrix which results when you delete the second row and the third column. This minor is therefore

$$\det \begin{pmatrix} 1 & 2 \\ 3 & 2 \end{pmatrix} = -4.$$

Therefore,

$$\text{cof}\,(A)_{23} = (-1)^{2+3} \det \begin{pmatrix} 1 & 2 \\ 3 & 2 \end{pmatrix} = (-1)^{2+3}(-4) = 4.$$

Similarly,

$$\text{cof}\,(A)_{22} = (-1)^{2+2} \det \begin{pmatrix} 1 & 3 \\ 3 & 1 \end{pmatrix} = -8.$$

Definition 2.4. The determinant of a 3×3 matrix A, is obtained by picking a row (column) and taking the product of each entry in that row (column) with its cofactor and adding these terms. This process, when applied to the ith row (column), is known as expanding the determinant along the ith row (column).

Example 2.4. Find the determinant of

$$A = \begin{pmatrix} 1 & 2 & 3 \\ 4 & 3 & 2 \\ 3 & 2 & 1 \end{pmatrix}.$$

Here is how it is done by "**expanding along the first column**".

$$\overbrace{1(-1)^{1+1}\begin{vmatrix} 3 & 2 \\ 2 & 1 \end{vmatrix}}^{\mathrm{cof}(A)_{11}} + \overbrace{4(-1)^{2+1}\begin{vmatrix} 2 & 3 \\ 2 & 1 \end{vmatrix}}^{\mathrm{cof}(A)_{21}} + \overbrace{3(-1)^{3+1}\begin{vmatrix} 2 & 3 \\ 3 & 2 \end{vmatrix}}^{\mathrm{cof}(A)_{31}} = 0.$$

You see, we just followed the rule in the above definition. We took the 1 in the first column and multiplied it by its cofactor, the 4 in the first column and multiplied it by its cofactor, and the 3 in the first column and multiplied it by its cofactor. Then we added these numbers together.

You could also expand the determinant along the second row as follows.

$$\overbrace{4(-1)^{2+1}\begin{vmatrix} 2 & 3 \\ 2 & 1 \end{vmatrix}}^{\mathrm{cof}(A)_{21}} + \overbrace{3(-1)^{2+2}\begin{vmatrix} 1 & 3 \\ 3 & 1 \end{vmatrix}}^{\mathrm{cof}(A)_{22}} + \overbrace{2(-1)^{2+3}\begin{vmatrix} 1 & 2 \\ 3 & 2 \end{vmatrix}}^{\mathrm{cof}(A)_{23}} = 0.$$

Observe this gives the same number. You should try expanding along other rows and columns. If you do not make any mistakes, you will always get the same answer.

What about a 4×4 matrix? You know now how to find the determinant of a 3×3 matrix. The pattern is the same.

Definition 2.5. Suppose A is a 4×4 matrix. The ij^{th} **minor** is the determinant of the 3×3 matrix you obtain when you delete the ith row and the jth column. The ij^{th} **cofactor**, $\mathrm{cof}\,(A)_{ij}$ is defined to be $(-1)^{i+j} \times \left(ij^{th}\text{ minor}\right)$. In words, you multiply $(-1)^{i+j}$ times the ij^{th} minor to get the ij^{th} cofactor.

Definition 2.6. The determinant of a 4×4 matrix A, is obtained by picking a row (column) and taking the product of each entry in that row (column) with its cofactor and adding these terms. This process when applied to the ith row (column) is known as expanding the determinant along the ith row (column).

Example 2.5. Find $\det(A)$ where

$$A = \begin{pmatrix} 1 & 2 & 3 & 4 \\ 5 & 4 & 2 & 3 \\ 1 & 3 & 4 & 5 \\ 3 & 4 & 3 & 2 \end{pmatrix}$$

As in the case of a 3×3 matrix you can expand this along any row or column. Lets pick the third column. $\det(A) =$

$$3\,(-1)^{1+3}\begin{vmatrix} 5 & 4 & 3 \\ 1 & 3 & 5 \\ 3 & 4 & 2 \end{vmatrix} + 2\,(-1)^{2+3}\begin{vmatrix} 1 & 2 & 4 \\ 1 & 3 & 5 \\ 3 & 4 & 2 \end{vmatrix} +$$

$$4\left(-1\right)^{3+3}\begin{vmatrix} 1 & 2 & 4 \\ 5 & 4 & 3 \\ 3 & 4 & 2 \end{vmatrix} + 3\left(-1\right)^{4+3}\begin{vmatrix} 1 & 2 & 4 \\ 5 & 4 & 3 \\ 1 & 3 & 5 \end{vmatrix}.$$

Now you know how to expand each of these 3×3 matrices along a row or a column. If you do so, you will get -12 assuming you make no mistakes. You could expand this matrix along any row or any column and assuming you make no mistakes, you will always get the same thing which is defined to be the determinant of the matrix A. This method of evaluating a determinant by expanding along a row or a column is called the **method of Laplace expansion**.

Note that each of the four terms above involves three terms consisting of determinants of 2×2 matrices and each of these will need 2 terms. Therefore, there will be $4 \times 3 \times 2 = 24$ terms to evaluate in order to find the determinant using the method of Laplace expansion. Suppose now you have a 10×10 matrix and you follow the above pattern for evaluating determinants. By analogy to the above, there will be $10! = 3,628,800$ terms involved in the evaluation of such a determinant by Laplace expansion along a row or column. This is a lot of terms.

In addition to the difficulties just discussed, you should regard the above claim that you always get the same answer by picking any row or column with considerable skepticism. It is incredible and not at all obvious. However, it requires a little effort to establish it. This is done in the appendix on the theory of the determinant. The above examples motivate the following definitions, the second of which is incredible.

Definition 2.7. Let $A = (a_{ij})$ be an $n \times n$ matrix and suppose the determinant of a $(n-1) \times (n-1)$ matrix has been defined. Then a new matrix called the **cofactor** matrix $\operatorname{cof}(A)$ is defined by $\operatorname{cof}(A) = (c_{ij})$ where to obtain c_{ij} delete the ith row and the jth column of A, take the determinant of the $(n-1) \times (n-1)$ matrix which results, (This is called the ij^{th} **minor** of A.) and then multiply this number by $(-1)^{i+j}$. Thus $(-1)^{i+j} \times$ (the ij^{th} minor) equals the ij^{th} cofactor. $\operatorname{cof}(A)_{ij}$ will denote the ij^{th} entry of the cofactor matrix.

With this definition of the cofactor matrix, here is how to define the determinant of an $n \times n$ matrix.

Definition 2.8. Let A be an $n \times n$ matrix where $n \geq 2$ and suppose the determinant of an $(n-1) \times (n-1)$ has been defined. Then

$$\det(A) = \sum_{j=1}^{n} a_{ij} \operatorname{cof}(A)_{ij} = \sum_{i=1}^{n} a_{ij} \operatorname{cof}(A)_{ij}. \tag{2.1}$$

The first formula consists of expanding the determinant along the ith row and the second expands the determinant along the jth column. This is called the method of Laplace expansion.

Theorem 2.1. *Expanding the $n \times n$ matrix along any row or column always gives the same answer so the above definition is a good definition.*

2.1.2 The Determinant Of A Triangular Matrix

Notwithstanding the difficulties involved in using the method of Laplace expansion, certain types of matrices are very easy to deal with.

Definition 2.9. A matrix M, is upper triangular if $M_{ij} = 0$ whenever $i > j$. Thus such a matrix equals zero below the main diagonal, the entries of the form M_{ii}, as shown.

$$\begin{pmatrix} * & * & \cdots & * \\ 0 & * & \ddots & \vdots \\ \vdots & \ddots & \ddots & * \\ 0 & \cdots & 0 & * \end{pmatrix}$$

A lower triangular matrix is defined similarly as a matrix for which all entries above the main diagonal are equal to zero.

You should verify the following using the above theorem on Laplace expansion.

Corollary 2.1. *Let M be an upper (lower) triangular matrix. Then $\det(M)$ is obtained by taking the product of the entries on the main diagonal.*

Example 2.6. Let

$$A = \begin{pmatrix} 1 & 2 & 3 & 77 \\ 0 & 2 & 6 & 7 \\ 0 & 0 & 3 & 33.7 \\ 0 & 0 & 0 & -1 \end{pmatrix}$$

Find $\det(A)$.

From the above corollary, it suffices to take the product of the diagonal entries. Thus $\det(A) = 1 \times 2 \times 3 \times (-1) = -6$. Without using the corollary, you could expand along the first column. This gives

$$1 \begin{vmatrix} 2 & 6 & 7 \\ 0 & 3 & 33.7 \\ 0 & 0 & -1 \end{vmatrix} + 0(-1)^{2+1} \begin{vmatrix} 2 & 3 & 77 \\ 0 & 3 & 33.7 \\ 0 & 0 & -1 \end{vmatrix} +$$

$$0(-1)^{3+1} \begin{vmatrix} 2 & 3 & 77 \\ 2 & 6 & 7 \\ 0 & 0 & -1 \end{vmatrix} + 0(-1)^{4+1} \begin{vmatrix} 2 & 3 & 77 \\ 2 & 6 & 7 \\ 0 & 3 & 33.7 \end{vmatrix}$$

and the only nonzero term in the expansion is

$$1 \begin{vmatrix} 2 & 6 & 7 \\ 0 & 3 & 33.7 \\ 0 & 0 & -1 \end{vmatrix}.$$

Now expand this along the first column to obtain

$$1 \times \left(2 \times \begin{vmatrix} 3 & 33.7 \\ 0 & -1 \end{vmatrix} + 0\,(-1)^{2+1} \begin{vmatrix} 6 & 7 \\ 0 & -1 \end{vmatrix} + 0\,(-1)^{3+1} \begin{vmatrix} 6 & 7 \\ 3 & 33.7 \end{vmatrix} \right)$$

$$= 1 \times 2 \times \begin{vmatrix} 3 & 33.7 \\ 0 & -1 \end{vmatrix}$$

Next expand this last determinant along the first column to obtain the above equals

$$1 \times 2 \times 3 \times (-1) = -6$$

which is just the product of the entries down the main diagonal of the original matrix.

2.1.3   *Properties Of Determinants*

There are many properties satisfied by determinants. Some of these properties have to do with row operations which were discussed earlier.

Definition 2.10. The **row operations** consist of the following

(1) Switch two rows.
(2) Multiply a row by a nonzero number.
(3) Replace a row by itself added to a multiple of another row.

Theorem 2.2. *Let A be an $n \times n$ matrix and let A_1 be a matrix which results from multiplying some row of A by a scalar c. Then $c \det(A) = \det(A_1)$.*

Example 2.7. Let $A = \begin{pmatrix} 1 & 2 \\ 3 & 4 \end{pmatrix}$, $A_1 = \begin{pmatrix} 2 & 4 \\ 3 & 4 \end{pmatrix}$. $\det(A) = -2$, $\det(A_1) = -4$.

Theorem 2.3. *Let A be an $n \times n$ matrix and let A_1 be a matrix which results from switching two rows of A. Then $\det(A) = -\det(A_1)$. Also, if one row of A is a multiple of another row of A, then $\det(A) = 0$.*

Example 2.8. Let $A = \begin{pmatrix} 1 & 2 \\ 3 & 4 \end{pmatrix}$ and let $A_1 = \begin{pmatrix} 3 & 4 \\ 1 & 2 \end{pmatrix}$. $\det A = -2$, $\det(A_1) = 2$.

Theorem 2.4. *Let A be an $n \times n$ matrix and let A_1 be a matrix which results from applying row operation 3. Thus a row $\mathbf{a}$ is replaced by $\mathbf{a} + c\mathbf{b}$ where $\mathbf{b}$ is some other row. Then $\det(A) = \det(A_1)$.*

Example 2.9. Let $A = \begin{pmatrix} 1 & 2 \\ 3 & 4 \end{pmatrix}$ and let $A_1 = \begin{pmatrix} 1 & 2 \\ 4 & 6 \end{pmatrix}$. Thus the second row of A_1 is one times the first row added to the second row. $\det(A) = -2$ and $\det(A_1) = -2$.

Theorem 2.5. *In Theorems 2.2 - 2.4 you can replace the word "row" with the word "column".*

There are two other major properties of determinants which do not involve row operations.

Theorem 2.6. *Let A and B be two $n \times n$ matrices. Then*

$$\boxed{\det(AB) = \det(A)\det(B).}$$

Also,

$$\boxed{\det(A) = \det(A^T).}$$

Example 2.10. Compare $\det(AB)$ and $\det(A)\det(B)$ for

$$A = \begin{pmatrix} 1 & 2 \\ -3 & 2 \end{pmatrix}, B = \begin{pmatrix} 3 & 2 \\ 4 & 1 \end{pmatrix}.$$

First

$$AB = \begin{pmatrix} 1 & 2 \\ -3 & 2 \end{pmatrix}\begin{pmatrix} 3 & 2 \\ 4 & 1 \end{pmatrix} = \begin{pmatrix} 11 & 4 \\ -1 & -4 \end{pmatrix}$$

and so

$$\det(AB) = \det\begin{pmatrix} 11 & 4 \\ -1 & -4 \end{pmatrix} = -40.$$

Now

$$\det(A) = \det\begin{pmatrix} 1 & 2 \\ -3 & 2 \end{pmatrix} = 8$$

and

$$\det(B) = \det\begin{pmatrix} 3 & 2 \\ 4 & 1 \end{pmatrix} = -5.$$

Thus $\det(A)\det(B) = 8 \times (-5) = -40$.

2.1.4 *Finding Determinants Using Row Operations*

Theorems 2.4 - 2.5 can be used to find determinants using row operations. As pointed out above, the method of Laplace expansion will not be practical for any matrix of large size. Here is an example in which all the row operations are used.

Example 2.11. Find the determinant of the matrix

$$A = \begin{pmatrix} 1 & 2 & 3 & 4 \\ 5 & 1 & 2 & 3 \\ 4 & 5 & 4 & 3 \\ 2 & 2 & -4 & 5 \end{pmatrix}$$

Replace the second row by (-5) times the first row added to it. Then replace the third row by (-4) times the first row added to it. Finally, replace the fourth row by (-2) times the first row added to it. This yields the matrix

$$B = \begin{pmatrix} 1 & 2 & 3 & 4 \\ 0 & -9 & -13 & -17 \\ 0 & -3 & -8 & -13 \\ 0 & -2 & -10 & -3 \end{pmatrix}$$

and from Theorem 2.4, it has the same determinant as A. Now using other row operations, $\det(B) = \left(\frac{-1}{3}\right) \det(C)$ where

$$C = \begin{pmatrix} 1 & 2 & 3 & 4 \\ 0 & 0 & 11 & 22 \\ 0 & -3 & -8 & -13 \\ 0 & 6 & 30 & 9 \end{pmatrix}.$$

The second row was replaced by (-3) times the third row added to the second row. By Theorem 2.4 this didn't change the value of the determinant. Then the last row was multiplied by (-3). By Theorem 2.2 the resulting matrix has a determinant which is (-3) times the determinant of the un-multiplied matrix. Therefore, we multiplied by $-1/3$ to retain the correct value. Now replace the last row with 2 times the third added to it. By Theorem 2.4, this does not change the value of the determinant. Finally switch the third and second rows. This causes the determinant to be multiplied by (-1). Thus $\det(C) = -\det(D)$ where

$$D = \begin{pmatrix} 1 & 2 & 3 & 4 \\ 0 & -3 & -8 & -13 \\ 0 & 0 & 11 & 22 \\ 0 & 0 & 14 & -17 \end{pmatrix}$$

You could do more row operations or you could note that this can be easily expanded along the first column followed by expanding the 3×3 matrix which results along its first column. Thus

$$\det(D) = 1(-3) \begin{vmatrix} 11 & 22 \\ 14 & -17 \end{vmatrix} = 1485$$

and so $\det(C) = -1485$ and $\det(A) = \det(B) = \left(\frac{-1}{3}\right)(-1485) = 495$.

Example 2.12. Find the determinant of the matrix

$$\begin{pmatrix} 1 & 2 & 3 & 2 \\ 1 & -3 & 2 & 1 \\ 2 & 1 & 2 & 5 \\ 3 & -4 & 1 & 2 \end{pmatrix}$$

Replace the second row by (-1) times the first row added to it. Next take -2 times the first row and add to the third and finally take -3 times the first row and add to the last row. This yields

$$\begin{pmatrix} 1 & 2 & 3 & 2 \\ 0 & -5 & -1 & -1 \\ 0 & -3 & -4 & 1 \\ 0 & -10 & -8 & -4 \end{pmatrix}.$$

By Theorem 2.4 this matrix has the same determinant as the original matrix. Remember you can work with the columns also. Take -5 times the last column and add to the second column. This yields

$$\begin{pmatrix} 1 & -8 & 3 & 2 \\ 0 & 0 & -1 & -1 \\ 0 & -8 & -4 & 1 \\ 0 & 10 & -8 & -4 \end{pmatrix}$$

By Theorem 2.5 this matrix has the same determinant as the original matrix. Now take (-1) times the third row and add to the top row. This gives.

$$\begin{pmatrix} 1 & 0 & 7 & 1 \\ 0 & 0 & -1 & -1 \\ 0 & -8 & -4 & 1 \\ 0 & 10 & -8 & -4 \end{pmatrix}$$

which by Theorem 2.4 has the same determinant as the original matrix. Let us expand it now along the first column. This yields the following for the determinant of the original matrix.

$$\det \begin{pmatrix} 0 & -1 & -1 \\ -8 & -4 & 1 \\ 10 & -8 & -4 \end{pmatrix}$$

which equals

$$8 \det \begin{pmatrix} -1 & -1 \\ -8 & -4 \end{pmatrix} + 10 \det \begin{pmatrix} -1 & -1 \\ -4 & 1 \end{pmatrix} = -82$$

Do not try to be fancy in using row operations. That is, stick mostly to the one which replaces a row or column with a multiple of another row or column added to it. Also note there is no way to check your answer other than working the problem more than one way. To be sure you have gotten it right you must do this.

2.2 Applications

2.2.1 *A Formula For The Inverse*

The definition of the determinant in terms of Laplace expansion along a row or column also provides a way to give a formula for the inverse of a matrix. Recall the definition of the inverse of a matrix in Definition 1.11 on Page 12. Also recall the definition of the cofactor matrix given in Definition 2.7 on Page 26. This cofactor matrix was just the matrix which results from replacing the ij^{th} entry of the matrix with the ij^{th} cofactor.

The following theorem says that to find the inverse, take the transpose of the cofactor matrix and divide by the determinant. The transpose of the cofactor matrix is called the **adjugate** or sometimes the **classical adjoint** of the matrix A. In other words, A^{-1} is equal to one divided by the determinant of A times the adjugate matrix of A. This is what the following theorem says with more precision. Its proof is in the appendix on the determinant, Theorem A.8 on page 313.

Theorem 2.7. A^{-1} *exists if and only if* $\det(A) \neq 0$. *If* $\det(A) \neq 0$, *then* $A^{-1} = \left(a_{ij}^{-1}\right)$ *where*

$$a_{ij}^{-1} = \det(A)^{-1} \operatorname{cof}(A)_{ji}$$

for $\operatorname{cof}(A)_{ij}$ *the* ij^{th} *cofactor of* A.

Example 2.13. Find the inverse of the matrix

$$A = \begin{pmatrix} 1 & 2 & 3 \\ 3 & 0 & 1 \\ 1 & 2 & 1 \end{pmatrix}$$

First find the determinant of this matrix. Using Theorems 2.4 - 2.5 on Page 29, the determinant of this matrix equals the determinant of the matrix

$$\begin{pmatrix} 1 & 2 & 3 \\ 0 & -6 & -8 \\ 0 & 0 & -2 \end{pmatrix}$$

which equals 12. The cofactor matrix of A is

$$\begin{pmatrix} -2 & -2 & 6 \\ 4 & -2 & 0 \\ 2 & 8 & -6 \end{pmatrix}.$$

Each entry of A was replaced by its cofactor. Therefore, from the above theorem, the inverse of A should equal

$$\frac{1}{12} \begin{pmatrix} -2 & -2 & 6 \\ 4 & -2 & 0 \\ 2 & 8 & -6 \end{pmatrix}^T = \begin{pmatrix} -\frac{1}{6} & \frac{1}{3} & \frac{1}{6} \\ -\frac{1}{6} & -\frac{1}{6} & \frac{2}{3} \\ \frac{1}{2} & 0 & -\frac{1}{2} \end{pmatrix}.$$

Does it work? You should check to see if it does. When the matrices are multiplied

$$\begin{pmatrix} -\frac{1}{6} & \frac{1}{3} & \frac{1}{6} \\ -\frac{1}{6} & -\frac{1}{6} & \frac{2}{3} \\ \frac{1}{2} & 0 & -\frac{1}{2} \end{pmatrix} \begin{pmatrix} 1 & 2 & 3 \\ 3 & 0 & 1 \\ 1 & 2 & 1 \end{pmatrix} = \begin{pmatrix} 1 & 0 & 0 \\ 0 & 1 & 0 \\ 0 & 0 & 1 \end{pmatrix}$$

and so it is correct.

Example 2.14. Find the inverse of the matrix

$$A = \begin{pmatrix} \frac{1}{2} & 0 & \frac{1}{2} \\ -\frac{1}{6} & \frac{1}{3} & -\frac{1}{2} \\ -\frac{5}{6} & \frac{2}{3} & -\frac{1}{2} \end{pmatrix}$$

First find its determinant. This determinant is $\frac{1}{6}$. The inverse is therefore equal to

$$6 \begin{pmatrix} \begin{vmatrix} \frac{1}{3} & -\frac{1}{2} \\ \frac{2}{3} & -\frac{1}{2} \end{vmatrix} & - \begin{vmatrix} -\frac{1}{6} & -\frac{1}{2} \\ -\frac{5}{6} & -\frac{1}{2} \end{vmatrix} & \begin{vmatrix} -\frac{1}{6} & \frac{1}{3} \\ -\frac{5}{6} & \frac{2}{3} \end{vmatrix} \\ - \begin{vmatrix} 0 & \frac{1}{2} \\ \frac{2}{3} & -\frac{1}{2} \end{vmatrix} & \begin{vmatrix} \frac{1}{2} & \frac{1}{2} \\ -\frac{5}{6} & -\frac{1}{2} \end{vmatrix} & - \begin{vmatrix} \frac{1}{2} & 0 \\ -\frac{5}{6} & \frac{2}{3} \end{vmatrix} \\ \begin{vmatrix} 0 & \frac{1}{2} \\ \frac{1}{3} & -\frac{1}{2} \end{vmatrix} & - \begin{vmatrix} \frac{1}{2} & \frac{1}{2} \\ -\frac{1}{6} & -\frac{1}{2} \end{vmatrix} & \begin{vmatrix} \frac{1}{2} & 0 \\ -\frac{1}{6} & \frac{1}{3} \end{vmatrix} \end{pmatrix}^T .$$

Expanding all the 2×2 determinants this yields

$$6 \begin{pmatrix} \frac{1}{6} & \frac{1}{3} & \frac{1}{6} \\ \frac{1}{3} & \frac{1}{6} & -\frac{1}{3} \\ -\frac{1}{6} & \frac{1}{6} & \frac{1}{6} \end{pmatrix}^T = \begin{pmatrix} 1 & 2 & -1 \\ 2 & 1 & 1 \\ 1 & -2 & 1 \end{pmatrix}$$

Always check your work.

$$\begin{pmatrix} 1 & 2 & -1 \\ 2 & 1 & 1 \\ 1 & -2 & 1 \end{pmatrix} \begin{pmatrix} \frac{1}{2} & 0 & \frac{1}{2} \\ -\frac{1}{6} & \frac{1}{3} & -\frac{1}{2} \\ -\frac{5}{6} & \frac{2}{3} & -\frac{1}{2} \end{pmatrix} = \begin{pmatrix} 1 & 0 & 0 \\ 0 & 1 & 0 \\ 0 & 0 & 1 \end{pmatrix}$$

and so it is correct. If the result of multiplying these matrices had been something other than the identity matrix, you would know there was an error. When this happens, you need to search for the mistake if you am interested in getting the right answer. A common mistake is to forget to take the transpose of the cofactor matrix.

This way of finding inverses is especially useful in the case where it is desired to find the inverse of a matrix whose entries are functions.

Example 2.15. Suppose

$$A(t) = \begin{pmatrix} e^t & 0 & 0 \\ 0 & \cos t & \sin t \\ 0 & -\sin t & \cos t \end{pmatrix}$$

Show that $A(t)^{-1}$ exists and then find it.

First note $\det(A(t)) = e^t \neq 0$ so $A(t)^{-1}$ exists. The cofactor matrix is

$$C(t) = \begin{pmatrix} 1 & 0 & 0 \\ 0 & e^t \cos t & e^t \sin t \\ 0 & -e^t \sin t & e^t \cos t \end{pmatrix}$$

and so the inverse is

$$\frac{1}{e^t} \begin{pmatrix} 1 & 0 & 0 \\ 0 & e^t \cos t & e^t \sin t \\ 0 & -e^t \sin t & e^t \cos t \end{pmatrix}^T = \begin{pmatrix} e^{-t} & 0 & 0 \\ 0 & \cos t & -\sin t \\ 0 & \sin t & \cos t \end{pmatrix}.$$

Example 2.16. Find the equation of the plane which contains the three points

$$(1, 2, 1), (3, -1, 2), \text{ and } (4, 2, 1).$$

You need to find numbers, a, b, c, d not all zero such that each of the given three points satisfies the equation

$$ax + by + cz = d.$$

Then you must have for (x, y, z) a point on this plane,

$$a + 2b + c - d = 0,$$
$$3a - b + 2c - d = 0,$$
$$4a + 2b + c - d = 0,$$
$$xa + yb + zc - d = 0.$$

You need a nonzero solution to the above system of four equations for the unknowns, $a, b, c,$ and d. Therefore, the determinant of the matrix of coefficients must equal 0 since this matrix cannot be one to one.

$$\det \begin{pmatrix} 1 & 2 & 1 & -1 \\ 3 & -1 & 2 & -1 \\ 4 & 2 & 1 & -1 \\ x & y & z & -1 \end{pmatrix} = 0$$

because the matrix sends a nonzero vector $(a, b, c, -d)$ to zero. Consequently from Theorem 2.7 on Page 32, its determinant equals zero. Hence, upon evaluating the determinant,

$$-15 + 9z + 3y = 0$$

which reduces to $3z + y = 5$.

2.2.2   *Cramer's Rule*

This formula for the inverse also implies a famous procedure known as **Cramer's rule**. Cramer's rule gives a formula for the solutions **x**, to a system of equations, $A\mathbf{x} = \mathbf{y}$ in the special case that A is a square matrix. Note this rule does not apply if you have a system of equations in which there is a different number of equations than variables.

In case you are solving a system of equations, $A\mathbf{x} = \mathbf{y}$ for **x**, it follows that if A^{-1} exists,

$$\mathbf{x} = \left(A^{-1}A\right)\mathbf{x} = A^{-1}\left(A\mathbf{x}\right) = A^{-1}\mathbf{y}$$

thus solving the system. Now in the case that A^{-1} exists, there is a formula for A^{-1} given above. Using this formula,

$$x_i = \sum_{j=1}^{n} a_{ij}^{-1} y_j = \sum_{j=1}^{n} \frac{1}{\det(A)} \operatorname{cof}(A)_{ji}\, y_j.$$

By the formula for the expansion of a determinant along a column,

$$x_i = \frac{1}{\det(A)} \det \begin{pmatrix} * & \cdots & y_1 & \cdots & * \\ \vdots & & \vdots & & \vdots \\ * & \cdots & y_n & \cdots & * \end{pmatrix},$$

where here the ith column of A is replaced with the column vector $(y_1, \cdots, y_n)^T$, and the determinant of this modified matrix is taken and divided by $\det(A)$. This formula is known as Cramer's rule.

Procedure 2.8. Suppose A is an $n \times n$ matrix and it is desired to solve the system $A\mathbf{x} = \mathbf{y}, \mathbf{y} = (y_1, \cdots, y_n)^T$ for $\mathbf{x} = (x_1, \cdots, x_n)^T$. Then Cramer's rule says

$$x_i = \frac{\det A_i}{\det A}$$

where A_i is obtained from A by replacing the ith column of A with the column $(y_1, \cdots, y_n)^T$.

Example 2.17. Find x, y if

$$\begin{pmatrix} 1 & 2 & 1 \\ 3 & 2 & 1 \\ 2 & -3 & 2 \end{pmatrix} \begin{pmatrix} x \\ y \\ z \end{pmatrix} = \begin{pmatrix} 1 \\ 2 \\ 3 \end{pmatrix}.$$

From Cramer's rule,

$$x = \frac{\begin{vmatrix} 1 & 2 & 1 \\ 2 & 2 & 1 \\ 3 & -3 & 2 \end{vmatrix}}{\begin{vmatrix} 1 & 2 & 1 \\ 3 & 2 & 1 \\ 2 & -3 & 2 \end{vmatrix}} = \frac{1}{2}$$

Now to find y, z

$$y = \frac{\begin{vmatrix} 1 & 1 & 1 \\ 3 & 2 & 1 \\ 2 & 3 & 2 \end{vmatrix}}{\begin{vmatrix} 1 & 2 & 1 \\ 3 & 2 & 1 \\ 2 & -3 & 2 \end{vmatrix}} = -\frac{1}{7}, \quad z = \frac{\begin{vmatrix} 1 & 2 & 1 \\ 3 & 2 & 2 \\ 2 & -3 & 3 \end{vmatrix}}{\begin{vmatrix} 1 & 2 & 1 \\ 3 & 2 & 1 \\ 2 & -3 & 2 \end{vmatrix}} = \frac{11}{14}.$$

You see the pattern. For large systems, Cramer's rule is less than useful if you want to find an answer. This is because to use it you must evaluate determinants. However, you have no practical way to evaluate determinants for large matrices other than row operations, and if you are using row operations, you might just as well use them to solve the system to begin with. It will be a lot less trouble. Nevertheless, there are situations in which Cramer's rule is useful.

Example 2.18. Solve for z if

$$\begin{pmatrix} 1 & 0 & 0 \\ 0 & e^t \cos t & e^t \sin t \\ 0 & -e^t \sin t & e^t \cos t \end{pmatrix} \begin{pmatrix} x \\ y \\ z \end{pmatrix} = \begin{pmatrix} 1 \\ t \\ t^2 \end{pmatrix}$$

You could do it by row operations but it might be easier in this case to use Cramer's rule because the matrix of coefficients does not consist of numbers but of functions. Thus

$$z = \frac{\begin{vmatrix} 1 & 0 & 1 \\ 0 & e^t \cos t & t \\ 0 & -e^t \sin t & t^2 \end{vmatrix}}{\begin{vmatrix} 1 & 0 & 0 \\ 0 & e^t \cos t & e^t \sin t \\ 0 & -e^t \sin t & e^t \cos t \end{vmatrix}} = t\left((\cos t)\, t + \sin t\right) e^{-t}.$$

You end up doing this sort of thing sometimes in ordinary differential equations in the method of variation of parameters.

2.3   Exercises

(1) Find the determinants of the following matrices.

(a) $\begin{pmatrix} 1 & 2 & 3 \\ 3 & 2 & 2 \\ 0 & 9 & 8 \end{pmatrix}$ (The answer is 31.)

(b) $\begin{pmatrix} 4 & 3 & 2 \\ 1 & 7 & 8 \\ 3 & -9 & 3 \end{pmatrix}$ (The answer is 375.)

(c) $\begin{pmatrix} 1 & 2 & 3 & 2 \\ 1 & 3 & 2 & 3 \\ 4 & 1 & 5 & 0 \\ 1 & 2 & 1 & 2 \end{pmatrix}$. (The answer is -2.)

(2) Find the following determinant by expanding along the first row and second column.

$$\begin{vmatrix} 1 & 2 & 1 \\ 2 & 1 & 3 \\ 2 & 1 & 1 \end{vmatrix}$$

(3) Find the following determinant by expanding along the first column and third row.

$$\begin{vmatrix} 1 & 2 & 1 \\ 1 & 0 & 1 \\ 2 & 1 & 1 \end{vmatrix}$$

(4) Find the following determinant by expanding along the second row and first column.

$$\begin{vmatrix} 1 & 2 & 1 \\ 2 & 1 & 3 \\ 2 & 1 & 1 \end{vmatrix}$$

(5) Compute the determinant by cofactor expansion. Pick the easiest row or column to use.

$$\begin{vmatrix} 1 & 0 & 0 & 1 \\ 2 & 1 & 1 & 0 \\ 0 & 0 & 0 & 2 \\ 2 & 1 & 3 & 1 \end{vmatrix}$$

(6) Find the determinant using row operations.

$$\begin{vmatrix} 1 & 2 & 1 \\ 2 & 3 & 2 \\ -4 & 1 & 2 \end{vmatrix}$$

(7) Find the determinant using row operations.

$$\begin{vmatrix} 2 & 1 & 3 \\ 2 & 4 & 2 \\ 1 & 4 & -5 \end{vmatrix}$$

(8) Find the determinant using row operations.

$$\begin{vmatrix} 1 & 2 & 1 & 2 \\ 3 & 1 & -2 & 3 \\ -1 & 0 & 3 & 1 \\ 2 & 3 & 2 & -2 \end{vmatrix}$$

(9) Find the determinant using row operations.

$$\begin{vmatrix} 1 & 4 & 1 & 2 \\ 3 & 2 & -2 & 3 \\ -1 & 0 & 3 & 3 \\ 2 & 1 & 2 & -2 \end{vmatrix}$$

(10) An operation is done to get from the first matrix to the second. Identify what was done and tell how it will affect the value of the determinant.

$$\begin{pmatrix} a & b \\ c & d \end{pmatrix}, \begin{pmatrix} a & c \\ b & d \end{pmatrix}$$

(11) An operation is done to get from the first matrix to the second. Identify what was done and tell how it will affect the value of the determinant.

$$\begin{pmatrix} a & b \\ c & d \end{pmatrix}, \begin{pmatrix} c & d \\ a & b \end{pmatrix}$$

(12) An operation is done to get from the first matrix to the second. Identify what was done and tell how it will affect the value of the determinant.

$$\begin{pmatrix} a & b \\ c & d \end{pmatrix}, \begin{pmatrix} a & b \\ a+c & b+d \end{pmatrix}$$

(13) An operation is done to get from the first matrix to the second. Identify what was done and tell how it will affect the value of the determinant.

$$\begin{pmatrix} a & b \\ c & d \end{pmatrix}, \begin{pmatrix} a & b \\ 2c & 2d \end{pmatrix}$$

(14) An operation is done to get from the first matrix to the second. Identify what was done and tell how it will affect the value of the determinant.

$$\begin{pmatrix} a & b \\ c & d \end{pmatrix}, \begin{pmatrix} b & a \\ d & c \end{pmatrix}$$

(15) Tell whether the statement is true or false.

(a) If A is a 3×3 matrix with a zero determinant, then one column must be a multiple of some other column.

(b) If any two columns of a square matrix are equal, then the determinant of the matrix equals zero.

(c) For A and B two $n \times n$ matrices $\det(A+B) = \det(A) + \det(B)$.

(d) For A an $n \times n$ matrix, $\det(3A) = 3\det(A)$

(e) If A^{-1} exists then $\det(A^{-1}) = \det(A)^{-1}$.

(f) If B is obtained by multiplying a single row of A by 4 then $\det(B) = 4\det(A)$.

(g) For A an $n \times n$ matrix, $\det(-A) = (-1)^n \det(A)$.

(h) If A is a real $n \times n$ matrix, then $\det(A^T A) \geq 0$.

(i) Cramer's rule is useful for finding solutions to systems of linear equations in which there is an infinite set of solutions.

(j) If $A^k = 0$ for some positive integer k, then $\det(A) = 0$.

(k) If $A\mathbf{x} = \mathbf{0}$ for some $\mathbf{x} \neq \mathbf{0}$, then $\det(A) = 0$.

(16) Verify an example of each property of determinants found in Theorems 2.4 - 2.5 for 2×2 matrices.

(17) A matrix is said to be **orthogonal** if $A^T A = I$. Thus the inverse of an orthogonal matrix is just its transpose. What are the possible values of $\det(A)$ if A is an orthogonal matrix?

(18) Fill in the missing entries to make the matrix orthogonal as in Problem 17.

$$\begin{pmatrix} \frac{-1}{\sqrt{2}} & \frac{1}{\sqrt{6}} & \frac{\sqrt{12}}{6} \\ \frac{1}{\sqrt{2}} & - & - \\ - & \frac{\sqrt{6}}{3} & - \end{pmatrix}.$$

(19) If A^{-1} exist, what is the relationship between $\det(A)$ and $\det(A^{-1})$. Explain your answer.

(20) Is it true that $\det(A + B) = \det(A) + \det(B)$? If this is so, explain why it is so and if it is not so, give a counter example.

(21) Let A be an $r \times r$ matrix and suppose there are $r - 1$ rows (columns) such that all rows (columns) are linear combinations of these $r - 1$ rows (columns). Show $\det(A) = 0$.

(22) Show $\det(aA) = a^n \det(A)$ where here A is an $n \times n$ matrix and a is a scalar.

(23) Suppose A is an upper triangular matrix. Show that A^{-1} exists if and only if all entries of the main diagonal are non zero. Is it true that A^{-1} will also be upper triangular? Explain. Is everything the same for lower triangular matrices?

(24) Let A and B be two $n \times n$ matrices. $A \sim B$ (A is **similar** to B) means there exists an invertible matrix S such that $A = S^{-1}BS$. Show that if $A \sim B$, then $B \sim A$. Show also that $A \sim A$ and that if $A \sim B$ and $B \sim C$, then $A \sim C$.

(25) In the context of Problem 24 show that if $A \sim B$, then $\det(A) = \det(B)$.

(26) Two $n \times n$ matrices A and B, are similar if $B = S^{-1}AS$ for some invertible $n \times n$ matrix S. Show that if two matrices are similar, they have the same characteristic polynomials. The characteristic polynomial of an $n \times n$ matrix M is the polynomial, $\det(\lambda I - M)$.

(27) Prove by doing computations that $\det(AB) = \det(A)\det(B)$ if A and B are 2×2 matrices.

(28) Illustrate with an example of 2×2 matrices that the determinant of a product equals the product of the determinants.

(29) An $n \times n$ matrix is called **nilpotent** if for some positive integer k it follows $A^k = 0$. If A is a nilpotent matrix and k is the smallest possible integer such that $A^k = 0$, what are the possible values of $\det(A)$?

(30) Use Cramer's rule to find the solution to
$$x + 2y = 1$$
$$2x - y = 2$$

(31) Use Cramer's rule to find the solution to

$$x + 2y + z = 1$$
$$2x - y - z = 2$$
$$x + z = 1$$

(32) Here is a matrix

$$\begin{pmatrix} 1 & 2 & 3 \\ 0 & 2 & 1 \\ 3 & 1 & 0 \end{pmatrix}$$

Determine whether the matrix has an inverse by finding whether the determinant is non zero.

(33) Here is a matrix

$$\begin{pmatrix} 1 & 0 & 0 \\ 0 & \cos t & -\sin t \\ 0 & \sin t & \cos t \end{pmatrix}$$

Does there exist a value of t for which this matrix fails to have an inverse? Explain.

(34) Here is a matrix

$$\begin{pmatrix} 1 & t & t^2 \\ 0 & 1 & 2t \\ t & 0 & 2 \end{pmatrix}$$

Does there exist a value of t for which this matrix fails to have an inverse? Explain.

(35) Here is a matrix

$$\begin{pmatrix} e^t & e^{-t}\cos t & e^{-t}\sin t \\ e^t & -e^{-t}\cos t - e^{-t}\sin t & -e^{-t}\sin t + e^{-t}\cos t \\ e^t & 2e^{-t}\sin t & -2e^{-t}\cos t \end{pmatrix}$$

Does there exist a value of t for which this matrix fails to have an inverse? Explain.

(36) Here is a matrix

$$\begin{pmatrix} e^t & \cosh t & \sinh t \\ e^t & \sinh t & \cosh t \\ e^t & \cosh t & \sinh t \end{pmatrix}$$

Does there exist a value of t for which this matrix fails to have an inverse? Explain.

(37) Use the formula for the inverse in terms of the cofactor matrix to find if possible the inverses of the matrices

$$\begin{pmatrix} 1 & 1 \\ 1 & 2 \end{pmatrix}, \begin{pmatrix} 1 & 2 & 3 \\ 0 & 2 & 1 \\ 4 & 1 & 1 \end{pmatrix}, \begin{pmatrix} 1 & 2 & 1 \\ 2 & 3 & 0 \\ 0 & 1 & 2 \end{pmatrix}.$$

If it is not possible to take the inverse, explain why.

(38) Use the formula for the inverse in terms of the cofactor matrix to find the inverse of the matrix

$$A = \begin{pmatrix} e^t & 0 & 0 \\ 0 & e^t \cos t & e^t \sin t \\ 0 & e^t \cos t - e^t \sin t & e^t \cos t + e^t \sin t \end{pmatrix}.$$

(39) Find the inverse if it exists of the matrix

$$\begin{pmatrix} e^t & \cos t & \sin t \\ e^t & -\sin t & \cos t \\ e^t & -\cos t & -\sin t \end{pmatrix}.$$

(40) *Let $F(t) = \det \begin{pmatrix} a(t) & b(t) \\ c(t) & d(t) \end{pmatrix}$. Verify

$$F'(t) = \det \begin{pmatrix} a'(t) & b'(t) \\ c(t) & d(t) \end{pmatrix} + \det \begin{pmatrix} a(t) & b(t) \\ c'(t) & d'(t) \end{pmatrix}.$$

Now suppose

$$F(t) = \det \begin{pmatrix} a(t) & b(t) & c(t) \\ d(t) & e(t) & f(t) \\ g(t) & h(t) & i(t) \end{pmatrix}.$$

Use Laplace expansion and the first part to verify $F'(t) =$

$$\det \begin{pmatrix} a'(t) & b'(t) & c'(t) \\ d(t) & e(t) & f(t) \\ g(t) & h(t) & i(t) \end{pmatrix} + \det \begin{pmatrix} a(t) & b(t) & c(t) \\ d'(t) & e'(t) & f'(t) \\ g(t) & h(t) & i(t) \end{pmatrix}$$

$$+ \det \begin{pmatrix} a(t) & b(t) & c(t) \\ d(t) & e(t) & f(t) \\ g'(t) & h'(t) & i'(t) \end{pmatrix}.$$

Conjecture a general result valid for $n \times n$ matrices and explain why it will be true. Can a similar thing be done with the columns?

(41) *Let $Ly = y^{(n)} + a_{n-1}(x) y^{(n-1)} + \cdots + a_1(x) y' + a_0(x) y$ where the a_i are given continuous functions defined on a closed interval (a, b) and y is some function which has n derivatives so it makes sense to write Ly. Suppose $Ly_k = 0$ for $k = 1, 2, \cdots, n$. The **Wronskian** of these functions y_i is defined as

$$W(y_1, \cdots, y_n)(x) \equiv \det \begin{pmatrix} y_1(x) & \cdots & y_n(x) \\ y_1'(x) & \cdots & y_n'(x) \\ \vdots & & \vdots \\ y_1^{(n-1)}(x) & \cdots & y_n^{(n-1)}(x) \end{pmatrix}$$

Show that for $W(x) = W(y_1, \cdots, y_n)(x)$ to save space,

$$W'(x) = \det \begin{pmatrix} y_1(x) & \cdots & y_n(x) \\ y_1'(x) & \cdots & y_n'(x) \\ \vdots & & \vdots \\ y_1^{(n)}(x) & \cdots & y_n^{(n)}(x) \end{pmatrix}.$$

Now use the differential equation $Ly = 0$ which is satisfied by each of these functions y_i and properties of determinants presented above to verify that $W' + a_{n-1}(x)W = 0$. Give an explicit solution of this linear differential equation **Abel's formula**, and use your answer to verify that the Wronskian of these solutions to the equation $Ly = 0$ either vanishes identically on (a, b) or never. **Hint:** To solve the differential equation, let $A'(x) = a_{n-1}(x)$ and multiply both sides of the differential equation by $e^{A(x)}$ and then argue the left side is the derivative of something.

(42) A determinant of the form

$$
\begin{vmatrix}
1 & 1 & \cdots & 1 \\
a_0 & a_1 & \cdots & a_n \\
a_0^2 & a_1^2 & \cdots & a_n^2 \\
\vdots & \vdots & & \vdots \\
a_0^{n-1} & a_1^{n-1} & \cdots & a_n^{n-1} \\
a_0^n & a_1^n & \cdots & a_n^n
\end{vmatrix}
$$

is called a Vandermonde determinant. Show this determinant equals

$$
\prod_{0 \le i < j \le n} (a_j - a_i)
$$

By this is meant to take the product of all terms of the form $(a_j - a_i)$ such that $j > i$. **Hint:** Show it works if $n = 1$ so you are looking at

$$
\begin{vmatrix}
1 & 1 \\
a_0 & a_1
\end{vmatrix}
$$

Then suppose it holds for $n - 1$ and consider the case n. Consider the following polynomial.

$$
p(t) \equiv
\begin{vmatrix}
1 & 1 & \cdots & 1 \\
a_0 & a_1 & \cdots & t \\
a_0^2 & a_1^2 & \cdots & t^2 \\
\vdots & \vdots & & \vdots \\
a_0^{n-1} & a_1^{n-1} & \cdots & t^{n-1} \\
a_0^n & a_1^n & \cdots & t^n
\end{vmatrix}.
$$

Explain why $p(a_j) = 0$ for $i = 0, \cdots, n - 1$. Thus

$$
p(t) = c \prod_{i=0}^{n-1} (t - a_i).
$$

Of course c is the coefficient of t^n. Find this coefficient from the above description of $p(t)$ and the induction hypothesis. Then plug in $t = a_n$ and observe you have the formula valid for n.

Chapter 3

Spectral Theory

3.1 Eigenvalues And Eigenvectors Of A Matrix

Spectral Theory refers to the study of eigenvalues and eigenvectors of a matrix. It is of fundamental importance in many areas.

3.1.1 Definition Of Eigenvectors And Eigenvalues

In this section, the field of scalars $\mathbb{F}$ will equal $\mathbb{C}$, the complex numbers, if nothing is stated to the contrary.

To illustrate the idea behind what will be discussed, consider the following example.

Example 3.1. Here is a matrix.

$$\begin{pmatrix} 0 & 5 & -10 \\ 0 & 22 & 16 \\ 0 & -9 & -2 \end{pmatrix}.$$

Multiply this matrix by the vector

$$\begin{pmatrix} -5 \\ -4 \\ 3 \end{pmatrix}$$

and see what happens. Then multiply it by

$$\begin{pmatrix} 1 \\ 0 \\ 0 \end{pmatrix}$$

and see what happens. Does this matrix act this way for some other vector?

First

$$\begin{pmatrix} 0 & 5 & -10 \\ 0 & 22 & 16 \\ 0 & -9 & -2 \end{pmatrix} \begin{pmatrix} -5 \\ -4 \\ 3 \end{pmatrix} = \begin{pmatrix} -50 \\ -40 \\ 30 \end{pmatrix} = 10 \begin{pmatrix} -5 \\ -4 \\ 3 \end{pmatrix}.$$

Next

$$\begin{pmatrix} 0 & 5 & -10 \\ 0 & 22 & 16 \\ 0 & -9 & -2 \end{pmatrix} \begin{pmatrix} 1 \\ 0 \\ 0 \end{pmatrix} = \begin{pmatrix} 0 \\ 0 \\ 0 \end{pmatrix} = 0 \begin{pmatrix} 1 \\ 0 \\ 0 \end{pmatrix}.$$

When you multiply the first vector by the given matrix, it stretched the vector, multiplying it by 10. When you multiplied the matrix by the second vector it sent it to the zero vector. Now consider

$$\begin{pmatrix} 0 & 5 & -10 \\ 0 & 22 & 16 \\ 0 & -9 & -2 \end{pmatrix} \begin{pmatrix} 1 \\ 1 \\ 1 \end{pmatrix} = \begin{pmatrix} -5 \\ 38 \\ -11 \end{pmatrix}.$$

In this case, multiplication by the matrix did not result in merely multiplying the vector by a number.

In the above example, the first two vectors are called eigenvectors and the numbers, 10 and 0 are called eigenvalues. Not every number is an eigenvalue and not every vector is an eigenvector.

Definition 3.1. Let M be an $n \times n$ matrix and let $\mathbf{x} \in \mathbb{C}^n$ be a **nonzero** **vector** for which

$$M\mathbf{x} = \lambda \mathbf{x} \tag{3.1}$$

for some scalar λ. Then $\mathbf{x}$ is called an **eigenvector** and λ is called an **eigenvalue** (**characteristic value**) of the matrix M.

> **Note: Eigenvectors are never equal to zero!**

The set of all eigenvalues of an $n \times n$ matrix M, is denoted by $\sigma(M)$ and is referred to as the **spectrum** of M.

The eigenvectors of a matrix M are those nonzero vectors $\mathbf{x}$ for which multiplication by M results in a scalar multiple of $\mathbf{x}$. As noted above, $\mathbf{0}$ is never allowed to be an eigenvector. How can eigenvectors be identified? Suppose $\mathbf{x}$ satisfies (3.1). Then

$$(M - \lambda I)\mathbf{x} = \mathbf{0}$$

for some $\mathbf{x} \neq \mathbf{0}$. (Equivalently, you could write $(\lambda I - M)\mathbf{x} = \mathbf{0}$.) Sometimes we will use

$$(\lambda I - M)\mathbf{x} = \mathbf{0}$$

and sometimes

$$(M - \lambda I)\mathbf{x} = \mathbf{0}.$$

It makes absolutely no difference because the two equations have the same solutions so you should use whichever you like better. Therefore, the matrix $M - \lambda I$ cannot have an inverse because if it did, the equation could be solved,

$$\mathbf{x} = \left((M - \lambda I)^{-1} (M - \lambda I) \right) \mathbf{x} = (M - \lambda I)^{-1} ((M - \lambda I) \mathbf{x}) = (M - \lambda I)^{-1} \mathbf{0} = \mathbf{0},$$

and this would require $\mathbf{x} = \mathbf{0}$, contrary to the requirement that $\mathbf{x} \neq \mathbf{0}$. By Theorem 2.7 on Page 32,

$$\det (M - \lambda I) = 0. \tag{3.2}$$

(Equivalently you could write $\det (\lambda I - M) = 0$.) The expression $\det (\lambda I - M)$ or equivalently, $\det (M - \lambda I)$ is a polynomial called the **characteristic polynomial** and the above equation is called the characteristic equation. For M an $n \times n$ matrix, it follows from the theorem on expanding a matrix by its cofactor, that $\det (M - \lambda I)$ is a polynomial of degree n. As such, the equation (3.2) has a solution $\lambda \in \mathbb{C}$ by the fundamental theorem of algebra. Is it actually an eigenvalue? The answer is yes from Theorem 2.7 which implies that $M - \lambda I$ has no inverse. From linear algebra, it follows $M - \lambda I$ is not one to one and so there exists a nonzero vector $\mathbf{x}$ such that $(M - \lambda I) \mathbf{x} = \mathbf{0}$. If you are interested in the proof of this assertion, see the appendix on the determinant. This proves the following corollary.

Corollary 3.1. *Let M be an $n \times n$ matrix and $\det (M - \lambda I) = 0$. Then there exists a nonzero vector $\mathbf{x} \in \mathbb{C}^n$ such that $(M - \lambda I) \mathbf{x} = \mathbf{0}$.*

3.1.2 Finding Eigenvectors And Eigenvalues

As an example, consider the following.

Example 3.2. Find the eigenvalues and eigenvectors for the matrix

$$A = \begin{pmatrix} 5 & -10 & -5 \\ 2 & 14 & 2 \\ -4 & -8 & 6 \end{pmatrix}.$$

You first need to identify the eigenvalues. Recall this requires the solution of the equation

$$\det (A - \lambda I) = 0.$$

In this case this equation is

$$\det \left(\begin{pmatrix} 5 & -10 & -5 \\ 2 & 14 & 2 \\ -4 & -8 & 6 \end{pmatrix} - \lambda \begin{pmatrix} 1 & 0 & 0 \\ 0 & 1 & 0 \\ 0 & 0 & 1 \end{pmatrix} \right) = 0$$

When you expand this determinant and simplify, you find the equation you need to solve is

$$(\lambda - 5) (\lambda^2 - 20\lambda + 100) = 0$$

and so the eigenvalues are

$$5, 10, 10.$$

10 is listed twice because it is a zero of multiplicity two due to

$$\lambda^2 - 20\lambda + 100 = (\lambda - 10)^2.$$

Having found the eigenvalues, it only remains to find the eigenvectors. First find the eigenvectors for $\lambda = 5$. As explained above, this requires you to solve the equation

$$\left(\begin{pmatrix} 5 & -10 & -5 \\ 2 & 14 & 2 \\ -4 & -8 & 6 \end{pmatrix} - 5 \begin{pmatrix} 1 & 0 & 0 \\ 0 & 1 & 0 \\ 0 & 0 & 1 \end{pmatrix} \right) \begin{pmatrix} x \\ y \\ z \end{pmatrix} = \begin{pmatrix} 0 \\ 0 \\ 0 \end{pmatrix}.$$

That is you need to find the solution to

$$\begin{pmatrix} 0 & -10 & -5 \\ 2 & 9 & 2 \\ -4 & -8 & 1 \end{pmatrix} \begin{pmatrix} x \\ y \\ z \end{pmatrix} = \begin{pmatrix} 0 \\ 0 \\ 0 \end{pmatrix}$$

By now this is an old problem. You set up the augmented matrix and row reduce to get the solution. Thus the matrix you must row reduce is

$$\begin{pmatrix} 0 & -10 & -5 & | & 0 \\ 2 & 9 & 2 & | & 0 \\ -4 & -8 & 1 & | & 0 \end{pmatrix}. \tag{3.3}$$

After row operations, you can obtain the following

$$\begin{pmatrix} 1 & 0 & -\frac{5}{4} & | & 0 \\ 0 & 1 & \frac{1}{2} & | & 0 \\ 0 & 0 & 0 & | & 0 \end{pmatrix}$$

and so the solution is any vector of the form

$$\begin{pmatrix} \frac{5}{4}t \\ \frac{-1}{2}t \\ t \end{pmatrix} = t \begin{pmatrix} \frac{5}{4} \\ \frac{-1}{2} \\ 1 \end{pmatrix}$$

where $t \in \mathbb{F}$. You would obtain the same collection of vectors if you replaced t with $4t$. Thus a simpler description for the solutions to this system of equations whose augmented matrix is in (3.3) is

$$t \begin{pmatrix} 5 \\ -2 \\ 4 \end{pmatrix} \tag{3.4}$$

where $t \in \mathbb{F}$. Now you need to remember that you cannot take $t = 0$ because this would result in the zero vector and

$$\boxed{\textbf{Eigenvectors are \underline{never} \underline{equal} \underline{to} \underline{zero}!}}$$

Other than this value, every other choice of t in (3.4) results in an eigenvector. It is a good idea to check your work! To do so, we will take the original matrix and multiply by this vector and see if we get 5 times this vector.

$$\begin{pmatrix} 5 & -10 & -5 \\ 2 & 14 & 2 \\ -4 & -8 & 6 \end{pmatrix} \begin{pmatrix} 5 \\ -2 \\ 4 \end{pmatrix} = \begin{pmatrix} 25 \\ -10 \\ 20 \end{pmatrix} = 5 \begin{pmatrix} 5 \\ -2 \\ 4 \end{pmatrix}$$

so it appears this is correct. Always check your work on these problems if you care about getting the answer right.

Next consider the eigenvectors for $\lambda = 10$. These vectors are solutions to the equation

$$\left(\begin{pmatrix} 5 & -10 & -5 \\ 2 & 14 & 2 \\ -4 & -8 & 6 \end{pmatrix} - 10 \begin{pmatrix} 1 & 0 & 0 \\ 0 & 1 & 0 \\ 0 & 0 & 1 \end{pmatrix} \right) \begin{pmatrix} x \\ y \\ z \end{pmatrix} = \begin{pmatrix} 0 \\ 0 \\ 0 \end{pmatrix}$$

That is you must find the solutions to

$$\begin{pmatrix} -5 & -10 & -5 \\ 2 & 4 & 2 \\ -4 & -8 & -4 \end{pmatrix} \begin{pmatrix} x \\ y \\ z \end{pmatrix} = \begin{pmatrix} 0 \\ 0 \\ 0 \end{pmatrix}$$

which reduces to consideration of the augmented matrix

$$\begin{pmatrix} -5 & -10 & -5 & | & 0 \\ 2 & 4 & 2 & | & 0 \\ -4 & -8 & -4 & | & 0 \end{pmatrix}$$

After row operations,

$$\begin{pmatrix} 1 & 2 & 1 & | & 0 \\ 0 & 0 & 0 & | & 0 \\ 0 & 0 & 0 & | & 0 \end{pmatrix}$$

and so the eigenvectors are of the form

$$\begin{pmatrix} -2s - t \\ s \\ t \end{pmatrix} = s \begin{pmatrix} -2 \\ 1 \\ 0 \end{pmatrix} + t \begin{pmatrix} -1 \\ 0 \\ 1 \end{pmatrix}.$$

You cannot pick t and s both equal to zero because this would result in the zero vector and

$$\boxed{\text{Eigenvectors } \underline{\text{are}} \ \underline{\text{never}} \ \underline{\text{equal}} \ \underline{\text{to}} \ \underline{\text{zero}}!}$$

However, every other choice of t and s does result in an eigenvector for the eigenvalue $\lambda = 10$. As in the case for $\lambda = 5$ you should check your work if you care about getting it right.

$$\begin{pmatrix} 5 & -10 & -5 \\ 2 & 14 & 2 \\ -4 & -8 & 6 \end{pmatrix} \begin{pmatrix} -1 \\ 0 \\ 1 \end{pmatrix} = \begin{pmatrix} -10 \\ 0 \\ 10 \end{pmatrix} = 10 \begin{pmatrix} -1 \\ 0 \\ 1 \end{pmatrix}$$

so it worked. The other vector will also work. Check it.

Here is another example.

Example 3.3. Let

$$A = \begin{pmatrix} 2 & 2 & -2 \\ 1 & 3 & -1 \\ -1 & 1 & 1 \end{pmatrix}$$

First find the eigenvalues.

$$\det\left(\begin{pmatrix} 2 & 2 & -2 \\ 1 & 3 & -1 \\ -1 & 1 & 1 \end{pmatrix} - \lambda \begin{pmatrix} 1 & 0 & 0 \\ 0 & 1 & 0 \\ 0 & 0 & 1 \end{pmatrix}\right) = 0$$

This reduces to $\lambda^3 - 6\lambda^2 + 8\lambda = 0$ and the solutions are 0, 2, and 4.

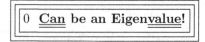

Now find the eigenvectors. For $\lambda = 0$ the augmented matrix for finding the solutions is

$$\begin{pmatrix} 2 & 2 & -2 & | & 0 \\ 1 & 3 & -1 & | & 0 \\ -1 & 1 & 1 & | & 0 \end{pmatrix}$$

and a reduced echelon form is

$$\begin{pmatrix} 1 & 0 & -1 & | & 0 \\ 0 & 1 & 0 & | & 0 \\ 0 & 0 & 0 & | & 0 \end{pmatrix}$$

Therefore, the eigenvectors are of the form

$$t \begin{pmatrix} 1 \\ 0 \\ 1 \end{pmatrix}$$

where $t \neq 0$.

Next find the eigenvectors for $\lambda = 2$. The augmented matrix for the system of equations needed to find these eigenvectors is

$$\begin{pmatrix} 0 & 2 & -2 & | & 0 \\ 1 & 1 & -1 & | & 0 \\ -1 & 1 & -1 & | & 0 \end{pmatrix}$$

and a reduced echelon form is

$$\begin{pmatrix} 1 & 0 & 0 & | & 0 \\ 0 & 1 & -1 & | & 0 \\ 0 & 0 & 0 & | & 0 \end{pmatrix}$$

and so the eigenvectors are of the form

$$t \begin{pmatrix} 0 \\ 1 \\ 1 \end{pmatrix}$$

where $t \neq 0$.

Finally find the eigenvectors for $\lambda = 4$. The augmented matrix for the system of equations needed to find these eigenvectors is

$$\begin{pmatrix} -2 & 2 & -2 & | & 0 \\ 1 & -1 & -1 & | & 0 \\ -1 & 1 & -3 & | & 0 \end{pmatrix}$$

and a reduced echelon form is

$$\begin{pmatrix} 1 & -1 & 0 & | & 0 \\ 0 & 0 & 1 & | & 0 \\ 0 & 0 & 0 & | & 0 \end{pmatrix}.$$

Therefore, the eigenvectors are of the form

$$t \begin{pmatrix} 1 \\ 1 \\ 0 \end{pmatrix}$$

where $t \neq 0$.

3.1.3 Volumes

The determinant and the concept of eigenvalues and eigenvectors provide a way to give a unified treatment of the concept of volumes in various dimensions. First here is a useful theorem which is of considerable interest for its own sake.

Theorem 3.1. *Let A be an $n \times n$ matrix. Then*

$$\det(A) = \prod_{i=1}^{n} \lambda_i$$

where λ_i are the eigenvalues of A. In words, the determinant of a matrix equals the product of its eigenvalues.

Proof: The characteristic polynomial is $\det(\lambda I - A) = \prod_{j=1}^{n}(\lambda - \lambda_j)$ where λ_i are the eigenvalues. This follows from the fundamental theorem of algebra which says every polynomial can be factored. Then, letting $\lambda = 0$ it follows $\det(-A) = (-1)^n \det(A) = \prod_{j=1}^{n}(0 - \lambda_j) = (-1)^n \prod_{j=1}^{n} \lambda_j$. ∎

Recall the geometric definition of the cross product of two vectors. The magnitude of the cross product of two vectors was the area of the parallelogram determined by the two vectors. There was also a coordinate description of the cross product.

The ith coordinate of the cross product is given by $\varepsilon_{ijk} u_j v_k$ where the two vectors are (u_1, u_2, u_3) and (v_1, v_2, v_3). Therefore, using the reduction identities,

$$
\begin{aligned}
|\mathbf{u} \times \mathbf{v}|^2 &= \varepsilon_{ijk} u_j v_k \varepsilon_{irs} u_r v_s \\
&= (\delta_{jr}\delta_{ks} - \delta_{kr}\delta_{js}) u_j v_k u_r v_s \\
&= u_j v_k u_j v_k - u_j v_k u_k v_j \\
&= (\mathbf{u} \cdot \mathbf{u})(\mathbf{v} \cdot \mathbf{v}) - (\mathbf{u} \cdot \mathbf{v})^2
\end{aligned}
$$

which equals

$$
\det \begin{pmatrix} \mathbf{u} \cdot \mathbf{u} & \mathbf{u} \cdot \mathbf{v} \\ \mathbf{u} \cdot \mathbf{v} & \mathbf{v} \cdot \mathbf{v} \end{pmatrix}.
$$

Now recall the box product and how the box product was $\pm$ the volume of the parallelepiped spanned by the three vectors. From the definition of the box product

$$
\mathbf{u} \times \mathbf{v} \cdot \mathbf{w} = \begin{vmatrix} \mathbf{i} & \mathbf{j} & \mathbf{k} \\ u_1 & u_2 & u_3 \\ v_1 & v_2 & v_3 \end{vmatrix} \cdot (w_1 \mathbf{i} + w_2 \mathbf{j} + w_3 \mathbf{k})
$$

$$
= \det \begin{pmatrix} w_1 & w_2 & w_3 \\ u_1 & u_2 & u_3 \\ v_1 & v_2 & v_3 \end{pmatrix}.
$$

Therefore,

$$
|\mathbf{u} \times \mathbf{v} \cdot \mathbf{w}|^2 = \det \begin{pmatrix} w_1 & w_2 & w_3 \\ u_1 & u_2 & u_3 \\ v_1 & v_2 & v_3 \end{pmatrix}^2
$$

which from the theory of determinants equals

$$
\det \begin{pmatrix} u_1 & u_2 & u_3 \\ v_1 & v_2 & v_3 \\ w_1 & w_2 & w_3 \end{pmatrix} \det \begin{pmatrix} u_1 & v_1 & w_1 \\ u_2 & v_2 & w_2 \\ u_3 & v_3 & w_3 \end{pmatrix} =
$$

$$
\det \left(\begin{pmatrix} u_1 & u_2 & u_3 \\ v_1 & v_2 & v_3 \\ w_1 & w_2 & w_3 \end{pmatrix} \begin{pmatrix} u_1 & v_1 & w_1 \\ u_2 & v_2 & w_2 \\ u_3 & v_3 & w_3 \end{pmatrix} \right) =
$$

$$
\det \begin{pmatrix} u_1^2 + u_2^2 + u_3^2 & u_1 v_1 + u_2 v_2 + u_3 v_3 & u_1 w_1 + u_2 w_2 + u_3 w_3 \\ u_1 v_1 + u_2 v_2 + u_3 v_3 & v_1^2 + v_2^2 + v_3^2 & v_1 w_1 + v_2 w_2 + v_3 w_3 \\ u_1 w_1 + u_2 w_2 + u_3 w_3 & v_1 w_1 + v_2 w_2 + v_3 w_3 & w_1^2 + w_2^2 + w_3^2 \end{pmatrix}
$$

$$
= \det \begin{pmatrix} \mathbf{u} \cdot \mathbf{u} & \mathbf{u} \cdot \mathbf{v} & \mathbf{u} \cdot \mathbf{w} \\ \mathbf{u} \cdot \mathbf{v} & \mathbf{v} \cdot \mathbf{v} & \mathbf{v} \cdot \mathbf{w} \\ \mathbf{u} \cdot \mathbf{w} & \mathbf{v} \cdot \mathbf{w} & \mathbf{w} \cdot \mathbf{w} \end{pmatrix}
$$

You see there is a definite pattern emerging here. These earlier cases were for a parallelepiped determined by either two or three vectors in $\mathbb{R}^3$. It makes sense to speak of a parallelepiped in any number of dimensions.

Definition 3.2. Let $\mathbf{u}_1, \cdots, \mathbf{u}_p$ be vectors in $\mathbb{R}^k$. The parallelepiped determined by these vectors will be denoted by $P(\mathbf{u}_1, \cdots, \mathbf{u}_p)$ and it is defined as

$$P(\mathbf{u}_1, \cdots, \mathbf{u}_p) \equiv \left\{ \sum_{j=1}^{p} s_j \mathbf{u}_j : s_j \in [0,1] \right\}.$$

The volume of this parallelepiped is defined as

$$\text{volume of } P(\mathbf{u}_1, \cdots, \mathbf{u}_p) \equiv \left(\det \left(\mathbf{u}_i \cdot \mathbf{u}_j \right) \right)^{1/2}.$$

In this definition, $\mathbf{u}_i \cdot \mathbf{u}_j$ is the ij^{th} entry of a $p \times p$ matrix. Note this definition agrees with all earlier notions of area and volume for parallelepipeds and it makes sense in any number of dimensions. However, it is important to verify the above determinant is nonnegative. After all, the above definition requires a square root of this determinant.

Lemma 3.1. *Let* $\mathbf{u}_1, \cdots, \mathbf{u}_p$ *be vectors in* $\mathbb{R}^k$ *for some* k. *Then* $\det \left(\mathbf{u}_i \cdot \mathbf{u}_j \right) \geq 0$.

Proof: Recall $\mathbf{v} \cdot \mathbf{w} = \mathbf{v}^T \mathbf{w}$. Therefore, in terms of matrix multiplication, the matrix $(\mathbf{u}_i \cdot \mathbf{u}_j)$ is just the following

$$\overbrace{\begin{pmatrix} \mathbf{u}_1^T \\ \vdots \\ \mathbf{u}_p^T \end{pmatrix}}^{p \times k} \overbrace{\begin{pmatrix} \mathbf{u}_1 & \cdots & \mathbf{u}_p \end{pmatrix}}^{k \times p}$$

which is of the form

$$U^T U.$$

Now the eigenvalues of the matrix $U^T U$ are all nonnegative. Here is why. Suppose $U^T U \mathbf{x} = \lambda \mathbf{x}$. Then

$$0 \leq \overline{\mathbf{x}}^T U^T U \mathbf{x} = \lambda \overline{\mathbf{x}}^T \mathbf{x} = \lambda \sum_k |x_k|^2.$$

Therefore, from Theorem 3.1, $\det \left(U^T U \right) \geq 0$ because it is the product of nonnegative numbers. ∎

Note it gives the right answer in the case where all the vectors are perpendicular. Here is why. Suppose $\{ \mathbf{u}_1, \cdots, \mathbf{u}_p \}$ are vectors which have the property that $\mathbf{u}_i \cdot \mathbf{u}_j = 0$ if $i \neq j$. Thus $P(\mathbf{u}_1, \cdots, \mathbf{u}_p)$ is a box which has all p sides perpendicular. What should its p dimensional volume be? Shouldn't it equal the product of the lengths of the sides? What does $\det \left(\mathbf{u}_i \cdot \mathbf{u}_j \right)$ give? The matrix $(\mathbf{u}_i \cdot \mathbf{u}_j)$ is a diagonal matrix having the squares of the magnitudes of the sides down the diagonal. Therefore, $\det \left(\mathbf{u}_i \cdot \mathbf{u}_j \right)^{1/2}$ equals the product of the lengths of the sides as it should.

The matrix $(\mathbf{u}_i \cdot \mathbf{u}_j)$ whose determinant gives the square of the volume of the parallelepiped spanned by the vectors $\{\mathbf{u}_1, \cdots, \mathbf{u}_p\}$ is called the Gramian matrix and sometimes the metric tensor.

These considerations are of great significance because they allow the computation in a systematic manner of k dimensional volumes of parallelepipeds which happen to be in $\mathbb{R}^n$ for $n \neq k$. Think for example of a plane in $\mathbb{R}^3$ and the problem of finding the two dimensional area of something on this plane.

Example 3.4. Find the equation of the plane containing the three points

$$(1, 2, 3), (0, 2, 1), (3, 1, 0).$$

These three points determine two vectors, the one from $(0, 2, 1)$ to $(1, 2, 3)$, $\mathbf{i} + 0\mathbf{j} + 2\mathbf{k}$, and the one from $(0, 2, 1)$ to $(3, 1, 0)$, $3\mathbf{i} + (-1)\mathbf{j} + (-1)\mathbf{k}$. If (x, y, z) denotes a point in the plane, then the volume of the parallelepiped spanned by the vector from $(0, 2, 1)$ to (x, y, z) and these other two vectors must be zero. Thus

$$\det \begin{pmatrix} x & y-2 & z-1 \\ 3 & -1 & -1 \\ 1 & 0 & 2 \end{pmatrix} = 0$$

Therefore, $-2x - 7y + 13 + z = 0$ is the equation of the plane. You should check it contains all three points.

Example 3.5. In the above example find the two dimensional volume of the parallelogram determined by the two vectors $\mathbf{i} + 0\mathbf{j} + 2\mathbf{k}$ and $3\mathbf{i} + (-1)\mathbf{j} + (-1)\mathbf{k}$.

This is easy. Just take the square root of the determinant of the Grammian matrix. Thus the area is

$$\sqrt{\det \begin{pmatrix} 5 & 1 \\ 1 & 11 \end{pmatrix}} = 3\sqrt{6}.$$

3.1.4   *Complex Eigenvalues*

Sometimes you have to consider eigenvalues which are complex numbers. This occurs in differential equations for example. You do these problems exactly the same way as you do the ones in which the eigenvalues are real. Here is an example.

Example 3.6. Find the eigenvalues and eigenvectors of the matrix

$$A = \begin{pmatrix} 1 & 0 & 0 \\ 0 & 2 & -1 \\ 0 & 1 & 2 \end{pmatrix}.$$

You need to find the eigenvalues. Solve

$$\det \left(\begin{pmatrix} 1 & 0 & 0 \\ 0 & 2 & -1 \\ 0 & 1 & 2 \end{pmatrix} - \lambda \begin{pmatrix} 1 & 0 & 0 \\ 0 & 1 & 0 \\ 0 & 0 & 1 \end{pmatrix} \right) = 0.$$

This reduces to $(\lambda - 1)\left(\lambda^2 - 4\lambda + 5\right) = 0$. The solutions are $\lambda = 1, \lambda = 2 + i, \lambda = 2 - i$.

There is nothing new about finding the eigenvectors for $\lambda = 1$ so consider the eigenvalue $\lambda = 2 + i$. You need to solve

$$\left((2+i) \begin{pmatrix} 1 & 0 & 0 \\ 0 & 1 & 0 \\ 0 & 0 & 1 \end{pmatrix} - \begin{pmatrix} 1 & 0 & 0 \\ 0 & 2 & -1 \\ 0 & 1 & 2 \end{pmatrix} \right) \begin{pmatrix} x \\ y \\ z \end{pmatrix} = \begin{pmatrix} 0 \\ 0 \\ 0 \end{pmatrix}$$

In other words, you must consider the augmented matrix

$$\begin{pmatrix} 1+i & 0 & 0 & | & 0 \\ 0 & i & 1 & | & 0 \\ 0 & -1 & i & | & 0 \end{pmatrix}$$

for the solution. Divide the top row by $(1+i)$ and then take $-i$ times the second row and add to the bottom. This yields

$$\begin{pmatrix} 1 & 0 & 0 & | & 0 \\ 0 & i & 1 & | & 0 \\ 0 & 0 & 0 & | & 0 \end{pmatrix}$$

Now multiply the second row by $-i$ to obtain

$$\begin{pmatrix} 1 & 0 & 0 & | & 0 \\ 0 & 1 & -i & | & 0 \\ 0 & 0 & 0 & | & 0 \end{pmatrix}$$

Therefore, the eigenvectors are of the form

$$t \begin{pmatrix} 0 \\ i \\ 1 \end{pmatrix}.$$

You should find the eigenvectors for $\lambda = 2 - i$. These are

$$t \begin{pmatrix} 0 \\ -i \\ 1 \end{pmatrix}.$$

As usual, if you want to get it right you had better check it.

$$\begin{pmatrix} 1 & 0 & 0 \\ 0 & 2 & -1 \\ 0 & 1 & 2 \end{pmatrix} \begin{pmatrix} 0 \\ -i \\ 1 \end{pmatrix} = \begin{pmatrix} 0 \\ -1-2i \\ 2-i \end{pmatrix} = (2-i) \begin{pmatrix} 0 \\ -i \\ 1 \end{pmatrix}$$

so it worked.

3.2   Exercises

(1) State the eigenvalue problem from an algebraic perspective.
(2) State the eigenvalue problem for a real eigenvalue from a geometric perspective.
(3) Suppose T is a nonzero linear transformation and it satisfies $T^2 = T$. Show that the eigenvalues are either 1 or 0. Then show that $\lambda = 1$ is an eigenvalue.
(4) Is it possible for a nonzero matrix to have only 0 as an eigenvalue?
(5) Show that if $A\mathbf{x} = \lambda \mathbf{x}$ and $A\mathbf{y} = \lambda \mathbf{y}$, then whenever a, b are scalars,

$$A\left(a\mathbf{x} + b\mathbf{y}\right) = \lambda \left(a\mathbf{x} + b\mathbf{y}\right).$$

Does this imply that $a\mathbf{x} + b\mathbf{y}$ is an eigenvector? Explain.
(6) Find the eigenvalues and eigenvectors of the matrix

$$\begin{pmatrix} -1 & -1 & 7 \\ -1 & 0 & 4 \\ -1 & -1 & 5 \end{pmatrix}.$$

(7) Find the eigenvalues and eigenvectors of the matrix

$$\begin{pmatrix} -3 & -7 & 19 \\ -2 & -1 & 8 \\ -2 & -3 & 10 \end{pmatrix}.$$

(8) Find the eigenvalues and eigenvectors of the matrix

$$\begin{pmatrix} -7 & -12 & 30 \\ -3 & -7 & 15 \\ -3 & -6 & 14 \end{pmatrix}.$$

(9) Find the eigenvalues and eigenvectors of the matrix

$$\begin{pmatrix} 7 & -2 & 0 \\ 8 & -1 & 0 \\ -2 & 4 & 6 \end{pmatrix}.$$

(10) Find the eigenvalues and eigenvectors of the matrix

$$\begin{pmatrix} 3 & -2 & -1 \\ 0 & 5 & 1 \\ 0 & 2 & 4 \end{pmatrix}.$$

(11) Find the eigenvalues and eigenvectors of the matrix

$$\begin{pmatrix} 6 & 8 & -23 \\ 4 & 5 & -16 \\ 3 & 4 & -12 \end{pmatrix}.$$

(12) Find the eigenvalues and eigenvectors of the matrix

$$\begin{pmatrix} 5 & 2 & -5 \\ 12 & 3 & -10 \\ 12 & 4 & -11 \end{pmatrix}.$$

(13) Find the eigenvalues and eigenvectors of the matrix

$$\begin{pmatrix} 20 & 9 & -18 \\ 6 & 5 & -6 \\ 30 & 14 & -27 \end{pmatrix}.$$

(14) Find the eigenvalues and eigenvectors of the matrix

$$\begin{pmatrix} 1 & 26 & -17 \\ 4 & -4 & 4 \\ -9 & -18 & 9 \end{pmatrix}.$$

(15) Find the eigenvalues and eigenvectors of the matrix

$$\begin{pmatrix} 3 & -1 & -2 \\ 11 & 3 & -9 \\ 8 & 0 & -6 \end{pmatrix}.$$

(16) Find the eigenvalues and eigenvectors of the matrix

$$\begin{pmatrix} -2 & 1 & 2 \\ -11 & -2 & 9 \\ -8 & 0 & 7 \end{pmatrix}.$$

(17) Find the eigenvalues and eigenvectors of the matrix

$$\begin{pmatrix} 2 & 1 & -1 \\ 2 & 3 & -2 \\ 2 & 2 & -1 \end{pmatrix}.$$

(18) Find the complex eigenvalues and eigenvectors of the matrix

$$\begin{pmatrix} 4 & -2 & -2 \\ 0 & 2 & -2 \\ 2 & 0 & 2 \end{pmatrix}.$$

(19) Let T be the linear transformation which reflects vectors about the x axis. Find a matrix for T and then find its eigenvalues and eigenvectors.

(20) Let T be the linear transformation which rotates all vectors in $\mathbb{R}^2$ counterclockwise through an angle of $\pi/2$. Find a matrix of T and then find eigenvalues and eigenvectors.

(21) Let T be the linear transformation which reflects all vectors in $\mathbb{R}^3$ through the plane determined by $z = 0$. Find a matrix for T and then obtain its eigenvalues and eigenvectors.

(22) Let A be the 2×2 matrix of a linear transformation which rotates all vectors through an angle of $\pi/3$. Explain from geometric reasoning why this matrix cannot have any real eigenvalues. For which angles will such a rotation matrix have a real eigenvalue?

(23) Here are three vectors in $\mathbb{R}^4$: $(1, 2, 0, 3)^T$, $(2, 1, -3, 2)^T$, $(0, 0, 1, 2)^T$. Find the volume of the parallelepiped determined by these three vectors.

(24) Here are two vectors in $\mathbb{R}^4$: $(1, 2, 0, 3)^T$, $(2, 1, -3, 2)^T$. Find the volume of the parallelepiped determined by these two vectors.

(25) Here are three vectors in $\mathbb{R}^2$: $(1, 2)^T$, $(2, 1)^T$, $(0, 1)^T$. Find the volume of the parallelepiped determined by these three vectors. Why should this volume equal zero?

(26) If there are $n + 1$ or more vectors in $\mathbb{R}^n$, Lemma 3.1 implies the parallelepiped determined by these $n + 1$ vectors must have zero volume. What is the geometric significance of this assertion?

Chapter 4

Vector Valued Functions

4.1 Vector Valued Functions

Vector valued functions have values in $\mathbb{R}^p$ where p is an integer at least as large as 1. Here are some examples.

Example 4.1. A rocket is launched from the rotating earth. You could define a function having values in $\mathbb{R}^3$ as $(r(t), \theta(t), \phi(t))$ where $r(t)$ is the distance of the center of mass of the rocket from the center of the earth, $\theta(t)$ is the longitude, and $\phi(t)$ is the latitude of the rocket.

Example 4.2. Let $\mathbf{f}(x, y) = (\sin xy, y^3 + x, x^4)$. Then $\mathbf{f}$ is a function defined on $\mathbb{R}^2$ which has values in $\mathbb{R}^3$. For example, $\mathbf{f}(1, 2) = (\sin 2, 9, 16)$.

As usual, $D(\mathbf{f})$ denotes the domain of the function $\mathbf{f}$ which is written in bold face because it will possibly have values in $\mathbb{R}^p$. When $D(\mathbf{f})$ is not specified, it will be understood that the domain of $\mathbf{f}$ consists of those things for which $\mathbf{f}$ makes sense.

Example 4.3. Let $\mathbf{f}(x, y, z) = \left(\frac{x+y}{z}, \sqrt{1 - x^2}, y\right)$. Then $D(\mathbf{f})$ would consist of the set of all (x, y, z) such that $|x| \leq 1$ and $z \neq 0$.

There are many ways to make new functions from old ones.

Definition 4.1. Let $\mathbf{f}, \mathbf{g}$ be functions with values in $\mathbb{R}^p$. Let a, b be points of $\mathbb{R}$ (scalars). Then $a\mathbf{f} + b\mathbf{g}$ is the name of a function whose domain is $D(\mathbf{f}) \cap D(\mathbf{g})$ which is defined as

$$(a\mathbf{f} + b\mathbf{g})(\mathbf{x}) = a\mathbf{f}(\mathbf{x}) + b\mathbf{g}(\mathbf{x}).$$

$\mathbf{f} \cdot \mathbf{g}$ or $(\mathbf{f}, \mathbf{g})$ is the name of a function whose domain is $D(\mathbf{f}) \cap D(\mathbf{g})$ which is defined as

$$(\mathbf{f}, \mathbf{g})(\mathbf{x}) \equiv \mathbf{f} \cdot \mathbf{g}(\mathbf{x}) \equiv \mathbf{f}(\mathbf{x}) \cdot \mathbf{g}(\mathbf{x}).$$

If $\mathbf{f}$ and $\mathbf{g}$ have values in $\mathbb{R}^3$, define a new function $\mathbf{f} \times \mathbf{g}$ by

$$\mathbf{f} \times \mathbf{g}(t) \equiv \mathbf{f}(t) \times \mathbf{g}(t).$$

If $\mathbf{f} : D(\mathbf{f}) \to X$ and $\mathbf{g} : X \to Y$, then $\mathbf{g} \circ \mathbf{f}$ is the name of a function whose domain is

$$\{\mathbf{x} \in D(\mathbf{f}) : \mathbf{f}(\mathbf{x}) \in D(\mathbf{g})\}$$

which is defined as

$$\mathbf{g} \circ \mathbf{f}(\mathbf{x}) \equiv \mathbf{g}(\mathbf{f}(\mathbf{x})).$$

This is called the composition of the two functions.

You should note that $\mathbf{f}(\mathbf{x})$ is not a function. It is the value of the function at the point $\mathbf{x}$. The name of the function is $\mathbf{f}$. Nevertheless, people often write $\mathbf{f}(\mathbf{x})$ to denote a function and it does not cause too many problems in beginning courses. When this is done, the variable, $\mathbf{x}$ should be considered as a generic variable free to be anything in $D(\mathbf{f})$. I will use this slightly sloppy abuse of notation whenever convenient.

Example 4.4. Let $\mathbf{f}(t) \equiv (t, 1 + t, 2)$ and $\mathbf{g}(t) \equiv (t^2, t, t)$. Then $\mathbf{f} \cdot \mathbf{g}$ is the name of the function satisfying

$$\mathbf{f} \cdot \mathbf{g}(t) = \mathbf{f}(t) \cdot \mathbf{g}(t) = t^3 + t + t^2 + 2t = t^3 + t^2 + 3t$$

Note that in this case is was assumed the domains of the functions consisted of all of $\mathbb{R}$ because this was the set on which the two both made sense. Also note that $\mathbf{f}$ and $\mathbf{g}$ map $\mathbb{R}$ into $\mathbb{R}^3$ but $\mathbf{f} \cdot \mathbf{g}$ maps $\mathbb{R}$ into $\mathbb{R}$.

Example 4.5. Suppose $\mathbf{f}(t) = (2t, 1 + t^2)$ and $g : \mathbb{R}^2 \to \mathbb{R}$ is given by $g(x, y) \equiv x + y$. Then $g \circ \mathbf{f} : \mathbb{R} \to \mathbb{R}$ and

$$g \circ \mathbf{f}(t) = g(\mathbf{f}(t)) = g(2t, 1 + t^2) = 1 + 2t + t^2.$$

4.2   Vector Fields

Some people find it useful to try and draw pictures to illustrate a vector valued function. This can be a very useful idea in the case where the function takes points in $D \subseteq \mathbb{R}^2$ and delivers a vector in $\mathbb{R}^2$. For many points $(x, y) \in D$, you draw an arrow of the appropriate length and direction with its tail at (x, y). The picture of all these arrows can give you an understanding of what is happening. For example if the vector valued function gives the velocity of a fluid at the point (x, y), the picture of these arrows can give an idea of the motion of the fluid. When they are long the fluid is moving fast, when they are short, the fluid is moving slowly. The direction of these arrows is an indication of the direction of motion. The only sensible way to produce such a picture is with a computer. Otherwise, it becomes a worthless exercise in busy work. Furthermore, it is of limited usefulness in three dimensions because in three dimensions such pictures are too cluttered to convey much insight.

Example 4.6. Draw a picture of the vector field $(-x, y)$ which gives the velocity of a fluid flowing in two dimensions.

You can see how the arrows indicate the motion of this fluid.

Example 4.7. Draw a picture of the vector field (y, x) for the velocity of a fluid flowing in two dimensions.

Here is another such example.

Example 4.8. Draw a picture of the vector field $(y \cos(x) + 1, x \sin(y) - 1)$ for the velocity of a fluid flowing in two dimensions.

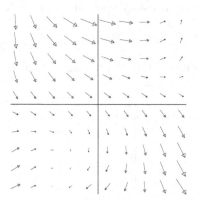

These pictures were drawn by maple. Note how they reveal both the direction and the magnitude of the vectors. However, if you try to draw these by hand, you will mainly waste time.

4.3   Exercises

(1) Here are some vector valued functions.

$$\mathbf{f}\left(x,y\right)=\left(x,y\right),\ \mathbf{g}\left(x,y\right)=\left(-\left(y-1\right),x\right),\ \mathbf{h}\left(x,y\right)=\left(x,-y\right).$$

Now here are the graphs of some vector fields. Match the function with the vector field.

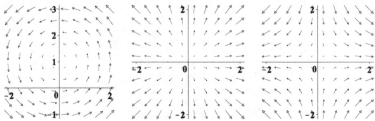

(2) Find $D\left(\mathbf{f}\right)$ for $\mathbf{f}\left(x,y,z,w\right)=\left(\frac{xy}{zw},\sqrt{6-x^{2}y^{2}}\right)$.

(3) Find $D\left(\mathbf{f}\right)$ for $\mathbf{f}\left(x,y,z\right)=\left(\frac{1}{1+x^{2}-y^{2}},\sqrt{4-\left(x^{2}+y^{2}+z^{2}\right)}\right)$.

(4) For $\mathbf{f}\left(x,y,z\right)=\left(x,y,xy\right),\mathbf{h}\left(x,y,z\right)=\left(y^{2},-x,z\right)$ and $\mathbf{g}\left(x,y,z\right)=\left(\frac{1}{x},yz,x^{2}-1\right)$, compute the following.

   (a) $\mathbf{f}\times\mathbf{g}$

   (b) $\mathbf{g}\times\mathbf{f}$

   (c) $\mathbf{f}\cdot\mathbf{g}$

   (d) $\mathbf{f}\times\mathbf{g}\cdot\mathbf{h}$

   (e) $\mathbf{f}\times\left(\mathbf{g}\times\mathbf{h}\right)$

   (f) $\left(\mathbf{f}\times\mathbf{g}\right)\cdot\left(\mathbf{g}\times\mathbf{h}\right)$

(5) Let $\mathbf{f}(x, y, z) = (y, z, x)$ and $\mathbf{g}(x, y, z) = (x^2 + y, z, x)$. Find $\mathbf{g} \circ \mathbf{f}(x, y, z)$.

(6) Let $\mathbf{f}(x, y, z) = (x, z, yz)$ and $\mathbf{g}(x, y, z) = (x, y, x^2 - 1)$. Find $\mathbf{g} \circ \mathbf{f}(x, y, z)$.

(7) For $\mathbf{f}, \mathbf{g}, \mathbf{h}$ vector valued functions and k, l scalar valued functions, which of the following make sense?

    (a) $\mathbf{f} \times \mathbf{g} \times \mathbf{h}$

    (b) $(k \times \mathbf{g}) \times \mathbf{h}$

    (c) $(\mathbf{f} \cdot \mathbf{g}) \times \mathbf{h}$

    (d) $(\mathbf{f} \times \mathbf{g}) \cdot \mathbf{h}$

    (e) $l\mathbf{g} \cdot k$

    (f) $\mathbf{f} \times (\mathbf{g} + \mathbf{h})$

(8) The Lotka Volterra system of differential equations, proposed in 1925 and 1926 by Lotka and Volterra respectively, is intended to model the interaction of predators and prey. An example of this situation is that of wolves and moose living on Isle Royal in the middle of Lake Superior. In these equations x is the number of prey and y is the number of predators. The equations are

$$x'(t) = x(t)(a - by(t)), \quad y'(t) = -y(t)(c - dx(t))$$

Written in terms of vectors,

$$(x', y') = (x(a - by), -y(c - dx))$$

The parameters a, b, c, d depend on the problem. The differential equations are saying that at a point (x, y), the population vector (x, y) moves in the direction of $(x(a - by), -y(c - dx))$. Here is the graph of the vector field which determines the Lotka Volterra system in the case where all the parameters equal 1 which is graphed near the point $(1,1)$. What conclusions seem to be true based on the graph of this vector field? What happens if you start with a population vector near the point $(1,1)$? Remember these vectors in the plane determine the directions of motion of the population vector.

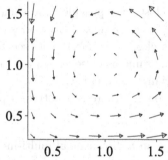

How did I know to graph the vector field near $(1,1)$?

4.4   Continuous Functions

What was done in one variable calculus for scalar functions is generalized here to include the case of a vector valued function of possibly many variables.

Definition 4.2. A function $\mathbf{f} : D(\mathbf{f}) \subseteq \mathbb{R}^p \to \mathbb{R}^q$ is continuous at $\mathbf{x} \in D(\mathbf{f})$ if for each $\varepsilon > 0$ there exists $\delta > 0$ such that whenever $\mathbf{y} \in D(\mathbf{f})$ and

$$|\mathbf{y} - \mathbf{x}| < \delta$$

it follows that

$$|\mathbf{f}(\mathbf{x}) - \mathbf{f}(\mathbf{y})| < \varepsilon.$$

$\mathbf{f}$ is continuous if it is continuous at every point of $D(\mathbf{f})$.

Note the total similarity to the scalar valued case.

4.4.1   *Sufficient Conditions For Continuity*

The next theorem is a fundamental result which allows less worry about the $\varepsilon\ \delta$ definition of continuity.

Theorem 4.1. *The following assertions are valid.*

(1) *The function* $a\mathbf{f} + b\mathbf{g}$ *is continuous at* $\mathbf{x}$ *whenever* $\mathbf{f}$, $\mathbf{g}$ *are continuous at* $\mathbf{x}$ $\in D(\mathbf{f}) \cap D(\mathbf{g})$ *and* $a, b \in \mathbb{R}$.

(2) *If* $\mathbf{f}$ *is continuous at* $\mathbf{x}$, $\mathbf{f}(\mathbf{x}) \in D(\mathbf{g}) \subseteq \mathbb{R}^p$, *and* $\mathbf{g}$ *is continuous at* $\mathbf{f}(\mathbf{x})$, *then* $\mathbf{g} \circ \mathbf{f}$ *is continuous at* $\mathbf{x}$.

(3) *If* $\mathbf{f} = (f_1, \cdots, f_q) : D(\mathbf{f}) \to \mathbb{R}^q$, *then* $\mathbf{f}$ *is continuous if and only if each* f_k *is a continuous real valued function.*

(4) *The function* $f : \mathbb{R}^p \to \mathbb{R}$, *given by* $f(\mathbf{x}) = |\mathbf{x}|$ *is continuous.*

The proof of this theorem is in the last section of this chapter. Its conclusions are not surprising. For example the first claim says that $(a\mathbf{f} + b\mathbf{g})(\mathbf{y})$ is close to $(a\mathbf{f} + b\mathbf{g})(\mathbf{x})$ when $\mathbf{y}$ is close to $\mathbf{x}$ provided the same can be said about $\mathbf{f}$ and $\mathbf{g}$. For the second claim, if $\mathbf{y}$ is close to $\mathbf{x}$, $\mathbf{f}(\mathbf{x})$ is close to $\mathbf{f}(\mathbf{y})$ and so by continuity of $\mathbf{g}$ at $\mathbf{f}(\mathbf{x})$, $\mathbf{g}(\mathbf{f}(\mathbf{y}))$ is close to $\mathbf{g}(\mathbf{f}(\mathbf{x}))$. To see the third claim is likely, note that closeness in $\mathbb{R}^p$ is the same as closeness in each coordinate. The fourth claim is immediate from the triangle inequality.

For functions defined on $\mathbb{R}^n$, there is a notion of polynomial just as there is for functions defined on $\mathbb{R}$.

Definition 4.3. Let α be an n dimensional multi-index. This means

$$\alpha = (\alpha_1, \cdots, \alpha_n)$$

where each α_i is a natural number or zero. Also, let

$$|\alpha| \equiv \sum_{i=1}^{n} |\alpha_i|$$

The symbol $\mathbf{x}^\alpha$ means

$$\mathbf{x}^\alpha \equiv x_1^{\alpha_1} x_2^{\alpha_2} \cdots x_3^{\alpha_n}.$$

An n dimensional polynomial of degree m is a function of the form

$$p(\mathbf{x}) = \sum_{|\alpha| \le m} d_\alpha \mathbf{x}^\alpha.$$

where the d_α are real numbers.

The above theorem implies that polynomials are all continuous.

4.5 Limits Of A Function

As in the case of scalar valued functions of one variable, a concept closely related to continuity is that of the **limit of a function**. The notion of limit of a function makes sense at points $\mathbf{x}$, which are limit points of $D(\mathbf{f})$ and this concept is defined next.

Definition 4.4. Let $A \subseteq \mathbb{R}^m$ be a set. A point $\mathbf{x}$, is a limit point of A if $B(\mathbf{x}, r)$ contains infinitely many points of A for every $r > 0$.

Definition 4.5. Let $\mathbf{f} : D(\mathbf{f}) \subseteq \mathbb{R}^p \to \mathbb{R}^q$ be a function and let $\mathbf{x}$ be a **limit point** of $D(\mathbf{f})$. Then

$$\lim_{\mathbf{y} \to \mathbf{x}} \mathbf{f}(\mathbf{y}) = \mathbf{L}$$

if and only if the following condition holds. For all $\varepsilon > 0$ there exists $\delta > 0$ such that if

$$0 < |\mathbf{y} - \mathbf{x}| < \delta, \text{ and } \mathbf{y} \in D(\mathbf{f})$$

then,

$$|\mathbf{L} - \mathbf{f}(\mathbf{y})| < \varepsilon.$$

Theorem 4.2. *If* $\lim_{\mathbf{y} \to \mathbf{x}} \mathbf{f}(\mathbf{y}) = \mathbf{L}$ *and* $\lim_{\mathbf{y} \to x} \mathbf{f}(\mathbf{y}) = \mathbf{L}_1$, *then* $\mathbf{L} = \mathbf{L}_1$.

Proof: Let $\varepsilon > 0$ be given. There exists $\delta > 0$ such that if $0 < |\mathbf{y} - \mathbf{x}| < \delta$ and $\mathbf{y} \in D(\mathbf{f})$, then

$$|\mathbf{f}(\mathbf{y}) - \mathbf{L}| < \varepsilon, \ |\mathbf{f}(\mathbf{y}) - \mathbf{L}_1| < \varepsilon.$$

Pick such a $\mathbf{y}$. There exists one because $\mathbf{x}$ is a limit point of $D(\mathbf{f})$. Then

$$|\mathbf{L} - \mathbf{L}_1| \le |\mathbf{L} - \mathbf{f}(\mathbf{y})| + |\mathbf{f}(\mathbf{y}) - \mathbf{L}_1| < \varepsilon + \varepsilon = 2\varepsilon.$$

Since $\varepsilon > 0$ was arbitrary, this shows $\mathbf{L} = \mathbf{L}_1$. ∎

As in the case of functions of one variable, one can define what it means for $\lim_{\mathbf{y} \to \mathbf{x}} f(\mathbf{x}) = \pm\infty$.

Definition 4.6. If $f(\mathbf{x}) \in \mathbb{R}$, $\lim_{\mathbf{y} \to \mathbf{x}} f(\mathbf{x}) = \infty$ if for every number l, there exists $\delta > 0$ such that whenever $|\mathbf{y} - \mathbf{x}| < \delta$ and $\mathbf{y} \in D(f)$, then $f(\mathbf{x}) > l$. $\lim_{\mathbf{y} \to \mathbf{x}} f(\mathbf{x}) = -\infty$ if for every number l, there exists $\delta > 0$ such that whenever $|\mathbf{y} - \mathbf{x}| < \delta$ and $\mathbf{y} \in D(f)$, then $f(\mathbf{x}) < l$.

The following theorem is just like the one variable version of calculus.

Theorem 4.3. *Suppose* $\mathbf{f} : D(\mathbf{f}) \to \mathbb{R}^q$. *Then for* $\mathbf{x}$ *a limit point of* $D(\mathbf{f})$,

$$\lim_{\mathbf{y} \to \mathbf{x}} \mathbf{f}(\mathbf{y}) = \mathbf{L} \tag{4.1}$$

if and only if

$$\lim_{\mathbf{y} \to \mathbf{x}} f_k(\mathbf{y}) = L_k \tag{4.2}$$

where $\mathbf{f}(\mathbf{y}) \equiv (f_1(\mathbf{y}), \cdots, f_p(\mathbf{y}))$ *and* $\mathbf{L} \equiv (L_1, \cdots, L_p)$. *Suppose*

$$\lim_{\mathbf{y} \to \mathbf{x}} \mathbf{f}(\mathbf{y}) = \mathbf{L}, \ \lim_{\mathbf{y} \to \mathbf{x}} \mathbf{g}(\mathbf{y}) = \mathbf{K}$$

where $\mathbf{K}, \mathbf{L} \in \mathbb{R}^q$. *Then if* $a, b \in \mathbb{R}$,

$$\lim_{\mathbf{y} \to \mathbf{x}} (a\mathbf{f}(\mathbf{y}) + b\mathbf{g}(\mathbf{y})) = a\mathbf{L} + b\mathbf{K}, \tag{4.3}$$

$$\lim_{y \to x} \mathbf{f} \cdot \mathbf{g}(y) = \mathbf{L} \cdot \mathbf{K} \tag{4.4}$$

In the case where $q = 3$ *and* $\lim_{\mathbf{y} \to \mathbf{x}} \mathbf{f}(\mathbf{y}) = \mathbf{L}$ *and* $\lim_{\mathbf{y} \to \mathbf{x}} \mathbf{g}(\mathbf{y}) = \mathbf{K}$, *then*

$$\lim_{\mathbf{y} \to \mathbf{x}} \mathbf{f}(\mathbf{y}) \times \mathbf{g}(\mathbf{y}) = \mathbf{L} \times \mathbf{K}. \tag{4.5}$$

If g *is scalar valued with* $\lim_{\mathbf{y} \to \mathbf{x}} g(\mathbf{y}) = K \neq 0$,

$$\lim_{\mathbf{y} \to \mathbf{x}} \mathbf{f}(\mathbf{y}) g(\mathbf{y}) = \mathbf{L}K. \tag{4.6}$$

Also, if $\mathbf{h}$ *is a continuous function defined near* $\mathbf{L}$, *then*

$$\lim_{\mathbf{y} \to \mathbf{x}} \mathbf{h} \circ \mathbf{f}(\mathbf{y}) = \mathbf{h}(\mathbf{L}). \tag{4.7}$$

Suppose $\lim_{\mathbf{y} \to \mathbf{x}} \mathbf{f}(\mathbf{y}) = \mathbf{L}$. *If* $|\mathbf{f}(\mathbf{y}) - \mathbf{b}| \leq r$ *for all* $\mathbf{y}$ *sufficiently close to* $\mathbf{x}$, *then* $|\mathbf{L} - \mathbf{b}| \leq r$ *also.*

Proof: Suppose (4.1). Then letting $\varepsilon > 0$ be given there exists $\delta > 0$ such that if $0 < |\mathbf{y} - \mathbf{x}| < \delta$, it follows

$$|f_k(\mathbf{y}) - L_k| \leq |\mathbf{f}(\mathbf{y}) - \mathbf{L}| < \varepsilon$$

which verifies (4.2).

Now suppose (4.2) holds. Then letting $\varepsilon > 0$ be given, there exists δ_k such that if $0 < |\mathbf{y} - \mathbf{x}| < \delta_k$, then

$$|f_k(\mathbf{y}) - L_k| < \frac{\varepsilon}{\sqrt{p}}.$$

Let $0 < \delta < \min(\delta_1, \cdots, \delta_p)$. Then if $0 < |\mathbf{y} - \mathbf{x}| < \delta$, it follows

$$\begin{aligned} |\mathbf{f}(\mathbf{y}) - \mathbf{L}| &= \left(\sum_{k=1}^{p} |f_k(\mathbf{y}) - L_k|^2 \right)^{1/2} \\ &< \left(\sum_{k=1}^{p} \frac{\varepsilon^2}{p} \right)^{1/2} = \varepsilon. \end{aligned}$$

Each of the remaining assertions follows immediately from the coordinate descriptions of the various expressions and the first part. However, I will give a different argument for these.

The proof of (4.3) is left for you. It is like a corresponding theorem for continuous functions. Now (4.4) is to be verified. Let $\varepsilon > 0$ be given. Then by the triangle inequality,

$$\begin{aligned} |\mathbf{f} \cdot \mathbf{g}(\mathbf{y}) - \mathbf{L} \cdot \mathbf{K}| &\leq |\mathbf{f} \cdot \mathbf{g}(\mathbf{y}) - \mathbf{f}(\mathbf{y}) \cdot \mathbf{K}| + |\mathbf{f}(\mathbf{y}) \cdot \mathbf{K} - \mathbf{L} \cdot \mathbf{K}| \\ &\leq |\mathbf{f}(\mathbf{y})| \, |\mathbf{g}(\mathbf{y}) - \mathbf{K}| + |\mathbf{K}| \, |\mathbf{f}(\mathbf{y}) - \mathbf{L}| . \end{aligned}$$

There exists δ_1 such that if $0 < |\mathbf{y} - \mathbf{x}| < \delta_1$ and $\mathbf{y} \in D(\mathbf{f})$, then

$$|\mathbf{f}(\mathbf{y}) - \mathbf{L}| < 1,$$

and so for such $\mathbf{y}$, the triangle inequality implies, $|\mathbf{f}(\mathbf{y})| < 1 + |\mathbf{L}|$. Therefore, for $0 < |\mathbf{y} - \mathbf{x}| < \delta_1$,

$$|\mathbf{f} \cdot \mathbf{g}(\mathbf{y}) - \mathbf{L} \cdot \mathbf{K}| \leq (1 + |\mathbf{K}| + |\mathbf{L}|) \left[|\mathbf{g}(\mathbf{y}) - \mathbf{K}| + |\mathbf{f}(\mathbf{y}) - \mathbf{L}| \right]. \qquad (4.8)$$

Now let $0 < \delta_2$ be such that if $\mathbf{y} \in D(\mathbf{f})$ and $0 < |\mathbf{x} - \mathbf{y}| < \delta_2$,

$$|\mathbf{f}(\mathbf{y}) - \mathbf{L}| < \frac{\varepsilon}{2(1 + |\mathbf{K}| + |\mathbf{L}|)}, \quad |\mathbf{g}(\mathbf{y}) - \mathbf{K}| < \frac{\varepsilon}{2(1 + |\mathbf{K}| + |\mathbf{L}|)}.$$

Then letting $0 < \delta \leq \min(\delta_1, \delta_2)$, it follows from (4.8) that

$$|\mathbf{f} \cdot \mathbf{g}(\mathbf{y}) - \mathbf{L} \cdot \mathbf{K}| < \varepsilon$$

and this proves (4.4).

Consider (4.5). Let δ_1 be as above. From the properties of the cross product,

$$|\mathbf{f}(\mathbf{y}) \times \mathbf{g}(\mathbf{y}) - \mathbf{L} \times \mathbf{K}| \leq |\mathbf{f}(\mathbf{y}) \times \mathbf{g}(\mathbf{y}) - \mathbf{f}(\mathbf{y}) \times \mathbf{K}| + |\mathbf{f}(\mathbf{y}) \times \mathbf{K} - \mathbf{L} \times \mathbf{K}|$$

$$= |\mathbf{f}(\mathbf{y}) \times (\mathbf{g}(\mathbf{y}) - \mathbf{K})| + |(\mathbf{f}(\mathbf{y}) - \mathbf{L}) \times \mathbf{K}|$$

Now from the geometric description of the cross product,

$$\leq |\mathbf{f}(\mathbf{y})| \, |\mathbf{g}(\mathbf{y}) - \mathbf{K}| + |\mathbf{f}(\mathbf{y}) - \mathbf{L}| \, |\mathbf{K}|$$

Then if $0 < |\mathbf{y} - \mathbf{x}| < \delta_1$, this is no larger than

$$(1 + |\mathbf{L}|) |\mathbf{g}(\mathbf{y}) - \mathbf{K}| + |\mathbf{f}(\mathbf{y}) - \mathbf{L}| |\mathbf{K}| \leq (1 + |\mathbf{K}| + |\mathbf{L}|) [|\mathbf{g}(\mathbf{y}) - \mathbf{K}| + |\mathbf{f}(\mathbf{y}) - \mathbf{L}|]$$

and now the conclusion follows as before in the case of the dot product.

The proof of (4.6) is left to you.

Consider (4.7). Since $\mathbf{h}$ is continuous near $\mathbf{L}$, it follows that for $\varepsilon > 0$ given, there exists $\eta > 0$ such that if $|\mathbf{y} - \mathbf{L}| < \eta$, then

$$|\mathbf{h}(\mathbf{y}) - \mathbf{h}(\mathbf{L})| < \varepsilon$$

Now since $\lim_{\mathbf{y} \to \mathbf{x}} \mathbf{f}(\mathbf{y}) = \mathbf{L}$, there exists $\delta > 0$ such that if $0 < |\mathbf{y} - \mathbf{x}| < \delta$, then

$$|\mathbf{f}(\mathbf{y}) - \mathbf{L}| < \eta.$$

Therefore, if $0 < |\mathbf{y} - \mathbf{x}| < \delta$,

$$|\mathbf{h}(\mathbf{f}(\mathbf{y})) - \mathbf{h}(\mathbf{L})| < \varepsilon.$$

It only remains to verify the last assertion. Assume $|\mathbf{f}(\mathbf{y}) - \mathbf{b}| \leq r$. It is required to show that $|\mathbf{L} - \mathbf{b}| \leq r$. If this is not true, then $|\mathbf{L} - \mathbf{b}| > r$. Consider $B(\mathbf{L}, |\mathbf{L} - \mathbf{b}| - r)$. Since $\mathbf{L}$ is the limit of $\mathbf{f}$, it follows $\mathbf{f}(\mathbf{y}) \in B(\mathbf{L}, |\mathbf{L} - \mathbf{b}| - r)$ whenever $\mathbf{y} \in D(\mathbf{f})$ is close enough to $\mathbf{x}$. Thus, by the triangle inequality,

$$|\mathbf{f}(\mathbf{y}) - \mathbf{L}| < |\mathbf{L} - \mathbf{b}| - r$$

and so

$$\begin{aligned} r &< |\mathbf{L} - \mathbf{b}| - |\mathbf{f}(\mathbf{y}) - \mathbf{L}| \leq ||\mathbf{b} - \mathbf{L}| - |\mathbf{f}(\mathbf{y}) - \mathbf{L}|| \\ &\leq |\mathbf{b} - \mathbf{f}(\mathbf{y})|, \end{aligned}$$

a contradiction to the assumption that $|\mathbf{b} - \mathbf{f}(\mathbf{y})| \leq r$. ∎

The relation between continuity and limits is as follows.

Theorem 4.4. *For* $\mathbf{f} : D(\mathbf{f}) \to \mathbb{R}^q$ *and* $\mathbf{x} \in D(\mathbf{f})$ *a limit point of* $D(\mathbf{f})$, $\mathbf{f}$ *is continuous at* $\mathbf{x}$ *if and only if*

$$\lim_{\mathbf{y} \to \mathbf{x}} \mathbf{f}(\mathbf{y}) = \mathbf{f}(\mathbf{x}).$$

Proof: First suppose $\mathbf{f}$ is continuous at $\mathbf{x}$ a limit point of $D(\mathbf{f})$. Then for every $\varepsilon > 0$ there exists $\delta > 0$ such that if $|\mathbf{y} - \mathbf{x}| < \delta$ and $\mathbf{y} \in D(\mathbf{f})$, then $|\mathbf{f}(\mathbf{x}) - \mathbf{f}(\mathbf{y})| < \varepsilon$. In particular, this holds if $0 < |\mathbf{x} - \mathbf{y}| < \delta$ and this is just the definition of the limit. Hence $\mathbf{f}(\mathbf{x}) = \lim_{\mathbf{y} \to \mathbf{x}} \mathbf{f}(\mathbf{y})$.

Next suppose $\mathbf{x}$ is a limit point of $D(\mathbf{f})$ and $\lim_{\mathbf{y} \to \mathbf{x}} \mathbf{f}(\mathbf{y}) = \mathbf{f}(\mathbf{x})$. This means that if $\varepsilon > 0$ there exists $\delta > 0$ such that for $0 < |\mathbf{x} - \mathbf{y}| < \delta$ and $\mathbf{y} \in D(\mathbf{f})$, it follows $|\mathbf{f}(\mathbf{y}) - \mathbf{f}(\mathbf{x})| < \varepsilon$. However, if $\mathbf{y} = \mathbf{x}$, then $|\mathbf{f}(\mathbf{y}) - \mathbf{f}(\mathbf{x})| = |\mathbf{f}(\mathbf{x}) - \mathbf{f}(\mathbf{x})| = 0$ and so whenever $\mathbf{y} \in D(\mathbf{f})$ and $|\mathbf{x} - \mathbf{y}| < \delta$, it follows $|\mathbf{f}(\mathbf{x}) - \mathbf{f}(\mathbf{y})| < \varepsilon$, showing $\mathbf{f}$ is continuous at $\mathbf{x}$. ∎

Example 4.9. Find $\lim_{(x,y) \to (3,1)} \left(\frac{x^2 - 9}{x - 3}, y \right)$.

It is clear that $\lim_{(x,y)\to(3,1)} \frac{x^2-9}{x-3} = 6$ and $\lim_{(x,y)\to(3,1)} y = 1$. Therefore, this limit equals $(6,1)$.

Example 4.10. Find $\lim_{(x,y)\to(0,0)} \frac{xy}{x^2+y^2}$.

First of all, observe the domain of the function is $\mathbb{R}^2 \setminus \{(0,0)\}$, every point in $\mathbb{R}^2$ except the origin. Therefore, $(0,0)$ is a limit point of the domain of the function so it might make sense to take a limit. However, just as in the case of a function of one variable, the limit may not exist. In fact, this is the case here. To see this, take points on the line $y = 0$. At these points, the value of the function equals 0. Now consider points on the line $y = x$ where the value of the function equals $1/2$. Since, arbitrarily close to $(0,0)$, there are points where the function equals $1/2$ and points where the function has the value 0, it follows there can be no limit. Just take $\varepsilon = 1/10$ for example. You cannot be within $1/10$ of $1/2$ and also within $1/10$ of 0 at the same time.

Note it is necessary to rely on the definition of the limit much more than in the case of a function of one variable and there are no easy ways to do limit problems for functions of more than one variable. It is what it is and you will not deal with these concepts without suffering and anguish.

4.6 Properties Of Continuous Functions

Functions of p variables have many of the same properties as functions of one variable. First there is a version of the extreme value theorem generalizing the one dimensional case.

Theorem 4.5. *Let C be closed and bounded and let $f : C \to \mathbb{R}$ be continuous. Then f achieves its maximum and its minimum on C. This means there exist, $\mathbf{x}_1, \mathbf{x}_2 \in C$ such that for all $\mathbf{x} \in C$,*

$$f(\mathbf{x}_1) \le f(\mathbf{x}) \le f(\mathbf{x}_2).$$

There is also the long technical theorem about sums and products of continuous functions. These theorems are proved later in this chapter.

Theorem 4.6. *The following assertions are valid.*

(1) The function $a\mathbf{f} + b\mathbf{g}$ is continuous at $\mathbf{x}$ when $\mathbf{f}$, $\mathbf{g}$ are continuous at $\mathbf{x} \in D(\mathbf{f}) \cap D(\mathbf{g})$ and $a, b \in \mathbb{R}$.

(2) If and f and g are each real valued functions continuous at $\mathbf{x}$, then fg is continuous at $\mathbf{x}$. If, in addition to this, $g(\mathbf{x}) \ne 0$, then f/g is continuous at $\mathbf{x}$.

(3) If $\mathbf{f}$ is continuous at $\mathbf{x}$, $\mathbf{f}(\mathbf{x}) \in D(\mathbf{g}) \subseteq \mathbb{R}^p$, and $\mathbf{g}$ is continuous at $\mathbf{f}(\mathbf{x})$, then $\mathbf{g} \circ \mathbf{f}$ is continuous at $\mathbf{x}$.

(4) If $\mathbf{f} = (f_1, \cdots, f_q) : D(\mathbf{f}) \to \mathbb{R}^q$, then $\mathbf{f}$ is continuous if and only if each f_k is a continuous real valued function.

(5) The function $f : \mathbb{R}^p \to \mathbb{R}$, given by $f(\mathbf{x}) = |\mathbf{x}|$ is continuous.

4.7   Exercises

(1) Let $\mathbf{f}(t) = \left(t, t^2 + 1, \frac{t}{t+1}\right)$ and let $\mathbf{g}(t) = \left(t+1, 1, \frac{t}{t^2+1}\right)$. Find $\mathbf{f} \cdot \mathbf{g}$.

(2) Let $\mathbf{f}, \mathbf{g}$ be given in the previous problem. Find $\mathbf{f} \times \mathbf{g}$.

(3) Let $\mathbf{f}(t) = \left(t, t^2, t^3\right), \mathbf{g}(t) = \left(1, t^2, t^2\right)$, and $\mathbf{h}(t) = (\sin t, t, 1)$. Find the time rate of change of the box product of the vectors $\mathbf{f}, \mathbf{g}$, and $\mathbf{h}$.

(4) Let $\mathbf{f}(t) = (t, \sin t)$. Show $\mathbf{f}$ is continuous at every point t.

(5) Suppose $|\mathbf{f}(\mathbf{x}) - \mathbf{f}(\mathbf{y})| \leq K |\mathbf{x} - \mathbf{y}|$ where K is a constant. Show that $\mathbf{f}$ is everywhere continuous. Functions satisfying such an inequality are called Lipschitz functions.

(6) Suppose $|\mathbf{f}(\mathbf{x}) - \mathbf{f}(\mathbf{y})| \leq K |\mathbf{x} - \mathbf{y}|^\alpha$ where K is a constant and $\alpha \in (0, 1)$. Show that $\mathbf{f}$ is everywhere continuous. Functions like this are called Holder continuous.

(7) Suppose $f : \mathbb{R}^3 \to \mathbb{R}$ is given by $f(\mathbf{x}) = 3x_1 x_2 + 2x_3^2$. Use Theorem 4.1 to verify that f is continuous. **Hint:** You should first verify that the function $\pi_k : \mathbb{R}^3 \to \mathbb{R}$ given by $\pi_k(\mathbf{x}) = x_k$ is a continuous function.

(8) Show that if $f : \mathbb{R}^q \to \mathbb{R}$ is a polynomial then it is continuous.

(9) State and prove a theorem which involves quotients of functions encountered in the previous problem.

(10) Let

$$f(x, y) \equiv \begin{cases} \frac{2x^2 - y^2}{x^2 + y^2} & \text{if } (x, y) \neq (0, 0) \\ 0 & \text{if } (x, y) = (0, 0) \end{cases}.$$

Find $\lim_{(x,y) \to (0,0)} f(x, y)$ if it exists. If it does not exist, tell why it does not exist. **Hint:** Consider along the line $y = x$ and along the line $y = 0$.

(11) Find the following limits if possible

(a) $\lim_{(x,y) \to (0,0)} \frac{x^2 - y^2}{x^2 + y^2}$.

(b) $\lim_{(x,y) \to (0,0)} \frac{x\left(x^2 - y^2\right)}{\left(x^2 + y^2\right)} = 0$.

(c) $\lim_{(x,y) \to (0,0)} \frac{\left(x^2 - y^4\right)^2}{\left(x^2 + y^4\right)^2}$. **Hint:** Consider along $y = 0$ and along $x = y^2$.

(d) $\lim_{(x,y) \to (0,0)} x \sin\left(\frac{1}{x^2 + y^2}\right)$.

(e) $\lim_{(x,y) \to (1,2)} \frac{-2yx^2 + 8yx + 34y + 3y^3 - 18y^2 + 6x^2 - 13x - 20 - xy^2 - x^3}{-y^2 + 4y - 5 - x^2 + 2x}$. **Hint:** It might help to write this in terms of the variables $(s, t) = (x - 1, y - 2)$.

(12) Suppose $\lim_{x \to 0} f(x, 0) = 0 = \lim_{y \to 0} f(0, y)$. Does it follow that

$$\lim_{(x,y) \to (0,0)} f(x, y) = 0?$$

Prove or give counter example.

(13) $\mathbf{f} : D \subseteq \mathbb{R}^p \to \mathbb{R}^q$ is Lipschitz continuous or just Lipschitz for short if there exists a constant K such that

$$|\mathbf{f}(\mathbf{x}) - \mathbf{f}(\mathbf{y})| \leq K |\mathbf{x} - \mathbf{y}|$$

for all $\mathbf{x}, \mathbf{y} \in D$. Show every Lipschitz function is uniformly continuous which means that given $\varepsilon > 0$ there exists $\delta > 0$ independent of $\mathbf{x}$ such that if $|\mathbf{x} - \mathbf{y}| < \delta$, then $|\mathbf{f}(\mathbf{x}) - \mathbf{f}(\mathbf{y})| < \varepsilon$.

(14) If $\mathbf{f}$ is uniformly continuous, does it follow that $|\mathbf{f}|$ is also uniformly continuous? If $|\mathbf{f}|$ is uniformly continuous does it follow that $\mathbf{f}$ is uniformly continuous? Answer the same questions with "uniformly continuous" replaced with "continuous". Explain why.

(15) Let f be defined on the positive integers. Thus $D(f) = \mathbb{N}$. Show that f is automatically continuous at every point of $D(f)$. Is it also uniformly continuous? What does this mean about the concept of continuous functions being those which can be graphed without taking the pencil off the paper?

(16) Let

$$f(x, y) = \frac{\left(x^2 - y^4\right)^2}{\left(x^2 + y^4\right)^2} \text{ if } (x, y) \neq (0, 0)$$

Show $\lim_{t \to 0} f(tx, ty) = 1$ for any choice of (x, y). Using Problem 11c, what does this tell you about limits existing just because the limit along any line exists.

(17) Let $f(x, y, z) = x^2 y + \sin(xyz)$. Does f achieve a maximum on the set

$$\{(x, y, z) : x^2 + y^2 + 2z^2 \leq 8\}?$$

Explain why.

(18) Suppose $\mathbf{x}$ is defined to be a limit point of a set A if and only if for all $r > 0$, $B(\mathbf{x}, r)$ contains a point of A different than $\mathbf{x}$. Show this is equivalent to the above definition of limit point.

(19) Give an example of an infinite set of points in $\mathbb{R}^3$ which has no limit points. Show that if $D(\mathbf{f})$ equals this set, then $\mathbf{f}$ is continuous. Show that more generally, if $\mathbf{f}$ is any function for which $D(\mathbf{f})$ has no limit points, then $\mathbf{f}$ is continuous.

(20) Let $\{\mathbf{x}_k\}_{k=1}^n$ be any finite set of points in $\mathbb{R}^p$. Show this set has no limit points.

(21) Suppose S is any set of points such that every pair of points is at least as far apart as 1. Show S has no limit points.

(22) Find $\lim_{\mathbf{x} \to 0} \frac{\sin(|\mathbf{x}|)}{|\mathbf{x}|}$ and prove your answer from the definition of limit.

(23) Suppose $\mathbf{g}$ is a continuous vector valued function of one variable defined on $[0, \infty)$. Prove

$$\lim_{\mathbf{x} \to \mathbf{x}_0} \mathbf{g}(|\mathbf{x}|) = \mathbf{g}(|\mathbf{x}_0|).$$

4.8   Open And Closed Sets

Eventually, one must consider functions which are defined on subsets of $\mathbb{R}^n$ and their properties. The next definition will end up being quite important. It describe a type of subset of $\mathbb{R}^n$ with the property that if $\mathbf{x}$ is in this set, then so is $\mathbf{y}$ whenever $\mathbf{y}$ is close enough to $\mathbf{x}$.

Definition 4.7. Recall that for $\mathbf{x}, \mathbf{y} \in \mathbb{R}^n$,

$$|\mathbf{x} - \mathbf{y}| = \left(\sum_{i=1}^{n} |x_i - y_i|^2 \right)^{1/2}.$$

Also let

$$B(\mathbf{x}, r) \equiv \{ \mathbf{y} \in \mathbb{R}^n : |\mathbf{x} - \mathbf{y}| < r \}$$

Let $U \subseteq \mathbb{R}^n$. U is an **open set** if whenever $\mathbf{x} \in U$, there exists $r > 0$ such that $B(\mathbf{x}, r) \subseteq U$. More generally, if U is any subset of $\mathbb{R}^n$, $\mathbf{x} \in U$ is an **interior point** of U if there exists $r > 0$ such that $\mathbf{x} \in B(\mathbf{x}, r) \subseteq U$. In other words U is an open set exactly when every point of U is an interior point of U.

If there is something called an open set, surely there should be something called a closed set and here is the definition of one.

Definition 4.8. A subset, C, of $\mathbb{R}^n$ is called a **closed set** if $\mathbb{R}^n \setminus C$ is an open set. They symbol $\mathbb{R}^n \setminus C$ denotes everything in $\mathbb{R}^n$ which is not in C. It is also called the **complement** of C. The symbol S^C is a short way of writing $\mathbb{R}^n \setminus S$.

To illustrate this definition, consider the following picture.

You see in this picture how the edges are dotted. This is because an open set, can not include the edges or the set would fail to be open. For example, consider what would happen if you picked a point out on the edge of U in the above picture. Every open ball centered at that point would have in it some points which are outside U. Therefore, such a point would violate the above definition. You also see the edges of $B(\mathbf{x}, r)$ dotted suggesting that $B(\mathbf{x}, r)$ ought to be an open set. This is intuitively clear but does require a proof. This will be done in the next theorem and will give examples of open sets. Also, you can see that if $\mathbf{x}$ is close to the edge of U, you might have to take r to be very small.

It is roughly the case that open sets do not have their skins while closed sets do. Here is a picture of a closed set, C.

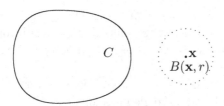

Note that $\mathbf{x} \notin C$ and since $\mathbb{R}^n \setminus C$ is open, there exists a ball, $B(\mathbf{x}, r)$ contained entirely in $\mathbb{R}^n \setminus C$. If you look at $\mathbb{R}^n \setminus C$, what would be its skin? It can't be in $\mathbb{R}^n \setminus C$ and so it must be in C. This is a rough heuristic explanation of what is going on with these definitions. Also note that $\mathbb{R}^n$ and $\emptyset$ are both open and closed. Here is why. If $\mathbf{x} \in \emptyset$, then there must be a ball centered at $\mathbf{x}$ which is also contained in $\emptyset$. This must be considered to be true because there is nothing in $\emptyset$ so there can be no example to show it false[1]. Therefore, from the definition, it follows $\emptyset$ is open. It is also closed because if $\mathbf{x} \notin \emptyset$, then $B(\mathbf{x}, 1)$ is also contained in $\mathbb{R}^n \setminus \emptyset = \mathbb{R}^n$. Therefore, $\emptyset$ is both open and closed. From this, it follows $\mathbb{R}^n$ is also both open and closed.

Theorem 4.7. *Let* $\mathbf{x} \in \mathbb{R}^n$ *and let* $r \geq 0$. *Then* $B(\mathbf{x}, r)$ *is an open set. Also,*

$$D(\mathbf{x}, r) \equiv \{\mathbf{y} \in \mathbb{R}^n : |\mathbf{y} - \mathbf{x}| \leq r\}$$

is a closed set.

Proof: Suppose $\mathbf{y} \in B(\mathbf{x}, r)$. It is necessary to show there exists $r_1 > 0$ such that $B(\mathbf{y}, r_1) \subseteq B(\mathbf{x}, r)$. Define $r_1 \equiv r - |\mathbf{x} - \mathbf{y}|$. Then if $|\mathbf{z} - \mathbf{y}| < r_1$, it follows from the above triangle inequality that

$$
\begin{aligned}
|\mathbf{z} - \mathbf{x}| &= |\mathbf{z} - \mathbf{y} + \mathbf{y} - \mathbf{x}| \\
&\leq |\mathbf{z} - \mathbf{y}| + |\mathbf{y} - \mathbf{x}| \\
&< r_1 + |\mathbf{y} - \mathbf{x}| = r - |\mathbf{x} - \mathbf{y}| + |\mathbf{y} - \mathbf{x}| = r.
\end{aligned}
$$

---

[1] To a mathematician, the statement: Whenever a pig is born with wings it can fly must be taken as true. We do not consider biological or aerodynamic considerations in such statements. There is no such thing as a winged pig and therefore, all winged pigs must be superb flyers since there can be no example of one which is not. On the other hand we would also consider the statement: Whenever a pig is born with wings it cannot possibly fly, as equally true. The point is, you can say anything you want about the elements of the empty set and no one can gainsay your statement. Therefore, such statements are considered as true by default. You may say this is a very strange way of thinking about truth and ultimately this is because mathematics is not about truth. It is more about consistency and logic.

Note that if $r = 0$ then $B(\mathbf{x}, r) = \emptyset$, the empty set. This is because if $\mathbf{y} \in \mathbb{R}^n$, $|\mathbf{x} - \mathbf{y}| \geq 0$ and so $\mathbf{y} \notin B(\mathbf{x}, 0)$. Since $\emptyset$ has no points in it, it must be open because every point in it, (There are none.) satisfies the desired property of being an interior point.

Now suppose $\mathbf{y} \notin D(\mathbf{x}, r)$. Then $|\mathbf{x} - \mathbf{y}| > r$ and defining $\delta \equiv |\mathbf{x} - \mathbf{y}| - r$, it follows that if $\mathbf{z} \in B(\mathbf{y}, \delta)$, then by the triangle inequality,

$$
\begin{aligned}
|\mathbf{x} - \mathbf{z}| &\geq |\mathbf{x} - \mathbf{y}| - |\mathbf{y} - \mathbf{z}| > |\mathbf{x} - \mathbf{y}| - \delta \\
&= |\mathbf{x} - \mathbf{y}| - (|\mathbf{x} - \mathbf{y}| - r) = r
\end{aligned}
$$

and this shows that $B(\mathbf{y}, \delta) \subseteq \mathbb{R}^n \setminus D(\mathbf{x}, r)$. Since $\mathbf{y}$ was an arbitrary point in $\mathbb{R}^n \setminus D(\mathbf{x}, r)$, it follows $\mathbb{R}^n \setminus D(\mathbf{x}, r)$ is an open set which shows, from the definition, that $D(\mathbf{x}, r)$ is a closed set as claimed. ∎

A picture which is descriptive of the conclusion of the above theorem which also implies the manner of proof is the following.

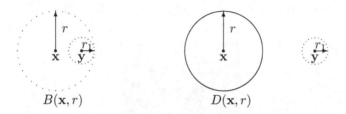

$$B(\mathbf{x}, r) \qquad\qquad D(\mathbf{x}, r)$$

Recall $\mathbb{R}^2$ consists of ordered pairs (x, y) such that $x \in \mathbb{R}$ and $y \in \mathbb{R}$. $\mathbb{R}^2$ is also written as $\mathbb{R} \times \mathbb{R}$. In general, the following definition holds.

Definition 4.9. The **Cartesian product** of two sets $A \times B$, means

$$\{(a, b) : a \in A, \ b \in B\}.$$

If you have n sets $A_1, A_2, \cdots, A_n$

$$\prod_{i=1}^{n} A_i = \{(x_1, x_2, \cdots, x_n) : \text{ each } x_i \in A_i\}.$$

Now suppose $A \subseteq \mathbb{R}^m$ and $B \subseteq \mathbb{R}^n$. Then if $(\mathbf{x}, \mathbf{y}) \in A \times B$, $\mathbf{x} = (x_1, \cdots, x_m)$ and $\mathbf{y} = (y_1, \cdots, y_n)$, the following identification will be made.

$$(\mathbf{x}, \mathbf{y}) = (x_1, \cdots, x_m, y_1, \cdots, y_n) \in \mathbb{R}^{n+m}.$$

Similarly, starting with something in $\mathbb{R}^{n+m}$, you can write it in the form $(\mathbf{x}, \mathbf{y})$ where $\mathbf{x} \in \mathbb{R}^m$ and $\mathbf{y} \in \mathbb{R}^n$. The following theorem has to do with the Cartesian product of two closed sets or two open sets. Also here is an important definition.

Definition 4.10. A set, $A \subseteq \mathbb{R}^n$ is said to be **bounded** if there exist finite intervals, $[a_i, b_i]$ such that

$$A \subseteq \prod_{i=1}^{n} [a_i, b_i].$$

Theorem 4.8. *Let U be an open set in $\mathbb{R}^m$ and let V be an open set in $\mathbb{R}^n$. Then $U \times V$ is an open set in $\mathbb{R}^{n+m}$. If C is a closed set in $\mathbb{R}^m$ and H is a closed set in $\mathbb{R}^n$, then $C \times H$ is a closed set in $\mathbb{R}^{n+m}$. If C and H are bounded, then so is $C \times H$.*

Proof: Let $(\mathbf{x}, \mathbf{y}) \in U \times V$. Since U is open, there exists $r_1 > 0$ such that $B(\mathbf{x}, r_1) \subseteq U$. Similarly, there exists $r_2 > 0$ such that $B(\mathbf{y}, r_2) \subseteq V$. Now

$$B((\mathbf{x}, \mathbf{y}), \delta) \equiv$$

$$\left\{ (\mathbf{s}, \mathbf{t}) \in \mathbb{R}^{n+m} : \sum_{k=1}^{m} |x_k - s_k|^2 + \sum_{j=1}^{n} |y_j - t_j|^2 < \delta^2 \right\}$$

Therefore, if $\delta \equiv \min(r_1, r_2)$ and $(\mathbf{s}, \mathbf{t}) \in B((\mathbf{x}, \mathbf{y}), \delta)$, then it follows that $\mathbf{s} \in B(\mathbf{x}, r_1) \subseteq U$ and that $\mathbf{t} \in B(\mathbf{y}, r_2) \subseteq V$ which shows that $B((\mathbf{x}, \mathbf{y}), \delta) \subseteq U \times V$. Hence $U \times V$ is open as claimed.

Next suppose $(\mathbf{x}, \mathbf{y}) \notin C \times H$. It is necessary to show there exists $\delta > 0$ such that $B((\mathbf{x}, \mathbf{y}), \delta) \subseteq \mathbb{R}^{n+m} \setminus (C \times H)$. Either $\mathbf{x} \notin C$ or $\mathbf{y} \notin H$ since otherwise $(\mathbf{x}, \mathbf{y})$ would be a point of $C \times H$. Suppose therefore, that $\mathbf{x} \notin C$. Since C is closed, there exists $r > 0$ such that $B(\mathbf{x}, r) \subseteq \mathbb{R}^m \setminus C$. Consider $B((\mathbf{x}, \mathbf{y}), r)$. If $(\mathbf{s}, \mathbf{t}) \in B((\mathbf{x}, \mathbf{y}), r)$, it follows that $\mathbf{s} \in B(\mathbf{x}, r)$ which is contained in $\mathbb{R}^m \setminus C$. Therefore, $B((\mathbf{x}, \mathbf{y}), r) \subseteq \mathbb{R}^{n+m} \setminus (C \times H)$ showing $C \times H$ is closed. A similar argument holds if $\mathbf{y} \notin H$.

If C is bounded, there exist $[a_i, b_i]$ such that $C \subseteq \prod_{i=1}^{m} [a_i, b_i]$ and if H is bounded, $H \subseteq \prod_{i=m+1}^{m+n} [a_i, b_i]$ for intervals $[a_{m+1}, b_{m+1}], \cdots, [a_{m+n}, b_{m+n}]$. Therefore, $C \times H \subseteq \prod_{i=1}^{m+n} [a_i, b_i]$. $\blacksquare$

4.9 Exercises

(1) Let $U = \{(x, y, z) \text{ such that } z > 0\}$. Determine whether U is open, closed or neither.

(2) Let $U = \{(x, y, z) \text{ such that } z \geq 0\}$. Determine whether U is open, closed or neither.

(3) Let $U = \left\{ (x, y, z) \text{ such that } \sqrt{x^2 + y^2 + z^2} < 1 \right\}$. Determine whether U is open, closed or neither.

(4) Let $U = \left\{ (x, y, z) \text{ such that } \sqrt{x^2 + y^2 + z^2} \leq 1 \right\}$. Determine whether U is open, closed or neither.

(5) Show carefully that $\mathbb{R}^n$ is both open and closed.

(6) Show that every open set in $\mathbb{R}^n$ is the union of open balls contained in it.

(7) Show the intersection of any two open sets is an open set.

(8) If S is a nonempty subset of $\mathbb{R}^p$, a point $\mathbf{x}$ is said to be a **limit point** of S if $B(\mathbf{x}, r)$ contains infinitely many points of S for each $r > 0$. Show this is

equivalent to saying that $B(\mathbf{x}, r)$ contains a point of S different than $\mathbf{x}$ for each $r > 0$.

(9) Closed sets were defined to be those sets which are complements of open sets. Show that a set is closed if and only if it contains all its limit points.

4.10 Some Fundamentals*

This section contains the proofs of the theorems which were stated without proof along with some other significant topics which will be useful later. These topics are of fundamental significance but are difficult.

4.10.1 *Combinations Of Continuous Functions*

Theorem 4.9. *The following assertions are valid.*

(1) The function $a\mathbf{f} + b\mathbf{g}$ is continuous at $\mathbf{x}$ when $\mathbf{f}$, $\mathbf{g}$ are continuous at $\mathbf{x} \in D(\mathbf{f}) \cap D(\mathbf{g})$ and $a, b \in \mathbb{R}$.

(2) If and f and g are each real valued functions continuous at $\mathbf{x}$, then fg is continuous at $\mathbf{x}$. If, in addition to this, $g(\mathbf{x}) \neq 0$, then f/g is continuous at $\mathbf{x}$.

(3) If $\mathbf{f}$ is continuous at $\mathbf{x}$, $\mathbf{f}(\mathbf{x}) \in D(\mathbf{g}) \subseteq \mathbb{R}^p$, and $\mathbf{g}$ is continuous at $\mathbf{f}(\mathbf{x})$, then $\mathbf{g} \circ \mathbf{f}$ is continuous at $\mathbf{x}$.

(4) If $\mathbf{f} = (f_1, \cdots, f_q) : D(\mathbf{f}) \to \mathbb{R}^q$, then $\mathbf{f}$ is continuous if and only if each f_k is a continuous real valued function.

(5) The function $f : \mathbb{R}^p \to \mathbb{R}$, given by $f(\mathbf{x}) = |\mathbf{x}|$ is continuous.

Proof: Begin with (1). Let $\varepsilon > 0$ be given. By assumption, there exist $\delta_1 > 0$ such that whenever $|\mathbf{x} - \mathbf{y}| < \delta_1$, it follows $|\mathbf{f}(\mathbf{x}) - \mathbf{f}(\mathbf{y})| < \frac{\varepsilon}{2(|a|+|b|+1)}$ and there exists $\delta_2 > 0$ such that whenever $|\mathbf{x} - \mathbf{y}| < \delta_2$, it follows that $|\mathbf{g}(\mathbf{x}) - \mathbf{g}(\mathbf{y})| < \frac{\varepsilon}{2(|a|+|b|+1)}$. Then let $0 < \delta \leq \min(\delta_1, \delta_2)$. If $|\mathbf{x} - \mathbf{y}| < \delta$, then everything happens at once. Therefore, using the triangle inequality

$$|a\mathbf{f}(\mathbf{x}) + b\mathbf{f}(\mathbf{x}) - (a\mathbf{g}(\mathbf{y}) + b\mathbf{g}(\mathbf{y}))|$$

$$\leq |a| \, |\mathbf{f}(\mathbf{x}) - \mathbf{f}(\mathbf{y})| + |b| \, |\mathbf{g}(\mathbf{x}) - \mathbf{g}(\mathbf{y})|$$

$$< |a| \left(\frac{\varepsilon}{2(|a| + |b| + 1)} \right) + |b| \left(\frac{\varepsilon}{2(|a| + |b| + 1)} \right) < \varepsilon.$$

Now begin on (2). There exists $\delta_1 > 0$ such that if $|\mathbf{y} - \mathbf{x}| < \delta_1$, then $|f(\mathbf{x}) - f(\mathbf{y})| < 1$. Therefore, for such $\mathbf{y}$,

$$|f(\mathbf{y})| < 1 + |f(\mathbf{x})|.$$

It follows that for such $\mathbf{y}$,

$$|fg(\mathbf{x}) - fg(\mathbf{y})| \le |f(\mathbf{x})g(\mathbf{x}) - g(\mathbf{x})f(\mathbf{y})| + |g(\mathbf{x})f(\mathbf{y}) - f(\mathbf{y})g(\mathbf{y})|$$

$$\le |g(\mathbf{x})||f(\mathbf{x}) - f(\mathbf{y})| + |f(\mathbf{y})||g(\mathbf{x}) - g(\mathbf{y})|$$
$$\le (1 + |g(\mathbf{x})| + |f(\mathbf{y})|)[|g(\mathbf{x}) - g(\mathbf{y})| + |f(\mathbf{x}) - f(\mathbf{y})|].$$

Now let $\varepsilon > 0$ be given. There exists δ_2 such that if $|\mathbf{x} - \mathbf{y}| < \delta_2$, then

$$|g(\mathbf{x}) - g(\mathbf{y})| < \frac{\varepsilon}{2(1 + |g(\mathbf{x})| + |f(\mathbf{y})|)},$$

and there exists δ_3 such that if $|\mathbf{x} - \mathbf{y}| < \delta_3$, then

$$|f(\mathbf{x}) - f(\mathbf{y})| < \frac{\varepsilon}{2(1 + |g(\mathbf{x})| + |f(\mathbf{y})|)}$$

Now let $0 < \delta \le \min(\delta_1, \delta_2, \delta_3)$. Then if $|\mathbf{x} - \mathbf{y}| < \delta$, all the above hold at once and

$$|fg(\mathbf{x}) - fg(\mathbf{y})| \le$$

$$(1 + |g(\mathbf{x})| + |f(\mathbf{y})|)[|g(\mathbf{x}) - g(\mathbf{y})| + |f(\mathbf{x}) - f(\mathbf{y})|]$$
$$< (1 + |g(\mathbf{x})| + |f(\mathbf{y})|)\left(\frac{\varepsilon}{2(1 + |g(\mathbf{x})| + |f(\mathbf{y})|)} + \frac{\varepsilon}{2(1 + |g(\mathbf{x})| + |f(\mathbf{y})|)}\right) = \varepsilon.$$

This proves the first part of (2). To obtain the second part, let δ_1 be as described above and let $\delta_0 > 0$ be such that for $|\mathbf{x} - \mathbf{y}| < \delta_0$,

$$|g(\mathbf{x}) - g(\mathbf{y})| < |g(\mathbf{x})|/2$$

and so by the triangle inequality,

$$-|g(\mathbf{x})|/2 \le |g(\mathbf{y})| - |g(\mathbf{x})| \le |g(\mathbf{x})|/2$$

which implies $|g(\mathbf{y})| \ge |g(\mathbf{x})|/2$, and $|g(\mathbf{y})| < 3|g(\mathbf{x})|/2$.
  Then if $|\mathbf{x} - \mathbf{y}| < \min(\delta_0, \delta_1)$,

$$\left|\frac{f(\mathbf{x})}{g(\mathbf{x})} - \frac{f(\mathbf{y})}{g(\mathbf{y})}\right| = \left|\frac{f(\mathbf{x})g(\mathbf{y}) - f(\mathbf{y})g(\mathbf{x})}{g(\mathbf{x})g(\mathbf{y})}\right|$$

$$\le \frac{|f(\mathbf{x})g(\mathbf{y}) - f(\mathbf{y})g(\mathbf{x})|}{\left(\frac{|g(\mathbf{x})|^2}{2}\right)}$$

$$= \frac{2|f(\mathbf{x})g(\mathbf{y}) - f(\mathbf{y})g(\mathbf{x})|}{|g(\mathbf{x})|^2}$$

$$\leq \frac{2}{|g(\mathbf{x})|^2}\left[|f(\mathbf{x})\,g(\mathbf{y}) - f(\mathbf{y})\,g(\mathbf{y}) + f(\mathbf{y})\,g(\mathbf{y}) - f(\mathbf{y})\,g(\mathbf{x})|\right]$$

$$\leq \frac{2}{|g(\mathbf{x})|^2}\left[|g(\mathbf{y})|\,|f(\mathbf{x}) - f(\mathbf{y})| + |f(\mathbf{y})|\,|g(\mathbf{y}) - g(\mathbf{x})|\right]$$

$$\leq \frac{2}{|g(\mathbf{x})|^2}\left[\frac{3}{2}\,|g(\mathbf{x})|\,|f(\mathbf{x}) - f(\mathbf{y})| + (1 + |f(\mathbf{x})|)\,|g(\mathbf{y}) - g(\mathbf{x})|\right]$$

$$\leq \frac{2}{|g(\mathbf{x})|^2}\,(1 + 2\,|f(\mathbf{x})| + 2\,|g(\mathbf{x})|)\,[|f(\mathbf{x}) - f(\mathbf{y})| + |g(\mathbf{y}) - g(\mathbf{x})|]$$

$$\equiv M\,[|f(\mathbf{x}) - f(\mathbf{y})| + |g(\mathbf{y}) - g(\mathbf{x})|]$$

where

$$M \equiv \frac{2}{|g(\mathbf{x})|^2}\,(1 + 2\,|f(\mathbf{x})| + 2\,|g(\mathbf{x})|)$$

Now let δ_2 be such that if $|\mathbf{x} - \mathbf{y}| < \delta_2$, then

$$|f(\mathbf{x}) - f(\mathbf{y})| < \frac{\varepsilon}{2}M^{-1}$$

and let δ_3 be such that if $|\mathbf{x} - \mathbf{y}| < \delta_3$, then

$$|g(\mathbf{y}) - g(\mathbf{x})| < \frac{\varepsilon}{2}M^{-1}.$$

Then if $0 < \delta \leq \min(\delta_0, \delta_1, \delta_2, \delta_3)$, and $|\mathbf{x} - \mathbf{y}| < \delta$, everything holds and

$$\left|\frac{f(\mathbf{x})}{g(\mathbf{x})} - \frac{f(\mathbf{y})}{g(\mathbf{y})}\right| \leq M\,[|f(\mathbf{x}) - f(\mathbf{y})| + |g(\mathbf{y}) - g(\mathbf{x})|]$$

$$< M\left[\frac{\varepsilon}{2}M^{-1} + \frac{\varepsilon}{2}M^{-1}\right] = \varepsilon.$$

This completes the proof of the second part of (2). Note that in these proofs no effort is made to find some sort of "best" δ. The problem is one which has a yes or a no answer. Either it is or it is not continuous.

Now begin on (3). If $\mathbf{f}$ is continuous at $\mathbf{x}$, $\mathbf{f}(\mathbf{x}) \in D(\mathbf{g}) \subseteq \mathbb{R}^p$, and $\mathbf{g}$ is continuous at $\mathbf{f}(\mathbf{x})$, then $\mathbf{g} \circ \mathbf{f}$ is continuous at $\mathbf{x}$. Let $\varepsilon > 0$ be given. Then there exists $\eta > 0$ such that if $|\mathbf{y} - \mathbf{f}(\mathbf{x})| < \eta$ and $\mathbf{y} \in D(\mathbf{g})$, it follows that $|\mathbf{g}(\mathbf{y}) - \mathbf{g}(\mathbf{f}(\mathbf{x}))| < \varepsilon$. It follows from continuity of $\mathbf{f}$ at $\mathbf{x}$ that there exists $\delta > 0$ such that if $|\mathbf{x} - \mathbf{z}| < \delta$ and $\mathbf{z} \in D(\mathbf{f})$, then $|\mathbf{f}(\mathbf{z}) - \mathbf{f}(\mathbf{x})| < \eta$. Then if $|\mathbf{x} - \mathbf{z}| < \delta$ and $\mathbf{z} \in D(\mathbf{g} \circ \mathbf{f}) \subseteq D(\mathbf{f})$, all the above hold and so

$$|\mathbf{g}(\mathbf{f}(\mathbf{z})) - \mathbf{g}(\mathbf{f}(\mathbf{x}))| < \varepsilon.$$

This proves part (3).

Part (4) says: If $\mathbf{f} = (f_1, \cdots, f_q) : D(\mathbf{f}) \to \mathbb{R}^q$, then $\mathbf{f}$ is continuous if and only if each f_k is a continuous real valued function. Then

$$|f_k(\mathbf{x}) - f_k(\mathbf{y})| \leq |\mathbf{f}(\mathbf{x}) - \mathbf{f}(\mathbf{y})| \equiv \left(\sum_{i=1}^{q}|f_i(\mathbf{x}) - f_i(\mathbf{y})|^2\right)^{1/2}$$

$$\leq \sum_{i=1}^{q} |f_i(\mathbf{x}) - f_i(\mathbf{y})|. \tag{4.9}$$

Suppose first that $\mathbf{f}$ is continuous at $\mathbf{x}$. Then there exists $\delta > 0$ such that if $|\mathbf{x} - \mathbf{y}| < \delta$, then $|\mathbf{f}(\mathbf{x}) - \mathbf{f}(\mathbf{y})| < \varepsilon$. The first part of the above inequality then shows that for each $k = 1, \cdots, q$, $|f_k(\mathbf{x}) - f_k(\mathbf{y})| < \varepsilon$. This shows the only if part. Now suppose each function f_k is continuous. Then if $\varepsilon > 0$ is given, there exists $\delta_k > 0$ such that whenever $|\mathbf{x} - \mathbf{y}| < \delta_k$

$$|f_k(\mathbf{x}) - f_k(\mathbf{y})| < \varepsilon/q.$$

Now let $0 < \delta \leq \min(\delta_1, \cdots, \delta_q)$. For $|\mathbf{x} - \mathbf{y}| < \delta$, the above inequality holds for all k and so the last part of (4.9) implies

$$|\mathbf{f}(\mathbf{x}) - \mathbf{f}(\mathbf{y})| \leq \sum_{i=1}^{q} |f_i(\mathbf{x}) - f_i(\mathbf{y})| < \sum_{i=1}^{q} \frac{\varepsilon}{q} = \varepsilon.$$

This proves part (4).

To verify part (5), let $\varepsilon > 0$ be given and let $\delta = \varepsilon$. Then if $|\mathbf{x} - \mathbf{y}| < \delta$, the triangle inequality implies

$$|f(\mathbf{x}) - f(\mathbf{y})| = ||\mathbf{x}| - |\mathbf{y}|| \leq |\mathbf{x} - \mathbf{y}| < \delta = \varepsilon.$$

This proves part (5) and completes the proof of the theorem. ∎

4.10.2 The Nested Interval Lemma

First, here is the one dimensional nested interval lemma.

Lemma 4.1. *Let $I_k = [a_k, b_k]$ be closed intervals, $a_k \leq b_k$, such that $I_k \supseteq I_{k+1}$ for all k. Then there exists a point c which is contained in all these intervals. If $\lim_{k \to \infty} (b_k - a_k) = 0$, then there is exactly one such point.*

Proof: Note that the $\{a_k\}$ are an increasing sequence and that $\{b_k\}$ is a decreasing sequence. Now note that if $m < n$, then

$$a_m \leq a_n \leq b_n$$

while if $m > n$,

$$b_n \geq b_m \geq a_m.$$

It follows that $a_m \leq b_n$ for any pair m, n. Therefore, each b_n is an upper bound for all the a_m and so if $c \equiv \sup\{a_k\}$, then for each n, it follows that $c \leq b_n$ and so for all, $a_n \leq c \leq b_n$ which shows that c is in all of these intervals.

If the condition on the lengths of the intervals holds, then if c, c' are in all the intervals, then if they are not equal, then eventually, for large enough k, they cannot both be contained in $[a_k, b_k]$ since eventually $b_k - a_k < |c - c'|$. This would be a contradiction. Hence $c = c'$. ∎

Definition 4.11. The **diameter** of a set S, is defined as

$$\text{diam}(S) \equiv \sup\{|\mathbf{x} - \mathbf{y}| : \mathbf{x}, \mathbf{y} \in S\}.$$

Thus $\text{diam}(S)$ is just a careful description of what you would think of as the diameter. It measures how stretched out the set is.

Here is a multidimensional version of the nested interval lemma.

Lemma 4.2. *Let* $I_k = \prod_{i=1}^{p} \left[a_i^k, b_i^k\right] \equiv \left\{\mathbf{x} \in \mathbb{R}^p : x_i \in \left[a_i^k, b_i^k\right]\right\}$ *and suppose that for all* $k = 1, 2, \cdots,$

$$I_k \supseteq I_{k+1}.$$

Then there exists a point $\mathbf{c} \in \mathbb{R}^p$ *which is an element of every* I_k. *If* $\lim_{k \to \infty} \text{diam}(I_k) = 0$, *then the point* $\mathbf{c}$ *is unique.*

Proof: For each $i = 1, \cdots, p$, $\left[a_i^k, b_i^k\right] \supseteq \left[a_i^{k+1}, b_i^{k+1}\right]$ and so, by Lemma 4.1, there exists a point $c_i \in \left[a_i^k, b_i^k\right]$ for all k. Then letting $\mathbf{c} \equiv (c_1, \cdots, c_p)$ it follows $\mathbf{c} \in I_k$ for all k. If the condition on the diameters holds, then the lengths of the intervals $\lim_{k \to \infty} \left[a_i^k, b_i^k\right] = 0$ and so by the same lemma, each c_i is unique. Hence $\mathbf{c}$ is unique. ∎

I will sometimes refer to the above Cartesian product of closed intervals as an interval to emphasize the analogy with one dimensions, and sometimes as a box.

4.10.3   *Convergent Sequences, Sequential Compactness*

A mapping $\mathbf{f} : \{k, k+1, k+2, \cdots\} \to \mathbb{R}^p$ is called a sequence. We usually write it in the form $\{\mathbf{a}_j\}$ where it is understood that $\mathbf{a}_j \equiv \mathbf{f}(j)$. In the same way as for sequences of real numbers, one can define what it means for convergence to take place.

Definition 4.12. *A sequence,* $\{\mathbf{a}_k\}$ *is said to* **converge** *to* $\mathbf{a}$ *if for every* $\varepsilon > 0$ *there exists* n_ε *such that if* $n > n_\varepsilon$, *then* $|\mathbf{a} - \mathbf{a}_n| < \varepsilon$. *The usual notation for this is* $\lim_{n \to \infty} \mathbf{a}_n = \mathbf{a}$ *although it is often written as* $\mathbf{a}_n \to \mathbf{a}$.

One can also define a subsequence in the same way as in the case of real valued sequences.

Definition 4.13. $\{\mathbf{a}_{n_k}\}$ *is a* **subsequence** *of* $\{\mathbf{a}_n\}$ *if* $n_1 < n_2 < \cdots$.

The following theorem says the limit, if it exists, is unique.

Theorem 4.10. *If a sequence,* $\{\mathbf{a}_n\}$ *converges to* $\mathbf{a}$ *and to* $\mathbf{b}$ *then* $\mathbf{a} = \mathbf{b}$.

Proof: There exists n_ε such that if $n > n_\varepsilon$ then $|\mathbf{a}_n - \mathbf{a}| < \frac{\varepsilon}{2}$ and if $n > n_\varepsilon$, then $|\mathbf{a}_n - \mathbf{b}| < \frac{\varepsilon}{2}$. Then pick such an n.

$$|\mathbf{a} - \mathbf{b}| < |\mathbf{a} - \mathbf{a}_n| + |\mathbf{a}_n - \mathbf{b}| < \frac{\varepsilon}{2} + \frac{\varepsilon}{2} = \varepsilon.$$

Since ε is arbitrary, this proves the theorem. ∎

The following is the definition of a Cauchy sequence in $\mathbb{R}^p$.

Definition 4.14. $\{\mathbf{a}_n\}$ is a Cauchy sequence if for all $\varepsilon > 0$, there exists n_ε such that whenever $n, m \geq n_\varepsilon$,

$$|\mathbf{a}_n - \mathbf{a}_m| < \varepsilon.$$

A sequence is Cauchy, means the terms are "bunching up to each other" as m, n get large.

Theorem 4.11. *The set of terms in a Cauchy sequence in $\mathbb{R}^p$ is bounded in the sense that for all n, $|\mathbf{a}_n| < M$ for some $M < \infty$.*

Proof: Let $\varepsilon = 1$ in the definition of a Cauchy sequence and let $n > n_1$. Then from the definition,

$$|\mathbf{a}_n - \mathbf{a}_{n_1}| < 1.$$

It follows that for all $n > n_1$,

$$|\mathbf{a}_n| < 1 + |\mathbf{a}_{n_1}|.$$

Therefore, for all n,

$$|\mathbf{a}_n| \leq 1 + |\mathbf{a}_{n_1}| + \sum_{k=1}^{n_1} |\mathbf{a}_k|. \qquad \blacksquare$$

Theorem 4.12. *If a sequence $\{\mathbf{a}_n\}$ in $\mathbb{R}^p$ converges, then the sequence is a Cauchy sequence. Also, if some subsequence of a Cauchy sequence converges, then the original sequence converges.*

Proof: Let $\varepsilon > 0$ be given and suppose $\mathbf{a}_n \to \mathbf{a}$. Then from the definition of convergence, there exists n_ε such that if $n > n_\varepsilon$, it follows that

$$|\mathbf{a}_n - \mathbf{a}| < \frac{\varepsilon}{2}$$

Therefore, if $m, n \geq n_\varepsilon + 1$, it follows that

$$|\mathbf{a}_n - \mathbf{a}_m| \leq |\mathbf{a}_n - \mathbf{a}| + |\mathbf{a} - \mathbf{a}_m| < \frac{\varepsilon}{2} + \frac{\varepsilon}{2} = \varepsilon$$

showing that, since $\varepsilon > 0$ is arbitrary, $\{\mathbf{a}_n\}$ is a Cauchy sequence. It remains to that the last claim.

Suppose then that $\{\mathbf{a}_n\}$ is a Cauchy sequence and $\mathbf{a} = \lim_{k \to \infty} \mathbf{a}_{n_k}$ where $\{\mathbf{a}_{n_k}\}_{k=1}^\infty$ is a subsequence. Let $\varepsilon > 0$ be given. Then there exists K such that if $k, l \geq K$, then $|\mathbf{a}_k - \mathbf{a}_l| < \frac{\varepsilon}{2}$. Then if $k > K$, it follows $n_k > K$ because $n_1, n_2, n_3, \cdots$ is strictly increasing as the subscript increases. Also, there exists K_1 such that if $k > K_1, |\mathbf{a}_{n_k} - \mathbf{a}| < \frac{\varepsilon}{2}$. Then letting $n > \max(K, K_1)$, pick $k > \max(K, K_1)$. Then

$$|\mathbf{a} - \mathbf{a}_n| \leq |\mathbf{a} - \mathbf{a}_{n_k}| + |\mathbf{a}_{n_k} - \mathbf{a}_n| < \frac{\varepsilon}{2} + \frac{\varepsilon}{2} = \varepsilon.$$

Therefore, the sequence converges. $\blacksquare$

Definition 4.15. A set K in $\mathbb{R}^p$ is said to be **sequentially compact** if every sequence in K has a subsequence which converges to a point of K.

Theorem 4.13. *If $I_0 = \prod_{i=1}^{p} [a_i, b_i]$ where $a_i \leq b_i$, then I_0 is sequentially compact.*

Proof: Let $\{\mathbf{a}_k\}_{k=1}^{\infty} \subseteq I_0$ and consider all sets of the form $\prod_{i=1}^{p} [c_i, d_i]$ where $[c_i, d_i]$ equals either $[a_i, \frac{a_i + b_i}{2}]$ or $[c_i, d_i] = [\frac{a_i + b_i}{2}, b_i]$. Thus there are 2^p of these sets because there are two choices for the ith slot for $i = 1, \cdots, p$. Also, if $\mathbf{x}$ and $\mathbf{y}$ are two points in one of these sets,

$$|x_i - y_i| \leq 2^{-1} |b_i - a_i|.$$

$\text{diam}(I_0) = \left(\sum_{i=1}^{p} |b_i - a_i|^2 \right)^{1/2},$

$$|\mathbf{x} - \mathbf{y}| = \left(\sum_{i=1}^{p} |x_i - y_i|^2 \right)^{1/2}$$

$$\leq 2^{-1} \left(\sum_{i=1}^{p} |b_i - a_i|^2 \right)^{1/2} \equiv 2^{-1} \text{diam}(I_0).$$

In particular, since $\mathbf{d} \equiv (d_1, \cdots, d_p)$ and $\mathbf{c} \equiv (c_1, \cdots, c_p)$ are two such points,

$$D_1 \equiv \left(\sum_{i=1}^{p} |d_i - c_i|^2 \right)^{1/2} \leq 2^{-1} \text{diam}(I_0)$$

Denote by $\{J_1, \cdots, J_{2^p}\}$ these sets determined above. Since the union of these sets equals all of $I_0 \equiv I$, it follows that for some J_k, the sequence, $\{\mathbf{a}_i\}$ is contained in J_k for infinitely many k. Let that one be called I_1. Next do for I_1 what was done for I_0 to get $I_2 \subseteq I_1$ such that the diameter is half that of I_1 and I_2 contains $\{\mathbf{a}_k\}$ for infinitely many values of k. Continue in this way obtaining a nested sequence $\{I_k\}$ such that $I_k \supseteq I_{k+1}$, and if $\mathbf{x}, \mathbf{y} \in I_k$, then $|\mathbf{x} - \mathbf{y}| \leq 2^{-k} \text{diam}(I_0)$, and I_n contains $\{\mathbf{a}_k\}$ for infinitely many values of k for each n. Then by the nested interval lemma, there exists $\mathbf{c}$ such that $\mathbf{c}$ is contained in each I_k. Pick $\mathbf{a}_{n_1} \in I_1$. Next pick $n_2 > n_1$ such that $\mathbf{a}_{n_2} \in I_2$. If $\mathbf{a}_{n_1}, \cdots, \mathbf{a}_{n_k}$ have been chosen, let $\mathbf{a}_{n_{k+1}} \in I_{k+1}$ and $n_{k+1} > n_k$. This can be done because in the construction, I_n contains $\{\mathbf{a}_k\}$ for infinitely many k. Thus the distance between $\mathbf{a}_{n_k}$ and $\mathbf{c}$ is no larger than $2^{-k} \text{diam}(I_0)$, and so $\lim_{k \to \infty} \mathbf{a}_{n_k} = \mathbf{c} \in I_0$. ∎

Corollary 4.1. *Let K be a closed and bounded set of points in $\mathbb{R}^p$. Then K is sequentially compact.*

Proof: Since K is closed and bounded, there exists a closed rectangle, $\prod_{k=1}^{p} [a_k, b_k]$ which contains K. Now let $\{\mathbf{x}_k\}$ be a sequence of points in K. By Theorem 4.13, there exists a subsequence $\{\mathbf{x}_{n_k}\}$ such that $\mathbf{x}_{n_k} \to \mathbf{x} \in \prod_{k=1}^{p} [a_k, b_k]$. However, K is closed and each of the points of the sequence is in K so $\mathbf{x} \in K$. If not, then since K^C is open, it would follow that eventually $\mathbf{x}_{n_k} \in K^C$ which is impossible. ∎

Theorem 4.14. *Every Cauchy sequence in $\mathbb{R}^p$ converges.*

Proof: Let $\{\mathbf{a}_k\}$ be a Cauchy sequence. By Theorem 4.11, there is some box $\prod_{i=1}^{p} [a_i, b_i]$ containing all the terms of $\{\mathbf{a}_k\}$. Therefore, by Theorem 4.13, a subsequence converges to a point of $\prod_{i=1}^{p} [a_i, b_i]$. By Theorem 4.12, the original sequence converges. ∎

4.10.4 *Continuity And The Limit Of A Sequence*

Just as in the case of a function of one variable, there is a very useful way of thinking of continuity in terms of limits of sequences found in the following theorem. In words, it says a function is continuous if it takes convergent sequences to convergent sequences whenever possible.

Theorem 4.15. *A function* $\mathbf{f} : D(\mathbf{f}) \to \mathbb{R}^q$ *is continuous at* $\mathbf{x} \in D(\mathbf{f})$ *if and only if, whenever* $\mathbf{x}_n \to \mathbf{x}$ *with* $\mathbf{x}_n \in D(\mathbf{f})$, *it follows* $\mathbf{f}(\mathbf{x}_n) \to \mathbf{f}(\mathbf{x})$.

Proof: Suppose first that $\mathbf{f}$ is continuous at $\mathbf{x}$ and let $\mathbf{x}_n \to \mathbf{x}$. Let $\varepsilon > 0$ be given. By continuity, there exists $\delta > 0$ such that if $|\mathbf{y} - \mathbf{x}| < \delta$, then $|\mathbf{f}(\mathbf{x}) - \mathbf{f}(\mathbf{y})| < \varepsilon$. However, there exists n_δ such that if $n \geq n_\delta$, then $|\mathbf{x}_n - \mathbf{x}| < \delta$, and so for all n this large,

$$|\mathbf{f}(\mathbf{x}) - \mathbf{f}(\mathbf{x}_n)| < \varepsilon$$

which shows $\mathbf{f}(\mathbf{x}_n) \to \mathbf{f}(\mathbf{x})$.

Now suppose the condition about taking convergent sequences to convergent sequences holds at $\mathbf{x}$. Suppose $\mathbf{f}$ fails to be continuous at $\mathbf{x}$. Then there exists $\varepsilon > 0$ and $\mathbf{x}_n \in D(f)$ such that $|\mathbf{x} - \mathbf{x}_n| < \frac{1}{n}$, yet

$$|\mathbf{f}(\mathbf{x}) - \mathbf{f}(\mathbf{x}_n)| \geq \varepsilon.$$

But this is clearly a contradiction because, although $\mathbf{x}_n \to \mathbf{x}$, $\mathbf{f}(\mathbf{x}_n)$ fails to converge to $\mathbf{f}(\mathbf{x})$. It follows $\mathbf{f}$ must be continuous after all. ∎

4.10.5 *The Extreme Value Theorem And Uniform Continuity*

Definition 4.16. A function $\mathbf{f}$ having values in $\mathbb{R}^p$ is said to be bounded if the set of values of $\mathbf{f}$ is a bounded set.

Lemma 4.3. *Let* $C \subseteq \mathbb{R}^p$ *be closed and bounded and let* $\mathbf{f} : C \to \mathbb{R}^s$ *be continuous. Then* $\mathbf{f}$ *is bounded.*

Proof: Suppose not. Then since $\mathbf{f}$ is not bounded, there exists $\mathbf{x}_n$ such that

$$\mathbf{f}(\mathbf{x}_n) \notin \prod_{i=1}^{s} (-n, n) \equiv R_n.$$

By Corollary 4.1, C is sequentially compact, and so there exists a subsequence $\{\mathbf{x}_{n_k}\}$ which converges to $\mathbf{x} \in C$. Now $\mathbf{f}(\mathbf{x}) \in R_m$ for large enough m. Hence, by continuity of $\mathbf{f}$, it follows $\mathbf{f}(\mathbf{x}_n) \in R_m$ for all n large enough, contradicting the construction. ∎

Here is a proof of the extreme value theorem.

Theorem 4.16. *Let C be closed and bounded and let $f : C \to \mathbb{R}$ be continuous. Then f achieves its maximum and its minimum on C. This means there exist $\mathbf{x}_1, \mathbf{x}_2 \in C$ such that for all $\mathbf{x} \in C$,*

$$f(\mathbf{x}_1) \leq f(\mathbf{x}) \leq f(\mathbf{x}_2).$$

Proof: Let $M = \sup\{f(\mathbf{x}) : \mathbf{x} \in C\}$. Then by Lemma 4.3, M is a finite number. Is $f(\mathbf{x}_2) = M$ for some x_2? If not, you could consider the function

$$g(\mathbf{x}) \equiv \frac{1}{M - f(\mathbf{x})}$$

and g would be a continuous and unbounded function defined on C, contrary to Lemma 4.3. Therefore, there exists $\mathbf{x}_2 \in C$ such that $f(\mathbf{x}_2) = M$. A similar argument applies to show the existence of $\mathbf{x}_1 \in C$ such that

$$f(\mathbf{x}_1) = \inf\{f(\mathbf{x}) : \mathbf{x} \in C\}. \qquad ∎$$

As in the case of a function of one variable, there is a concept of uniform continuity.

Definition 4.17. A function $\mathbf{f} : D(\mathbf{f}) \to \mathbb{R}^q$ is uniformly continuous if for every $\varepsilon > 0$ there exists $\delta > 0$ such that whenever $\mathbf{x}, \mathbf{y}$ are points of $D(\mathbf{f})$ such that $|\mathbf{x} - \mathbf{y}| < \delta$, it follows $|\mathbf{f}(\mathbf{x}) - \mathbf{f}(\mathbf{y})| < \varepsilon$.

Theorem 4.17. *Let $\mathbf{f} : K \to \mathbb{R}^q$ be continuous at every point of K where K is a closed and bounded set in $\mathbb{R}^p$. Then $\mathbf{f}$ is uniformly continuous.*

Proof: Suppose not. Then there exists $\varepsilon > 0$ and sequences $\{\mathbf{x}_j\}$ and $\{\mathbf{y}_j\}$ of points in K such that

$$|\mathbf{x}_j - \mathbf{y}_j| < \frac{1}{j}$$

but $|\mathbf{f}(\mathbf{x}_j) - \mathbf{f}(\mathbf{y}_j)| \geq \varepsilon$. Then by Corollary 4.1 on Page 80 which says K is sequentially compact, there is a subsequence $\{\mathbf{x}_{n_k}\}$ of $\{\mathbf{x}_j\}$ which converges to a point $\mathbf{x} \in K$. Then since $|\mathbf{x}_{n_k} - \mathbf{y}_{n_k}| < \frac{1}{k}$, it follows that $\{\mathbf{y}_{n_k}\}$ also converges to $\mathbf{x}$. Therefore,

$$\varepsilon \leq \lim_{k \to \infty} |\mathbf{f}(\mathbf{x}_{n_k}) - \mathbf{f}(\mathbf{y}_{n_k})| = |\mathbf{f}(\mathbf{x}) - \mathbf{f}(\mathbf{x})| = 0,$$

a contradiction. Therefore, $\mathbf{f}$ is uniformly continuous as claimed. ∎

Some of the following exercises have been essentially done in the above discussion. Try doing them yourself. There are also some new topics.

4.11 Exercises

(1) Suppose $\{\mathbf{x}_n\}$ is a sequence contained in a closed set C such that $\lim_{n\to\infty} \mathbf{x}_n = \mathbf{x}$. Show that $\mathbf{x} \in C$. **Hint:** Recall that a set is closed if and only if the complement of the set is open. That is if and only if $\mathbb{R}^n \setminus C$ is open.

(2) Show using Problem 1 and Theorem 4.13 that every closed and bounded set is sequentially compact. **Hint:** If C is such a set, then $C \subseteq I_0 \equiv \prod_{i=1}^{n} [a_i, b_i]$. Now if $\{\mathbf{x}_n\}$ is a sequence in C, it must also be a sequence in I_0. Apply Problem 1 and Theorem 4.13.

(3) Prove the extreme value theorem, a continuous function achieves its maximum and minimum on any closed and bounded set C, using the result of Problem 2. **Hint:** Suppose $\lambda = \sup \{f(\mathbf{x}) : \mathbf{x} \in C\}$. Then there exists $\{\mathbf{x}_n\} \subseteq C$ such that $f(\mathbf{x}_n) \to \lambda$. Now select a convergent subsequence using Problem 2. Do the same for the minimum.

(4) Let C be a closed and bounded set and suppose $\mathbf{f} : C \to \mathbb{R}^m$ is continuous. Show that $\mathbf{f}$ must also be **uniformly continuous.** This means: For every $\varepsilon > 0$ there exists $\delta > 0$ such that whenever $\mathbf{x}, \mathbf{y} \in C$ and $|\mathbf{x} - \mathbf{y}| < \delta$, it follows $|\mathbf{f}(\mathbf{x}) - \mathbf{f}(\mathbf{y})| < \varepsilon$. This is a good time to review the definition of continuity so you will see the difference. **Hint:** Suppose it is not so. Then there exists $\varepsilon > 0$ and $\{\mathbf{x}_k\}$ and $\{\mathbf{y}_k\}$ such that $|\mathbf{x}_k - \mathbf{y}_k| < \frac{1}{k}$ but $|\mathbf{f}(\mathbf{x}_k) - \mathbf{f}(\mathbf{y}_k)| \geq \varepsilon$. Now use Problem 2 to obtain a convergent subsequence.

(5) From Problem 2 every closed and bounded set is sequentially compact. Are these the only sets which are sequentially compact? Explain.

(6) A set whose elements are open sets $\mathcal{C}$ is called an **open cover** of H if $\cup \mathcal{C} \supseteq H$. In other words, $\mathcal{C}$ is an open cover of H if every point of H is in at least one set of $\mathcal{C}$. Show that if $\mathcal{C}$ is an open cover of a closed and bounded set H then there exists $\delta > 0$ such that whenever $\mathbf{x} \in H$, $B(\mathbf{x}, \delta)$ is contained in some set of $\mathcal{C}$. This number δ is called a **Lebesgue number**. **Hint:** If there is no Lebesgue number for H, let $H \subseteq I = \prod_{i=1}^{n} [a_i, b_i]$. Use the process of chopping the intervals in half to get a sequence of nested intervals, I_k contained in I where $\operatorname{diam}(I_k) \leq 2^{-k} \operatorname{diam}(I)$ and there is no Lebesgue number for the open cover on $H_k \equiv H \cap I_k$. Now use the nested interval theorem to get $\mathbf{c}$ in all these H_k. For some $r > 0$ it follows $B(\mathbf{c}, r)$ is contained in some open set of U. But for large k, it must be that $H_k \subseteq B(\mathbf{c}, r)$ which contradicts the construction. You fill in the details.

(7) A set is **compact** if for every open cover of the set, there exists a finite subset of the open cover which also covers the set. Show every closed and bounded set in $\mathbb{R}^p$ is compact. Next show that if a set in $\mathbb{R}^p$ is compact, then it must be closed and bounded. This is called the Heine Borel theorem. **Hint:** To show closed and bounded is compact, you might use the technique of chopping into small pieces of the above problem.

(8) Suppose S is a nonempty set in $\mathbb{R}^p$. Define

$$\text{dist}\,(\mathbf{x},S) \equiv \inf\,\{|\mathbf{x} - \mathbf{y}| : \mathbf{y} \in S\}.$$

Show that

$$|\text{dist}\,(\mathbf{x},S) - \text{dist}\,(\mathbf{y},S)| \le |\mathbf{x} - \mathbf{y}|.$$

Hint: Suppose $\text{dist}\,(\mathbf{x}, S) < \text{dist}\,(\mathbf{y}, S)$. If these are equal there is nothing to show. Explain why there exists $\mathbf{z} \in S$ such that $|\mathbf{x} - \mathbf{z}| < \text{dist}\,(\mathbf{x},S) + \varepsilon$. Now explain why

$$|\text{dist}\,(\mathbf{x},S) - \text{dist}\,(\mathbf{y},S)| = \text{dist}\,(\mathbf{y},S) - \text{dist}\,(\mathbf{x},S) \le |\mathbf{y} - \mathbf{z}| - (|\mathbf{x} - \mathbf{z}| - \varepsilon)$$

Now use the triangle inequality and observe that ε is arbitrary.

(9) Suppose H is a closed set and $H \subseteq U \subseteq \mathbb{R}^p$, an open set. Show there exists a continuous function defined on $\mathbb{R}^p$, f such that $f\,(\mathbb{R}^p) \subseteq [0,1]$, $f\,(\mathbf{x}) = 0$ if $\mathbf{x} \notin U$ and $f\,(\mathbf{x}) = 1$ if $\mathbf{x} \in H$. **Hint:** Try something like

$$\frac{\text{dist}\,(\mathbf{x}, U^C)}{\text{dist}\,(\mathbf{x}, U^C) + \text{dist}\,(\mathbf{x}, H)},$$

where $U^C \equiv \mathbb{R}^p \setminus U$, a closed set. You need to explain why the denominator is never equal to zero. The rest is supplied by Problem 8. This is a special case of a major theorem called Urysohn's lemma.

Chapter 5

Vector Valued Functions Of One Variable

5.1 Limits Of A Vector Valued Function Of One Real Variable

As in the case of a scalar valued function of one variable, the derivative is defined as

$$\lim_{h \to 0} \frac{\mathbf{f}(t_0 + h) - \mathbf{f}(t_0)}{h}.$$

Thus the derivative of a function of one variable involves a limit. The following is the definition of what is meant by a limit. The new topic is the case of one sided limits although there is really nothing essentially new from what was done earlier. Here is the definition.

Definition 5.1. In the case where $D(\mathbf{f})$ is only assumed to satisfy $D(\mathbf{f}) \supseteq (t, t + r)$,

$$\lim_{s \to t+} \mathbf{f}(s) = \mathbf{L}$$

if and only if for all $\varepsilon > 0$ there exists $\delta > 0$ such that if

$$0 < s - t < \delta,$$

then

$$|\mathbf{f}(s) - \mathbf{L}| < \varepsilon.$$

In the case where $D(\mathbf{f})$ is only assumed to satisfy $D(\mathbf{f}) \supseteq (t - r, t)$,

$$\lim_{s \to t-} \mathbf{f}(s) = \mathbf{L}$$

if and only if for all $\varepsilon > 0$ there exists $\delta > 0$ such that if

$$0 < t - s < \delta,$$

then

$$|\mathbf{f}(s) - \mathbf{L}| < \varepsilon.$$

One can also consider limits as a variable "approaches" infinity. Of course nothing is "close" to infinity and so this requires a slightly different definition.

$$\lim_{t \to \infty} \mathbf{f}(t) = \mathbf{L}$$

if for every $\varepsilon > 0$ there exists l such that whenever $t > l$,

$$|\mathbf{f}(t) - \mathbf{L}| < \varepsilon \tag{5.1}$$

and

$$\lim_{t \to -\infty} \mathbf{f}(t) = \mathbf{L}$$

if for every $\varepsilon > 0$ there exists l such that whenever $t < l$, (5.1) holds.

Note that in all of this the definitions are identical to the case of scalar valued functions. The only difference is that here $|\cdot|$ refers to the norm or length in $\mathbb{R}^p$ where maybe $p > 1$.

Example 5.1. Let $\mathbf{f}(t) = (\cos t, \sin t, t^2 + 1, \ln(t))$. Find $\lim_{t \to \pi/2} \mathbf{f}(t)$.

Use Theorem 4.3 on Page 64 and the continuity of the functions to write this limit equals

$$\left(\lim_{t \to \pi/2} \cos t, \ \lim_{t \to \pi/2} \sin t, \ \lim_{t \to \pi/2} \left(t^2 + 1 \right), \ \lim_{t \to \pi/2} \ln(t) \right)$$

$$= \left(0, 1, \ln\left(\frac{\pi^2}{4} + 1 \right), \ln\left(\frac{\pi}{2} \right) \right).$$

Example 5.2. Let $\mathbf{f}(t) = \left(\frac{\sin t}{t}, t^2, t + 1 \right)$. Find $\lim_{t \to 0} \mathbf{f}(t)$.

Recall that $\lim_{t \to 0} \frac{\sin t}{t} = 1$. Then from Theorem 4.3 on Page 64, $\lim_{t \to 0} \mathbf{f}(t) = (1, 0, 1)$.

5.2   The Derivative And Integral

The following definition is on the derivative and integral of a vector valued function of one variable.

Definition 5.2. The derivative of a function $\mathbf{f}'(t)$, is defined as the following limit whenever the limit exists. If the limit does not exist, then neither does $\mathbf{f}'(t)$.

$$\lim_{h \to 0} \frac{\mathbf{f}(t+h) - \mathbf{f}(t)}{h} \equiv \mathbf{f}'(t)$$

As before,

$$\mathbf{f}'(t) = \lim_{s \to t} \frac{\mathbf{f}(s) - \mathbf{f}(t)}{s - t}.$$

The function of h on the left is called the difference quotient just as it was for a scalar valued function. If $\mathbf{f}(t) = (f_1(t), \cdots, f_p(t))$ and $\int_a^b f_i(t) \, dt$ exists for each $i = 1, \cdots, p$, then $\int_a^b \mathbf{f}(t) \, dt$ is defined as the vector

$$\left(\int_a^b f_1(t) \, dt, \cdots, \int_a^b f_p(t) \, dt \right).$$

This is what is meant by saying $\mathbf{f} \in R([a, b])$.

Here is a simple proposition which is useful to have.

Proposition 5.1. Let $a \leq b$, $\mathbf{f} = (f_1, \cdots, f_n)$ is vector valued and each f_i is continuous, then

$$\left| \int_a^b \mathbf{f}(t)\, dt \right| \leq \sqrt{n} \int_a^b |\mathbf{f}(t)|\, dt.$$

Proof: This follows from the Cauchy Schwarz inequality.

$$\left| \int_a^b \mathbf{f}(t)\, dt \right| = \left| \left(\int_a^b f_1(t)\, dt, \cdots, \int_a^b f_n(t)\, dt \right) \right|$$

$$= \left(\left| \int_a^b f_1(t)\, dt \right|, \cdots, \left| \int_a^b f_n(t)\, dt \right| \right) \leq \left(\int_a^b |f_1(t)|\, dt, \cdots, \int_a^b |f_n(t)|\, dt \right)$$

$$\leq \left(\int_a^b |\mathbf{f}(t)|\, dt, \cdots, \int_a^b |\mathbf{f}(t)|\, dt \right) = \sqrt{n} \int_a^b |\mathbf{f}(t)|\, dt. \quad \blacksquare$$

As in the case of a scalar valued function differentiability implies continuity but not the other way around.

Theorem 5.1. *If $\mathbf{f}'(t)$ exists, then $\mathbf{f}$ is continuous at t.*

Proof: Suppose $\varepsilon > 0$ is given and choose $\delta_1 > 0$ such that if $|h| < \delta_1$,

$$\left| \frac{\mathbf{f}(t+h) - \mathbf{f}(t)}{h} - \mathbf{f}'(t) \right| < 1.$$

then for such h, the triangle inequality implies $|\mathbf{f}(t+h) - \mathbf{f}(t)| < |h| + |\mathbf{f}'(t)| |h|$. Now letting $\delta < \min \left(\delta_1, \frac{\varepsilon}{1+|\mathbf{f}'(x)|} \right)$ it follows if $|h| < \delta$, then $|\mathbf{f}(t+h) - \mathbf{f}(t)| < \varepsilon$. Letting $y = h + t$, this shows that if $|y - t| < \delta, |\mathbf{f}(y) - \mathbf{f}(t)| < \varepsilon$ which proves $\mathbf{f}$ is continuous at t. $\blacksquare$

As in the scalar case, there is a fundamental theorem of calculus.

Theorem 5.2. *If $\mathbf{f} \in R([a, b])$ and if $\mathbf{f}$ is continuous at $t \in (a, b)$, then*

$$\frac{d}{dt} \left(\int_a^t \mathbf{f}(s)\, ds \right) = \mathbf{f}(t).$$

Proof: Say $\mathbf{f}(t) = (f_1(t), \cdots, f_p(t))$. Then it follows

$$\frac{1}{h} \int_a^{t+h} \mathbf{f}(s)\, ds - \frac{1}{h} \int_a^t \mathbf{f}(s)\, ds = \left(\frac{1}{h} \int_t^{t+h} f_1(s)\, ds, \cdots, \frac{1}{h} \int_t^{t+h} f_p(s)\, ds \right)$$

and $\lim_{h \to 0} \frac{1}{h} \int_t^{t+h} f_i(s)\, ds = f_i(t)$ for each $i = 1, \cdots, p$ from the fundamental theorem of calculus for scalar valued functions. Therefore,

$$\lim_{h \to 0} \frac{1}{h} \int_a^{t+h} \mathbf{f}(s)\, ds - \frac{1}{h} \int_a^t \mathbf{f}(s)\, ds = (f_1(t), \cdots, f_p(t)) = \mathbf{f}(t). \quad \blacksquare$$

Example 5.3. Let $\mathbf{f}(x) = \mathbf{c}$ where $\mathbf{c}$ is a constant. Find $\mathbf{f}'(x)$.

The difference quotient,

$$\frac{\mathbf{f}(x+h) - \mathbf{f}(x)}{h} = \frac{\mathbf{c} - \mathbf{c}}{h} = \mathbf{0}$$

Therefore,

$$\lim_{h \to 0} \frac{\mathbf{f}(x+h) - \mathbf{f}(x)}{h} = \lim_{h \to 0} \mathbf{0} = \mathbf{0}$$

Example 5.4. Let $\mathbf{f}(t) = (at, bt)$ where a, b are constants. Find $\mathbf{f}'(t)$.

From the above discussion this derivative is just the vector valued functions whose components consist of the derivatives of the components of $\mathbf{f}$. Thus $\mathbf{f}'(t) = (a, b)$.

5.2.1 *Geometric And Physical Significance Of The Derivative*

Suppose $\mathbf{r}$ is a vector valued function of a parameter t not necessarily time and consider the following picture of the points traced out by $\mathbf{r}$.

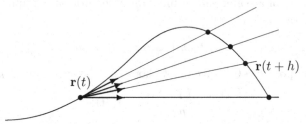

In this picture there are unit vectors in the direction of the vector from $\mathbf{r}(t)$ to $\mathbf{r}(t+h)$. You can see that it is reasonable to suppose these unit vectors, if they converge, converge to a unit vector $\mathbf{T}$ which is tangent to the curve at the point $\mathbf{r}(t)$. Now each of these unit vectors is of the form

$$\frac{\mathbf{r}(t+h) - \mathbf{r}(t)}{|\mathbf{r}(t+h) - \mathbf{r}(t)|} \equiv \mathbf{T}_h.$$

Thus $\mathbf{T}_h \to \mathbf{T}$, a unit tangent vector to the curve at the point $\mathbf{r}(t)$. Therefore,

$$\mathbf{r}'(t) \equiv \lim_{h \to 0} \frac{\mathbf{r}(t+h) - \mathbf{r}(t)}{h} = \lim_{h \to 0} \frac{|\mathbf{r}(t+h) - \mathbf{r}(t)|}{h} \frac{\mathbf{r}(t+h) - \mathbf{r}(t)}{|\mathbf{r}(t+h) - \mathbf{r}(t)|}$$

$$= \lim_{h \to 0} \frac{|\mathbf{r}(t+h) - \mathbf{r}(t)|}{h} \mathbf{T}_h = |\mathbf{r}'(t)| \mathbf{T}.$$

In the case that t is time, the expression $|\mathbf{r}(t+h) - \mathbf{r}(t)|$ is a good approximation for the distance traveled by the object on the time interval $[t, t+h]$. The real distance would be the length of the curve joining the two points but if h is very small, this is essentially equal to $|\mathbf{r}(t+h) - \mathbf{r}(t)|$ as suggested by the picture below.

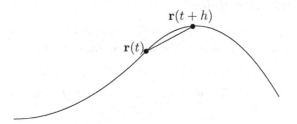

Therefore, $\frac{|\mathbf{r}(t+h)-\mathbf{r}(t)|}{h}$ gives for small h, the approximate distance travelled on the time interval $[t, t+h]$ divided by the length of time h. Therefore, this expression is really the average speed of the object on this small time interval and so the limit as $h \to 0$, deserves to be called the instantaneous speed of the object. Thus $|\mathbf{r}'(t)|\,\mathbf{T}$ represents the speed times a unit direction vector $\mathbf{T}$ which defines the direction in which the object is moving. Thus $\mathbf{r}'(t)$ is the velocity of the object. This is the physical significance of the derivative when t is time. In general, $\mathbf{r}'(t)$ and $\mathbf{T}(\mathbf{t})$ are vectors tangent to the curve which point in the direction of motion.

How do you go about computing $\mathbf{r}'(t)$? Letting $\mathbf{r}(t) = (r_1(t), \cdots, r_q(t))$, the expression

$$\frac{\mathbf{r}(t_0 + h) - \mathbf{r}(t_0)}{h} \tag{5.2}$$

is equal to

$$\left(\frac{r_1(t_0 + h) - r_1(t_0)}{h}, \cdots, \frac{r_q(t_0 + h) - r_q(t_0)}{h} \right).$$

Then as h converges to 0, (5.2) converges to $\mathbf{v} \equiv (v_1, \cdots, v_q)$ where $v_k = r_k'(t)$. This is because of Theorem 4.3 on Page 64, which says that the term in (5.2) gets close to a vector $\mathbf{v}$ if and only if all the coordinate functions of the term in (5.2) get close to the corresponding coordinate functions of $\mathbf{v}$.

In the case where t is time, this simply says the velocity vector equals the vector whose components are the derivatives of the components of the displacement vector $\mathbf{r}(t)$.

Example 5.5. Let $\mathbf{r}(t) = (\sin t, t^2, t + 1)$ for $t \in [0, 5]$. Find a tangent line to the curve parameterized by $\mathbf{r}$ at the point $\mathbf{r}(2)$.

From the above discussion, a direction vector has the same direction as $\mathbf{r}'(2)$. Therefore, it suffices to simply use $\mathbf{r}'(2)$ as a direction vector for the line. $\mathbf{r}'(2) = (\cos 2, 4, 1)$. Therefore, a parametric equation for the tangent line is

$$(\sin 2, 4, 3) + t(\cos 2, 4, 1) = (x, y, z).$$

Example 5.6. Let $\mathbf{r}(t) = (\sin t, t^2, t + 1)$ for $t \in [0, 5]$. Find the velocity vector when $t = 1$.

From the above discussion, this is simply $\mathbf{r}'(1) = (\cos 1, 2, 1)$.

5.2.2 Differentiation Rules

There are rules which relate the derivative to the various operations done with vectors such as the dot product, the cross product, vector addition, and scalar multiplication.

Theorem 5.3. *Let $a, b \in \mathbb{R}$ and suppose $\mathbf{f}'(t)$ and $\mathbf{g}'(t)$ exist. Then the following formulas are obtained.*

$$(a\mathbf{f} + b\mathbf{g})'(t) = a\mathbf{f}'(t) + b\mathbf{g}'(t). \tag{5.3}$$

$$(\mathbf{f} \cdot \mathbf{g})'(t) = \mathbf{f}'(t) \cdot \mathbf{g}(t) + \mathbf{f}(t) \cdot \mathbf{g}'(t) \tag{5.4}$$

If $\mathbf{f}, \mathbf{g}$ have values in $\mathbb{R}^3$, then

$$(\mathbf{f} \times \mathbf{g})'(t) = \mathbf{f}(t) \times \mathbf{g}'(t) + \mathbf{f}'(t) \times \mathbf{g}(t) \tag{5.5}$$

The formulas, (5.4), and (5.5) are referred to as the product rule.

Proof: The first formula is left for you to prove. Consider the second, (5.4).

$$\lim_{h \to 0} \frac{\mathbf{f} \cdot \mathbf{g}(t+h) - \mathbf{fg}(t)}{h}$$

$$= \lim_{h \to 0} \frac{\mathbf{f}(t+h) \cdot \mathbf{g}(t+h) - \mathbf{f}(t+h) \cdot \mathbf{g}(t)}{h} + \frac{\mathbf{f}(t+h) \cdot \mathbf{g}(t) - \mathbf{f}(t) \cdot \mathbf{g}(t)}{h}$$

$$= \lim_{h \to 0} \left(\mathbf{f}(t+h) \cdot \frac{(\mathbf{g}(t+h) - \mathbf{g}(t))}{h} + \frac{(\mathbf{f}(t+h) - \mathbf{f}(t))}{h} \cdot \mathbf{g}(t) \right)$$

$$= \lim_{h \to 0} \sum_{k=1}^{n} f_k(t+h) \frac{(g_k(t+h) - g_k(t))}{h} + \sum_{k=1}^{n} \frac{(f_k(t+h) - f_k(t))}{h} g_k(t)$$

$$= \sum_{k=1}^{n} f_k(t) g_k'(t) + \sum_{k=1}^{n} f_k'(t) g_k(t) = \mathbf{f}'(t) \cdot \mathbf{g}(t) + \mathbf{f}(t) \cdot \mathbf{g}'(t).$$

Formula (5.5) is left as an exercise which follows from the product rule and the definition of the cross product. ∎

Example 5.7. Let $\mathbf{r}(t) = (t^2, \sin t, \cos t)$ and let $\mathbf{p}(t) = (t, \ln(t+1), 2t)$. Find $(\mathbf{r}(t) \times \mathbf{p}(t))'$.

From (5.5) this equals $(2t, \cos t, -\sin t) \times (t, \ln(t+1), 2t) + (t^2, \sin t, \cos t) \times \left(1, \frac{1}{t+1}, 2\right)$.

Example 5.8. Let $\mathbf{r}(t) = (t^2, \sin t, \cos t)$ Find $\int_0^\pi \mathbf{r}(t)\, dt$.

This equals $\left(\int_0^\pi t^2\, dt, \int_0^\pi \sin t\, dt, \int_0^\pi \cos t\, dt \right) = \left(\frac{1}{3}\pi^3, 2, 0 \right)$.

Example 5.9. An object has position $\mathbf{r}(t) = \left(t^3, \frac{t}{1+t}, \sqrt{t^2 + 2} \right)$ kilometers where t is given in hours. Find the velocity of the object in kilometers per hour when $t = 1$.

Recall the velocity at time t was $\mathbf{r}'(t)$. Therefore, find $\mathbf{r}'(t)$ and plug in $t = 1$ to find the velocity.

$$\mathbf{r}'(t) = \left(3t^2, \frac{1(1+t) - t}{(1+t)^2}, \frac{1}{2}(t^2 + 2)^{-1/2} 2t \right) = \left(3t^2, \frac{1}{(1+t)^2}, \frac{1}{\sqrt{(t^2+2)}} t \right)$$

When $t = 1$, the velocity is

$$\mathbf{r}'(1) = \left(3, \frac{1}{4}, \frac{1}{\sqrt{3}} \right) \quad \text{kilometers per hour.}$$

Obviously, this can be continued. That is, you can consider the possibility of taking the derivative of the derivative and then the derivative of that and so forth. The main thing to consider about this is the notation, and it is exactly like it was in the case of a scalar valued function presented earlier. Thus $\mathbf{r}''(t)$ denotes the second derivative.

When you are given a vector valued function of one variable, sometimes it is possible to give a simple description of the curve which results. Usually it is not possible to do this!

Example 5.10. Describe the curve which results from the vector valued function $\mathbf{r}(t) = (\cos 2t, \sin 2t, t)$ where $t \in \mathbb{R}$.

The first two components indicate that for $\mathbf{r}(t) = (x(t), y(t), z(t))$, the pair, $(x(t), y(t))$ traces out a circle. While it is doing so, $z(t)$ is moving at a steady rate in the positive direction. Therefore, the curve which results is a cork screw shaped thing called a helix.

As an application of the theorems for differentiating curves, here is an interesting application. It is also a situation where the curve can be identified as something familiar.

Example 5.11. Sound waves have the angle of incidence equal to the angle of reflection. Suppose you are in a large room and you make a sound. The sound waves spread out and you would expect your sound to be inaudible very far away. But what if the room were shaped so that the sound is reflected off the wall toward a single point, possibly far away from you? Then you might have the interesting phenomenon of someone far away hearing what you said quite clearly. How should the room be designed?

Suppose you are located at the point $\mathbf{P}_0$ and the point where your sound is to be reflected is $\mathbf{P}_1$. Consider a plane which contains the two points and let $\mathbf{r}(t)$ denote a parametrization of the intersection of this plane with the walls of the room. Then the condition that the angle of reflection equals the angle of incidence reduces to saying the angle between $\mathbf{P}_0 - \mathbf{r}(t)$ and $-\mathbf{r}'(t)$ equals the angle between $\mathbf{P}_1 - \mathbf{r}(t)$ and $\mathbf{r}'(t)$. Draw a picture to see this. Therefore,

$$\frac{(\mathbf{P}_0 - \mathbf{r}(t)) \cdot (-\mathbf{r}'(t))}{|\mathbf{P}_0 - \mathbf{r}(t)| \, |\mathbf{r}'(t)|} = \frac{(\mathbf{P}_1 - \mathbf{r}(t)) \cdot (\mathbf{r}'(t))}{|\mathbf{P}_1 - \mathbf{r}(t)| \, |\mathbf{r}'(t)|}.$$

This reduces to

$$\frac{(\mathbf{r}(t) - \mathbf{P}_0) \cdot (-\mathbf{r}'(t))}{|\mathbf{r}(t) - \mathbf{P}_0|} = \frac{(\mathbf{r}(t) - \mathbf{P}_1) \cdot (\mathbf{r}'(t))}{|\mathbf{r}(t) - \mathbf{P}_1|} \qquad (5.6)$$

Now

$$\frac{(\mathbf{r}(t) - \mathbf{P}_1) \cdot (\mathbf{r}'(t))}{|\mathbf{r}(t) - \mathbf{P}_1|} = \frac{d}{dt} |\mathbf{r}(t) - \mathbf{P}_1|$$

and a similar formula holds for $\mathbf{P}_1$ replaced with $\mathbf{P}_0$. This is because

$$|\mathbf{r}(t) - \mathbf{P}_1| = \sqrt{(\mathbf{r}(t) - \mathbf{P}_1) \cdot (\mathbf{r}(t) - \mathbf{P}_1)}$$

and so using the chain rule and product rule,

$$\begin{aligned} \frac{d}{dt} |\mathbf{r}(t) - \mathbf{P}_1| &= \frac{1}{2} ((\mathbf{r}(t) - \mathbf{P}_1) \cdot (\mathbf{r}(t) - \mathbf{P}_1))^{-1/2} \, 2 \, ((\mathbf{r}(t) - \mathbf{P}_1) \cdot \mathbf{r}'(t)) \\ &= \frac{(\mathbf{r}(t) - \mathbf{P}_1) \cdot (\mathbf{r}'(t))}{|\mathbf{r}(t) - \mathbf{P}_1|}. \end{aligned}$$

Therefore, from (5.6),

$$\frac{d}{dt} (|\mathbf{r}(t) - \mathbf{P}_1|) + \frac{d}{dt} (|\mathbf{r}(t) - \mathbf{P}_0|) = 0$$

showing that $|\mathbf{r}(t) - \mathbf{P}_1| + |\mathbf{r}(t) - \mathbf{P}_0| = C$ for some constant C. This implies the curve of intersection of the plane with the room is an ellipse having $\mathbf{P}_0$ and $\mathbf{P}_1$ as the foci.

5.2.3 Leibniz's Notation

Leibniz's notation also generalizes routinely. For example, $\frac{d\mathbf{y}}{dt} = \mathbf{y}'(t)$ with other similar notations holding.

5.3 Exercises

(1) Find the following limits if possible

(a) $\lim_{x \to 0+} \left(\frac{|x|}{x}, \sin x / x, \cos x \right)$

(b) $\lim_{x \to 0+} \left(\frac{x}{|x|}, \sec x, e^x \right)$

(c) $\lim_{x \to 4} \left(\frac{x^2 - 16}{x + 4}, x + 7, \frac{\tan 4x}{5x} \right)$

(d) $\lim_{x \to \infty} \left(\frac{x}{1 + x^2}, \frac{x^2}{1 + x^2}, \frac{\sin x^2}{x} \right)$

(2) Find

$$\lim_{x \to 2} \left(\frac{x^2 - 4}{x + 2}, x^2 + 2x - 1, \frac{x^2 - 4}{x - 2} \right).$$

(3) Prove from the definition that $\lim_{x \to a} (\sqrt[3]{x}, x + 1) = (\sqrt[3]{a}, a + 1)$ for all $a \in \mathbb{R}$. **Hint:** You might want to use the formula for the difference of two cubes,

$$a^3 - b^3 = (a - b)(a^2 + ab + b^2).$$

(4) Let

$$\mathbf{r}(t) = \left(4 + t^2, \sqrt{t^2 + 1}t^3, t^3\right)$$

describe the position of an object in $\mathbb{R}^3$ as a function of t where t is measured in seconds and $\mathbf{r}(t)$ is measured in meters. Is the velocity of this object ever equal to zero? If so, find the value of t at which this occurs and the point in $\mathbb{R}^3$ at which the velocity is zero.

(5) Let $\mathbf{r}(t) = (\sin 2t, t^2, 2t + 1)$ for $t \in [0, 4]$. Find a tangent line to the curve parameterized by $\mathbf{r}$ at the point $\mathbf{r}(2)$.

(6) Let $\mathbf{r}(t) = (t, \sin t^2, t + 1)$ for $t \in [0, 5]$. Find a tangent line to the curve parameterized by $\mathbf{r}$ at the point $\mathbf{r}(2)$.

(7) Let $\mathbf{r}(t) = (\sin t, t^2, \cos(t^2))$ for $t \in [0, 5]$. Find a tangent line to the curve parameterized by $\mathbf{r}$ at the point $\mathbf{r}(2)$.

(8) Let $\mathbf{r}(t) = (\sin t, \cos(t^2), t + 1)$ for $t \in [0, 5]$. Find the velocity when $t = 3$.

(9) Let $\mathbf{r}(t) = (\sin t, t^2, t + 1)$ for $t \in [0, 5]$. Find the velocity when $t = 3$.

(10) Let $\mathbf{r}(t) = (t, \ln(t^2 + 1), t + 1)$ for $t \in [0, 5]$. Find the velocity when $t = 3$.

(11) Suppose an object has position $\mathbf{r}(t) \in \mathbb{R}^3$ where $\mathbf{r}$ is differentiable and suppose also that $|\mathbf{r}(t)| = c$ where c is a constant.

    (a) Show first that this condition does not require $\mathbf{r}(t)$ to be a constant. **Hint:** You can do this either mathematically or by giving a physical example.

    (b) Show that you can conclude that $\mathbf{r}'(t) \cdot \mathbf{r}(t) = 0$. That is, the velocity is always perpendicular to the displacement.

(12) Prove (5.5) from the component description of the cross product.

(13) Prove (5.5) from the formula $(\mathbf{f} \times \mathbf{g})_i = \varepsilon_{ijk} f_j g_k$.

(14) Prove (5.5) directly from the definition of the derivative without considering components.

(15) A Bezier curve in $\mathbb{R}^p$ is a vector valued function of the form

$$\mathbf{y}(t) = \sum_{k=0}^{n} \binom{n}{k} \mathbf{x}_k (1 - t)^{n-k} t^k$$

where here the $\binom{n}{k}$ are the binomial coefficients and $\mathbf{x}_k$ are $n + 1$ points in $\mathbb{R}^n$. Show that $\mathbf{y}(0) = \mathbf{x}_0$, $\mathbf{y}(1) = \mathbf{x}_n$, and find $\mathbf{y}'(0)$ and $\mathbf{y}'(1)$. Recall that $\binom{n}{0} = \binom{n}{n} = 1$ and $\binom{n}{n-1} = \binom{n}{1} = n$. Curves of this sort are important in various computer programs.

(16) Suppose $\mathbf{r}(t)$, $\mathbf{s}(t)$, and $\mathbf{p}(t)$ are three differentiable functions of t which have values in $\mathbb{R}^3$. Find a formula for $(\mathbf{r}(t) \times \mathbf{s}(t) \cdot \mathbf{p}(t))'$.

(17) If $\mathbf{r}'(t) = \mathbf{0}$ for all $t \in (a, b)$, show that there exists a constant vector $\mathbf{c}$ such that $\mathbf{r}(t) = \mathbf{c}$ for all $t \in (a, b)$.

(18) If $\mathbf{F}'(t) = \mathbf{f}(t)$ for all $t \in (a, b)$ and $\mathbf{F}$ is continuous on $[a, b]$, show that $\int_a^b \mathbf{f}(t)\, dt = \mathbf{F}(b) - \mathbf{F}(a)$.

(19) Verify that if $\mathbf{\Omega} \times \mathbf{u} = \mathbf{0}$ for all $\mathbf{u}$, then $\mathbf{\Omega} = \mathbf{0}$.

5.4   Line Integrals

The concept of the integral can be extended to functions which are not defined on an interval of the real line but on some curve in $\mathbb{R}^n$. This is done by defining things in such a way that the more general concept reduces to the earlier notion. First it is necessary to consider what is meant by arc length.

5.4.1   *Arc Length And Orientations*

The application of the integral considered here is the concept of the **length of a curve**.

Definition 5.3. C is a **smooth curve** in $\mathbb{R}^n$ if there exists an interval $[a, b] \subseteq \mathbb{R}$ and functions $x_i : [a, b] \to \mathbb{R}$ such that the following conditions hold

(1) x_i is continuous on $[a, b]$.

(2) x_i' exists and is continuous and bounded on $[a, b]$, with $x_i'(a)$ defined as the derivative from the right,

$$\lim_{h \to 0+} \frac{x_i(a + h) - x_i(a)}{h},$$

and $x_i'(b)$ defined similarly as the derivative from the left.

(3) For $\mathbf{p}(t) \equiv (x_1(t), \cdots, x_n(t))$, $t \to \mathbf{p}(t)$ is one to one on (a, b).

(4) $|\mathbf{p}'(t)| \equiv \left(\sum_{i=1}^n |x_i'(t)|^2 \right)^{1/2} \neq 0$ for all $t \in [a, b]$.

(5) $C = \cup \{(x_1(t), \cdots, x_n(t)) : t \in [a, b]\}$.

The functions $x_i(t)$, defined above are giving the coordinates of a point in $\mathbb{R}^n$ and the list of these functions is called a **parametrization** for the smooth curve. Note the natural direction of the interval also gives a direction for moving along the curve. Such a direction is called an orientation. The integral is used to define what is meant by the length of such a smooth curve. Consider such a smooth curve having parametrization $(x_1, \cdots, x_n)$. Forming a partition of $[a, b]$, $a = t_0 < \cdots < t_n = b$ and letting $\mathbf{p}_i = (x_1(t_i), \cdots, x_n(t_i))$, you could consider the polygon formed by lines from $\mathbf{p}_0$ to $\mathbf{p}_1$ and from $\mathbf{p}_1$ to $\mathbf{p}_2$ and from $\mathbf{p}_3$ to $\mathbf{p}_4$ etc. to be an approximation to the curve C. The following picture illustrates what is meant by this.

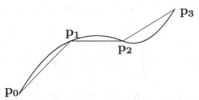

Now consider what happens when the partition is refined by including more points. You can see from the following picture that the polygonal approximation would appear to be even better and that as more points are added in the partition, the sum of the lengths of the line segments seems to get close to something which deserves to be defined as the length of the curve C.

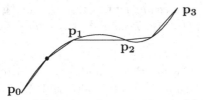

Thus the length of the curve is approximated by

$$\sum_{k=1}^{n} |\mathbf{p}(t_k) - \mathbf{p}(t_{k-1})|.$$

Since the functions in the parametrization are differentiable, it is reasonable to expect this to be close to

$$\sum_{k=1}^{n} |\mathbf{p}'(t_{k-1})| (t_k - t_{k-1})$$

which is seen to be a Riemann sum for the integral $\int_a^b |\mathbf{p}'(t)|\, dt$ and it is this integral which is **defined** as the length of the curve.

Definition 5.4. Let $\mathbf{p}(t)$, $t \in [a,b]$ be a parametrization for a smooth curve. Then the length of this curve is defined as $\int_a^b |\mathbf{p}'(t)|\, dt$.

Would the same length be obtained if another parametrization were used? This is a very important question because the length of the curve should depend only on the curve itself and not on the method used to trace out the curve. The answer to this question is that the length of the curve does not depend on parametrization. The proof is somewhat technical so is given in the last section of this chapter.

Does the definition of length given above correspond to the usual definition of length in the case when the curve is a line segment? It is easy to see that it does so by considering two points in $\mathbb{R}^n$ $\mathbf{p}$ and $\mathbf{q}$. A parametrization for the line segment joining these two points is

$$f_i(t) \equiv t p_i + (1 - t) q_i,\ t \in [0, 1].$$

Using the definition of length of a smooth curve just given, the length according to this definition is

$$\int_0^1 \left(\sum_{i=1}^n (p_i - q_i)^2 \right)^{1/2} dt = |\mathbf{p} - \mathbf{q}| .$$

Thus this new definition which is valid for smooth curves which may not be straight line segments gives the usual length for straight line segments.

The proof that curve length is well defined for a smooth curve contains a result which deserves to be stated as a corollary. It is proved in Lemma 5.3 on Page 105 but the proof is mathematically fairly advanced so it is presented later.

Corollary 5.1. *Let C be a smooth curve and let $\mathbf{f} : [a, b] \to C$ and $\mathbf{g} : [c, d] \to C$ be two parameterizations satisfying (1) - (5). Then $\mathbf{g}^{-1} \circ \mathbf{f}$ is either strictly increasing or strictly decreasing.*

Definition 5.5. If $\mathbf{g}^{-1} \circ \mathbf{f}$ is increasing, then $\mathbf{f}$ and $\mathbf{g}$ are said to be equivalent parameterizations and this is written as $\mathbf{f} \sim \mathbf{g}$. It is also said that the two parameterizations give the same orientation for the curve when $\mathbf{f} \sim \mathbf{g}$.

When the parameterizations are equivalent, they preserve the direction of motion along the curve, and this also shows there are exactly two orientations of the curve since either $\mathbf{g}^{-1} \circ \mathbf{f}$ is increasing or it is decreasing. This is not hard to believe. In simple language, the message is that there are exactly two directions of motion along a curve. The difficulty is in proving this is actually the case.

Lemma 5.1. *The following hold for $\sim$.*

$$\mathbf{f} \sim \mathbf{f}; \tag{5.7}$$

$$\text{If } \mathbf{f} \sim \mathbf{g} \text{ then } \mathbf{g} \sim \mathbf{f}; \tag{5.8}$$

$$\text{If } \mathbf{f} \sim \mathbf{g} \text{ and } \mathbf{g} \sim \mathbf{h}, \text{ then } \mathbf{f} \sim \mathbf{h}. \tag{5.9}$$

Proof: Formula (5.7) is obvious because $\mathbf{f}^{-1} \circ \mathbf{f}(t) = t$ so it is clearly an increasing function. If $\mathbf{f} \sim \mathbf{g}$ then $\mathbf{f}^{-1} \circ \mathbf{g}$ is increasing. Now $\mathbf{g}^{-1} \circ \mathbf{f}$ must also be increasing because it is the inverse of $\mathbf{f}^{-1} \circ \mathbf{g}$. This verifies (5.8). To see (5.9), $\mathbf{f}^{-1} \circ \mathbf{h} = (\mathbf{f}^{-1} \circ \mathbf{g}) \circ (\mathbf{g}^{-1} \circ \mathbf{h})$ and so since both of these functions are increasing, it follows $\mathbf{f}^{-1} \circ \mathbf{h}$ is also increasing. ∎

The symbol $\sim$ is called an equivalence relation. If C is such a smooth curve just described, and if $\mathbf{f} : [a, b] \to C$ is a parametrization of C, consider $\mathbf{g}(t) \equiv \mathbf{f}((a + b) - t)$, also a parametrization of C. Now by Corollary 5.1, if $\mathbf{h}$ is a parametrization, then if $\mathbf{f}^{-1} \circ \mathbf{h}$ is not increasing, it must be the case that $\mathbf{g}^{-1} \circ \mathbf{h}$ is increasing. Consequently, either $\mathbf{h} \sim \mathbf{g}$ or $\mathbf{h} \sim \mathbf{f}$. These parameterizations, $\mathbf{h}$, which satisfy $\mathbf{h} \sim \mathbf{f}$ are called the equivalence class determined by $\mathbf{f}$ and those $\mathbf{h} \sim \mathbf{g}$ are called the equivalence class determined by $\mathbf{g}$. These two classes are called **orientations** of C. They give the direction of motion on C. You see that going from $\mathbf{f}$ to $\mathbf{g}$ corresponds to tracing out the curve in the opposite direction.

Sometimes people wonder why it is required, in the definition of a smooth curve that $\mathbf{p}'(t) \neq \mathbf{0}$. Imagine t is time and $\mathbf{p}(t)$ gives the location of a point in space. If $\mathbf{p}'(t)$ is allowed to equal zero, the point can stop and change directions abruptly, producing a pointy place in C. Here is an example.

Example 5.12. Graph the curve (t^3, t^2) for $t \in [-1, 1]$.

In this case, $t = x^{1/3}$ and so $y = x^{2/3}$. Thus the graph of this curve looks like the picture below. Note the pointy place. Such a curve should not be considered smooth.

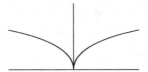

So what is the thing to remember from all this? First, there are certain conditions which must be satisfied for a curve to be smooth. These are listed above. Next, if you have any curve, there are two directions you can move over this curve, each called an orientation. This is illustrated in the following picture.

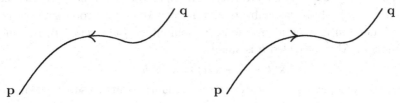

Either you move from $\mathbf{p}$ to $\mathbf{q}$ or you move from $\mathbf{q}$ to $\mathbf{p}$.

Definition 5.6. A curve C is piecewise smooth if there exist points on this curve $\mathbf{p}_0, \mathbf{p}_1, \cdots, \mathbf{p}_n$ such that, denoting $C_{\mathbf{p}_{k-1}\mathbf{p}_k}$ the part of the curve joining $\mathbf{p}_{k-1}$ and $\mathbf{p}_k$, it follows $C_{\mathbf{p}_{k-1}\mathbf{p}_k}$ is a smooth curve and $\cup_{k=1}^n C_{\mathbf{p}_{k-1}\mathbf{p}_k} = C$. In other words, it is piecewise smooth if it consists of a finite number of smooth curves linked together.

Note that Example 5.12 is an example of a piecewise smooth curve although it is not smooth.

5.4.2 Line Integrals And Work

Let C be a smooth curve contained in $\mathbb{R}^p$. A curve C is an "**oriented curve**" if the only parameterizations considered are those which lie in exactly one of the two equivalence classes, each of which is called an "**orientation**". In simple language, orientation specifies a direction over which motion along the curve is to take place.

Thus, it specifies the order in which the points of C are encountered. The pair of concepts consisting of the set of points making up the curve along with a direction of motion along the curve is called an **oriented curve.**

Definition 5.7. Suppose $\mathbf{F}(\mathbf{x}) \in \mathbb{R}^p$ is given for each $\mathbf{x} \in C$ where C is a smooth oriented curve and suppose $\mathbf{x} \to \mathbf{F}(\mathbf{x})$ is continuous. The mapping $\mathbf{x} \to \mathbf{F}(\mathbf{x})$ is called a **vector field.** In the case that $\mathbf{F}(\mathbf{x})$ is a force, it is called a **force field.**

Next the concept of work done by a force field $\mathbf{F}$ on an object as it moves along the curve C, in the direction determined by the given orientation of the curve will be defined. This is new. Earlier the work done by a force which acts on an object moving in a straight line was discussed but here the object moves over a curve. In order to define what is meant by the work, consider the following picture.

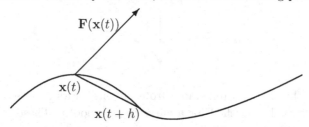

In this picture, the work done by a constant force $\mathbf{F}$ on an object which moves from the point $\mathbf{x}(t)$ to the point $\mathbf{x}(t+h)$ along the straight line shown would equal $\mathbf{F}\cdot(\mathbf{x}(t+h) - \mathbf{x}(t))$. It is reasonable to assume this would be a good approximation to the work done in moving along the curve joining $\mathbf{x}(t)$ and $\mathbf{x}(t+h)$ provided h is small enough. Also, provided h is small,

$$\mathbf{x}(t+h) - \mathbf{x}(t) \approx \mathbf{x}'(t)\,h$$

where the wriggly equal sign indicates the two quantities are close. In the notation of Leibniz, one writes dt for h and

$$dW = \mathbf{F}(\mathbf{x}(t)) \cdot \mathbf{x}'(t)\,dt$$

or in other words,

$$\frac{dW}{dt} = \mathbf{F}(\mathbf{x}(t)) \cdot \mathbf{x}'(t).$$

Defining the total work done by the force at $t = 0$, corresponding to the first endpoint of the curve, to equal zero, the work would satisfy the following initial value problem.

$$\frac{dW}{dt} = \mathbf{F}(\mathbf{x}(t)) \cdot \mathbf{x}'(t),\ W(a) = 0.$$

This motivates the following definition of work.

Definition 5.8. Let $\mathbf{F}(\mathbf{x})$ be given above. Then the work done by this force field on an object moving over the curve C in the direction determined by the specified orientation is defined as

$$\int_C \mathbf{F} \cdot d\mathbf{R} \equiv \int_a^b \mathbf{F}(\mathbf{x}(t)) \cdot \mathbf{x}'(t)\,dt$$

where the function **x** is one of the allowed parameterizations of C in the given orientation of C. In other words, there is an interval $[a, b]$ and as t goes from a to b, **x** (t) moves in the direction determined from the given orientation of the curve.

Theorem 5.4. *The symbol* $\int_C \mathbf{F} \cdot d\mathbf{R}$, *is well defined in the sense that every parametrization in the given orientation of C gives the same value for* $\int_C \mathbf{F} \cdot d\mathbf{R}$.

Proof: Suppose **g** $: [c, d] \rightarrow C$ is another allowed parametrization. Thus $\mathbf{g}^{-1} \circ \mathbf{f}$ is an increasing function ϕ. Then since ϕ is increasing, it follows from the change of variables formula that

$$\int_c^d \mathbf{F}\left(\mathbf{g}\left(s\right)\right) \cdot \mathbf{g}'\left(s\right)\, ds = \int_a^b \mathbf{F}\left(\mathbf{g}\left(\phi\left(t\right)\right)\right) \cdot \mathbf{g}'\left(\phi\left(t\right)\right) \phi'\left(t\right)\, dt$$

$$= \int_a^b \mathbf{F}\left(\mathbf{f}\left(t\right)\right) \cdot \frac{d}{dt}\left(\mathbf{g}\left(\mathbf{g}^{-1} \circ \mathbf{f}\left(t\right)\right)\right)\, dt = \int_a^b \mathbf{F}\left(\mathbf{f}\left(t\right)\right) \cdot \mathbf{f}'\left(t\right)\, dt. \qquad \blacksquare$$

Regardless the physical interpretation of **F**, this is called the **line integral**. When **F** is interpreted as a force, the line integral measures the extent to which the motion over the curve in the indicated direction is aided by the force. If the net effect of the force on the object is to impede rather than to aid the motion, this will show up as the work being negative.

Does the concept of work as defined here coincide with the earlier concept of work when the object moves over a straight line when acted on by a constant force?

Let **p** and **q** be two points in $\mathbb{R}^n$ and suppose **F** is a constant force acting on an object which moves from **p** to **q** along the straight line joining these points. Then the work done is $\mathbf{F} \cdot (\mathbf{q} - \mathbf{p})$. Is the same thing obtained from the above definition? Let **x** $(t) \equiv \mathbf{p} + t\,(\mathbf{q} - \mathbf{p})$, $t \in [0, 1]$ be a parametrization for this oriented curve, thestraight line in the direction from **p** to **q**. Then $\mathbf{x}'(t) = \mathbf{q} - \mathbf{p}$ and $\mathbf{F}\left(\mathbf{x}\left(t\right)\right) = \mathbf{F}$. Therefore, the above definition yields

$$\int_0^1 \mathbf{F} \cdot (\mathbf{q} - \mathbf{p})\, dt = \mathbf{F} \cdot (\mathbf{q} - \mathbf{p}).$$

Therefore, the new definition adds to but does not contradict the old one.

Example 5.13. Suppose for $t \in [0, \pi]$ the position of an object is given by **r** $(t) = t\mathbf{i} + \cos(2t)\mathbf{j} + \sin(2t)\mathbf{k}$. Also suppose there is a force field defined on $\mathbb{R}^3$, $\mathbf{F}(x, y, z) \equiv 2xy\mathbf{i} + x^2\mathbf{j} + \mathbf{k}$. Find $\int_C \mathbf{F} \cdot d\mathbf{R}$ where C is the curve traced out by this object which has the orientation determined by the direction of increasing t.

To find this line integral use the above definition and write

$$\int_C \mathbf{F} \cdot d\mathbf{R} = \int_0^\pi \left(2t\left(\cos\left(2t\right)\right), t^2, 1\right) \cdot \left(1, -2\sin\left(2t\right), 2\cos\left(2t\right)\right)\, dt$$

In evaluating this replace the x in the formula for **F** with t, the y in the formula for **F** with $\cos(2t)$ and the z in the formula for **F** with $\sin(2t)$ because these are the values of these variables which correspond to the value of t. Taking the dot product, this equals the following integral.

$$\int_0^\pi \left(2t \cos 2t - 2 (\sin 2t) t^2 + 2 \cos 2t\right) dt = \pi^2$$

Example 5.14. Let C denote the oriented curve obtained by $\mathbf{r}(t) = (t, \sin t, t^3)$ where the orientation is determined by increasing t for $t \in [0, 2]$. Also let $\mathbf{F} = (x, y, xz + z)$. Find $\int_C \mathbf{F} \cdot d\mathbf{R}$.

You use the definition.

$$\int_C \mathbf{F} \cdot d\mathbf{R} = \int_0^2 \left(t, \sin(t), (t+1) t^3\right) \cdot \left(1, \cos(t), 3t^2\right) dt$$

$$= \int_0^2 \left(t + \sin(t) \cos(t) + 3 (t+1) t^5\right) dt = \frac{1251}{14} - \frac{1}{2} \cos^2(2).$$

Suppose you have a curve specified by $\mathbf{r}(s) = (x(s), y(s), z(s))$ and it has the property that $|\mathbf{r}'(s)| = 1$ for all $s \in [0, b]$. Then the length of this curve for s between 0 and s_1 is $\int_0^{s_1} |\mathbf{r}'(s)| \, ds = \int_0^{s_1} 1 \, ds = s_1$. This parameter is therefore called arc length because the length of the curve up to s equals s. Now you can always change the parameter to be arc length.

Proposition 5.2. Suppose C is an oriented smooth curve parameterized by $\mathbf{r}(t)$ for $t \in [a, b]$. Then letting l denote the total length of C, there exists $\mathbf{R}(s)$, $s \in [0, l]$ another parametrization for this curve which preserves the orientation and such that $|\mathbf{R}'(s)| = 1$ so that s is arc length.

Prove: Let $\phi(t) \equiv \int_a^t |\mathbf{r}'(\tau)| \, d\tau \equiv s$. Then s is an increasing function of t because

$$\frac{ds}{dt} = \phi'(t) = |\mathbf{r}'(t)| > 0.$$

Now define $\mathbf{R}(s) \equiv \mathbf{r}\left(\phi^{-1}(s)\right)$. Then

$$\mathbf{R}'(s) = \mathbf{r}'\left(\phi^{-1}(s)\right) \left(\phi^{-1}\right)'(s) = \frac{\mathbf{r}'\left(\phi^{-1}(s)\right)}{\left|\mathbf{r}'\left(\phi^{-1}(s)\right)\right|}$$

and so $|\mathbf{R}'(s)| = 1$ as claimed. $\mathbf{R}(l) = \mathbf{r}\left(\phi^{-1}(l)\right) = \mathbf{r}\left(\phi^{-1}\left(\int_a^b |\mathbf{r}'(\tau)| \, d\tau\right)\right) = \mathbf{r}(b)$ and $\mathbf{R}(0) = \mathbf{r}\left(\phi^{-1}(0)\right) = \mathbf{r}(a)$ and $\mathbf{R}$ delivers the same set of points in the same order as $\mathbf{r}$ because $\frac{ds}{dt} > 0$. ∎

The arc length parameter is just like any other parameter, in so far as considerations of line integrals are concerned, because it was shown above that line integrals are independent of parametrization. However, when things are defined in terms of the arc length parametrization, it is clear they depend only on geometric properties of the curve itself and for this reason, the arc length parametrization is important in differential geometry.

5.4.3 *Another Notation For Line Integrals*

Definition 5.9. Let $\mathbf{F}(x, y, z) = (P(x, y, z), Q(x, y, z), R(x, y, z))$ and let C be an oriented curve. Then another way to write $\int_C \mathbf{F} \cdot d\mathbf{R}$ is

$$\int_C Pdx + Qdy + Rdz$$

This last is referred to as the integral of a **differential form**, $Pdx + Qdy + Rdz$. The study of differential forms is important. Formally, $d\mathbf{R} = (dx, dy, dz)$ and so the integrand in the above is formally $\mathbf{F} \cdot d\mathbf{R}$. Other occurrences of this notation are handled similarly in 2 or higher dimensions.

5.5 **Exercises**

(1) Let $\mathbf{r}(t) = \left(\ln(t), \frac{t^2}{2}, \sqrt{2}t \right)$ for $t \in [1, 2]$. Find the length of this curve.

(2) Let $\mathbf{r}(t) = \left(\frac{2}{3}t^{3/2}, t, t \right)$ for $t \in [0, 1]$. Find the length of this curve.

(3) Let $\mathbf{r}(t) = (t, \cos(3t), \sin(3t))$ for $t \in [0, 1]$. Find the length of this curve.

(4) Suppose for $t \in [0, \pi]$ the position of an object is given by $\mathbf{r}(t) = t\mathbf{i} + \cos(2t)\mathbf{j} + \sin(2t)\mathbf{k}$. Also suppose there is a force field defined on $\mathbb{R}^3$, $\mathbf{F}(x, y, z) \equiv 2xy\mathbf{i} + (x^2 + 2zy)\mathbf{j} + y^2\mathbf{k}$. Find the work $\int_C \mathbf{F} \cdot d\mathbf{R}$ where C is the curve traced out by this object having the orientation determined by the direction of increasing t.

(5) In the following, a force field is specified followed by the parametrization of a curve. Find the work.

    (a) $\mathbf{F} = (x, y, z), \mathbf{r}(t) = (t, t^2, t + 1), t \in [0, 1]$

    (b) $\mathbf{F} = (x - y, y + z, z), \mathbf{r}(t) = (\cos(t), t, \sin(t)), t \in [0, \pi]$

    (c) $\mathbf{F} = (x^2, y^2, z + x), \mathbf{r}(t) = (t, 2t, t + t^2), t \in [0, 1]$

    (d) $\mathbf{F} = (z, y, x), \mathbf{r}(t) = (t^2, 2t, t), t \in [0, 1]$

(6) The curve consists of straight line segments which go from $(0, 0, 0)$ to $(1, 1, 1)$ and finally to $(1, 2, 3)$. Find the work done if the force field is

    (a) $\mathbf{F} = (2xy, x^2 + 2y, 1)$

    (b) $\mathbf{F} = (yz^2, xz^2, 2xyz + 1)$

    (c) $\mathbf{F} = (\cos x, -\sin y, 1)$

    (d) $\mathbf{F} = (2x \sin y, x^2 \cos y, 1)$

(7) *Read ahead about the gradient in Definition 8.5. Show the vector fields in the preceding problems are respectively $\nabla(x^2y + y^2 + z)$, $\nabla(xyz^2 + z)$, $\nabla(\sin x + \cos y + z - 1)$, and $\nabla(x^2 \sin y + z)$. Thus each of these vector fields is of the form ∇f where f is a function of three variables. For each f in the above, compute $f(1, 2, 3) - f(0, 0, 0)$ and compare with your solutions to the above line integrals. You should get the same thing from $f(1, 2, 3) - f(0, 0, 0)$. This is not a coincidence and will be fully discussed later. Such vector fields are called **conservative**.

(8) Here is a vector field $(y, x + z^2, 2yz)$ and here is the parametrization of a curve C. $\mathbf{R}(t) = (\cos 2t, 2\sin 2t, t)$ where t goes from 0 to $\pi/4$. Find $\int_C \mathbf{F} \cdot d\mathbf{R}$.

(9) If f and g are both increasing functions, show that $f \circ g$ is an increasing function also. Assume anything you like about the domains of the functions.

(10) Suppose for $t \in [0, 3]$ the position of an object is given by $\mathbf{r}(t) = t\mathbf{i} + t\mathbf{j} + t\mathbf{k}$. Also suppose there is a force field defined on $\mathbb{R}^3$, $\mathbf{F}(x, y, z) \equiv yz\mathbf{i} + xz\mathbf{j} + xy\mathbf{k}$. Find $\int_C \mathbf{F} \cdot d\mathbf{R}$ where C is the curve traced out by this object which has the orientation determined by the direction of increasing t. Repeat the problem for $\mathbf{r}(t) = t\mathbf{i} + t^2\mathbf{j} + t\mathbf{k}$.

(11) Suppose for $t \in [0, 1]$ the position of an object is given by $\mathbf{r}(t) = t\mathbf{i} + t\mathbf{j} + t\mathbf{k}$. Also suppose there is a force field defined on $\mathbb{R}^3$, $\mathbf{F}(x, y, z) \equiv z\mathbf{i} + xz\mathbf{j} + xy\mathbf{k}$. Find $\int_C \mathbf{F} \cdot d\mathbf{R}$ where C is the curve traced out by this object which has the orientation determined by the direction of increasing t. Repeat the problem for $\mathbf{r}(t) = t\mathbf{i} + t^2\mathbf{j} + t\mathbf{k}$.

(12) Let $\mathbf{F}(x, y, z)$ be a given force field and suppose it acts on an object having mass m on a curve with parametrization, $(x(t), y(t), z(t))$ for $t \in [a, b]$. Show directly that the work done equals the difference in the kinetic energy. **Hint:**

$$\int_a^b \mathbf{F}(x(t), y(t), z(t)) \cdot (x'(t), y'(t), z'(t)) \, dt =$$

$$\int_a^b m(x''(t), y''(t), z''(t)) \cdot (x'(t), y'(t), z'(t)) \, dt,$$

etc.

(13) Suppose for $t \in [0, 2\pi]$ the position of an object is given by

$$\mathbf{r}(t) = 2t\mathbf{i} + \cos(t)\mathbf{j} + \sin(t)\mathbf{k}.$$

Also suppose there is a force field defined on $\mathbb{R}^3$,

$$\mathbf{F}(x, y, z) \equiv 2xy\mathbf{i} + (x^2 + 2zy)\mathbf{j} + y^2\mathbf{k}.$$

Find the work $\int_C \mathbf{F} \cdot d\mathbf{R}$ where C is the curve traced out by this object which has the orientation determined by the direction of increasing t.

(14) Here is a vector field $(y, x^2 + z, 2yz)$ and here is the parametrization of a curve C. $\mathbf{R}(t) = (\cos 2t, 2\sin 2t, t)$ where t goes from 0 to $\pi/4$. Find $\int_C \mathbf{F} \cdot d\mathbf{R}$.

(15) Suppose for $t \in [0, 1]$ the position of an object is given by $\mathbf{r}(t) = t\mathbf{i} + t\mathbf{j} + t\mathbf{k}$. Also suppose there is a force field defined on $\mathbb{R}^3$, $\mathbf{F}(x, y, z) \equiv yz\mathbf{i} + xz\mathbf{j} + xy\mathbf{k}$. Find $\int_C \mathbf{F} \cdot d\mathbf{R}$ where C is the curve traced out by this object which has the orientation determined by the direction of increasing t. Repeat the problem for $\mathbf{r}(t) = t\mathbf{i} + t^2\mathbf{j} + t\mathbf{k}$. You should get the same answer in this case. This is because the vector field happens to be conservative. (More on this later.)

5.6 Independence Of Parametrization*

Recall that if $\mathbf{p}(t) : t \in [a, b]$ was a parametrization of a smooth curve C, the length of C is defined as $\int_a^b |\mathbf{p}'(t)|\, dt$. If some other parametrization were used to trace out C, would the same answer be obtained? To answer this question in a satisfactory manner requires some hard calculus.

5.6.1 *Hard Calculus*

Recall Theorem 4.15 about continuity and convergent sequences. It said roughly that a function $\mathbf{f}$ is continuous at $\mathbf{x}$ if and only if whenever $\mathbf{x}_k \to \mathbf{x}$, then $\mathbf{f}(\mathbf{x}_k) \to \mathbf{f}(\mathbf{x})$. Also recall the following Lemma from Volume 1, whose proof is summarized below for convenience.

Lemma 5.2. *Let $\phi : [a, b] \to \mathbb{R}$ be a continuous function and suppose ϕ is $1 - 1$ on (a, b). Then ϕ is either strictly increasing or strictly decreasing on $[a, b]$. Furthermore, ϕ^{-1} is continuous.*

Proof: First it is shown that ϕ is either strictly increasing or strictly decreasing on (a, b).

If ϕ is not strictly decreasing on (a, b), then there exists $x_1 < y_1$, $x_1, y_1 \in (a, b)$ such that

$$(\phi(y_1) - \phi(x_1))(y_1 - x_1) > 0.$$

If for some other pair of points $x_2 < y_2$ with $x_2, y_2 \in (a, b)$, the above inequality does not hold, then since ϕ is $1 - 1$,

$$(\phi(y_2) - \phi(x_2))(y_2 - x_2) < 0.$$

Let $x_t \equiv t x_1 + (1 - t) x_2$ and $y_t \equiv t y_1 + (1 - t) y_2$. It follows that $x_t < y_t$ for all $t \in [0, 1]$. Now define

$$h(t) \equiv (\phi(y_t) - \phi(x_t))(y_t - x_t).$$

Then $h(0) < 0$, $h(1) > 0$ but by assumption, $h(t) \neq 0$ for any $t \in (0, 1)$, a contradiction.

This property of being either strictly increasing or strictly decreasing on (a, b) carries over to $[a, b]$ by the continuity of ϕ.

It only remains to verify ϕ^{-1} is continuous. If not, there exists $s_n \to s$ where s_n and s are points of $\phi([a, b])$ but $|\phi^{-1}(s_n) - \phi^{-1}(s)| \geq \varepsilon$. By sequential compactness of $[a, b]$, there is a subsequence, still denoted by n, such that $\left|\phi^{-1}(s_n) - t_1\right| \to 0$. Thus $s_n \to \phi(t_1)$, so $s = \phi(t_1)$, and $t_1 = \phi^{-1}(s)$, a contradiction. ∎

Corollary 5.2. *Let $f : (a, b) \to \mathbb{R}$ be one to one and continuous. Then $f(a, b)$ is an open interval (c, d) and $f^{-1} : (c, d) \to (a, b)$ is continuous.*

Proof: Since f is either strictly increasing or strictly decreasing, it follows that $f(a, b)$ is an open interval (c, d). Assume f is decreasing. Now let $x \in (a, b)$. Why is f^{-1} is continuous at $f(x)$? Since f is decreasing, if $f(x) < f(y)$, then $y \equiv f^{-1}(f(y)) < x \equiv f^{-1}(f(x))$ and so f^{-1} is also decreasing. Let $\varepsilon > 0$ be given. Let $\varepsilon > \eta > 0$ and $(x - \eta, x + \eta) \subseteq (a, b)$. Then $f(x) \in (f(x + \eta), f(x - \eta))$. Let
$$\delta = \min\left(f(x) - f(x + \eta), f(x - \eta) - f(x)\right).$$
Then if $\left|f(z) - f(x)\right| < \delta$, it follows
$$z \equiv f^{-1}(f(z)) \in (x - \eta, x + \eta) \subseteq (x - \varepsilon, x + \varepsilon)$$
which implies
$$\left|f^{-1}(f(z)) - x\right| = \left|f^{-1}(f(z)) - f^{-1}(f(x))\right| < \varepsilon.$$
This proves the theorem in the case where f is strictly decreasing. The case where f is increasing is similar. ∎

Theorem 5.5. *Let $f : [a, b] \to \mathbb{R}$ be continuous and one to one. Suppose $f'(x_1)$ exists for some $x_1 \in [a, b]$ and $f'(x_1) \neq 0$. Then $\left(f^{-1}\right)'(f(x_1))$ exists and is given by the formula $\left(f^{-1}\right)'(f(x_1)) = \frac{1}{f'(x_1)}$.*

Proof: By Lemma 5.2 f is either strictly increasing or strictly decreasing and f^{-1} is continuous on $[a, b]$. Therefore there exists $\eta > 0$ such that if $0 < \left|f(x_1) - f(x)\right| < \eta$, then
$$0 < \left|x_1 - x\right| = \left|f^{-1}(f(x_1)) - f^{-1}(f(x))\right| < \delta$$
where δ is small enough that for $0 < \left|x_1 - x\right| < \delta$,
$$\left|\frac{x - x_1}{f(x) - f(x_1)} - \frac{1}{f'(x_1)}\right| < \varepsilon.$$
It follows that if $0 < \left|f(x_1) - f(x)\right| < \eta$,
$$\left|\frac{f^{-1}(f(x)) - f^{-1}(f(x_1))}{f(x) - f(x_1)} - \frac{1}{f'(x_1)}\right| = \left|\frac{x - x_1}{f(x) - f(x_1)} - \frac{1}{f'(x_1)}\right| < \varepsilon$$
Therefore, since $\varepsilon > 0$ is arbitrary,
$$\lim_{y \to f(x_1)} \frac{f^{-1}(y) - f^{-1}(f(x_1))}{y - f(x_1)} = \frac{1}{f'(x_1)}. \qquad ∎$$
The following obvious corollary comes from the above by not bothering with end points.

Corollary 5.3. *Let $f : (a, b) \to \mathbb{R}$ be continuous and one to one. Suppose $f'(x_1)$ exists for some $x_1 \in (a, b)$ and $f'(x_1) \neq 0$. Then $\left(f^{-1}\right)'(f(x_1))$ exists and is given by the formula $\left(f^{-1}\right)'(f(x_1)) = \frac{1}{f'(x_1)}$.*

Proof: From the definition of the derivative and continuity of f^{-1},
$$\lim_{f(x) \to f(x_1)} \frac{f^{-1}(f(x)) - f^{-1}(f(x_1))}{f(x) - f(x_1)} = \lim_{x \to x_1} \frac{x - x_1}{f(x) - f(x_1)} = \frac{1}{f'(x_1)}. \qquad ∎$$

5.6.2   *Independence Of Parametrization*

Theorem 5.6. *Let $\phi : [a,b] \to [c,d]$ be one to one and suppose ϕ' exists and is continuous on $[a,b]$. Then if f is a continuous function defined on $[c,d]$ which is Riemann integrable*[1],

$$\int_c^d f(s)\, ds = \int_a^b f(\phi(t)) \left|\phi'(t)\right| dt$$

Proof: Let $F'(s) = f(s)$. (For example, let $F(s) = \int_a^s f(r)\, dr$.) Then the first integral equals $F(d) - F(c)$ by the fundamental theorem of calculus. Since ϕ is one to one, it follows from Lemma 5.2 above that ϕ is either strictly increasing or strictly decreasing. Suppose ϕ is strictly decreasing. Then $\phi(a) = d$ and $\phi(b) = c$. Therefore, $\phi' \leq 0$ and the second integral equals

$$-\int_a^b f(\phi(t))\,\phi'(t)\, dt = \int_b^a \frac{d}{dt}\left(F(\phi(t))\right) dt = F(\phi(a)) - F(\phi(b)) = F(d) - F(c).$$

The case when ϕ is increasing is similar but easier. ∎

Lemma 5.3. *Let $\mathbf{f} : [a,b] \to C$, $\mathbf{g} : [c,d] \to C$ be parameterizations of a smooth curve which satisfy conditions (1) - (5). Then $\phi(t) \equiv \mathbf{g}^{-1} \circ \mathbf{f}(t)$ is $1-1$ on (a,b), continuous on $[a,b]$, and either strictly increasing or strictly decreasing on $[a,b]$.*

Proof: It is obvious ϕ is $1-1$ on (a,b) from the conditions $\mathbf{f}$ and $\mathbf{g}$ satisfy. It only remains to verify continuity on $[a,b]$ because then the final claim follows from Lemma 5.2. If ϕ is not continuous on $[a,b]$, then there exists a sequence, $\{t_n\} \subseteq [a,b]$ such that $t_n \to t$ but $\phi(t_n)$ fails to converge to $\phi(t)$. Therefore, for some $\varepsilon > 0$, there exists a subsequence, still denoted by n such that $|\phi(t_n) - \phi(t)| \geq \varepsilon$. By sequential compactness of $[c,d]$, (See Theorem 4.13 on Page 80.) there is a further subsequence, still denoted by n, such that $\{\phi(t_n)\}$ converges to a point, s, of $[c,d]$ which is not equal to $\phi(t)$. Thus $\mathbf{g}^{-1} \circ \mathbf{f}(t_n) \to s$ while $t_n \to t$. Therefore, the continuity of $\mathbf{f}$ and $\mathbf{g}$ imply $\mathbf{f}(t_n) \to \mathbf{g}(s)$ and $\mathbf{f}(t_n) \to \mathbf{f}(t)$. Thus, $\mathbf{g}(s) = \mathbf{f}(t)$, so $s = \mathbf{g}^{-1} \circ \mathbf{f}(t) = \phi(t)$, a contradiction. Therefore, ϕ is continuous as claimed. ∎

Theorem 5.7. *The length of a smooth curve is not dependent on which parametrization is used.*

Proof: Let C be the curve and suppose $\mathbf{f} : [a,b] \to C$ and $\mathbf{g} : [c,d] \to C$ both satisfy conditions (1) - (5). Is it true that $\int_a^b |\mathbf{f}'(t)|\, dt = \int_c^d |\mathbf{g}'(s)|\, ds$?

Let $\phi(t) \equiv \mathbf{g}^{-1} \circ \mathbf{f}(t)$ for $t \in [a,b]$. I want to show that ϕ is C^1 on an interval of the form $[a + \delta, b - \delta]$. By the above lemma, ϕ is either strictly increasing or strictly decreasing on $[a,b]$. Suppose for the sake of simplicity that it is strictly increasing. The decreasing case is handled similarly.

Let $s_0 \in \phi([a + \delta, b - \delta]) \subset (c,d)$. Then by assumption 4 for smooth curves, $g_i'(s_0) \neq 0$ for some i. By continuity of g_i', it follows $g_i'(s) \neq 0$ for all $s \in I$ where I is

---

[1]Recall that all continuous functions of this sort are Riemann integrable.

an open interval contained in $[c, d]$ which contains s_0. It follows from the mean value theorem that on this interval g_i is either strictly increasing or strictly decreasing. Therefore, $J \equiv g_i(I)$ is also an open interval and you can define a differentiable function $h_i : J \to I$ by

$$h_i(g_i(s)) = s.$$

This implies that for $s \in I$,

$$h_i'(g_i(s)) = \frac{1}{g_i'(s)}. \tag{5.10}$$

Now letting $s = \phi(t)$ for $s \in I$, it follows $t \in J_1$, an open interval. Also, for s and t related this way, $\mathbf{f}(t) = \mathbf{g}(s)$ and so in particular, for $s \in I$, $g_i(s) = f_i(t)$. Consequently,

$$s = h_i(g_i(s)) = h_i(f_i(t)) = \phi(t)$$

and so, for $t \in J_1$,

$$\phi'(t) = h_i'(f_i(t)) f_i'(t) = h_i'(g_i(s)) f_i'(t) = \frac{f_i'(t)}{g_i'(\phi(t))} \tag{5.11}$$

which shows that ϕ' exists and is continuous on J_1, an open interval containing $\phi^{-1}(s_0)$. Since s_0 is arbitrary, this shows ϕ' exists on $[a + \delta, b - \delta]$ and is continuous there.

Now $\mathbf{f}(t) = \mathbf{g} \circ (\mathbf{g}^{-1} \circ \mathbf{f})(t) = \mathbf{g}(\phi(t))$, and it was just shown that ϕ' is a continuous function on $[a - \delta, b + \delta]$. It follows from the chain rule, $\mathbf{f}'(t) = \mathbf{g}'(\phi(t)) \phi'(t)$ and so, by Theorem 5.6,

$$\int_{\phi(a+\delta)}^{\phi(b-\delta)} |\mathbf{g}'(s)| \, ds = \int_{a+\delta}^{b-\delta} |\mathbf{g}'(\phi(t))| \, |\phi'(t)| \, dt = \int_{a+\delta}^{b-\delta} |\mathbf{f}'(t)| \, dt.$$

Now using the continuity of $\phi, \mathbf{g}'$, and $\mathbf{f}'$ on $[a, b]$ and letting $\delta \to 0+$ in the above, yields

$$\int_c^d |\mathbf{g}'(s)| \, ds = \int_a^b |\mathbf{f}'(t)| \, dt. \qquad \blacksquare$$

Chapter 6

Motion On A Space Curve

6.1 Space Curves

A fly buzzing around the room, a person riding a roller coaster, and a satellite orbiting the earth all have something in common. They are moving over some sort of curve in three dimensions.

Denote by $\mathbf{R}(t)$ the position vector of the point on the curve which occurs at time t. Assume that $\mathbf{R}', \mathbf{R}''$ exist and are continuous. Thus $\mathbf{R}' = \mathbf{v}$, the velocity and $\mathbf{R}'' = \mathbf{a}$ is defined as the acceleration.

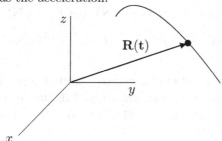

Lemma 6.1. *Define* $\mathbf{T}(t) \equiv \mathbf{R}'(t) / |\mathbf{R}'(t)|$. *Then* $|\mathbf{T}(t)| = 1$ *and if* $\mathbf{T}'(t) \neq 0$, *then there exists a unit vector* $\mathbf{N}(t)$ *perpendicular to* $\mathbf{T}(t)$ *and a scalar valued function* $\kappa(t)$, *with* $\mathbf{T}'(t) = \kappa(t) |\mathbf{v}| \mathbf{N}(t)$.

Proof: It follows from the definition that $|\mathbf{T}| = 1$. Therefore, $\mathbf{T} \cdot \mathbf{T} = 1$ and so, upon differentiating both sides,

$$\mathbf{T}' \cdot \mathbf{T} + \mathbf{T} \cdot \mathbf{T}' = 2\mathbf{T}' \cdot \mathbf{T} = 0.$$

Therefore, $\mathbf{T}'$ is perpendicular to $\mathbf{T}$. Let $\mathbf{N}(t) \equiv \frac{\mathbf{T}'}{|\mathbf{T}'|}$. Then letting $|\mathbf{T}'| \equiv \kappa(t) |\mathbf{v}(t)|$, it follows

$$\mathbf{T}'(t) = \kappa(t) |\mathbf{v}(t)| \mathbf{N}(t).$$

If $|\mathbf{T}'| = 0$, you could let $\mathbf{N}$ be any unit vector and $\kappa(t) = 0$. ∎

Definition 6.1. *The vector* $\mathbf{T}(t)$ *is called the* **unit tangent vector** *and the vector* $\mathbf{N}(t)$ *is called the* **principal normal.** *The function* $\kappa(t)$ *in the above lemma is* called the **curvature**. The **radius of curvature** is defined as $\rho = 1/\kappa$. The plane determined by the two vectors $\mathbf{T}$ and $\mathbf{N}$ in the case where $\mathbf{T}' \neq \mathbf{0}$ is called the **osculating[1] plane**. It identifies a particular plane which is in a sense tangent to this space curve.

The important thing about this is that it is possible to write the acceleration as the sum of two vectors, one perpendicular to the direction of motion and the other in the direction of motion.

Theorem 6.1. *For* $\mathbf{R}(t)$ *the position vector of a space curve, the acceleration is given by the formula*

$$\mathbf{a} = \frac{d\,|\mathbf{v}|}{dt}\mathbf{T} + \kappa\,|\mathbf{v}|^2\,\mathbf{N} \equiv a_T\mathbf{T} + a_N\mathbf{N}.$$

Furthermore, $a_T^2 + a_N^2 = |\mathbf{a}|^2.$

Proof:

$$\mathbf{a} = \frac{d\mathbf{v}}{dt} = \frac{d}{dt}(\mathbf{R}') = \frac{d}{dt}(|\mathbf{v}|\,\mathbf{T}) = \frac{d\,|\mathbf{v}|}{dt}\mathbf{T} + |\mathbf{v}|\,\mathbf{T}' = \frac{d\,|\mathbf{v}|}{dt}\mathbf{T} + |\mathbf{v}|^2\,\kappa\mathbf{N}.$$

This proves the first part.

For the second part,

$$|\mathbf{a}|^2 = (a_T\mathbf{T} + a_N\mathbf{N}) \cdot (a_T\mathbf{T} + a_N\mathbf{N})$$

$$= a_T^2\mathbf{T}\cdot\mathbf{T} + 2a_Na_T\mathbf{T}\cdot\mathbf{N} + a_N^2\mathbf{N}\cdot\mathbf{N} = a_T^2 + a_N^2$$

because $\mathbf{T}\cdot\mathbf{N} = 0$. ∎

Finally, it is good to point out that the curvature is a property of the curve itself, and does not depend on the parametrization of the curve. If the curve is given by two different vector valued functions $\mathbf{R}(t)$ and $\mathbf{R}(\tau)$, then from the formula above for the curvature,

$$\kappa(t) = \frac{|\mathbf{T}'(t)|}{|\mathbf{v}(t)|} = \frac{\left|\frac{d\mathbf{T}}{d\tau}\frac{d\tau}{dt}\right|}{\left|\frac{d\mathbf{R}}{d\tau}\frac{d\tau}{dt}\right|} = \frac{\left|\frac{d\mathbf{T}}{d\tau}\right|}{\left|\frac{d\mathbf{R}}{d\tau}\right|} \equiv \kappa(\tau).$$

From this, it is possible to give an important formula from physics. Suppose an object orbits a point at constant speed v. In the above notation, $|\mathbf{v}| = v$. What is the centripetal acceleration of this object? You may know from a physics class that the answer is v^2/r where r is the radius. This follows from the above quite easily. The parametrization of the object which is as described is

$$\mathbf{R}(t) = \left(r\cos\left(\frac{v}{r}t\right), r\sin\left(\frac{v}{r}t\right)\right).$$

Therefore, $\mathbf{T} = \left(-\sin\left(\frac{v}{r}t\right), \cos\left(\frac{v}{r}t\right)\right)$ and $\mathbf{T}' = \left(-\frac{v}{r}\cos\left(\frac{v}{r}t\right), -\frac{v}{r}\sin\left(\frac{v}{r}t\right)\right)$. Thus,

$$\kappa = |\mathbf{T}'(t)|/v = \frac{1}{r}.$$

---

[1]To osculate means to kiss. Thus this plane could be called the kissing plane. However, that does not sound formal enough so we call it the osculating plane.

It follows

$$\mathbf{a} = \frac{dv}{dt}\mathbf{T} + v^2 \kappa \mathbf{N} = \frac{v^2}{r}\mathbf{N}.$$

The vector $\mathbf{N}$ points from the object toward the center of the circle because it is a positive multiple of the vector $\left(-\frac{v}{r}\cos\left(\frac{v}{r}t\right), -\frac{v}{r}\sin\left(\frac{v}{r}t\right)\right)$.

This formula for the acceleration also yields an easy way to find the curvature. Take the cross product of both sides with $\mathbf{v}$, the velocity. Then

$$\mathbf{a}\times\mathbf{v} = \frac{d\,|\mathbf{v}|}{dt}\mathbf{T}\times\mathbf{v} + |\mathbf{v}|^2\,\kappa\mathbf{N}\times\mathbf{v} = \frac{d\,|\mathbf{v}|}{dt}\mathbf{T}\times\mathbf{v} + |\mathbf{v}|^3\,\kappa\mathbf{N}\times\mathbf{T}$$

Now $\mathbf{T}$ and $\mathbf{v}$ have the same direction so the first term on the right equals zero. Taking the magnitude of both sides, and using the fact that $\mathbf{N}$ and $\mathbf{T}$ are two perpendicular unit vectors, $|\mathbf{a}\times\mathbf{v}| = |\mathbf{v}|^3\,\kappa$ and so

$$\kappa = \frac{|\mathbf{a}\times\mathbf{v}|}{|\mathbf{v}|^3}. \tag{6.1}$$

Example 6.1. Let $\mathbf{R}(t) = \left(\cos(t), t, t^2\right)$ for $t \in [0,3]$. Find the speed, velocity, curvature, and write the acceleration in terms of normal and tangential components.

First of all, $\mathbf{v}(t) = (-\sin t, 1, 2t)$ and so the speed is given by $|\mathbf{v}| = \sqrt{\sin^2(t) + 1 + 4t^2}$. Therefore,

$$a_T = \frac{d}{dt}\left(\sqrt{\sin^2(t) + 1 + 4t^2}\right) = \frac{\sin(t)\cos(t) + 4t}{\sqrt{(2 + 4t^2 - \cos^2 t)}}.$$

It remains to find a_N. To do this, you can find the curvature first if you like.

$$\mathbf{a}(t) = \mathbf{R}''(t) = (-\cos t, 0, 2).$$

Then

$$\kappa = \frac{|(-\cos t, 0, 2)\times(-\sin t, 1, 2t)|}{\left(\sqrt{\sin^2(t) + 1 + 4t^2}\right)^3} = \frac{\sqrt{4 + (-2\sin(t) + 2(\cos(t))t)^2 + \cos^2(t)}}{\left(\sqrt{\sin^2(t) + 1 + 4t^2}\right)^3}$$

Then $a_N = \kappa\,|\mathbf{v}|^2$

$$= \frac{\sqrt{4 + (-2\sin(t) + 2(\cos(t))t)^2 + \cos^2(t)}}{\left(\sqrt{\sin^2(t) + 1 + 4t^2}\right)^3}\left(\sin^2(t) + 1 + 4t^2\right)$$

$$= \frac{\sqrt{4 + (-2\sin(t) + 2(\cos(t))t)^2 + \cos^2(t)}}{\sqrt{\sin^2(t) + 1 + 4t^2}}.$$

You can observe the formula $a_N^2 + a_T^2 = |\mathbf{a}|^2$ holds. Indeed $a_N^2 + a_T^2 =$

$$\left(\frac{\sqrt{4 + (-2\sin(t) + 2(\cos(t))t)^2 + \cos^2(t)}}{\sqrt{\sin^2(t) + 1 + 4t^2}}\right)^2 + \left(\frac{\sin(t)\cos(t) + 4t}{\sqrt{(2 + 4t^2 - \cos^2 t)}}\right)^2$$

$$= \frac{4 + (-2\sin t + 2(\cos t)t)^2 + \cos^2 t}{\sin^2 t + 1 + 4t^2} + \frac{(\sin t\cos t + 4t)^2}{2 + 4t^2 - \cos^2 t} = \cos^2 t + 4 = |\mathbf{a}|^2$$

6.1.1   *Some Simple Techniques*

Recall the formula for acceleration is

$$\mathbf{a} = a_T \mathbf{T} + a_N \mathbf{N} \qquad\qquad (6.2)$$

where $a_T = \frac{d|\mathbf{v}|}{dt}$ and $a_N = \kappa |\mathbf{v}|^2$. Of course one way to find a_T and a_N is to just find $|\mathbf{v}|, \frac{d|\mathbf{v}|}{dt}$ and κ and plug in. However, there is another way which might be easier. Take the dot product of both sides with $\mathbf{T}$. This gives,

$$\mathbf{a} \cdot \mathbf{T} = a_T \mathbf{T} \cdot \mathbf{T} + a_N \mathbf{N} \cdot \mathbf{T} = a_T.$$

Thus

$$\mathbf{a} = (\mathbf{a} \cdot \mathbf{T}) \mathbf{T} + a_N \mathbf{N}$$

and so

$$\mathbf{a} - (\mathbf{a} \cdot \mathbf{T}) \mathbf{T} = a_N \mathbf{N} \qquad\qquad (6.3)$$

and taking norms of both sides,

$$|\mathbf{a} - (\mathbf{a} \cdot \mathbf{T}) \mathbf{T}| = a_N.$$

Also from (6.3),

$$\frac{\mathbf{a} - (\mathbf{a} \cdot \mathbf{T}) \mathbf{T}}{|\mathbf{a} - (\mathbf{a} \cdot \mathbf{T}) \mathbf{T}|} = \frac{a_N \mathbf{N}}{a_N |\mathbf{N}|} = \mathbf{N}.$$

Also recall

$$\kappa = \frac{|\mathbf{a} \times \mathbf{v}|}{|\mathbf{v}|^3}, \quad a_T^2 + a_N^2 = |\mathbf{a}|^2$$

This is usually easier than computing $\mathbf{T}'/|\mathbf{T}'|$. To illustrate the use of these simple observations, consider the example worked above which was fairly messy. I will make it easier by selecting a value of t and by using the above simplifying techniques.

Example 6.2. Let $\mathbf{R}(t) = \left(\cos(t), t, t^2\right)$ for $t \in [0, 3]$. Find the speed, velocity, curvature, and write the acceleration in terms of normal and tangential components when $t = 0$. Also find $\mathbf{N}$ at the point where $t = 0$.

First I need to find the velocity and acceleration. Thus

$$\mathbf{v} = (-\sin t, 1, 2t), \quad \mathbf{a} = (-\cos t, 0, 2)$$

and consequently, $\mathbf{T} = \frac{(-\sin t, 1, 2t)}{\sqrt{\sin^2(t) + 1 + 4t^2}}$. When $t = 0$, this reduces to

$$\mathbf{v}(0) = (0, 1, 0), \quad \mathbf{a} = (-1, 0, 2), \quad |\mathbf{v}(0)| = 1, \quad \mathbf{T} = (0, 1, 0).$$

Then the tangential component of acceleration when $t = 0$ is

$$a_T = (-1, 0, 2) \cdot (0, 1, 0) = 0$$

Now $|\mathbf{a}|^2 = 5$ and so $a_N = \sqrt{5}$ because $a_T^2 + a_N^2 = |\mathbf{a}|^2$. Thus $\sqrt{5} = \kappa |\mathbf{v}(0)|^2 = \kappa \cdot 1 = \kappa$. Next lets find $\mathbf{N}$. From $\mathbf{a} = a_T \mathbf{T} + a_N \mathbf{N}$ it follows

$$(-1, 0, 2) = 0 \cdot \mathbf{T} + \sqrt{5} \mathbf{N}$$

and so

$$\mathbf{N} = \frac{1}{\sqrt{5}} (-1, 0, 2).$$

This was pretty easy.

Example 6.3. Find a formula for the curvature of the curve given by the graph of $y = f(x)$ for $x \in [a, b]$. Assume whatever you like about smoothness of f.

You need to write this as a parametric curve. This is most easily accomplished by letting $t = x$. Thus a parametrization is $(t, f(t), 0) : t \in [a, b]$. Then you can use the formula given above. The acceleration is $(0, f''(t), 0)$ and the velocity is $(1, f'(t), 0)$. Therefore,

$$\mathbf{a} \times \mathbf{v} = (0, f''(t), 0) \times (1, f'(t), 0) = (0, 0, -f''(t)).$$

Therefore, the curvature is given by

$$\frac{|\mathbf{a} \times \mathbf{v}|}{|\mathbf{v}|^3} = \frac{|f''(t)|}{\left(1 + f'(t)^2\right)^{3/2}}.$$

Sometimes curves do not come to you parametrically. This is unfortunate when it occurs but you can sometimes find a parametric description of such curves. It should be emphasized that it is only sometimes when you can actually find a parametrization. General systems of nonlinear equations cannot be solved using algebra.

Example 6.4. Find a parametrization for the intersection of the surfaces

$$y + 3z = 2x^2 + 4 \text{ and } y + 2z = x + 1.$$

You need to solve for x and y in terms of x. This yields

$$z = 2x^2 - x + 3, \ y = -4x^2 + 3x - 5.$$

Therefore, letting $t = x$, the parametrization is

$$(x, y, z) = \left(t, -4t^2 - 5 + 3t, -t + 3 + 2t^2\right).$$

Example 6.5. Find a parametrization for the straight line joining $(3, 2, 4)$ and $(1, 10, 5)$.

$(x, y, z) = (3, 2, 4) + t(-2, 8, 1) = (3 - 2t, 2 + 8t, 4 + t)$ where $t \in [0, 1]$. Note where this came from. The vector $(-2, 8, 1)$ is obtained from $(1, 10, 5) - (3, 2, 4)$. Now you should check to see this works.

6.2 Geometry Of Space Curves*

If you are interested in more on space curves, you should read this section. Otherwise, proceed to the exercises. Denote by $\mathbf{R}(s)$ the function which takes s to a point on this curve where s is arc length. Thus $\mathbf{R}(s)$ equals the point on the curve which occurs when you have traveled a distance of s along the curve from one end. This is known as the parametrization of the curve in terms of arc length. Note also that it incorporates an orientation on the curve because there are exactly two ends you could begin measuring length from. In this section, assume anything about smoothness and continuity to make the following manipulations valid. In particular, assume that $\mathbf{R}'$ exists and is continuous.

Lemma 6.2. *Define* $\mathbf{T}(s) \equiv \mathbf{R}'(s)$. *Then* $|\mathbf{T}(s)| = 1$ *and if* $\mathbf{T}'(s) \neq 0$, *then there exists a unit vector* $\mathbf{N}(s)$ *perpendicular to* $\mathbf{T}(s)$ *and a scalar valued function* $\kappa(s)$ *with* $\mathbf{T}'(s) = \kappa(s) \mathbf{N}(s)$.

Proof: First, $s = \int_0^s |\mathbf{R}'(r)|\, dr$ because of the definition of arc length. Therefore, from the fundamental theorem of calculus, $1 = |\mathbf{R}'(s)| = |\mathbf{T}(s)|$. Therefore, $\mathbf{T} \cdot \mathbf{T} = 1$ and so upon differentiating this on both sides, yields $\mathbf{T}' \cdot \mathbf{T} + \mathbf{T} \cdot \mathbf{T}' = 0$ which shows $\mathbf{T} \cdot \mathbf{T}' = 0$. Therefore, the vector $\mathbf{T}'$ is perpendicular to the vector $\mathbf{T}$. In case $\mathbf{T}'(s) \neq \mathbf{0}$, let $\mathbf{N}(s) = \frac{\mathbf{T}'(s)}{|\mathbf{T}'(s)|}$ and so $\mathbf{T}'(s) = |\mathbf{T}'(s)| \mathbf{N}(s)$, showing the scalar valued function is $\kappa(s) = |\mathbf{T}'(s)|$. ∎

The radius of curvature is defined as $\rho = \frac{1}{\kappa}$. Thus at points where there is a lot of curvature, the radius of curvature is small and at points where the curvature is small, the radius of curvature is large. The plane determined by the two vectors $\mathbf{T}$ and $\mathbf{N}$ is called the osculating plane. It identifies a particular plane which is in a sense tangent to this space curve. In the case where $|\mathbf{T}'(s)| = 0$ near the point of interest, $\mathbf{T}(s)$ equals a constant and so the space curve is a straight line which it would be supposed has no curvature. Also, the principal normal is undefined in this case. This makes sense because if there is no curving going on, there is no special direction normal to the curve at such points which could be distinguished from any other direction normal to the curve. In the case where $|\mathbf{T}'(s)| = 0$, $\kappa(s) = 0$ and the radius of curvature would be considered infinite.

Definition 6.2. The vector $\mathbf{T}(s)$ is called the unit tangent vector and the vector $\mathbf{N}(s)$ is called the **principal normal**. The function $\kappa(s)$ in the above lemma is called the **curvature**. When $\mathbf{T}'(s) \neq 0$ so the principal normal is defined, the vector $\mathbf{B}(s) \equiv \mathbf{T}(s) \times \mathbf{N}(s)$ is called the **binormal.**

The binormal is normal to the osculating plane and $\mathbf{B}'$ tells how fast this vector changes. Thus it measures the rate at which the curve twists.

Lemma 6.3. *Let* $\mathbf{R}(s)$ *be a parametrization of a space curve with respect to arc length and let the vectors* $\mathbf{T}, \mathbf{N}$, *and* $\mathbf{B}$ *be as defined above. Then* $\mathbf{B}' = \mathbf{T} \times \mathbf{N}'$ *and there exists a scalar function* $\tau(s)$ *such that* $\mathbf{B}' = \tau \mathbf{N}$.

Proof: From the definition of $\mathbf{B} = \mathbf{T} \times \mathbf{N}$, and you can differentiate both sides and get $\mathbf{B}' = \mathbf{T}' \times \mathbf{N} + \mathbf{T} \times \mathbf{N}'$. Now recall that $\mathbf{T}'$ is a multiple called curvature multiplied by $\mathbf{N}$ so the vectors $\mathbf{T}'$ and $\mathbf{N}$ have the same direction, so $\mathbf{B}' = \mathbf{T} \times \mathbf{N}'$. Therefore, $\mathbf{B}'$ is either zero or is perpendicular to $\mathbf{T}$. But also, from the definition of $\mathbf{B}, \mathbf{B}$ is a unit vector and so $\mathbf{B}(s) \cdot \mathbf{B}(s) = 1$. Differentiating this, $\mathbf{B}'(s) \cdot \mathbf{B}(s) + \mathbf{B}(s) \cdot \mathbf{B}'(s) = 0$ showing that $\mathbf{B}'$ is perpendicular to $\mathbf{B}$ also. Therefore, $\mathbf{B}'$ is a vector which is perpendicular to both vectors $\mathbf{T}$ and $\mathbf{B}$ and since this is in three dimensions, $\mathbf{B}'$ must be some scalar multiple of $\mathbf{N}$, and this multiple is called τ. Thus $\mathbf{B}' = \tau\mathbf{N}$ as claimed. ∎

Lets go over this last claim a little more. The following situation is obtained. There are two vectors $\mathbf{T}$ and $\mathbf{B}$ which are perpendicular to each other and both $\mathbf{B}'$ and $\mathbf{N}$ are perpendicular to these two vectors, hence perpendicular to the plane determined by them. Therefore, $\mathbf{B}'$ must be a multiple of $\mathbf{N}$. Take a piece of paper, draw two unit vectors on it which are perpendicular. Then you can see that any two vectors which are perpendicular to this plane must be multiples of each other.

The scalar function τ is called the torsion. In case $\mathbf{T}' = 0$, none of this is defined because in this case there is not a well defined osculating plane. The conclusion of the following theorem is called the Serret Frenet formulas.

Theorem 6.2. *(Serret Frenet) Let $\mathbf{R}(s)$ be the parametrization with respect to arc length of a space curve and $\mathbf{T}(s) = \mathbf{R}'(s)$ is the unit tangent vector. Suppose $|\mathbf{T}'(s)| \neq 0$ so the principal normal $\mathbf{N}(s) = \frac{\mathbf{T}'(s)}{|\mathbf{T}'(s)|}$ is defined. The binormal is the vector $\mathbf{B} \equiv \mathbf{T} \times \mathbf{N}$ so $\mathbf{T}, \mathbf{N}, \mathbf{B}$ forms a right handed system of unit vectors each of which is perpendicular to every other. Then the following system of differential equations holds in $\mathbb{R}^9$.*

$$\mathbf{B}' = \tau\mathbf{N}, \ \mathbf{T}' = \kappa\mathbf{N}, \ \mathbf{N}' = -\kappa\mathbf{T} - \tau\mathbf{B}$$

*where κ is the curvature and is nonnegative and τ is the **torsion**.*

Proof: $\kappa \geq 0$ because $\kappa = |\mathbf{T}'(s)|$. The first two equations are already established. To get the third, note that $\mathbf{B} \times \mathbf{T} = \mathbf{N}$ which follows because $\mathbf{T}, \mathbf{N}, \mathbf{B}$ is given to form a right handed system of unit vectors each perpendicular to the others. (Use your right hand.) Now take the derivative of this expression. thus

$$\mathbf{N}' = \mathbf{B}' \times \mathbf{T} + \mathbf{B} \times \mathbf{T}' = \tau\mathbf{N} \times \mathbf{T} + \kappa\mathbf{B} \times \mathbf{N}.$$

Now recall again that $\mathbf{T}, \mathbf{N}, \mathbf{B}$ is a right hand system. Thus

$$\mathbf{N} \times \mathbf{T} = -\mathbf{B}, \ \mathbf{B} \times \mathbf{N} = -\mathbf{T}.$$

This establishes the Frenet Serret formulas. ∎

This is an important example of a system of differential equations in $\mathbb{R}^9$. It is a remarkable result because it says that from knowledge of the two scalar functions τ and κ, and initial values for $\mathbf{B}, \mathbf{T}$, and $\mathbf{N}$ when $s = 0$ you can obtain the binormal, unit tangent, and principal normal vectors. It is just the solution of an initial value problem although this is for a vector valued rather than scalar valued function.

Having done this, you can reconstruct the entire space curve starting at some point $\mathbf{R}_0$ because $\mathbf{R}'(s) = \mathbf{T}(s)$ and so $\mathbf{R}(s) = \mathbf{R}_0 + \int_0^s \mathbf{T}(r)\, dr$.

The vectors $\mathbf{B}, \mathbf{T}$, and $\mathbf{N}$ are vectors which are functions of position on the space curve. Often, especially in applications, you deal with a space curve which is parameterized by a function of t where t is time. Thus a value of t would correspond to a point on this curve and you could let $\mathbf{B}(t), \mathbf{T}(t)$, and $\mathbf{N}(t)$ be the binormal, unit tangent, and principal normal at this point of the curve. The following example is typical.

Example 6.6. Given the circular helix, $\mathbf{R}(t) = (a\cos t)\,\mathbf{i} + (a\sin t)\,\mathbf{j} + (bt)\,\mathbf{k}$, find the arc length $s(t)$, the unit tangent vector $\mathbf{T}(t)$, the principal normal $\mathbf{N}(t)$, the binormal $\mathbf{B}(t)$, the curvature $\kappa(t)$, and the torsion, $\tau(t)$. Here $t \in [0, T]$.

The arc length is $s(t) = \int_0^t \left(\sqrt{a^2 + b^2}\right)\, dr = \left(\sqrt{a^2 + b^2}\right) t$. Now the tangent vector is obtained using the chain rule as

$$\mathbf{T} = \frac{d\mathbf{R}}{ds} = \frac{d\mathbf{R}}{dt}\frac{dt}{ds} = \frac{1}{\sqrt{a^2+b^2}}\mathbf{R}'(t) = \frac{1}{\sqrt{a^2+b^2}}\left((-a\sin t)\,\mathbf{i} + (a\cos t)\,\mathbf{j} + b\mathbf{k}\right)$$

The principal normal:

$$\frac{d\mathbf{T}}{ds} = \frac{d\mathbf{T}}{dt}\frac{dt}{ds} = \frac{1}{a^2+b^2}\left((-a\cos t)\,\mathbf{i} + (-a\sin t)\,\mathbf{j} + 0\mathbf{k}\right)$$

and so

$$\mathbf{N} = \frac{d\mathbf{T}}{ds} \Big/ \left|\frac{d\mathbf{T}}{ds}\right| = -\left((\cos t)\,\mathbf{i} + (\sin t)\,\mathbf{j}\right)$$

The binormal:

$$\mathbf{B} = \frac{1}{\sqrt{a^2+b^2}}\begin{vmatrix} \mathbf{i} & \mathbf{j} & \mathbf{k} \\ -a\sin t & a\cos t & b \\ -\cos t & -\sin t & 0 \end{vmatrix} = \frac{1}{\sqrt{a^2+b^2}}\left((b\sin t)\,\mathbf{i} - b\cos t\,\mathbf{j} + a\mathbf{k}\right)$$

Now the curvature $\kappa(t) = \left|\frac{d\mathbf{T}}{ds}\right| = \sqrt{\left(\frac{a\cos t}{a^2+b^2}\right)^2 + \left(\frac{a\sin t}{a^2+b^2}\right)^2} = \frac{a}{a^2+b^2}$. Note the curvature is constant in this example. The final task is to find the torsion. Recall that $\mathbf{B}' = \tau\mathbf{N}$ where the derivative on $\mathbf{B}$ is taken with respect to arc length. Therefore, remembering that t is a function of s,

$$\mathbf{B}'(s) = \frac{1}{\sqrt{a^2+b^2}}\left((b\cos t)\,\mathbf{i} + (b\sin t)\,\mathbf{j}\right)\frac{dt}{ds} = \frac{1}{a^2+b^2}\left((b\cos t)\,\mathbf{i} + (b\sin t)\,\mathbf{j}\right)$$

$$= \tau\left(-(\cos t)\,\mathbf{i} - (\sin t)\,\mathbf{j}\right) = \tau\mathbf{N}$$

and it follows $-b/\left(a^2 + b^2\right) = \tau$.

An important application of the usefulness of these ideas involves the decomposition of the acceleration in terms of these vectors of an object moving over a space curve.

Corollary 6.1. *Let $\mathbf{R}(t)$ be a space curve and denote by $\mathbf{v}(t)$ the velocity, $\mathbf{v}(t) = \mathbf{R}'(t)$, let $v(t) \equiv |\mathbf{v}(t)|$ denote the speed, and let $\mathbf{a}(t)$ denote the acceleration. Then $\mathbf{v} = v\mathbf{T}$ and $\mathbf{a} = \frac{dv}{dt}\mathbf{T} + \kappa v^2\mathbf{N}$.*

Proof: $\mathbf{T} = \frac{d\mathbf{R}}{ds} = \frac{d\mathbf{R}}{dt}\frac{dt}{ds} = \mathbf{v}\frac{dt}{ds}$. Also, $s = \int_0^t v(r)\,dr$ and so $\frac{ds}{dt} = v$ which implies $\frac{dt}{ds} = \frac{1}{v}$. Therefore, $\mathbf{T} = \mathbf{v}/v$ which implies $\mathbf{v} = v\mathbf{T}$ as claimed.

Now the acceleration is just the derivative of the velocity and so by the Serrat Frenet formulas,

$$\mathbf{a} = \frac{dv}{dt}\mathbf{T} + v\frac{d\mathbf{T}}{dt} = \frac{dv}{dt}\mathbf{T} + v\frac{d\mathbf{T}}{ds}v = \frac{dv}{dt}\mathbf{T} + v^2\kappa\mathbf{N}$$

Note how this decomposes the acceleration into a component tangent to the curve and one which is normal to it. Also note that from the above, $v|\mathbf{T}'|\frac{\mathbf{T}'(t)}{|\mathbf{T}'|} = v^2\kappa\mathbf{N}$ and so $\frac{|\mathbf{T}'|}{v} = \kappa$ and $\mathbf{N} = \frac{\mathbf{T}'(t)}{|\mathbf{T}'|}$. ∎

6.3 Exercises

(1) Find a parametrization for the intersection of the planes $2x + y + 3z = -2$ and $3x - 2y + z = -4$.

(2) Find a parametrization for the intersection of the plane $3x + y + z = -3$ and the circular cylinder $x^2 + y^2 = 1$.

(3) Find a parametrization for the intersection of the plane $4x + 2y + 3z = 2$ and the elliptic cylinder $x^2 + 4z^2 = 9$.

(4) Find a parametrization for the straight line joining $(1, 2, 1)$ and $(-1, 4, 4)$.

(5) Find a parametrization for the intersection of the surfaces $3y + 3z = 3x^2 + 2$ and $3y + 2z = 3$.

(6) Find a formula for the curvature of the curve $y = \sin x$ in the xy plane.

(7) An object moves over the curve (t, e^t, at) where $t \in \mathbb{R}$ and a is a positive constant. Find the value of t at which the normal component of acceleration is largest if there is such a point.

(8) Find a formula for the curvature of the space curve in $\mathbb{R}^2$, $(x(t), y(t))$.

(9) An object moves over the helix, $(\cos 3t, \sin 3t, 5t)$. Find the normal and tangential components of the acceleration of this object as a function of t and write the acceleration in the form $a_T\mathbf{T} + a_N\mathbf{N}$.

(10) An object moves over the helix, $(\cos t, \sin t, t)$. Find the normal and tangential components of the acceleration of this object as a function of t and write the acceleration in the form $a_T\mathbf{T} + a_N\mathbf{N}$.

(11) An object moves in $\mathbb{R}^3$ according to the formula $(\cos 3t, \sin 3t, t^2)$. Find the normal and tangential components of the acceleration of this object as a function of t and write the acceleration in the form $a_T\mathbf{T} + a_N\mathbf{N}$.

(12) An object moves over the helix, $(\cos t, \sin t, 2t)$. Find the osculating plane at the point of the curve corresponding to $t = \pi/4$.

(13) An object moves over a circle of radius r according to the formula

$$\mathbf{r}(t) = (r\cos(\omega t), r\sin(\omega t))$$

where $v = r\omega$. Show that the speed of the object is constant and equals to v. Tell why $a_T = 0$ and find a_N, $\mathbf{N}$.

(14) Suppose $|\mathbf{R}(t)| = c$ where c is a constant$\mathbf{R}(t)$. Show the velocity, $\mathbf{R}'(t)$ is always perpendicular to $\mathbf{R}(t)$.

(15) An object moves in three dimensions and the only force on the object is a central force. This means that if $\mathbf{r}(t)$ is the position of the object, $\mathbf{a}(t) = k(\mathbf{r}(t))\mathbf{r}(t)$ where k is some function. Show that if this happens, then the motion of the object must be in a plane. **Hint:** First argue that $\mathbf{a} \times \mathbf{r} = \mathbf{0}$. Next show that $(\mathbf{a} \times \mathbf{r}) = (\mathbf{v} \times \mathbf{r})'$. Therefore, $(\mathbf{v} \times \mathbf{r})' = \mathbf{0}$. Explain why this requires $\mathbf{v} \times \mathbf{r} = \mathbf{c}$ for some vector $\mathbf{c}$ which does not depend on t. Then explain why $\mathbf{c} \cdot \mathbf{r} = 0$. This implies the motion is in a plane. Why? What are some examples of central forces?

(16) Let $\mathbf{R}(t) = (\cos t)\mathbf{i} + (\cos t)\mathbf{j} + (\sqrt{2}\sin t)\mathbf{k}$. Find the arc length, s as a function of the parameter t, if $t = 0$ is taken to correspond to $s = 0$.

(17) Let $\mathbf{R}(t) = 2\mathbf{i} + (4t + 2)\mathbf{j} + 4t\mathbf{k}$. Find the arc length, s as a function of the parameter t, if $t = 0$ is taken to correspond to $s = 0$.

(18) Let $\mathbf{R}(t) = e^{5t}\mathbf{i} + e^{-5t}\mathbf{j} + 5\sqrt{2}t\mathbf{k}$. Find the arc length, s as a function of the parameter t, if $t = 0$ is taken to correspond to $s = 0$.

(19) Consider the curve obtained from the graph of $y = f(x)$. Find a formula for the curvature.

(20) Consider the curve in the plane $y = e^x$. Find the point on this curve at which the curvature is a maximum.

(21) An object moves along the x axis toward $(0, 0)$ and then along the curve $y = x^2$ in the direction of increasing x at constant speed. Is the force acting on the object a continuous function? Explain. Is there any physically reasonable way to make this force continuous by relaxing the requirement that the object move at constant speed? If the curve were part of a railroad track, what would happen at the point where $x = 0$?

(22) An object of mass m moving over a space curve is acted on by a force $\mathbf{F}$. Show the work done by this force equals ma_T (length of the curve). In other words, it is only the tangential component of the force which does work.

(23) The edge of an elliptical skating rink represented in the following picture has a light at its left end and satisfies the equation $\frac{x^2}{900} + \frac{y^2}{256} = 1$. (Distances measured in yards.)

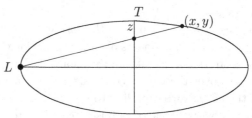

A hockey puck slides from the point T towards the center of the rink at the rate of 2 yards per second. What is the speed of its shadow along the wall when $z = 8$? **Hint:** You need to find $\sqrt{x'^2 + y'^2}$ at the instant described.

Chapter 7

Functions Of Many Variables

7.1 The Graph Of A Function Of Two Variables

In calculus, we are permitted and even required to think in a meaningful way about things which cannot be drawn. However, it is certainly interesting to consider some things which can be visualized and this will help to formulate and understand more general notions which make sense in higher dimensions. One of these is the concept of a scalar valued function of two variables.

Let $f(x, y)$ denote a scalar valued function of two variables evaluated at the point (x, y). Its graph consists of the set of points (x, y, z) such that $z = f(x, y)$. How does one go about depicting such a graph? The usual way is to fix one of the variables, say x and consider the function $z = f(x, y)$ where y is allowed to vary and x is fixed. Graphing this would give a curve which lies in the surface to be depicted. Then do the same thing for other values of x and the result would depict the desired graph. Computers do this very well. The following is the graph of the function $z = \cos(x)\sin(2x + y)$ drawn using Maple, a computer algebra system.[1]

Notice how elaborate this picture is. The lines in the drawing correspond to taking one of the variables constant and graphing the curve which results. The computer did this drawing in seconds but you could not do it as well if you spent all day on it. I used a grid consisting of 70 choices for x and 70 choices for y.

Sometimes attempts are made to understand three dimensional objects like the

---

[1]I used Maple and exported the graph as a file which I then imported into this document.

above graph by looking at contour graphs in two dimensions. The contour graph of the above three dimensional graph is below and comes from using the computer algebra system again.

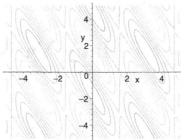

This is in two dimensions and the different lines in two dimensions correspond to points on the three dimensional graph which have the same z value. If you have looked at a weather map, these lines are called isotherms or isobars depending on whether the function involved is temperature or pressure. In a contour geographic map, the contour lines represent constant altitude. If many contour lines are close to each other, this indicates rapid change in the altitude, temperature, pressure, or whatever else may be measured.

A scalar function of three variables, cannot be visualized because four dimensions are required. However, some people like to try and visualize even these examples. This is done by looking at level surfaces in $\mathbb{R}^3$ which are defined as surfaces where the function assumes a constant value. They play the role of contour lines for a function of two variables. As a simple example, consider $f(x, y, z) = x^2 + y^2 + z^2$. The level surfaces of this function would be concentric spheres centered at 0. (Why?) Another way to visualize objects in higher dimensions involves the use of color and animation. However, there really are limits to what you can accomplish in this direction. So much for art.

However, the concept of level curves is quite useful because these can be drawn.

Example 7.1. Determine from a contour map where the function $f(x, y) = \sin\left(x^2 + y^2\right)$ is steepest.

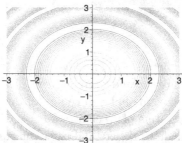

In the picture, the steepest places are where the contour lines are close together because they correspond to various values of the function. You can look at the picture and see where they are close and where they are far. This is the advantage of a contour map.

7.2 Review Of Limits

Recall the concept of limit of a function of many variables. When $\mathbf{f} : D(\mathbf{f}) \to \mathbb{R}^q$ one can only consider in a meaningful way limits at limit points of the set $D(\mathbf{f})$.

Definition 7.1. Let A denote a nonempty subset of $\mathbb{R}^p$. A point $\mathbf{x}$ is said to be a **limit point** of the set A if for every $r > 0, B(\mathbf{x}, r)$ contains infinitely many points of A.

Example 7.2. Let S denote the set $\{(x, y, z) \in \mathbb{R}^3 : x, y, z \text{ are all in } \mathbb{N}\}$. Which points are limit points?

This set does not have any because any two of these points are at least as far apart as 1. Therefore, if $\mathbf{x}$ is any point of $\mathbb{R}^3, B(\mathbf{x}, 1/4)$ contains at most one point.

Example 7.3. Let U be an open set in $\mathbb{R}^3$. Which points of U are limit points of U?

They all are. From the definition of U being open, if $\mathbf{x} \in U$, There exists $B(\mathbf{x}, r) \subseteq U$ for some $r > 0$. Now consider the line segment $\mathbf{x} + t r \mathbf{e}_1$ where $t \in [0, 1/2]$. This describes infinitely many points and they are all in $B(\mathbf{x}, r)$ because $|\mathbf{x} + t r \mathbf{e}_1 - \mathbf{x}| = tr < r$. Therefore, every point of U is a limit point of U.

The case where U is open will be the one of most interest, but many other sets have limit points.

Definition 7.2. Let $\mathbf{f} : D(\mathbf{f}) \subseteq \mathbb{R}^p \to \mathbb{R}^q$ where $q, p \geq 1$ be a function and let $\mathbf{x}$ be a limit point of $D(\mathbf{f})$. Then

$$\lim_{\mathbf{y} \to \mathbf{x}} \mathbf{f}(\mathbf{y}) = \mathbf{L}$$

if and only if the following condition holds. For all $\varepsilon > 0$ there exists $\delta > 0$ such that if

$$0 < |\mathbf{y} - \mathbf{x}| < \delta \text{ and } \mathbf{y} \in D(\mathbf{f})$$

then,

$$|\mathbf{L} - \mathbf{f}(\mathbf{y})| < \varepsilon.$$

The condition that $\mathbf{x}$ must be a limit point of $D(\mathbf{f})$ if you are to take a limit at $\mathbf{x}$ is what makes the limit well defined.

Proposition 7.1. Let $\mathbf{f} : D(\mathbf{f}) \subseteq \mathbb{R}^p \to \mathbb{R}^q$ where $q, p \geq 1$ be a function and let $\mathbf{x}$ be a limit point of $D(\mathbf{f})$. Then if $\lim_{\mathbf{y} \to \mathbf{x}} \mathbf{f}(\mathbf{y})$ exists, it must be unique.

Proof: Suppose $\lim_{\mathbf{y}\to\mathbf{x}} \mathbf{f}(\mathbf{y}) = \mathbf{L}_1$ and $\lim_{\mathbf{y}\to\mathbf{x}} \mathbf{f}(\mathbf{y}) = \mathbf{L}_2$. Then for $\varepsilon > 0$ given, let $\delta_i > 0$ correspond to $\mathbf{L}_i$ in the definition of the limit and let $\delta = \min(\delta_1, \delta_2)$. Since $\mathbf{x}$ is a limit point, there exists $\mathbf{y} \in B(\mathbf{x}, \delta) \cap D(\mathbf{f})$. Therefore,

$$|\mathbf{L}_1 - \mathbf{L}_2| \leq |\mathbf{L}_1 - \mathbf{f}(\mathbf{y})| + |\mathbf{f}(\mathbf{y}) - \mathbf{L}_2| < \varepsilon + \varepsilon = 2\varepsilon.$$

Since $\varepsilon > 0$ is arbitrary, this shows $\mathbf{L}_1 = \mathbf{L}_2$. $\blacksquare$

The following theorem summarized many important interactions involving continuity. Most of this theorem has been proved in Theorem 4.3 on Page 64.

Theorem 7.1. *Suppose* $\mathbf{x}$ *is a limit point of* $D(\mathbf{f})$ *and*

$$\lim_{\mathbf{y}\to\mathbf{x}} \mathbf{f}(\mathbf{y}) = \mathbf{L}, \quad \lim_{\mathbf{y}\to\mathbf{x}} \mathbf{g}(\mathbf{y}) = \mathbf{K}$$

where $\mathbf{K}$ *and* $\mathbf{L}$ *are vectors in* $\mathbb{R}^p$ *for* $p \geq 1$. *Then if* $a, b \in \mathbb{R}$,

$$\lim_{\mathbf{y}\to\mathbf{x}} a\mathbf{f}(\mathbf{y}) + b\mathbf{g}(\mathbf{y}) = a\mathbf{L} + b\mathbf{K}, \tag{7.1}$$

$$\lim_{\mathbf{y}\to\mathbf{x}} \mathbf{f} \cdot \mathbf{g}(\mathbf{y}) = \mathbf{L} \cdot \mathbf{K} \tag{7.2}$$

Also, if $\mathbf{h}$ *is a continuous function defined near* $\mathbf{L}$, *then*

$$\lim_{\mathbf{y}\to\mathbf{x}} \mathbf{h} \circ \mathbf{f}(\mathbf{y}) = \mathbf{h}(\mathbf{L}). \tag{7.3}$$

For a vector valued function

$$\mathbf{f}(\mathbf{y}) = (f_1(\mathbf{y}), \cdots, f_q(\mathbf{y}))^T,$$

$\lim_{\mathbf{y}\to\mathbf{x}} \mathbf{f}(\mathbf{y}) = \mathbf{L} = (L_1 \cdots, L_k)^T$ *if and only if*

$$\lim_{\mathbf{y}\to\mathbf{x}} f_k(\mathbf{y}) = L_k \tag{7.4}$$

for each $k = 1, \cdots, p$.

In the case where $\mathbf{f}$ *and* $\mathbf{g}$ *have values in* $\mathbb{R}^3$

$$\lim_{\mathbf{y}\to\mathbf{x}} \mathbf{f}(\mathbf{y}) \times \mathbf{g}(\mathbf{y}) = \mathbf{L} \times \mathbf{K}. \tag{7.5}$$

Also recall Theorem 4.4 on Page 66.

Theorem 7.2. *For* $\mathbf{f} : D(\mathbf{f}) \to \mathbb{R}^q$ *and* $\mathbf{x} \in D(\mathbf{f})$ *such that* $\mathbf{x}$ *is a limit point of* $D(\mathbf{f})$, *it follows* $\mathbf{f}$ *is continuous at* $\mathbf{x}$ *if and only if* $\lim_{\mathbf{y}\to\mathbf{x}} \mathbf{f}(\mathbf{y}) = \mathbf{f}(\mathbf{x})$.

7.3 Exercises

(1) Sketch the contour graph of the function of two variables $f(x,y) = (x-1)^2 + (y-2)^2$.

(2) Which of the following functions could correspond to the following contour graphs? $z = x^2 + 3y^2$, $z = 3x^2 + y^2$, $z = x^2 - y^2$, $z = x + y$.

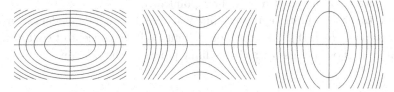

(3) Which of the following functions could correspond to the following contour graphs? $z = x^2 - 3y^2$, $z = y^2 + 3x^2$, $z = x - y$, $z = x + y$.

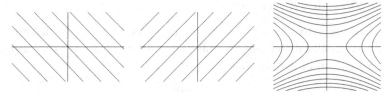

(4) Which of the following functions could correspond to the following contour graphs? $z = \sin(x + y)$, $z = x + y$, $z = (x + y)^2$, $z = x^2 - y$.

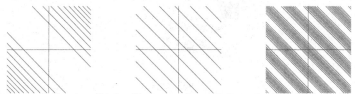

(5) Find the following limits if they exist. If they do not exist, explain why.

(a) $\lim_{(x,y)\to(0,0)} \frac{x^2 - y^2}{x^2 + y^2}$

(b) $\lim_{(x,y)\to(0,0)} \frac{2x^3 + xy^2 - x^2 - 2y^2}{x^2 + 2y^2}$

(c) $\lim_{(x,y)\to(0,0)} \frac{\sin(x^2 + y^2)}{x^2 + y^2}$

(d) $\lim_{(x,y)\to(0,0)} \frac{\sin(x^2 + 2y^2)}{x^2 + 2y^2}$

(e) $\lim_{(x,y)\to(0,0)} \frac{\sin(x^2 + 2y^2)}{2x^2 + y^2}$

(f) $\lim_{(x,y)\to(0,0)} \frac{(x^2 - y^4)^2}{(x^2 + y^4)^2}$

(6) Find the following limits if they exist. If they do not exist, tell why.

(a) $\lim_{(x,y)\to(0,0)} x \frac{(x^2 - y^4)^2}{(x^2 + y^4)^2}$

(b) $\lim_{(x,y)\to(0,0)} \frac{x \sin(x^2 + 2y^2)}{2x^2 + y^2}$

(c) $\lim_{(x,y)\to(0,0)} \frac{xy}{x^2 + y^2}$

(d) $\lim_{(x,y)\to(1,0)} \frac{x^3-3x^2+3x-1-y^2x+y^2}{x^2-2x+1+y^2}$

(7) *Suppose f is a function defined on a set D and that $\mathbf{a} \in D$ is not a limit point of D. Show that if I define the notion of limit in the same way as above, then $\lim_{\mathbf{x}\to\mathbf{a}} f(\mathbf{x}) = 5$. Show that it is also the case that $\lim_{\mathbf{x}\to\mathbf{a}} f(\mathbf{x}) = 7$. In other words, the concept of limit is totally meaningless. This is why the insistence that the point $\mathbf{a}$ be a limit point of D.

(8) *Show that the definition of continuity at $\mathbf{a} \in D(\mathbf{f})$ is not dependent on $\mathbf{a}$ being a limit point of $D(\mathbf{f})$. The concept of limit and the concept of continuity are related at those points $\mathbf{a}$ which are limit points of the domain.

7.4 The Directional Derivative And Partial Derivatives

7.4.1 *The Directional Derivative*

The directional derivative is just what its name suggests. It is the derivative of a function in a particular direction. The following picture illustrates the situation in the case of a function of two variables.

$$z = f(x, y)$$

$$(x_0, y_0)$$

In this picture, $\mathbf{v} \equiv (v_1, v_2)$ is a unit vector in the xy plane and $\mathbf{x}_0 \equiv (x_0, y_0)$ is a point in the xy plane. When (x, y) moves in the direction of $\mathbf{v}$, this results in a change in $z = f(x, y)$ as shown in the picture. The directional derivative in this direction is defined as

$$\lim_{t\to 0} \frac{f(x_0 + tv_1, y_0 + tv_2) - f(x_0, y_0)}{t}.$$

It tells how fast z is changing in this direction. If you looked at it from the side, you would be getting the slope of the indicated tangent line. A simple example of this is a person climbing a mountain. He could go various directions, some steeper than others. The directional derivative is just a measure of the steepness in a given direction. This motivates the following general definition of the directional derivative.

Definition 7.3. Let $f : U \to \mathbb{R}$ where U is an open set in $\mathbb{R}^n$ and let $\mathbf{v}$ be a unit vector. For $\mathbf{x} \in U$, define the **directional derivative** of f in the direction $\mathbf{v}$, at the point $\mathbf{x}$ as

$$D_{\mathbf{v}}f(\mathbf{x}) \equiv \lim_{t \to 0} \frac{f(\mathbf{x}+t\mathbf{v}) - f(\mathbf{x})}{t}.$$

Example 7.4. Find the directional derivative of the function $f(x,y) = x^2 y$ in the direction of $\mathbf{i} + \mathbf{j}$ at the point $(1,2)$.

First you need a unit vector which has the same direction as the given vector. This unit vector is $\mathbf{v} \equiv \left(\frac{1}{\sqrt{2}}, \frac{1}{\sqrt{2}} \right)$. Then to find the directional derivative from the definition, write the difference quotient described above. Thus $f(\mathbf{x}+t\mathbf{v}) = \left(1 + \frac{t}{\sqrt{2}} \right)^2 \left(2 + \frac{t}{\sqrt{2}} \right)$ and $f(\mathbf{x}) = 2$. Therefore,

$$\frac{f(\mathbf{x}+t\mathbf{v}) - f(\mathbf{x})}{t} = \frac{\left(1 + \frac{t}{\sqrt{2}} \right)^2 \left(2 + \frac{t}{\sqrt{2}} \right) - 2}{t},$$

and to find the directional derivative, you take the limit of this as $t \to 0$. However, this difference quotient equals $\frac{1}{4}\sqrt{2}\left(10 + 4t\sqrt{2} + t^2 \right)$ and so, letting $t \to 0, D_{\mathbf{v}} f(1,2) = \left(\frac{5}{2}\sqrt{2} \right)$.

There is something you must keep in mind about this. The direction vector must always be a unit vector[2].

7.4.2 *Partial Derivatives*

There are some special unit vectors which come to mind immediately. These are the vectors $\mathbf{e}_i$ where

$$\mathbf{e}_i = (0, \cdots, 0, 1, 0, \cdots 0)^T$$

and the 1 is in the ith position.

Thus in case of a function of two variables, the directional derivative in the direction $\mathbf{i} = \mathbf{e}_1$ is the slope of the indicated straight line in the following picture.

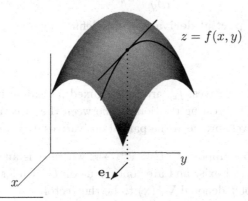

---

[2]Actually, there is a more general formulation of the notion of directional derivative known as the Gateaux derivative in which the length of $\mathbf{v}$ is not one but it is not considered here.

As in the case of a general directional derivative, you fix y and take the derivative of the function $x \to f(x, y)$. More generally, even in situations which cannot be drawn, the definition of a partial derivative is as follows.

Definition 7.4. Let U be an open subset of $\mathbb{R}^n$ and let $f : U \to \mathbb{R}$. Then letting $\mathbf{x} = (x_1, \cdots, x_n)^T$ be a typical element of $\mathbb{R}^n$,

$$\frac{\partial f}{\partial x_i} (\mathbf{x}) \equiv D_{\mathbf{e}_i} f(\mathbf{x}).$$

This is called the **partial derivative** of f. Thus,

$$\frac{\partial f}{\partial x_i}(\mathbf{x}) \equiv \lim_{t \to 0} \frac{f(\mathbf{x} + t\mathbf{e}_i) - f(\mathbf{x})}{t}$$

$$= \lim_{t \to 0} \frac{f(x_1, \cdots, x_i + t, \cdots x_n) - f(x_1, \cdots, x_i, \cdots x_n)}{t},$$

and to find the partial derivative, differentiate with respect to the variable of interest and regard all the others as constants. Other notation for this partial derivative is $f_{x_i}, f_{,i}$, or $D_i f$. If $y = f(\mathbf{x})$, the partial derivative of f with respect to x_i may also be denoted by $\frac{\partial y}{\partial x_i}$ or y_{x_i}.

Example 7.5. Find $\frac{\partial f}{\partial x}, \frac{\partial f}{\partial y}$, and $\frac{\partial f}{\partial z}$ if $f(x, y) = y \sin x + x^2 y + z$.

From the definition above, $\frac{\partial f}{\partial x} = y \cos x + 2xy$, $\frac{\partial f}{\partial y} = \sin x + x^2$, and $\frac{\partial f}{\partial z} = 1$. Having taken one partial derivative, there is no reason to stop doing it. Thus, one could take the partial derivative with respect to y of the partial derivative with respect to x, denoted by $\frac{\partial^2 f}{\partial y \partial x}$ or f_{xy}. In the above example,

$$\frac{\partial^2 f}{\partial y \partial x} = f_{xy} = \cos x + 2x.$$

Also observe that

$$\frac{\partial^2 f}{\partial x \partial y} = f_{yx} = \cos x + 2x.$$

Higher order partial derivatives are defined by analogy to the above. Thus in the above example,

$$f_{yxx} = -\sin x + 2.$$

These partial derivatives, f_{xy} are called mixed partial derivatives.

There is an interesting relationship between the directional derivatives and the partial derivatives, provided the partial derivatives exist and are continuous.

Definition 7.5. Suppose $f : U \subseteq \mathbb{R}^n \to \mathbb{R}$ where U is an open set and the partial derivatives of f all exist and are continuous on U. Under these conditions, define the **gradient** of f denoted $\nabla f(\mathbf{x})$ to be the vector

$$\nabla f(\mathbf{x}) = (f_{x_1}(\mathbf{x}), f_{x_2}(\mathbf{x}), \cdots, f_{x_n}(\mathbf{x}))^T.$$

Proposition 7.2. In the situation of Definition 7.5 and for $\mathbf{v}$ a unit vector

$$D_{\mathbf{v}}f(\mathbf{x}) = \nabla f(\mathbf{x}) \cdot \mathbf{v}.$$

This proposition will be proved in a more general setting later. For now, you can use it to compute directional derivatives.

Example 7.6. Find the directional derivative of the function $f(x,y) = \sin(2x^2 + y^3)$ at $(1,1)$ in the direction $\left(\frac{1}{\sqrt{2}}, \frac{1}{\sqrt{2}}\right)^T$.

First find the gradient.

$$\nabla f(x,y) = \left(4x \cos\left(2x^2 + y^3\right), 3y^2 \cos\left(2x^2 + y^3\right)\right)^T.$$

Therefore,

$$\nabla f(1,1) = \left(4\cos(3), 3\cos(3)\right)^T$$

The directional derivative is therefore,

$$\left(4\cos(3), 3\cos(3)\right)^T \cdot \left(\frac{1}{\sqrt{2}}, \frac{1}{\sqrt{2}}\right)^T = \frac{7}{2}(\cos 3)\sqrt{2}.$$

Another important observation is that the gradient gives the direction in which the function changes most rapidly. The following proposition will be proved later.

Proposition 7.3. In the situation of Definition 7.5, suppose $\nabla f(\mathbf{x}) \neq \mathbf{0}$. Then the direction in which f increases most rapidly, that is the direction in which the directional derivative is largest, is the direction of the gradient. Thus $\mathbf{v} = \nabla f(\mathbf{x})/|\nabla f(\mathbf{x})|$ is the unit vector which maximizes $D_{\mathbf{v}}f(\mathbf{x})$ and this maximum value is $|\nabla f(\mathbf{x})|$. Similarly, $\mathbf{v} = -\nabla f(\mathbf{x})/|\nabla f(\mathbf{x})|$ is the unit vector which minimizes $D_{\mathbf{v}}f(\mathbf{x})$ and this minimum value is $-|\nabla f(\mathbf{x})|$.

The concept of a **directional derivative for a vector valued function** is also easy to define although the geometric significance expressed in pictures is not.

Definition 7.6. Let $\mathbf{f} : U \to \mathbb{R}^p$ where U is an open set in $\mathbb{R}^n$ and let $\mathbf{v}$ be a unit vector. For $\mathbf{x} \in U$, define the directional derivative of $\mathbf{f}$ in the direction $\mathbf{v}$, at the point $\mathbf{x}$ as

$$D_{\mathbf{v}}\mathbf{f}(\mathbf{x}) \equiv \lim_{t \to 0} \frac{\mathbf{f}(\mathbf{x}+t\mathbf{v}) - \mathbf{f}(\mathbf{x})}{t}.$$

Example 7.7. Let $\mathbf{f}(x,y) = \left(xy^2, yx\right)^T$. Find the directional derivative in the direction $(1,2)^T$ at the point (x,y).

First, a unit vector in this direction is $\left(1/\sqrt{5}, 2/\sqrt{5}\right)^T$ and from the definition, the desired limit is

$$\lim_{t \to 0} \frac{\left(\left(x + t\left(1/\sqrt{5}\right)\right)\left(y + t\left(2/\sqrt{5}\right)\right)^2 - xy^2, \left(x + t\left(1/\sqrt{5}\right)\right)\left(y + t\left(2/\sqrt{5}\right)\right) - xy\right)}{t}$$

$$= \lim_{t \to 0} \left(\frac{4}{5}xy\sqrt{5} + \frac{4}{5}xt + \frac{1}{5}\sqrt{5}y^2 + \frac{4}{5}ty + \frac{4}{25}t^2\sqrt{5}, \frac{2}{5}x\sqrt{5} + \frac{1}{5}y\sqrt{5} + \frac{2}{5}t \right)$$

$$= \left(\frac{4}{5}xy\sqrt{5} + \frac{1}{5}\sqrt{5}y^2, \frac{2}{5}x\sqrt{5} + \frac{1}{5}y\sqrt{5} \right).$$

You see from this example and the above definition that all you have to do is to form the vector which is obtained by replacing each component of the vector with its directional derivative. In particular, you can take partial derivatives of vector valued functions and use the same notation.

Example 7.8. Find the partial derivative with respect to x of the function $\mathbf{f}(x, y, z, w) = \left(xy^2, z\sin(xy), z^3x \right)^T$.

From the above definition, $\mathbf{f}_x(x, y, z) = D_1\mathbf{f}(x, y, z) = \left(y^2, zy\cos(xy), z^3 \right)^T$.

7.5　Exercises

(1) Find the directional derivative of $f(x, y, z) = x^2y + z^4$ in the direction of the vector $(1, 3, -1)$ when $(x, y, z) = (1, 1, 1)$.

(2) Find the directional derivative of $f(x, y, z) = \sin(x + y^2) + z$ in the direction of the vector $(1, 2, -1)$ when $(x, y, z) = (1, 1, 1)$.

(3) Find the directional derivative of $f(x, y, z) = \ln(x + y^2) + z^2$ in the direction of the vector $(1, 1, -1)$ when $(x, y, z) = (1, 1, 1)$.

(4) Using the conclusion of Proposition 7.2, prove Proposition 7.3 from the geometric description of the dot product, the one which says the dot product is the product of the lengths of the vectors and the cosine of the included angle which is no larger than π.

(5) Find the largest value of the directional derivative of $f(x, y, z) = \ln(x + y^2) + z^2$ at the point $(1, 1, 1)$.

(6) Find the smallest value of the directional derivative of $f(x, y, z) = x\sin(4xy^2) + z^2$ at the point $(1, 1, 1)$.

(7) An ant falls to the top of a stove having temperature $T(x, y) = x^2\sin(x + y)$ at the point $(2, 3)$. In what direction should the ant go to minimize the temperature? In what direction should he go to maximize the temperature?

(8) Find the partial derivative with respect to y of the function $\mathbf{f}(x, y, z, w) = \left(y^2, z^2\sin(xy), z^3x \right)^T$.

(9) Find the partial derivative with respect to x of the function $\mathbf{f}(x, y, z, w) = \left(wx, zx\sin(xy), z^3x \right)^T$.

(10) Find $\frac{\partial f}{\partial x}, \frac{\partial f}{\partial y}$, and $\frac{\partial f}{\partial z}$ for $f =$

　(a) $x^2y^2z + w$

　(b) $e^2 + xy + z^2$

　(c) $\sin(z^2) + \cos(xy)$

(d) $\ln\left(x^2 + y^2 + 1\right) + e^z$

(e) $\sin\left(xyz\right) + \cos\left(xy\right)$

(11) Find $\frac{\partial f}{\partial x}$, $\frac{\partial f}{\partial y}$, and $\frac{\partial f}{\partial z}$ for $f =$

(a) $x^2 y + \cos\left(xy\right) + z^3 y$

(b) $e^{x^2 + y^2} z \sin\left(x + y\right)$

(c) $z^2 \sin^3\left(e^{x^2 + y^3}\right)$

(d) $x^2 \cos\left(\sin\left(\tan\left(z^2 + y^2\right)\right)\right)$

(e) $x^{y^2 + z}$

(12) Suppose

$$f\left(x, y\right) = \begin{cases} \frac{2xy + 6x^3 + 12xy^2 + 18yx^2 + 36y^3 + \sin\left(x^3\right) + \tan\left(3y^3\right)}{3x^2 + 6y^2} & \text{if } \left(x, y\right) \neq (0, 0) \\ 0 \text{ if } \left(x, y\right) = (0, 0). \end{cases}$$

Find $\frac{\partial f}{\partial x}\left(0, 0\right)$ and $\frac{\partial f}{\partial y}\left(0, 0\right)$.

(13) Why must the vector in the definition of the directional derivative be a unit vector? **Hint:** Suppose not. Would the directional derivative be a correct manifestation of steepness?

7.6 Mixed Partial Derivatives

Under certain conditions the **mixed partial derivatives** will always be equal. This astonishing fact may have been known to Euler in 1734.

Theorem 7.3. *Suppose $f : U \subseteq \mathbb{R}^2 \to \mathbb{R}$ where U is an open set on which f_x, f_y, f_{xy} and f_{yx} exist. Then if f_{xy} and f_{yx} are continuous at the point $(x, y) \in U$, it follows*

$$f_{xy}\left(x, y\right) = f_{yx}\left(x, y\right).$$

Proof: Since U is open, there exists $r > 0$ such that $B\left(\left(x, y\right), r\right) \subseteq U$. Now let $|t|, |s| < r/2$ and consider

$$\Delta\left(s, t\right) \equiv \frac{1}{st}\{\overbrace{f\left(x + t, y + s\right) - f\left(x + t, y\right)}^{h(t)} - \overbrace{\left(f\left(x, y + s\right) - f\left(x, y\right)\right)}^{h(0)}\}. \quad (7.6)$$

Note that $(x + t, y + s) \in U$ because

$$\left|(x + t, y + s) - (x, y)\right| = \left|(t, s)\right| = \left(t^2 + s^2\right)^{1/2}$$

$$\leq \left(\frac{r^2}{4} + \frac{r^2}{4}\right)^{1/2} = \frac{r}{\sqrt{2}} < r.$$

As implied above, $h\left(t\right) \equiv f\left(x + t, y + s\right) - f\left(x + t, y\right)$. Therefore, by the mean value theorem from calculus and the (one variable) chain rule,

$$\Delta\left(s, t\right) = \frac{1}{st}\left(h\left(t\right) - h\left(0\right)\right) = \frac{1}{st} h'\left(\alpha t\right) t$$

$$= \frac{1}{s}\left(f_x\left(x + \alpha t, y + s\right) - f_x\left(x + \alpha t, y\right)\right)$$

for some $\alpha \in (0,1)$. Applying the mean value theorem again,

$$\Delta(s,t) = f_{xy}(x + \alpha t, y + \beta s)$$

where $\alpha, \beta \in (0,1)$.

If the terms $f(x+t,y)$ and $f(x,y+s)$ are interchanged in (7.6), $\Delta(s,t)$ is also unchanged and the above argument shows there exist $\gamma, \delta \in (0,1)$ such that

$$\Delta(s,t) = f_{yx}(x + \gamma t, y + \delta s).$$

Letting $(s,t) \to (0,0)$ and using the continuity of f_{xy} and f_{yx} at (x,y),

$$\lim_{(s,t)\to(0,0)} \Delta(s,t) = f_{xy}(x,y) = f_{yx}(x,y). \qquad \blacksquare$$

The following is obtained from the above by simply fixing all the variables except for the two of interest.

Corollary 7.1. *Suppose U is an open subset of $\mathbb{R}^n$ and $f : U \to \mathbb{R}$ has the property that for two indices k, l, f_{x_k}, f_{x_l}, $f_{x_l x_k}$, and $f_{x_k x_l}$ exist on U and $f_{x_k x_l}$ and $f_{x_l x_k}$ are both continuous at $\mathbf{x} \in U$. Then $f_{x_k x_l}(\mathbf{x}) = f_{x_l x_k}(\mathbf{x})$.*

It is necessary to assume the mixed partial derivatives are continuous in order to assert they are equal. The following is a well known example [3].

Example 7.9. Let

$$f(x,y) = \begin{cases} \frac{xy(x^2 - y^2)}{x^2 + y^2} & \text{if } (x,y) \neq (0,0) \\ 0 \text{ if } (x,y) = (0,0) \end{cases}$$

Here is a picture of the graph of this function. It looks innocuous but isn't.

From the definition of partial derivatives it follows immediately that $f_x(0,0) = f_y(0,0) = 0$. Using the standard rules of differentiation, for $(x,y) \neq (0,0)$,

$$f_x = y\frac{x^4 - y^4 + 4x^2 y^2}{(x^2 + y^2)^2}, \quad f_y = x\frac{x^4 - y^4 - 4x^2 y^2}{(x^2 + y^2)^2}$$

Now

$$f_{xy}(0,0) \equiv \lim_{y\to 0} \frac{f_x(0,y) - f_x(0,0)}{y} = \lim_{y\to 0} \frac{-y^4}{(y^2)^2} = -1$$

while

$$f_{yx}(0,0) \equiv \lim_{x\to 0} \frac{f_y(x,0) - f_y(0,0)}{x} = \lim_{x\to 0} \frac{x^4}{(x^2)^2} = 1$$

showing that, although the mixed partial derivatives do exist at $(0,0)$, they are not equal there.

7.7 Partial Differential Equations

Partial differential equations are equations which involve the partial derivatives of some function. The most famous partial differential equations involve the **Laplacian**, named after Laplace[3].

Definition 7.7. Let u be a function of n variables. Then $\Delta u \equiv \sum_{k=1}^{n} u_{x_k x_k}$. This is also written as $\nabla^2 u$. The symbol Δ or ∇^2 is called the Laplacian. When $\Delta u = 0$ the function u is called **harmonic.Laplace's equation** is $\Delta u = 0$. The **heat equation** is $u_t - \Delta u = 0$ and the **wave equation** is $u_{tt} - \Delta u = 0$.

Example 7.10. Find the Laplacian of $u(x, y) = x^2 - y^2$.

$u_{xx} = 2$ while $u_{yy} = -2$. Therefore, $\Delta u = u_{xx} + u_{yy} = 2 - 2 = 0$. Thus this function is harmonic, $\Delta u = 0$.

Example 7.11. Find $u_t - \Delta u$ where $u(t, x, y) = e^{-t} \cos x$.

In this case, $u_t = -e^{-t} \cos x$ while $u_{yy} = 0$ and $u_{xx} = -e^{-t} \cos x$ therefore, $u_t - \Delta u = 0$ and so u solves the heat equation $u_t - \Delta u = 0$.

Example 7.12. Let $u(t, x) = \sin t \cos x$. Find $u_{tt} - \Delta u$.

In this case, $u_{tt} = -\sin t \cos x$ while $\Delta u = -\sin t \cos x$. Therefore, u is a solution of the wave equation $u_{tt} - \Delta u = 0$.

7.8 Exercises

(1) Find $f_x, f_y, f_z, f_{xy}, f_{yx}, f_{xz}, f_{zx}, f_{zy}, f_{yz}$ for the following. Verify the mixed partial derivatives are equal.

    (a) $x^2 y^3 z^4 + \sin(xyz)$
    (b) $\sin(xyz) + x^2 yz$
    (c) $z \ln |x^2 + y^2 + 1|$
    (d) $e^{x^2 + y^2 + z^2}$
    (e) $\tan(xyz)$

(2) Suppose f is a continuous function and $f : U \to \mathbb{R}$ where U is an open set and suppose that $\mathbf{x} \in U$ has the property that for all $\mathbf{y}$ near $\mathbf{x}$, $f(\mathbf{x}) \le f(\mathbf{y})$. Prove that if f has all of its partial derivatives at $\mathbf{x}$, then $f_{x_i}(\mathbf{x}) = 0$ for each x_i. **Hint:** This is just a repeat of the usual one variable theorem seen in beginning calculus. You just do this one variable argument for each variable to get the conclusion.

---

[3]Laplace was a great physicist and mathematician of the 1700's. He made fundamental contributions to mechanics and astronomy.

(3) As an important application of Problem 2 consider the following. Experiments are done at n times, $t_1, t_2, \cdots, t_n$ and at each time there results a collection of numerical outcomes. Denote by $\{(t_i, x_i)\}_{i=1}^p$ the set of all such pairs and try to find numbers a and b such that the line $x = at + b$ approximates these ordered pairs as well as possible in the sense that out of all choices of a and b, $\sum_{i=1}^p (at_i + b - x_i)^2$ is as small as possible. In other words, you want to minimize the function of two variables $f(a, b) \equiv \sum_{i=1}^p (at_i + b - x_i)^2$. Find a formula for a and b in terms of the given ordered pairs. You will be finding the formula for the least squares regression line.

(4) Show that if $v(x, y) = u(\alpha x, \beta y)$, then $v_x = \alpha u_x$ and $v_y = \beta u_y$. State and prove a generalization to any number of variables.

(5) Let f be a function which has continuous derivatives. Show that $u(t, x) = f(x - ct)$ solves the wave equation $u_{tt} - c^2 \Delta u = 0$. What about $u(x, t) = f(x + ct)$?

(6) D'Alembert found a formula for the solution to the wave equation $u_{tt} = c^2 u_{xx}$ along with the initial conditions $u(x, 0) = f(x)$, $u_t(x, 0) = g(x)$. Here is how he did it. He looked for a solution of the form $u(x, t) = h(x + ct) + k(x - ct)$ and then found h and k in terms of the given functions f and g. He ended up with something like

$$u(x, t) = \frac{1}{2c} \int_{x-ct}^{x+ct} g(r)\, dr + \frac{1}{2} \left(f(x + ct) + f(x - ct) \right).$$

Fill in the details.

(7) Determine which of the following functions satisfy Laplace's equation.

   (a) $x^3 - 3xy^2$
   (b) $3x^2y - y^3$
   (c) $x^3 - 3xy^2 + 2x^2 - 2y^2$
   (d) $3x^2y - y^3 + 4xy$
   (e) $3x^2 - y^3 + 4xy$
   (f) $3x^2y - y^3 + 4y$
   (g) $x^3 - 3x^2y^2 + 2x^2 - 2y^2$

(8) Show that $z = \sqrt{x^2 + y^2}$ is a solution to $x\frac{\partial z}{\partial x} + y\frac{\partial z}{\partial y} = z$.

(9) Show that if $\Delta u = \lambda u$ where u is a function of only x, then $e^{\lambda t}u$ solves the heat equation $u_t - \Delta u = 0$.

(10) Show that if a, b are scalars and u, v are functions which satisfy Laplace's equation then $au + bv$ also satisfies Laplace's equation. Verify a similar statement for the heat and wave equations.

(11) Show that $u(x, t) = \frac{1}{\sqrt{t}} e^{-x^2/4c^2 t}$ solves the heat equation $u_t = c^2 u_{xx}$.

Chapter 8

The Derivative Of A Function Of Many Variables

8.1 The Derivative Of Functions Of One Variable, $o(v)$

First consider the notion of the derivative of a function of one variable.

Observation 8.1. Suppose a function f of one variable has a derivative at x. Then

$$\lim_{h \to 0} \frac{|f(x+h) - f(x) - f'(x)h|}{|h|} = 0.$$

This observation follows from the definition of the derivative of a function of one variable, namely

$$f'(x) \equiv \lim_{h \to 0} \frac{f(x+h) - f(x)}{h}.$$

Thus

$$\lim_{h \to 0} \frac{|f(x+h) - f(x) - f'(x)h|}{|h|} = \lim_{h \to 0} \left| \frac{f(x+h) - f(x)}{h} - f'(x) \right| = 0$$

Definition 8.1. A vector valued function of a vector $\mathbf{v}$ is called $\mathbf{o}(\mathbf{v})$ (referred to as "little o of $\mathbf{v}$") if

$$\lim_{|\mathbf{v}| \to 0} \frac{\mathbf{o}(\mathbf{v})}{|\mathbf{v}|} = \mathbf{0}. \tag{8.1}$$

Thus for a function of one variable, the function $f(x+h) - f(x) - f'(x)h$ is $o(h)$. When we say a function is $o(h)$, it is used like an adjective. It is like saying the function is white or black or green or fat or thin. The term is used very imprecisely. Thus in general,

$$\mathbf{o}(\mathbf{v}) = \mathbf{o}(\mathbf{v}) + \mathbf{o}(\mathbf{v}), \, \mathbf{o}(\mathbf{v}) = 45 \times \mathbf{o}(\mathbf{v}), \, \mathbf{o}(\mathbf{v}) = \mathbf{o}(\mathbf{v}) - \mathbf{o}(\mathbf{v}), \text{etc.}$$

When you add two functions with the property of the above definition, you get another one having that same property. When you multiply by 45, the property is also retained, as it is when you subtract two such functions. How could something so sloppy be useful? The notation is useful precisely because it prevents you from obsessing over things which are not relevant and should be ignored.

Theorem 8.1. *Let $f : (a, b) \to \mathbb{R}$ be a function of one variable. Then $f'(x)$ exists if and only if there exists p such that*

$$f(x+h) - f(x) = ph + o(h) \tag{8.2}$$

In this case, $p = f'(x)$.

Proof: From the above observation it follows that if $f'(x)$ does exist, then (8.2) holds.

Suppose then that (8.2) is true. Then

$$\frac{f(x+h) - f(x)}{h} - p = \frac{o(h)}{h}.$$

Taking a limit, you see that

$$p = \lim_{h \to 0} \frac{f(x+h) - f(x)}{h}$$

and that in fact this limit exists which shows that $p = f'(x)$. ■

This theorem shows that one way to define $f'(x)$ is as the number p, if there is one, which has the property that

$$f(x+h) = f(x) + ph + o(h).$$

You should think of p as the linear transformation resulting from multiplication by the 1×1 matrix (p).

Example 8.1. Let $f(x) = x^3$. Find $f'(x)$.

$$
\begin{aligned}
f(x+h) &= (x+h)^3 = x^3 + 3x^2h + 3xh^2 + h^3 \\
&= f(x) + 3x^2h + (3xh + h^2)h.
\end{aligned}
$$

Since $(3xh + h^2)h = o(h)$, it follows $f'(x) = 3x^2$.

Example 8.2. Let $f(x) = \sin(x)$. Find $f'(x)$.

$$
\begin{aligned}
f(x+h) - f(x) &= \sin(x+h) - \sin(x) = \sin(x)\cos(h) + \cos(x)\sin(h) - \sin(x) \\
&= \cos(x)\sin(h) + \sin(x)\frac{(\cos(h) - 1)}{h}h \\
&= \cos(x)h + \cos(x)\frac{(\sin(h) - h)}{h}h + \sin x \frac{(\cos(h) - 1)}{h}h.
\end{aligned}
$$

Now

$$\cos(x)\frac{(\sin(h) - h)}{h}h + \sin x\frac{(\cos(h) - 1)}{h}h = o(h). \tag{8.3}$$

Remember the fundamental limits which allowed you to find the derivative of $\sin(x)$ were

$$\lim_{h \to 0} \frac{\sin(h)}{h} = 1, \quad \lim_{h \to 0} \frac{\cos(h) - 1}{h} = 0. \tag{8.4}$$

These same limits are what is needed to verify (8.3).

How can you tell whether a function of two variables (u, v) is $\mathbf{o}\begin{pmatrix} u \\ v \end{pmatrix}$? In general, there is no substitute for the definition, but you can often identify this property by observing that the expression involves only "higher order terms". These are terms like $u^2 v, uv, v^4$, etc. If you sum the exponents on the u and the v you get something larger than 1. For example,

$$\left| \frac{vu}{\sqrt{u^2 + v^2}} \right| \le \frac{1}{2} \left(u^2 + v^2 \right) \frac{1}{\sqrt{u^2 + v^2}} = \frac{1}{2} \sqrt{u^2 + v^2}$$

and this converges to 0 as $(u, v) \to (0, 0)$. This follows from the inequality $|uv| \le \frac{1}{2} \left(u^2 + v^2 \right)$ which you can verify from $(u - v)^2 \ge 0$. Similar considerations apply in higher dimensions also. In general, this is a hard question because it involves a limit of a function of many variables. Furthermore, there is really no substitute for answering this question, because its resolution involves the definition of whether a function is differentiable. That may be why we spend most of our time on one dimensional considerations which involve taking the partial derivatives. The following exercises should help give you an idea of how to determine whether something is o.

8.2   Exercises

(1) Determine which of the following functions are $o(h)$.

(a) h^2
(b) $h \sin(h)$
(c) $|h|^{3/2} \ln(|h|)$
(d) $h^2 x + y h^3$
(e) $\sin(h^2)$
(f) $\sin(h)$
(g) $xh \sin\left(\sqrt{|h|}\right) + x^5 h^2$
(h) $\exp\left(-1/|h|^2\right)$

(2) Here are some scalar valued functions of several variables. Determine which of these functions are $o(\mathbf{v})$. Here $\mathbf{v}$ is a vector in $\mathbb{R}^n$, $\mathbf{v} = (v_1, \cdots, v_n)$.

(a) $v_1 v_2$
(b) $v_2 \sin(v_1)$
(c) $v_1^2 + v_2$

    (d) $v_2 \sin (v_1 + v_2)$

    (e) $v_1 (v_1 + v_2 + xv_3)$

    (f) $(e^{v_1} - 1 - v_1)$

    (g) $(\mathbf{x} \cdot \mathbf{v}) |\mathbf{v}|$

(3) Here are some vector valued functions of $\mathbf{v} \in \mathbb{R}^n$. Determine which ones are $o(\mathbf{v})$.

    (a) $(\mathbf{x} \cdot \mathbf{v}) \mathbf{v}$

    (b) $\sin (v_1) \mathbf{v}$

    (c) $\sqrt{|(\mathbf{x} \cdot \mathbf{v})|} |\mathbf{v}|^{2/3}$

    (d) $\sqrt{|(\mathbf{x} \cdot \mathbf{v})|} |\mathbf{v}|^{1/2}$

    (e) $\left(\sin \left(\sqrt{|\mathbf{x} \cdot \mathbf{v}|} \right) - \sqrt{|\mathbf{x} \cdot \mathbf{v}|} \right) \cdot$
        $|\mathbf{v}|^{-1/4}$

    (f) $\exp \left(-1/|\mathbf{v}|^2 \right)$

    (g) $\mathbf{v}^T A \mathbf{v}$ where A is an $n \times n$ matrix.

(4) Show that if $f(x) = o(x)$, then $f'(0) = 0$.

(5) Show that if $\lim_{h \to 0} f(x) = 0$ then $xf(x) = o(x)$.

(6) Show that if $f'(0)$ exists and $f(0) = 0$, then $f(|x|^p) = o(x)$ whenever $p > 1$.

8.3   The Derivative Of Functions Of Many Variables

This way of thinking about the derivative is exactly what is needed to define the derivative of a function of n variables. Recall the following definition.

Definition 8.2. A function T which maps $\mathbb{R}^n$ to $\mathbb{R}^p$ is called a linear transformation if for every pair of scalars, a, b and vectors $\mathbf{x}, \mathbf{y} \in \mathbb{R}^n$, it follows that $T(a\mathbf{x} + b\mathbf{y}) = aT(\mathbf{x}) + bT(\mathbf{y})$.

Recall that from the properties of matrix multiplication, if A is an $p \times n$ matrix, and if $\mathbf{x}, \mathbf{y}$ are vectors in $\mathbb{R}^n$, then $A(a\mathbf{x} + b\mathbf{y}) = aA(\mathbf{x}) + bA(\mathbf{y})$. Thus you can define a linear transformation by multiplying by a matrix. Of course the simplest example is that of a 1×1 matrix or number. You can think of the number 3 as a linear transformation T mapping $\mathbb{R}$ to $\mathbb{R}$ according to the rule $Tx = 3x$. It satisfies the properties needed for a linear transformation because $3(ax + by) = a3x + b3y = aTx + bTy$. The case of the derivative of a scalar valued function of one variable is of this sort. You get a number for the derivative. However, you can think of this number as a linear transformation. Of course it might not be worth the fuss to think of it this way for a function of one variable but this is the way you must think of it for a function of n variables.

Definition 8.3. Let $\mathbf{f} : U \to \mathbb{R}^p$ where U is an open set in $\mathbb{R}^n$ for $n, p \geq 1$ and let $\mathbf{x} \in U$ be given. Then $\mathbf{f}$ is defined to be **differentiable** at $\mathbf{x} \in U$ if and only if there exist column vectors $\mathbf{v}_i$ such that for $\mathbf{h} = (h_1 \cdots , h_n)^T$,

$$\mathbf{f}(\mathbf{x} + \mathbf{h}) = \mathbf{f}(\mathbf{x}) + \sum_{i=1}^{n} \mathbf{v}_i h_i + \mathbf{o}(\mathbf{h}). \tag{8.5}$$

The derivative of the function $\mathbf{f}$, denoted by $D\mathbf{f}(\mathbf{x})$, is the linear transformation defined by multiplying by the matrix whose columns are the $p \times 1$ vectors $\mathbf{v}_i$. Thus if $\mathbf{h}$ is a vector in $\mathbb{R}^n$,

$$D\mathbf{f}(\mathbf{x})\,\mathbf{h} \equiv \left(\begin{array}{ccc} | & & | \\ \mathbf{v}_1 & \cdots & \mathbf{v}_n \\ | & & | \end{array} \right) \mathbf{h}$$

and one can write $\mathbf{f}$ is differentiable if and only if there is a matrix $D\mathbf{f}(\mathbf{x})$ such that
$$\mathbf{f}(\mathbf{x} + \mathbf{h}) = \mathbf{f}(\mathbf{x}) + D\mathbf{f}(\mathbf{x})\,\mathbf{h} + \mathbf{o}(\mathbf{h}).$$
If $\mathbf{h} = \mathbf{x} - \mathbf{x}_0$, this takes the form
$$\mathbf{f}(\mathbf{x}) = \mathbf{f}(\mathbf{x}_0) + D\mathbf{f}(\mathbf{x}_0)(\mathbf{x} - \mathbf{x}_0) + \mathbf{o}(\mathbf{x} - \mathbf{x}_0)$$

If you deleted the $\mathbf{o}(\mathbf{x} - \mathbf{x}_0)$ term and considered the function of $\mathbf{x}$ given by what is left, this is called the linear approximation to the function at the point $\mathbf{x}_0$. In the case where $\mathbf{x} \in \mathbb{R}^2$ and f has values in $\mathbb{R}$ one can draw a picture to illustrate this.

The linear approximation is graphed as that plane which is tangent to the graph of the function. This is what is going on geometrically.

It is common to think of this matrix as the derivative but strictly speaking, this is incorrect. The derivative is a "linear transformation" determined by multiplication by this matrix, called the **standard matrix** because it is based on the standard basis vectors for $\mathbb{R}^n$. The subtle issues involved in a thorough exploration of this issue will be avoided for now. It will be fine to think of the above matrix as the derivative. Other notations which are often used for this matrix or the linear transformation are $\mathbf{f}'(\mathbf{x})$, $J(\mathbf{x})$, and even $\frac{\partial \mathbf{f}}{\partial \mathbf{x}}$ or $\frac{d\mathbf{f}}{d\mathbf{x}}$.

The next theorem gives a description of this matrix and shows it is unique if it exists.

Theorem 8.2. *Suppose $\mathbf{f}$ is differentiable, as given above in (8.5). Then*

$$\mathbf{v}_k = \lim_{h \to 0} \frac{\mathbf{f}(\mathbf{x} + h\mathbf{e}_k) - \mathbf{f}(\mathbf{x})}{h} \equiv \frac{\partial \mathbf{f}}{\partial x_k}(\mathbf{x}),$$

the kth partial derivative.

Proof: Let $\mathbf{h} = (0, \cdots, h, 0, \cdots, 0)^T = h\mathbf{e}_k$ where the h is in the kth slot. Then (8.5) reduces to

$$\mathbf{f}(\mathbf{x} + \mathbf{h}) = \mathbf{f}(\mathbf{x}) + \mathbf{v}_k h + \mathbf{o}(h).$$

Therefore, dividing by h

$$\frac{\mathbf{f}(\mathbf{x} + h\mathbf{e}_k) - \mathbf{f}(\mathbf{x})}{h} = \mathbf{v}_k + \frac{\mathbf{o}(h)}{h}$$

and taking the limit,

$$\lim_{h \to 0} \frac{\mathbf{f}(\mathbf{x} + h\mathbf{e}_k) - \mathbf{f}(\mathbf{x})}{h} = \lim_{h \to 0} \left(\mathbf{v}_k + \frac{\mathbf{o}(h)}{h} \right) = \mathbf{v}_k$$

and so, the above limit exists. ∎

The above description of the derivative can be used to approximate a nonlinear function. Since the $\mathbf{o}(\mathbf{x} - \mathbf{x}_0)$ term is so small, it is a good approximation to write

$$\mathbf{f}(\mathbf{x}) \approx \mathbf{f}(\mathbf{x}_0) + D\mathbf{f}(\mathbf{x}_0)(\mathbf{x} - \mathbf{x}_0)$$

$$= \mathbf{f}(\mathbf{x}_0) + \left(\frac{\partial \mathbf{f}}{\partial x_1}(\mathbf{x}_0) \quad \frac{\partial \mathbf{f}}{\partial x_2}(\mathbf{x}_0) \quad \cdots \quad \frac{\partial \mathbf{f}}{\partial x_n}(\mathbf{x}_0) \right) \begin{pmatrix} h_1 \\ h_2 \\ \vdots \\ h_n \end{pmatrix}$$

$$= \mathbf{f}(\mathbf{x}_0) + \sum_{i=1}^{n} \frac{\partial \mathbf{f}}{\partial x_i}(\mathbf{x}_0) h_i$$

Here is an example of a scalar valued nonlinear function.

Example 8.3. Suppose $f(x, y) = \sqrt{xy}$. Find the approximate change in f if x goes from 1 to 1.01 and y goes from 4 to 3.99.

We can do this by noting that

$$f(1.01, 3.99) - f(1, 4) \approx f_x(1, 2)(.01) + f_y(1, 2)(-.01)$$

$$= 1(.01) + \frac{1}{4}(-.01) = 7.5 \times 10^{-3}.$$

Of course the exact value is

$$\sqrt{(1.01)(3.99)} - \sqrt{4} = 7.461\,083\,1 \times 10^{-3}.$$

Notation 8.1. When f is a scalar valued function of n variables, the following is often written to express the idea that a small change in f due to small changes in the variables can be expressed in the form

$$df(\mathbf{x}) = f_{x_1}(\mathbf{x})\, dx_1 + \cdots + f_{x_n}(\mathbf{x})\, dx_n$$

where the small change in x_i is denoted as dx_i. As explained above, df is the approximate change in the function f. Sometimes df is referred to as the differential of f.

Let $\mathbf{f} : U \to \mathbb{R}^q$ where U is an open subset of $\mathbb{R}^p$ and $\mathbf{f}$ is differentiable. It was just shown that

$$\mathbf{f}(\mathbf{x} + \mathbf{v}) = \mathbf{f}(\mathbf{x}) + \sum_{j=1}^{p} \frac{\partial \mathbf{f}(\mathbf{x})}{\partial x_j} v_j + o(\mathbf{v}).$$

Taking the ith coordinate of the above equation yields

$$f_i(\mathbf{x} + \mathbf{v}) = f_i(\mathbf{x}) + \sum_{j=1}^{p} \frac{\partial f_i(\mathbf{x})}{\partial x_j} v_j + o(\mathbf{v}),$$

and it follows that the term with a sum is nothing more than the ith component of $J(\mathbf{x})\mathbf{v}$ where $J(\mathbf{x})$ is the $q \times p$ matrix

$$\begin{pmatrix} \frac{\partial f_1}{\partial x_1} & \frac{\partial f_1}{\partial x_2} & \cdots & \frac{\partial f_1}{\partial x_p} \\ \frac{\partial f_2}{\partial x_1} & \frac{\partial f_2}{\partial x_2} & \cdots & \frac{\partial f_2}{\partial x_p} \\ \vdots & \vdots & \ddots & \vdots \\ \frac{\partial f_q}{\partial x_1} & \frac{\partial f_q}{\partial x_2} & \cdots & \frac{\partial f_q}{\partial x_p} \end{pmatrix}.$$

Thus

$$\mathbf{f}(\mathbf{x} + \mathbf{v}) = \mathbf{f}(\mathbf{x}) + J(\mathbf{x})\mathbf{v} + o(\mathbf{v}), \qquad (8.6)$$

and to reiterate, the linear transformation which results by multiplication by this $q \times p$ matrix is known as the derivative.

Sometimes x, y, z is written instead of $x_1, x_2,$ and x_3. This is to save on notation and is easier to write and to look at although it lacks generality. When this is done it is understood that $x = x_1, y = x_2,$ and $z = x_3$. Thus the derivative is the linear transformation determined by

$$\begin{pmatrix} f_{1x} & f_{1y} & f_{1z} \\ f_{2x} & f_{2y} & f_{2z} \\ f_{3x} & f_{3y} & f_{3z} \end{pmatrix}.$$

Example 8.4. Let A be a constant $m \times n$ matrix and consider $\mathbf{f}(\mathbf{x}) = A\mathbf{x}$. Find $D\mathbf{f}(\mathbf{x})$ if it exists.

$$\mathbf{f}(\mathbf{x} + \mathbf{h}) - \mathbf{f}(\mathbf{x}) = A(\mathbf{x} + \mathbf{h}) - A(\mathbf{x}) = A\mathbf{h} = A\mathbf{h} + o(\mathbf{h}).$$

In fact in this case, $o(\mathbf{h}) = \mathbf{0}$. Therefore, $D\mathbf{f}(\mathbf{x}) = A$. Note that this looks the same as the case in one variable, $f(x) = ax$.

Example 8.5. Let $f(x, y, z) = xy + z^2 x$. Find $Df(x, y, z)$.

Consider $f(x+h, y+k, z+l) - f(x, y, z)$. This is something which is easily computed from the definition of the function. It equals

$$(x+h)(y+k) + (z+l)^2(x+h) - (xy + z^2x)$$

Multiply everything together and collect the terms. This yields

$$(z^2 + y)h + xk + 2zxl + (hk + +2zlh + l^2x + l^2h)$$

It follows easily the last term at the end is $o(h, k, l)$ and so the derivative of this function is the linear transformation coming from multiplication by the matrix $((z^2 + y), x, 2zx)$ and so this is the derivative. It follows from this and the description of the derivative in terms of partial derivatives that

$$\frac{\partial f}{\partial x}(x, y, z) = z^2 + y, \quad \frac{\partial f}{\partial y}(x, y, z) = x, \quad \frac{\partial f}{\partial z}(x, y, z) = 2xz.$$

Of course you could compute these partial derivatives directly.

Given a function of many variables, how can you tell if it is differentiable? In other words, when you make the linear approximation, how can you tell easily that what is left over is $o(\mathbf{v})$. Sometimes you have to go directly to the definition and verify it is differentiable from the definition. For example, you may have seen the following important example in one variable calculus.

Example 8.6. Let $f(x) = \begin{cases} x^2 \sin\left(\frac{1}{x}\right) & \text{if } x \neq 0 \\ 0 & \text{if } x = 0 \end{cases}$. Find $Df(0)$.

$$f(h) - f(0) = 0h + h^2 \sin\left(\frac{1}{h}\right) = o(h),$$

and so $Df(0) = 0$. If you find the derivative for $x \neq 0$, it is totally useless information if what you want is $Df(0)$. This is because the derivative turns out to be discontinuous. Try it. Find the derivative for $x \neq 0$ and try to obtain $Df(0)$ from it. You see, in this example you had to revert to the definition to find the derivative.

It is not really too hard to use the definition even for more ordinary examples.

Example 8.7. Let $\mathbf{f}(x, y) = \begin{pmatrix} x^2y + y^2 \\ y^3x \end{pmatrix}$. Find $D\mathbf{f}(1, 2)$.

First of all, note that the thing you are after is a 2×2 matrix.

$$\mathbf{f}(1, 2) = \begin{pmatrix} 6 \\ 8 \end{pmatrix}.$$

Then

$$\mathbf{f}(1 + h_1, 2 + h_2) - \mathbf{f}(1, 2)$$

$$= \begin{pmatrix} (1 + h_1)^2(2 + h_2) + (2 + h_2)^2 \\ (2 + h_2)^3(1 + h_1) \end{pmatrix} - \begin{pmatrix} 6 \\ 8 \end{pmatrix}$$

$$= \left(\begin{array}{c} 5h_2 + 4h_1 + 2h_1h_2 + 2h_1^2 + h_1^2h_2 + h_2^2 \\ 8h_1 + 12h_2 + 12h_1h_2 + 6h_2^2 + 6h_2^2h_1 + h_2^3 + h_2^3h_1 \end{array} \right)$$

$$= \left(\begin{array}{cc} 4 & 5 \\ 8 & 12 \end{array} \right) \left(\begin{array}{c} h_1 \\ h_2 \end{array} \right) + \left(\begin{array}{c} 2h_1h_2 + 2h_1^2 + h_1^2h_2 + h_2^2 \\ 12h_1h_2 + 6h_2^2 + 6h_2^2h_1 + h_2^3 + h_2^3h_1 \end{array} \right)$$

$$= \left(\begin{array}{cc} 4 & 5 \\ 8 & 12 \end{array} \right) \left(\begin{array}{c} h_1 \\ h_2 \end{array} \right) + \mathbf{o}\left(\mathbf{h} \right).$$

Therefore, the standard matrix of the derivative is $\left(\begin{array}{cc} 4 & 5 \\ 8 & 12 \end{array} \right)$.

Example 8.8. Let $\mathbf{f}\left(x, y \right) = \left(\begin{array}{c} x^3 y + y^2 \\ xy^2 + 1 \end{array} \right)$. Find $D\mathbf{f}\left(x, y \right)$.

You know that if there is a derivative, its standard matrix is of the form

$$\left(\begin{array}{cc} f_{1x}\left(x, y \right) & f_{1y}\left(x, y \right) \\ f_{2x}\left(x, y \right) & f_{2y}\left(x, y \right) \end{array} \right) = \left(\begin{array}{cc} 3x^2 y & x^3 + 2y \\ y^2 & 2xy \end{array} \right)$$

Does it work? Is

$$\left(\begin{array}{c} \left(x + u \right)^3 \left(y + v \right) + \left(y + v \right)^2 \\ \left(x + u \right)\left(y + v \right)^2 + 1 \end{array} \right) - \left(\begin{array}{c} x^3 y + y^2 \\ xy^2 + 1 \end{array} \right)$$

$$- \left(\begin{array}{cc} 3x^2 y & x^3 + 2y \\ y^2 & 2xy \end{array} \right) \left(\begin{array}{c} u \\ v \end{array} \right) = \mathbf{o}\left(\begin{array}{c} u \\ v \end{array} \right)?$$

Doing the computations, it follows the left side of the equal sign is of the form

$$\left(\begin{array}{c} 3x^2 uv + 3xu^2 y + 3xu^2 v + u^3 y + u^3 v + v^2 \\ xv^2 + 2uyv + uv^2 \end{array} \right)$$

This is $\mathbf{o}\left(\begin{array}{c} u \\ v \end{array} \right)$ because it involves terms like uv, u^2v, etc. Each term being of degree 2 or more.

8.4   Exercises

(1) Use the definition of the derivative to find the 1×1 matrix which is the derivative of the following functions.

(a) $f\left(t \right) = t^2 + t$.

(b) $f\left(t \right) = t^3$.

(c) $f\left(t \right) = t \sin\left(t \right)$.

(d) $f\left(t \right) = \ln\left(t^2 + 1 \right)$.

(e) $f\left(t \right) = t\left| t \right|$.

(2) Show that if f is a real valued function defined on (a, b) and it achieves a local maximum at $x \in (a, b)$, then $Df(x) = 0$.

(3) Use the above definition of the derivative to prove the product rule for functions of 1 variable.

(4) Let $f(x, y) = x \sin(y)$. Compute the derivative directly from the definition.

(5) Let $f(x, y) = x^2 \sin(y)$. Compute the derivative directly from the definition.

(6) Let $\mathbf{f}(x, y) = \begin{pmatrix} x^2 + y \\ y^2 \end{pmatrix}$. Compute the derivative directly from the definition.

(7) Let $\mathbf{f}(x, y) = \begin{pmatrix} x^2 y \\ x + y^2 \end{pmatrix}$. Compute the derivative directly from the definition.

(8) Let $f(x, y) = x^\alpha y^\beta$. Show $Df(x, y) = \begin{pmatrix} \alpha x^{\alpha-1} y^\beta & x^\alpha \beta y^{\beta-1} \end{pmatrix}$.

(9) Let $\mathbf{f}(x, y) = \begin{pmatrix} x^2 \sin(y) \\ x^2 + y \end{pmatrix}$. Find $Df(x, y)$.

(10) Let $f(x, y) = \sqrt{x} \sqrt[3]{y}$. Find the approximate change in f when (x, y) goes from $(4, 8)$ to $(4.01, 7.99)$.

(11) Suppose f is differentiable and $\mathbf{g}$ is also differentiable, $\mathbf{g}$ having values in $\mathbb{R}^3$ and f having values in $\mathbb{R}$. Find $D(f\mathbf{g})$ directly from the definition. Assume both functions are defined on an open subset of $\mathbb{R}^n$.

(12) Show, using the above definition, that if f is differentiable, then so is $t \to f(t)^n$ for any positive integer and in fact the derivative of this function is $nf(t)^{n-1} f'(t)$.

(13) Suppose f is a scalar valued function of two variables which is differentiable. Show that $(x, y) \to (f(x, y))^n$ is also differentiable and its derivative equals

$$nf(x, y)^{n-1} Df(x, y)$$

(14) Let $f(x, y)$ be defined on $\mathbb{R}^2$ as follows. $f(x, x^2) = 1$ if $x \neq 0$. Define $f(0, 0) = 0$, and $f(x, y) = 0$ if $y \neq x^2$. Show that f is not continuous at $(0, 0)$ but that

$$\lim_{h \to 0} \frac{f(ha, hb) - f(0, 0)}{h} = 0$$

for (a, b) an arbitrary unit vector. Thus the directional derivative exists at $(0, 0)$ in every direction but f is not even continuous there.

8.5 C^1 Functions

Most of the time, there is an easier way to conclude a derivative exists and to find it. It involves the notion of a C^1 function.

Definition 8.4. When $\mathbf{f} : U \to \mathbb{R}^p$ for U an open subset of $\mathbb{R}^n$ and the vector valued functions $\frac{\partial \mathbf{f}}{\partial x_i}$ are all continuous, (equivalently each $\frac{\partial f_i}{\partial x_j}$ is continuous), the function is said to be $C^1(U)$. If all the partial derivatives up to order k exist and are continuous, then the function is said to be C^k.

It turns out that for a C^1 function, all you have to do is write the matrix described in Theorem 8.2 and this will be the derivative. There is no question of existence for the derivative for such functions. This is the importance of the next theorem.

Theorem 8.3. *Suppose* $\mathbf{f} : U \to \mathbb{R}^p$ *where* U *is an open set in* $\mathbb{R}^n$. *Suppose also that all partial derivatives of* $\mathbf{f}$ *exist on* U *and are continuous. Then* $\mathbf{f}$ *is differentiable at every point of* U.

Proof: If you fix all the variables but one, you can apply the fundamental theorem of calculus as follows.

$$\mathbf{f}(\mathbf{x}+v_k\mathbf{e}_k) - \mathbf{f}(\mathbf{x}) = \int_0^1 \frac{\partial \mathbf{f}}{\partial x_k}(\mathbf{x}+tv_k\mathbf{e}_k)\,v_k dt. \tag{8.7}$$

Here is why. Let $\mathbf{h}(t) = \mathbf{f}(\mathbf{x}+tv_k\mathbf{e}_k)$. Then

$$\frac{\mathbf{h}(t+h) - \mathbf{h}(t)}{h} = \frac{\mathbf{f}(\mathbf{x}+tv_k\mathbf{e}_k+hv_k\mathbf{e}_k) - \mathbf{f}(\mathbf{x}+tv_k\mathbf{e}_k)}{hv_k}v_k$$

and so, taking the limit as $h \to 0$ yields

$$\mathbf{h}'(t) = \frac{\partial \mathbf{f}}{\partial x_k}(\mathbf{x}+tv_k\mathbf{e}_k)\,v_k$$

Therefore,

$$\mathbf{f}(\mathbf{x}+v_k\mathbf{e}_k) - \mathbf{f}(\mathbf{x}) = \mathbf{h}(1) - \mathbf{h}(0) = \int_0^1 \mathbf{h}'(t)\,dt = \int_0^1 \frac{\partial \mathbf{f}}{\partial x_k}(\mathbf{x}+tv_k\mathbf{e}_k)\,v_k dt.$$

Now I will use this observation to prove the theorem. Let $\mathbf{v} = (v_1, \cdots, v_n)$ with $|\mathbf{v}|$ sufficiently small. Thus $\mathbf{v} = \sum_{k=1}^n v_k\mathbf{e}_k$. For the purposes of this argument, define

$$\sum_{k=n+1}^n v_k\mathbf{e}_k \equiv \mathbf{0}.$$

Then with this convention,

$$\mathbf{f}(\mathbf{x}+\mathbf{v}) - \mathbf{f}(\mathbf{x}) = \sum_{i=1}^n \left(\mathbf{f}\left(\mathbf{x}+\sum_{k=i}^n v_k\mathbf{e}_k\right) - \mathbf{f}\left(\mathbf{x}+\sum_{k=i+1}^n v_k\mathbf{e}_k\right) \right)$$

$$= \sum_{i=1}^n \int_0^1 \frac{\partial \mathbf{f}}{\partial x_i}\left(\mathbf{x}+\sum_{k=i+1}^n v_k\mathbf{e}_k+tv_i\mathbf{e}_i\right) v_i dt$$

$$= \sum_{i=1}^n \int_0^1 \left(\frac{\partial \mathbf{f}}{\partial x_i}\left(\mathbf{x}+\sum_{k=i+1}^n v_k\mathbf{e}_k+tv_i\mathbf{e}_i\right) v_i - \frac{\partial \mathbf{f}}{\partial x_i}(\mathbf{x})\,v_i \right) dt$$

$$+ \sum_{i=1}^n \int_0^1 \frac{\partial \mathbf{f}}{\partial x_i}(\mathbf{x})\,v_i dt = \sum_{i=1}^n \frac{\partial \mathbf{f}}{\partial x_i}(\mathbf{x})\,v_i$$

$$+ \sum_{i=1}^{n} \int_{0}^{1} \left(\frac{\partial \mathbf{f}}{\partial x_i} \left(\mathbf{x} + \sum_{k=i+1}^{n} v_k \mathbf{e}_k + t v_i \mathbf{e}_i \right) - \frac{\partial \mathbf{f}}{\partial x_i} (\mathbf{x}) \right) v_i dt$$

$$= \sum_{i=1}^{n} \frac{\partial \mathbf{f}}{\partial x_i} (\mathbf{x}) \, v_i + o(\mathbf{v})$$

and this shows $\mathbf{f}$ is differentiable at $\mathbf{x}$.

Some explanation of the step to the last line is in order. The messy thing at the end is $o(\mathbf{v})$ because of the continuity of the partial derivatives. To see this, consider one term. By Proposition 5.1,

$$\left| \int_{0}^{1} \left(\frac{\partial \mathbf{f}}{\partial x_i} \left(\mathbf{x} + \sum_{k=i+1}^{n} v_k \mathbf{e}_k + t v_i \mathbf{e}_i \right) - \frac{\partial \mathbf{f}}{\partial x_i} (\mathbf{x}) \right) v_i dt \right|$$

$$\leq \sqrt{p} \int_{0}^{1} \left| \frac{\partial \mathbf{f}}{\partial x_i} \left(\mathbf{x} + \sum_{k=i+1}^{n} v_k \mathbf{e}_k + t v_i \mathbf{e}_i \right) - \frac{\partial \mathbf{f}}{\partial x_i} (\mathbf{x}) \right| dt \, |\mathbf{v}|$$

Thus, dividing by $|\mathbf{v}|$ and taking a limit as $|\mathbf{v}| \to 0$, this converges to 0 due to continuity of the partial derivatives of $\mathbf{f}$. The messy term is thus a finite sum of $o(\mathbf{v})$ terms and is therefore $o(\mathbf{v})$. ∎

Here is an example to illustrate.

Example 8.9. Let $\mathbf{f}(x, y) = \begin{pmatrix} x^2 y + y^2 \\ y^3 x \end{pmatrix}$. Find $D\mathbf{f}(x, y)$.

From Theorem 8.3 this function is differentiable because all possible partial derivatives are continuous. Thus

$$D\mathbf{f}(x, y) = \begin{pmatrix} 2xy & x^2 + 2y \\ y^3 & 3y^2 x \end{pmatrix}.$$

In particular,

$$D\mathbf{f}(1, 2) = \begin{pmatrix} 4 & 5 \\ 8 & 12 \end{pmatrix}.$$

Here is another example.

Example 8.10. Let $\mathbf{f}(x_1, x_2, x_3) = \begin{pmatrix} x_1^2 x_2 + x_2^2 \\ x_2 x_1 + x_3 \\ \sin(x_1 x_2 x_3) \end{pmatrix}$. Find $D\mathbf{f}(x_1, x_2, x_3)$.

All possible partial derivatives are continuous, so the function is differentiable. The matrix for this derivative is therefore the following 3×3 matrix

$$\begin{pmatrix} 2x_1 x_2 & x_1^2 + 2x_2 & 0 \\ x_2 & x_1 & 1 \\ x_2 x_3 \cos(x_1 x_2 x_3) & x_1 x_3 \cos(x_1 x_2 x_3) & x_1 x_2 \cos(x_1 x_2 x_3) \end{pmatrix}$$

Example 8.11. Suppose $f(x, y, z) = xy + z^2$. Find $Df(1, 2, 3)$.

Taking the partial derivatives of f, $f_x = y$, $f_y = x$, $f_z = 2z$. These are all continuous. Therefore, the function has a derivative and $f_x\left(1,2,3\right) = 1$, $f_y\left(1,2,3\right) = 2$, and $f_z\left(1,2,3\right) = 6$. Therefore, $Df\left(1,2,3\right)$ is given by

$$Df\left(1,2,3\right) = \left(1,2,6\right).$$

Also, for (x,y,z) close to $(1,2,3)$,

$$\begin{aligned} f\left(x,y,z\right) &\approx f\left(1,2,3\right) + 1\left(x-1\right) + 2\left(y-2\right) + 6\left(z-3\right) \\ &= 11 + 1\left(x-1\right) + 2\left(y-2\right) + 6\left(z-3\right) = -12 + x + 2y + 6z \end{aligned}$$

When a function is differentiable at $\mathbf{x}_0$, it follows the function must be continuous there. This is the content of the following important lemma.

Lemma 8.1. *Let* $\mathbf{f} : U \to \mathbb{R}^q$ *where U is an open subset of $\mathbb{R}^p$. If $\mathbf{f}$ is differentiable, then $\mathbf{f}$ is continuous at $\mathbf{x}_0$. Furthermore, if $C \geq \max\left\{\left|\frac{\partial \mathbf{f}}{\partial x_i}\left(\mathbf{x}_0\right)\right|, i = 1, \cdots, p\right\}$, then whenever $\left|\mathbf{x} - \mathbf{x}_0\right|$ is small enough,*

$$\left|\mathbf{f}\left(\mathbf{x}\right) - \mathbf{f}\left(\mathbf{x}_0\right)\right| \leq \left(Cp + 1\right)\left|\mathbf{x} - \mathbf{x}_0\right| \tag{8.8}$$

Proof: Suppose $\mathbf{f}$ is differentiable. Since $\mathbf{o}\left(\mathbf{v}\right)$ satisfies (8.1), there exists $\delta_1 > 0$ such that if $\left|\mathbf{x} - \mathbf{x}_0\right| < \delta_1$, then $\left|\mathbf{o}\left(\mathbf{x} - \mathbf{x}_0\right)\right| < \left|\mathbf{x} - \mathbf{x}_0\right|$. But also, by the triangle inequality, there exists a constant C depending on the partial derivatives of $\mathbf{f}$ such that

$$\left|\sum_{i=1}^{p} \frac{\partial \mathbf{f}}{\partial x_i}\left(\mathbf{x}_0\right)\left(x_i - x_{0i}\right)\right| \leq C \sum_{i=1}^{p} \left|x_i - x_{0i}\right| \leq Cp\left|\mathbf{x} - \mathbf{x}_0\right|$$

Therefore, if $\left|\mathbf{x} - \mathbf{x}_0\right| < \delta_1$,

$$\left|\mathbf{f}\left(\mathbf{x}\right) - \mathbf{f}\left(\mathbf{x}_0\right)\right| \leq \left|\sum_{i=1}^{p} \frac{\partial \mathbf{f}}{\partial x_i}\left(\mathbf{x}_0\right)\left(x_i - x_{0i}\right)\right| + \left|\mathbf{x} - \mathbf{x}_0\right| < \left(Cp + 1\right)\left|\mathbf{x} - \mathbf{x}_0\right|$$

which verifies (8.8). Now letting $\varepsilon > 0$ be given, let $\delta = \min\left(\delta_1, \frac{\varepsilon}{Cp+1}\right)$. Then for $\left|\mathbf{x} - \mathbf{x}_0\right| < \delta$,

$$\left|\mathbf{f}\left(\mathbf{x}\right) - \mathbf{f}\left(\mathbf{x}_0\right)\right| < \left(Cp + 1\right)\left|\mathbf{x} - \mathbf{x}_0\right| < \left(Cp + 1\right)\frac{\varepsilon}{Cp + 1} = \varepsilon$$

showing $\mathbf{f}$ is continuous at $\mathbf{x}_0$. ∎

There have been quite a few terms defined. First there was the concept of continuity. Next the concept of partial or directional derivative. Next there was the concept of differentiability and the derivative being a linear transformation determined by a certain matrix. Finally, it was shown that if a function is C^1, then it has a derivative. To give a rough idea of the relationships of these topics, here is a picture.

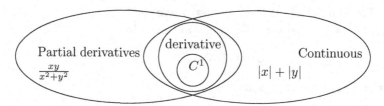

You might ask whether there are examples of functions which are differentiable but not C^1. Of course there are. In fact, Example 8.6 is just such an example as explained earlier. Then you should verify that $f'(x)$ exists for all $x \in \mathbb{R}$ but f' fails to be continuous at $x = 0$. Thus the function is differentiable at every point of $\mathbb{R}$ but fails to be C^1 because the derivative is not continuous at 0.

Example 8.12. Find an example of a function which is not differentiable at $(0,0)$ even though both partial derivatives exist at this point and the function is continuous at this point.

Here is a simple example.

$$f(x,y) \equiv \begin{cases} x \sin\left(\frac{1}{xy}\right) & \text{if } xy \neq 0 \\ 0 \text{ if } xy = 0 \end{cases}$$

To see this works, note that f is defined everywhere and

$$|f(x,y)| \leq |x|$$

so clearly f is continuous at $(0,0)$.

$$\frac{f(x,0) - f(0,0)}{x} = \frac{0 - 0}{x} = 0$$

and so $f_x(0,0) = 0$. Similarly,

$$\frac{f(0,y) - f(0,0)}{y} = \frac{0 - 0}{y} = 0$$

and so $f_y(0,0) = 0$. Thus the partial derivatives exist. However, the function is not differentiable at $(0,0)$ because

$$\lim_{(x,y)\to(0,0)} \frac{x \sin\left(\frac{1}{xy}\right)}{|(x,y)|}$$

does not even exist, much less equals 0. To see this, let $x = y$ and let $x \to 0$.

8.6   The Chain Rule

8.6.1   *The Chain Rule For Functions Of One Variable*

First recall the chain rule for a function of one variable. Consider the following picture.

$$I \xrightarrow{g} J \xrightarrow{f} \mathbb{R}$$

Here I and J are open intervals and it is assumed that $g(I) \subseteq J$. The chain rule says that if $f'(g(x))$ exists and $g'(x)$ exists for $x \in I$, then the composition, $f \circ g$ also has a derivative at x and

$$(f \circ g)'(x) = f'(g(x)) g'(x).$$

Recall that $f \circ g$ is the name of the function defined by $f \circ g(x) \equiv f(g(x))$. In the notation of this chapter, the chain rule is written as

$$Df(g(x)) Dg(x) = D(f \circ g)(x). \qquad (8.9)$$

8.6.2   The Chain Rule For Functions Of Many Variables

Let $U \subseteq \mathbb{R}^n$ and $V \subseteq \mathbb{R}^p$ be open sets and let $\mathbf{f}$ be a function defined on V having values in $\mathbb{R}^q$ while $\mathbf{g}$ is a function defined on U such that $\mathbf{g}(U) \subseteq V$ as in the following picture.

$$U \xrightarrow{\mathbf{g}} V \xrightarrow{\mathbf{f}} \mathbb{R}^q$$

The chain rule says that if the linear transformations (matrices) on the left in (8.9) both exist then the same formula holds in this more general case. Thus

$$D\mathbf{f}(\mathbf{g}(\mathbf{x})) D\mathbf{g}(\mathbf{x}) = D(\mathbf{f} \circ \mathbf{g})(\mathbf{x})$$

Note this all makes sense because $D\mathbf{f}(\mathbf{g}(\mathbf{x}))$ is a $q \times p$ matrix and $D\mathbf{g}(\mathbf{x})$ is a $p \times n$ matrix. Remember it is all right to do $(q \times p)(p \times n)$. The middle numbers match. More precisely,

Theorem 8.4. *(Chain rule) Let U be an open set in $\mathbb{R}^n$, let V be an open set in $\mathbb{R}^p$, let $\mathbf{g} : U \to \mathbb{R}^p$ be such that $\mathbf{g}(U) \subseteq V$, and let $\mathbf{f} : V \to \mathbb{R}^q$. Suppose $D\mathbf{g}(\mathbf{x})$ exists for some $\mathbf{x} \in U$ and that $D\mathbf{f}(\mathbf{g}(\mathbf{x}))$ exists. Then $D(\mathbf{f} \circ \mathbf{g})(\mathbf{x})$ exists and furthermore,*

$$D(\mathbf{f} \circ \mathbf{g})(\mathbf{x}) = D\mathbf{f}(\mathbf{g}(\mathbf{x})) D\mathbf{g}(\mathbf{x}). \qquad (8.10)$$

In particular,

$$\frac{\partial(\mathbf{f} \circ \mathbf{g})(\mathbf{x})}{\partial x_j} = \sum_{i=1}^{p} \frac{\partial \mathbf{f}(\mathbf{g}(\mathbf{x}))}{\partial y_i} \frac{\partial g_i(\mathbf{x})}{\partial x_j}. \qquad (8.11)$$

There is an easy way to remember this in terms of the repeated index summation convention presented earlier in Volume 1. Let $\mathbf{y} = \mathbf{g}(\mathbf{x})$ and $\mathbf{z} = \mathbf{f}(\mathbf{y})$. Then the above says

$$\frac{\partial \mathbf{z}}{\partial y_i} \frac{\partial y_i}{\partial x_k} = \frac{\partial \mathbf{z}}{\partial x_k}. \qquad (8.12)$$

Remember there is a sum on the repeated index. In particular, for each index r,

$$\frac{\partial z_r}{\partial y_i} \frac{\partial y_i}{\partial x_k} = \frac{\partial z_r}{\partial x_k}.$$

The proof of this major theorem will be given at the end of this section. It will include the chain rule for functions of one variable as a special case. First here are some examples.

Example 8.13. Let $f(u, v) = \sin(uv)$ and let $u(x, y, t) = t \sin x + \cos y$ and $v(x, y, t, s) = s \tan x + y^2 + ts$. Letting $z = f(u, v)$ where u, v are as just described, find $\frac{\partial z}{\partial t}$ and $\frac{\partial z}{\partial x}$.

From (8.12), $\frac{\partial z}{\partial t} = \frac{\partial z}{\partial u}\frac{\partial u}{\partial t} + \frac{\partial z}{\partial v}\frac{\partial v}{\partial t} = v \cos(uv) \sin(x) + us \cos(uv)$. Here $y_1 = u, y_2 = v, t = x_k$. Also,

$$\frac{\partial z}{\partial x} = \frac{\partial z}{\partial u}\frac{\partial u}{\partial x} + \frac{\partial z}{\partial v}\frac{\partial v}{\partial x} = v \cos(uv) t \cos(x) + us \sec^2(x) \cos(uv).$$

Clearly you can continue in this way, taking partial derivatives with respect to any of the other variables.

Example 8.14. Let $w = f(u_1, u_2) = u_2 \sin(u_1)$ and $u_1 = x^2 y + z, u_2 = \sin(xy)$. Find $\frac{\partial w}{\partial x}, \frac{\partial w}{\partial y}$, and $\frac{\partial w}{\partial z}$.

The derivative of f is of the form (w_x, w_y, w_z) and so it suffices to find the derivative of f using the chain rule. You need to find $Df(u_1, u_2) Dg(x, y, z)$ where

$$\mathbf{g}(x, y) = \begin{pmatrix} x^2 y + z \\ \sin(xy) \end{pmatrix}.$$

Then

$$Dg(x, y, z) = \begin{pmatrix} 2xy & x^2 & 1 \\ y \cos(xy) & x \cos(xy) & 0 \end{pmatrix}.$$

Also $Df(u_1, u_2) = (u_2 \cos(u_1), \sin(u_1))$. Therefore, the derivative is

$$Df(u_1, u_2) Dg(x, y, z)$$

$$= (u_2 \cos(u_1), \sin(u_1)) \begin{pmatrix} 2xy & x^2 & 1 \\ y \cos(xy) & x \cos(xy) & 0 \end{pmatrix}$$

$$= (2u_2 (\cos u_1) xy + (\sin u_1) y \cos xy, u_2 (\cos u_1) x^2$$
$$+ (\sin u_1) x \cos xy, u_2 \cos u_1)$$

$$= (w_x, w_y, w_z)$$

Thus

$$\frac{\partial w}{\partial x} = 2u_2 (\cos u_1) xy + (\sin u_1) y \cos xy$$

$$= 2 (\sin(xy)) (\cos(x^2 y + z)) xy$$
$$+ (\sin(x^2 y + z)) y \cos xy.$$

Similarly, you can find the other partial derivatives of w in terms of substituting in for u_1 and u_2 in the above. Note

$$\frac{\partial w}{\partial x} = \frac{\partial w}{\partial u_1}\frac{\partial u_1}{\partial x} + \frac{\partial w}{\partial u_2}\frac{\partial u_2}{\partial x}.$$

In fact, in general if you have $w = f(u_1, u_2)$ and $\mathbf{g}(x, y, z) = \begin{pmatrix} u_1(x, y, z) \\ u_2(x, y, z) \end{pmatrix}$, then $D(f \circ \mathbf{g})(x, y, z)$ is of the form

$$\begin{pmatrix} w_{u_1} & w_{u_2} \end{pmatrix} \begin{pmatrix} u_{1x} & u_{1y} & u_{1z} \\ u_{2x} & u_{2y} & u_{2z} \end{pmatrix}$$

$$= \begin{pmatrix} w_{u_1} u_x + w_{u_2} u_{2x} & w_{u_1} u_y + w_{u_2} u_{2y} & w_{u_1} u_z + w_{u_2} u_{2z} \end{pmatrix}.$$

Example 8.15. Let $w = f(u_1, u_2, u_3) = u_1^2 + u_3 + u_2$ and $\mathbf{g}(x, y, z) = \begin{pmatrix} u_1 \\ u_2 \\ u_3 \end{pmatrix} =$

$\begin{pmatrix} x + 2yz \\ x^2 + y \\ z^2 + x \end{pmatrix}$. Find $\frac{\partial w}{\partial x}$ and $\frac{\partial w}{\partial z}$.

By the chain rule,

$$(w_x, w_y, w_z) = \begin{pmatrix} w_{u_1} & w_{u_2} & w_{u_3} \end{pmatrix} \begin{pmatrix} u_{1x} & u_{1y} & u_{1z} \\ u_{2x} & u_{2y} & u_{2z} \\ u_{3x} & u_{3y} & u_{3z} \end{pmatrix} =$$

$$(w_{u_1} u_{1x} + w_{u_2} u_{2x} + w_{u_3} u_{3x}, w_{u_1} u_{1y} + w_{u_2} u_{2y} + w_{u_3} u_{3y},$$
$$w_{u_1} u_{1z} + w_{u_2} u_{2z} + w_{u_3} u_{3z})$$

Note the pattern,

$$\begin{aligned} w_x &= w_{u_1} u_{1x} + w_{u_2} u_{2x} + w_{u_3} u_{3x}, \\ w_y &= w_{u_1} u_{1y} + w_{u_2} u_{2y} + w_{u_3} u_{3y}, \\ w_z &= w_{u_1} u_{1z} + w_{u_2} u_{2z} + w_{u_3} u_{3z}. \end{aligned}$$

Therefore,

$$w_x = 2u_1(1) + 1(2x) + 1(1) = 2(x + 2yz) + 2x + 1 = 4x + 4yz + 1$$

and

$$w_z = 2u_1(2y) + 1(0) + 1(2z) = 4(x + 2yz)y + 2z = 4yx + 8y^2z + 2z.$$

Of course to find all the partial derivatives at once, you just use the chain rule. Thus you would get

$$\begin{pmatrix} w_x & w_y & w_z \end{pmatrix}$$

$$= \begin{pmatrix} 2u_1 & 1 & 1 \end{pmatrix} \begin{pmatrix} 1 & 2z & 2y \\ 2x & 1 & 0 \\ 1 & 0 & 2z \end{pmatrix}$$

$$= \begin{pmatrix} 2u_1 + 2x + 1 & 4u_1z + 1 & 4u_1y + 2z \end{pmatrix}$$

$$= \begin{pmatrix} 4x + 4yz + 1 & 4zx + 8yz^2 + 1 & 4yx + 8y^2z + 2z \end{pmatrix}$$

Example 8.16. Let $\mathbf{f}(u_1, u_2) = \begin{pmatrix} u_1^2 + u_2 \\ \sin(u_2) + u_1 \end{pmatrix}$ and

$$\mathbf{g}(x_1, x_2, x_3) = \begin{pmatrix} u_1(x_1, x_2, x_3) \\ u_2(x_1, x_2, x_3) \end{pmatrix} = \begin{pmatrix} x_1 x_2 + x_3 \\ x_2^2 + x_1 \end{pmatrix}.$$

Find $D(\mathbf{f} \circ \mathbf{g})(x_1, x_2, x_3)$.

To do this,

$$Df(u_1, u_2) = \begin{pmatrix} 2u_1 & 1 \\ 1 & \cos u_2 \end{pmatrix},$$

$$Dg(x_1, x_2, x_3) = \begin{pmatrix} x_2 & x_1 & 1 \\ 1 & 2x_2 & 0 \end{pmatrix}.$$

Then

$$Df(\mathbf{g}(x_1, x_2, x_3)) = \begin{pmatrix} 2(x_1 x_2 + x_3) & 1 \\ 1 & \cos(x_2^2 + x_1) \end{pmatrix}$$

and so by the chain rule,

$$D(\mathbf{f} \circ \mathbf{g})(x_1, x_2, x_3)$$

$$= \overbrace{\begin{pmatrix} 2(x_1 x_2 + x_3) & 1 \\ 1 & \cos(x_2^2 + x_1) \end{pmatrix}}^{Df(\mathbf{g}(\mathbf{x}))} \overbrace{\begin{pmatrix} x_2 & x_1 & 1 \\ 1 & 2x_2 & 0 \end{pmatrix}}^{Dg(\mathbf{x})}$$

$$= \begin{pmatrix} (2x_1 x_2 + 2x_3) x_2 + 1 & (2x_1 x_2 + 2x_3) x_1 + 2x_2 & 2x_1 x_2 + 2x_3 \\ x_2 + \cos(x_2^2 + x_1) & x_1 + 2x_2 (\cos(x_2^2 + x_1)) & 1 \end{pmatrix}$$

Therefore, in particular,

$$\frac{\partial f_1 \circ \mathbf{g}}{\partial x_1}(x_1, x_2, x_3) = (2x_1 x_2 + 2x_3) x_2 + 1,$$

$$\frac{\partial f_2 \circ \mathbf{g}}{\partial x_3}(x_1, x_2, x_3) = 1, \frac{\partial f_2 \circ \mathbf{g}}{\partial x_2}(x_1, x_2, x_3) = x_1 + 2x_2 (\cos(x_2^2 + x_1)).$$

etc.

In different notation, let $\begin{pmatrix} z_1 \\ z_2 \end{pmatrix} = \mathbf{f}(u_1, u_2) = \begin{pmatrix} u_1^2 + u_2 \\ \sin(u_2) + u_1 \end{pmatrix}$. Then

$$\frac{\partial z_1}{\partial x_1} = \frac{\partial z_1}{\partial u_1} \frac{\partial u_1}{\partial x_1} + \frac{\partial z_1}{\partial u_2} \frac{\partial u_2}{\partial x_1}$$

$$= 2u_1 x_2 + 1 = 2(x_1 x_2 + x_3) x_2 + 1.$$

Example 8.17. Let

$$\mathbf{f}(u_1, u_2, u_3) = \begin{pmatrix} z_1 \\ z_2 \\ z_3 \end{pmatrix} = \begin{pmatrix} u_1^2 + u_2 u_3 \\ u_1^2 + u_2^3 \\ \ln(1 + u_3^2) \end{pmatrix}$$

and let

$$\mathbf{g}(x_1, x_2, x_3, x_4) = \begin{pmatrix} u_1 \\ u_2 \\ u_3 \end{pmatrix} = \begin{pmatrix} x_1 + x_2^2 + \sin(x_3) + \cos(x_4) \\ x_4^2 - x_1 \\ x_3^2 + x_4 \end{pmatrix}.$$

Find $(\mathbf{f} \circ \mathbf{g})'(\mathbf{x})$.

$$Df\left(u\right)=\begin{pmatrix} 2u_1 & u_3 & u_2 \\ 2u_1 & 3u_2^2 & 0 \\ 0 & 0 & \frac{2u_3}{\left(1+u_3^2\right)} \end{pmatrix}$$

Similarly,

$$Dg\left(x\right)=\begin{pmatrix} 1 & 2x_2 & \cos\left(x_3\right) & -\sin\left(x_4\right) \\ -1 & 0 & 0 & 2x_4 \\ 0 & 0 & 2x_3 & 1 \end{pmatrix}.$$

Then by the chain rule, $D\left(f\circ g\right)\left(x\right)=Df\left(u\right)Dg\left(x\right)$ where $u=g\left(x\right)$ as described above. Thus $D\left(f\circ g\right)\left(x\right)=$

$$\begin{pmatrix} 2u_1 & u_3 & u_2 \\ 2u_1 & 3u_2^2 & 0 \\ 0 & 0 & \frac{2u_3}{\left(1+u_3^2\right)} \end{pmatrix}\begin{pmatrix} 1 & 2x_2 & \cos\left(x_3\right) & -\sin\left(x_4\right) \\ -1 & 0 & 0 & 2x_4 \\ 0 & 0 & 2x_3 & 1 \end{pmatrix}$$

$$=\begin{pmatrix} 2u_1-u_3 & 4u_1x_2 & 2u_1\cos x_3+2u_2x_3 & -2u_1\sin x_4+2u_3x_4+u_2 \\ 2u_1-3u_2^2 & 4u_1x_2 & 2u_1\cos x_3 & -2u_1\sin x_4+6u_2^2x_4 \\ 0 & 0 & 4\frac{u_3}{1+u_3^2}x_3 & 2\frac{u_3}{1+u_3^2} \end{pmatrix} \quad (8.13)$$

where each u_i is given by the above formulas. Thus $\frac{\partial z_1}{\partial x_1}$ equals

$$\begin{aligned} 2u_1-u_3 &= 2\left(x_1+x_2^2+\sin\left(x_3\right)+\cos\left(x_4\right)\right)-\left(x_3^2+x_4\right) \\ &= 2x_1+2x_2^2+2\sin x_3+2\cos x_4-x_3^2-x_4. \end{aligned}$$

while $\frac{\partial z_2}{\partial x_4}$ equals

$$-2u_1\sin x_4+6u_2^2x_4=-2\left(x_1+x_2^2+\sin\left(x_3\right)+\cos\left(x_4\right)\right)\sin\left(x_4\right)+6\left(x_4^2-x_1\right)^2 x_4.$$

If you wanted $\frac{\partial z}{\partial x_2}$ it would be the second column of the above matrix in (8.13). Thus $\frac{\partial z}{\partial x_2}$ equals

$$\begin{pmatrix} \frac{\partial z_1}{\partial x_2} \\ \frac{\partial z_2}{\partial x_2} \\ \frac{\partial z_3}{\partial x_2} \end{pmatrix}=\begin{pmatrix} 4u_1x_2 \\ 4u_1x_2 \\ 0 \end{pmatrix}=\begin{pmatrix} 4\left(x_1+x_2^2+\sin\left(x_3\right)+\cos\left(x_4\right)\right)x_2 \\ 4\left(x_1+x_2^2+\sin\left(x_3\right)+\cos\left(x_4\right)\right)x_2 \\ 0 \end{pmatrix}$$

I hope that by now it is clear that all the information you could desire about various partial derivatives is available and it all reduces to matrix multiplication and the consideration of entries of the matrix obtained by multiplying the two derivatives.

8.7 Exercises

(1) Let $z = f(x_1, \cdots, x_n)$ be as given and let $x_i = g_i(t_1, \cdots, t_m)$ as given. Find $\frac{\partial z}{\partial t_i}$ which is indicated.

    (a) $z = x_1^3 + x_2,\ x_1 = \sin(t_1) + \cos(t_2),\ x_2 = t_1 t_2^2$. Find $\frac{\partial z}{\partial t_1}$

    (b) $z = x_1 x_2^2,\ x_1 = t_1 t_2^2 t_3,\ x_2 = t_1 t_2^2$. Find $\frac{\partial z}{\partial t_1}$.

    (c) $z = x_1 x_2^2,\ x_1 = t_1 t_2^2 t_3,\ x_2 = t_1 t_2^2$. Find $\frac{\partial z}{\partial t_1}$.

    (d) $z = x_1 x_2^2,\ x_1 = t_1 t_2^2 t_3,\ x_2 = t_1 t_2^2$. Find $\frac{\partial z}{\partial t_3}$.

    (e) $z = x_1^2 x_2^2,\ x_1 = t_1 t_2^2 t_3,\ x_2 = t_1 t_2^2$. Find $\frac{\partial z}{\partial t_2}$.

    (f) $z = x_1^2 x_2 + x_3^2,\ x_1 = t_1 t_2,\ x_2 = t_1 t_2 t_4,\ x_3 = \sin(t_3)$. Find $\frac{\partial z}{\partial t_2}$.

    (g) $z = x_1^2 x_2 + x_3^2,\ x_1 = t_1 t_2,\ x_2 = t_1 t_2 t_4,\ x_3 = \sin(t_3)$. Find $\frac{\partial z}{\partial t_3}$.

    (h) $z = x_1^2 x_2 + x_3^2,\ x_1 = t_1 t_2,\ x_2 = t_1 t_2 t_4,\ x_3 = \sin(t_3)$. Find $\frac{\partial z}{\partial t_1}$.

(2) Let $z = f(\mathbf{y}) = \left(y_1^2 + \sin y_2 + \tan y_3\right)$ and

$$\mathbf{y} = \mathbf{g}(\mathbf{x}) \equiv \begin{pmatrix} x_1 + x_2 \\ x_2^2 - x_1 + x_2 \\ x_2^2 + x_1 + \sin x_2 \end{pmatrix}.$$

    Find $D(f \circ \mathbf{g})(\mathbf{x})$. Use to write $\frac{\partial z}{\partial x_i}$ for $i = 1, 2$.

(3) Let $z = f(\mathbf{y}) = \left(y_1^2 + \cot y_2 + \sin y_3\right)$ and $\mathbf{y} = \mathbf{g}(\mathbf{x}) \equiv \begin{pmatrix} x_1 + x_4 + x_3 \\ x_2^2 - x_1 + x_2 \\ x_2^2 + x_1 + \sin x_4 \end{pmatrix}.$

    Find $D(f \circ \mathbf{g})(\mathbf{x})$. Use to write $\frac{\partial z}{\partial x_i}$ for $i = 1, 2, 3, 4$.

(4) Let

$$z = f(\mathbf{y}) = \left(y_1^2 + y_2^2 + \sin y_3 + y_4\right) \text{ and } \mathbf{y} = \mathbf{g}(\mathbf{x}) \equiv \begin{pmatrix} x_1 + x_4 + x_3 \\ x_2^2 - x_1 + x_2 \\ x_2^2 + x_1 + \sin x_4 \\ x_4 + x_2 \end{pmatrix}.$$

    Find $D(f \circ \mathbf{g})(\mathbf{x})$. Use to write $\frac{\partial z}{\partial x_i}$ for $i = 1, 2, 3, 4$.

(5) Let

$$\mathbf{z} = \mathbf{f}(\mathbf{y}) = \begin{pmatrix} y_1^2 + \sin y_2 + \tan y_3 \\ y_1^2 y_2 + y_3 \end{pmatrix}$$

and $\mathbf{y} = \mathbf{g}(\mathbf{x}) \equiv \begin{pmatrix} x_1 + x_2 \\ x_2^2 - x_1 + x_2 \\ x_2^2 + x_1 + \sin x_2 \end{pmatrix}$. Find $D(\mathbf{f} \circ \mathbf{g})(\mathbf{x})$. Use to write $\frac{\partial z_k}{\partial x_i}$

for $i = 1, 2$ and $k = 1, 2$. Recall this will be of the form $\begin{pmatrix} z_{1x_1} & z_{1x_2} & z_{1x_3} \\ z_{2x_1} & z_{2x_2} & z_{2x_3} \end{pmatrix}.$

(6) Let $z = \mathbf{f}\left(\mathbf{y}\right) = \begin{pmatrix} y_1^2 + \sin y_2 + \tan y_3 \\ y_1^2 y_2 + y_3 \\ \cos\left(y_1^2\right) + y_2^3 y_3 \end{pmatrix}$ and

$$\mathbf{y} = \mathbf{g}\left(\mathbf{x}\right) \equiv \begin{pmatrix} x_1 + x_4 \\ x_2^2 - x_1 + x_3 \\ x_3^2 + x_1 + \sin x_2 \end{pmatrix}.$$

Find $D\left(\mathbf{f} \circ \mathbf{g}\right)\left(\mathbf{x}\right)$. Use to write $\frac{\partial z_k}{\partial x_i}$ for $i = 1, 2, 3, 4$ and $k = 1, 2, 3$.

(7) Give a version of the chain rule which involves three functions $\mathbf{f}, \mathbf{g}, \mathbf{h}$.

(8) If $\mathbf{f} : U \to V$ and $\mathbf{f}^{-1} : V \to U$ for U, V open sets such that $\mathbf{f}, \mathbf{f}^{-1}$ are both differentiable, show that

$$\det\left(D\mathbf{f}\left(\mathbf{f}^{-1}\left(\mathbf{y}\right)\right)\right) \det\left(D\mathbf{f}^{-1}\left(\mathbf{y}\right)\right) = 1$$

8.7.1 Related Rates Problems

Sometimes several variables are related and, given information about how one variable is changing, you want to find how the others are changing. The following law is discussed later in the book, on Page 257.

Example 8.18. Bernoulli's law states that in an incompressible fluid,

$$\frac{v^2}{2g} + z + \frac{P}{\gamma} = C$$

where C is a constant. Here v is the speed, P is the pressure, and z is the height above some reference point. The constants g and γ are the acceleration of gravity and the weight density of the fluid. Suppose measurements indicate that $\frac{dv}{dt} = -3$, and $\frac{dz}{dt} = 2$. Find $\frac{dP}{dt}$ when $v = 7$ and $z = 8$ in terms of g and γ.

This is just an exercise in using the chain rule. Differentiate the two sides with respect to t.

$$\frac{1}{g} v \frac{dv}{dt} + \frac{dz}{dt} + \frac{1}{\gamma} \frac{dP}{dt} = 0.$$

Then when $v = 7$ and $z = 8$, finding $\frac{dP}{dt}$ involves nothing more than solving the following for $\frac{dP}{dt}$.

$$\frac{7}{g}\left(-3\right) + 2 + \frac{1}{\gamma} \frac{dP}{dt} = 0$$

Thus

$$\frac{dP}{dt} = \gamma\left(\frac{21}{g} - 2\right)$$

at this instant in time.

Example 8.19. In Bernoulli's law above, each of v, z, and P are functions of (x, y, z), the position of a point in the fluid. Find a formula for $\frac{\partial P}{\partial x}$ in terms of the partial derivatives of the other variables.

This is an example of the chain rule. Differentiate both sides with respect to x.

$$\frac{v}{g}v_x + z_x + \frac{1}{\gamma}P_x = 0$$

and so

$$P_x = -\left(\frac{vv_x + z_x g}{g}\right)\gamma$$

Example 8.20. Suppose a level curve is of the form $f(x, y) = C$ and that near a point on this level curve y is a differentiable function of x. Find $\frac{dy}{dx}$.

This is an example of the chain rule. Differentiate both sides with respect to x. This gives

$$f_x + f_y\frac{dy}{dx} = 0.$$

Solving for $\frac{dy}{dx}$ gives

$$\frac{dy}{dx} = \frac{-f_x(x, y)}{f_y(x, y)}.$$

Example 8.21. Suppose a level surface is of the form $f(x, y, z) = C$. and that near a point (x, y, z) on this level surface z is a C^1 function of x and y. Find a formula for z_x.

This is an example of the use of the chain rule. Differentiate both sides of the equation with respect to x. Since $y_x = 0$,

$$f_x + f_z z_x = 0.$$

Then solving for z_x,

$$z_x = \frac{-f_x(x, y, z)}{f_z(x, y, z)}$$

Example 8.22. Polar coordinates are

$$x = r\cos\theta, \ y = r\sin\theta. \tag{8.14}$$

Thus if f is a C^1 scalar valued function you could ask to express f_x in terms of the variables r and θ. Do so.

This is an example of the chain rule. Abusing notation slightly, regard f as a function of position in the plane. This position can be described with any set of coordinates. Thus $f(x, y) = f(r, \theta)$ and so

$$f_x = f_r r_x + f_\theta \theta_x.$$

This will be done if you can find r_x and θ_x. However you must find these in terms of r and θ, not in terms of x and y. Using the chain rule on the two equations for the transformation in (8.14),

$$1 = r_x \cos\theta - (r\sin\theta)\theta_x, \ 0 = r_x \sin\theta + (r\cos\theta)\theta_x$$

Solving these using Cramer's rule,

$$r_x = \cos(\theta), \ \theta_x = \frac{-\sin(\theta)}{r}$$

Hence f_x in polar coordinates is

$$f_x = f_r(r, \theta)\cos(\theta) - f_\theta(r, \theta)\left(\frac{\sin(\theta)}{r}\right)$$

8.7.2 The Derivative Of The Inverse Function

Example 8.23. Let $\mathbf{f} : U \to V$ where U and V are open sets in $\mathbb{R}^n$ and $\mathbf{f}$ is one to one and onto. Suppose also that $\mathbf{f}$ and $\mathbf{f}^{-1}$ are both differentiable. How are $D\mathbf{f}^{-1}$ and $D\mathbf{f}$ related?

This can be done as follows. From the assumptions, $\mathbf{x} = \mathbf{f}^{-1}(\mathbf{f}(\mathbf{x}))$. Let $I\mathbf{x} = \mathbf{x}$. Then by Example 8.4 on Page 137 $DI = I$. By the chain rule,

$$I = DI = D\mathbf{f}^{-1}(\mathbf{f}(\mathbf{x}))(D\mathbf{f}(\mathbf{x})), \ I = DI = D\mathbf{f}(\mathbf{f}^{-1}(\mathbf{y}))D\mathbf{f}^{-1}(\mathbf{y})$$

Letting $\mathbf{y} = \mathbf{f}(\mathbf{x})$, the second yields

$$I = D\mathbf{f}(\mathbf{x})D\mathbf{f}^{-1}(\mathbf{f}(\mathbf{x})).$$

Therefore,

$$D\mathbf{f}(\mathbf{x})^{-1} = D\mathbf{f}^{-1}(\mathbf{f}(\mathbf{x})).$$

This is equivalent to

$$D\mathbf{f}(\mathbf{f}^{-1}(\mathbf{y}))^{-1} = D\mathbf{f}^{-1}(\mathbf{y})$$

or

$$D\mathbf{f}(\mathbf{x})^{-1} = D\mathbf{f}^{-1}(\mathbf{y}), \mathbf{y} = \mathbf{f}(\mathbf{x}).$$

This is just like a similar situation for functions of one variable. Remember

$$(f^{-1})'(f(x)) = 1/f'(x).$$

In terms of the repeated index summation convention, suppose $\mathbf{y} = \mathbf{f}(\mathbf{x})$ so that $\mathbf{x} = \mathbf{f}^{-1}(\mathbf{y})$. Then the above can be written as

$$\delta_{ij} = \frac{\partial x_i}{\partial y_k}(\mathbf{f}(\mathbf{x}))\frac{\partial y_k}{\partial x_j}(\mathbf{x}).$$

8.7.3 Proof Of The Chain Rule

As in the case of a function of one variable, it is important to consider the derivative of a composition of two functions. As in the case of a function of one variable, this rule is called the chain rule. Its proof depends on the following fundamental lemma. This proof will include the one dimensional case.

Lemma 8.2. *Let* $\mathbf{g} : U \to \mathbb{R}^p$ *where* U *is an open set in* $\mathbb{R}^n$ *and suppose* $\mathbf{g}$ *has a derivative at* $\mathbf{x} \in U$. *Then* $o(\mathbf{g}(\mathbf{x}+\mathbf{v}) - \mathbf{g}(\mathbf{x})) = o(\mathbf{v})$.

Proof: It is necessary to show that

$$\lim_{\mathbf{v}\to 0}\frac{|o(\mathbf{g}(\mathbf{x}+\mathbf{v}) - \mathbf{g}(\mathbf{x}))|}{|\mathbf{v}|} = 0. \tag{8.15}$$

From Lemma 8.1, there exists a constant K and $\delta > 0$ such that if $|\mathbf{v}| < \delta$, then

$$|\mathbf{g}(\mathbf{x}+\mathbf{v}) - \mathbf{g}(\mathbf{x})| \le K|\mathbf{v}|. \tag{8.16}$$

Now let $\varepsilon > 0$ be given. There exists $\eta > 0$ such that if $|g(x+v) - g(x)| < \eta$, then

$$|o(g(x+v) - g(x))| < \frac{\varepsilon}{K} |g(x+v) - g(x)| \tag{8.17}$$

Let $|v| < \min\left(\delta, \frac{\eta}{K}\right)$. For such v, $|g(x+v) - g(x)| \leq \eta$, which implies

$$\begin{aligned} |o(g(x+v) - g(x))| &< \frac{\varepsilon}{K} |g(x+v) - g(x)| \\ &< \left(\frac{\varepsilon}{K}\right) K |v| = \varepsilon |v| \end{aligned}$$

and so

$$\frac{|o(g(x+v) - g(x))|}{|v|} < \varepsilon$$

which establishes (8.15). ∎

Recall the notation $f \circ g(x) \equiv f(g(x))$. Thus $f \circ g$ is the name of a function, and this function is defined by what was just written. The following theorem is known as the **chain rule**.

Theorem 8.5. *(Chain rule) Let U be an open set in $\mathbb{R}^n$, let V be an open set in $\mathbb{R}^p$, let $g : U \to \mathbb{R}^p$ be such that $g(U) \subseteq V$, and let $f : V \to \mathbb{R}^q$. Suppose $Dg(x)$ exists for some $x \in U$ and that $Df(g(x))$ exists. Then $D(f \circ g)(x)$ exists and furthermore,*

$$D(f \circ g)(x) = Df(g(x)) Dg(x). \tag{8.18}$$

In particular, If $y = g(x)$ so $y_i = g_i(x)$,

$$\frac{\partial(f \circ g)(x)}{\partial x_j} = \sum_{i=1}^{p} \frac{\partial f(g(x))}{\partial y_i} \frac{\partial g_i(x)}{\partial x_j}. \tag{8.19}$$

Proof: From the assumption that $Df(g(x))$ exists,

$$f(g(x+v)) = f(g(x)) + Df(g(x))(g(x+v) - g(x)) + o(g(x+v) - g(x))$$

$$= f(g(x)) + Df(g(x))(Dg(x)v + o(v)) + o(g(x+v) - g(x))$$

which by Lemma 8.2 equals

$$\begin{aligned} &= f(g(x)) + Df(g(x)) Dg(x) v + Df(g(x)) o(v) + o(v) \\ &= f(g(x)) + Df(g(x)) Dg(x) v + o(v) \end{aligned}$$

and this shows

$$D(f \circ g)(x) = Df(g(x)) Dg(x)$$

from the definition of the derivative and its uniqueness established in Theorem 8.2 on Page 135. ∎

8.8 Exercises

(1) Suppose $\mathbf{f} : U \to \mathbb{R}^q$ and let $\mathbf{x} \in U$ and $\mathbf{v}$ be a unit vector. Show that $D_{\mathbf{v}}\mathbf{f}(\mathbf{x}) = D\mathbf{f}(\mathbf{x})\mathbf{v}$. Recall that

$$D_{\mathbf{v}}\mathbf{f}(\mathbf{x}) \equiv \lim_{t \to 0} \frac{\mathbf{f}(\mathbf{x} + t\mathbf{v}) - \mathbf{f}(\mathbf{x})}{t}.$$

(2) Let $f(x, y) = \begin{cases} xy \sin\left(\frac{1}{x}\right) & \text{if } x \neq 0 \\ 0 & \text{if } x = 0 \end{cases}$. Find where f is differentiable and compute the derivative at all these points.

(3) Let

$$f(x, y) = \begin{cases} x & \text{if } |y| > |x| \\ -x & \text{if } |y| \leq |x| \end{cases}.$$

Show that f is continuous at $(0, 0)$ and that the partial derivatives exist at $(0, 0)$ but the function is not differentiable at $(0, 0)$.

(4) Let

$$\mathbf{f}(x, y, z) = \begin{pmatrix} x^2 \sin y + z^3 \\ \sin(x + y) + z^3 \cos x \end{pmatrix}.$$

Find $D\mathbf{f}(1, 2, 3)$.

(5) Let

$$\mathbf{f}(x, y, z) = \begin{pmatrix} x \tan y + z^3 \\ \cos(x + y) + z^3 \cos x \end{pmatrix}.$$

Find $D\mathbf{f}(x, y, z)$.

(6) Let

$$\mathbf{f}(x, y, z) = \begin{pmatrix} x \sin y + z^3 \\ \sin(x + y) + z^3 \cos x \\ x^5 + y^2 \end{pmatrix}.$$

Find $D\mathbf{f}(x, y, z)$.

(7) Let

$$f(x, y) = \begin{cases} \frac{(x^2 - y^4)^2}{(x^2 + y^4)^2} & \text{if } (x, y) \neq (0, 0) \\ 1 & \text{if } (x, y) = (0, 0) \end{cases}.$$

Show that all directional derivatives of f exist at $(0, 0)$, and are all equal to zero but the function is not even continuous at $(0, 0)$. Therefore, it is not differentiable. Why?

(8) In the example of Problem 7 show that the partial derivatives exist but are not continuous.

(9) A certain building is shaped like the top half of the ellipsoid, $\frac{x^2}{900} + \frac{y^2}{900} + \frac{z^2}{400} = 1$ determined by letting $z \geq 0$. Here dimensions are measured in feet. The building needs to be painted. The paint, when applied is about .005 feet thick. About how many cubic feet of paint will be needed. **Hint:** This is going to replace the numbers, 900 and 400 with slightly larger numbers when the ellipsoid is fattened slightly by the paint. The volume of the top half of the ellipsoid, $x^2/a^2 + y^2/b^2 + z^2/c^2 \leq 1, z \geq 0$ is $(2/3)\pi abc$.

(10) Suppose $\mathbf{r}_1(t) = (\cos t, \sin t, t)$, $\mathbf{r}_2(t) = (t, 2t, 1)$, and $\mathbf{r}_3(t) = (1, t, 1)$. Find the rate of change with respect to t of the volume of the parallelepiped determined by these three vectors when $t = 1$.

(11) A trash compactor is compacting a rectangular block of trash. The width is changing at the rate of -1 inches per second, the length is changing at the rate of -2 inches per second and the height is changing at the rate of -3 inches per second. How fast is the volume changing when the length is 20, the height is 10, and the width is 10?

(12) A trash compactor is compacting a rectangular block of trash. The width is changing at the rate of -2 inches per second, the length is changing at the rate of -1 inches per second and the height is changing at the rate of -4 inches per second. How fast is the surface area changing when the length is 20, the height is 10, and the width is 10?

(13) The ideal gas law is $PV = kT$ where k is a constant which depends on the number of moles and on the gas being considered. If V is changing at the rate of 2 cubic cm. per second and T is changing at the rate of 3 degrees Kelvin per second, how fast is the pressure changing when $T = 300$ and V equals 400 cubic cm.?

(14) Let S denote a level surface of the form $f(x_1, x_2, x_3) = C$. Show that any smooth curve in the level surface is perpendicular to the gradient.

(15) Suppose $\mathbf{f}$ is a C^1 function which maps U, an open subset of $\mathbb{R}^n$ one to one and onto V, an open set in $\mathbb{R}^m$ such that the inverse map, $\mathbf{f}^{-1}$ is also C^1. What must be true of m and n? Why? **Hint:** Consider Example 8.23 on Page 153. Also you can use the fact that if A is an $m \times n$ matrix which maps $\mathbb{R}^n$ onto $\mathbb{R}^m$, then $m \leq n$.

(16) Finish Example 8.22 by finding f_y in terms of θ, r. Show that $f_y = \sin(\theta) f_r + \frac{\cos(\theta)}{r} f_\theta$.

(17) *Think of ∂_x as a differential operator which takes functions and differentiates them with respect to x. Thus $\partial_x f \equiv f_x$. In the context of Example 8.22, which is on polar coordinates, and Problem 16, explain how

$$\partial_x = \cos(\theta)\partial_r - \frac{\sin(\theta)}{r}\partial_\theta$$

$$\partial_y = \sin(\theta)\partial_r + \frac{\cos(\theta)}{r}\partial_\theta$$

The Laplacian of a function u is defined as $\Delta u = u_{xx} + u_{yy}$. Use the above observation to give a formula Δu in terms of r and θ. You should get $u_{rr} + \frac{1}{r}u_r + \frac{1}{r^2}u_{\theta\theta}$. This is the formula for the Laplacian in polar coordinates.

8.9 The Gradient

Here we review the concept of the gradient and the directional derivative and prove the formula for the directional derivative discussed earlier.

Let $f : U \to \mathbb{R}$ where U is an open subset of $\mathbb{R}^n$ and suppose f is differentiable on U. Thus if $\mathbf{x} \in U$,

$$f(\mathbf{x}+\mathbf{v}) = f(\mathbf{x}) + \sum_{j=1}^{n} \frac{\partial f(\mathbf{x})}{\partial x_i} v_i + o(\mathbf{v}). \tag{8.20}$$

Now we can prove the formula for the directional derivative in terms of the gradient.

Proposition 8.1. If f is differentiable at $\mathbf{x}$ and for $\mathbf{v}$ a unit vector

$$D_{\mathbf{v}}f(\mathbf{x}) = \nabla f(\mathbf{x}) \cdot \mathbf{v}. \tag{8.21}$$

Proof:

$$\frac{f(\mathbf{x}+t\mathbf{v}) - f(\mathbf{x})}{t} = \frac{1}{t}\left(f(\mathbf{x}) + \sum_{j=1}^{n} \frac{\partial f(\mathbf{x})}{\partial x_i} tv_i + o(t\mathbf{v}) - f(\mathbf{x}) \right)$$

$$= \frac{1}{t}\left(\sum_{j=1}^{n} \frac{\partial f(\mathbf{x})}{\partial x_i} tv_i + o(t\mathbf{v}) \right) = \sum_{j=1}^{n} \frac{\partial f(\mathbf{x})}{\partial x_i} v_i + \frac{o(t\mathbf{v})}{t}$$

Now $\lim_{t\to 0} \frac{o(t\mathbf{v})}{t} = 0$ and so

$$D_{\mathbf{v}}f(\mathbf{x}) = \lim_{t\to 0} \frac{f(\mathbf{x}+t\mathbf{v}) - f(\mathbf{x})}{t} = \sum_{j=1}^{n} \frac{\partial f(\mathbf{x})}{\partial x_i} v_i = \nabla f(\mathbf{x}) \cdot \mathbf{v}$$

as claimed. ∎

Example 8.24. Let $f(x, y, z) = x^2 + \sin(xy) + z$. Find $D_{\mathbf{v}}f(1, 0, 1)$ where

$$\mathbf{v} = \left(\frac{1}{\sqrt{3}}, \frac{1}{\sqrt{3}}, \frac{1}{\sqrt{3}} \right).$$

Note this vector which is given is already a unit vector. Therefore, from the above, it is only necessary to find $\nabla f(1, 0, 1)$ and take the dot product.

$$\nabla f(x, y, z) = (2x + (\cos xy)y, (\cos xy)x, 1).$$

Therefore, $\nabla f(1, 0, 1) = (2, 1, 1)$. Therefore, the directional derivative is

$$(2, 1, 1) \cdot \left(\frac{1}{\sqrt{3}}, \frac{1}{\sqrt{3}}, \frac{1}{\sqrt{3}} \right) = \frac{4}{3}\sqrt{3}.$$

Because of (8.21) it is easy to find the largest possible directional derivative and the smallest possible directional derivative. That which follows is a more algebraic treatment of an earlier result with the trigonometry removed.

Proposition 8.2. Let $f : U \to \mathbb{R}$ be a differentiable function and let $\mathbf{x} \in U$. Then

$$\max \{ D_{\mathbf{v}} f(\mathbf{x}) : |\mathbf{v}| = 1 \} = |\nabla f(x)| \tag{8.22}$$

and

$$\min \{ D_{\mathbf{v}} f(\mathbf{x}) : |\mathbf{v}| = 1 \} = - |\nabla f(x)| . \tag{8.23}$$

Furthermore, the maximum in (8.22) occurs when $\mathbf{v} = \nabla f(\mathbf{x}) / |\nabla f(\mathbf{x})|$ and the minimum in (8.23) occurs when $\mathbf{v} = -\nabla f(\mathbf{x}) / |\nabla f(\mathbf{x})|$.

Proof: From (8.21) and the Cauchy Schwarz inequality,

$$|D_{\mathbf{v}} f(\mathbf{x})| \leq |\nabla f(\mathbf{x})|$$

and so for any choice of $\mathbf{v}$ with $|\mathbf{v}| = 1$,

$$- |\nabla f(\mathbf{x})| \leq D_{\mathbf{v}} f(\mathbf{x}) \leq |\nabla f(\mathbf{x})| .$$

The proposition is proved by noting that if $\mathbf{v} = -\nabla f(\mathbf{x}) / |\nabla f(\mathbf{x})|$, then

$$\begin{aligned} D_{\mathbf{v}} f(\mathbf{x}) &= \nabla f(\mathbf{x}) \cdot (-\nabla f(\mathbf{x}) / |\nabla f(\mathbf{x})|) \\ &= - |\nabla f(\mathbf{x})|^2 / |\nabla f(\mathbf{x})| = - |\nabla f(\mathbf{x})| \end{aligned}$$

while if $\mathbf{v} = \nabla f(\mathbf{x}) / |\nabla f(\mathbf{x})|$, then

$$\begin{aligned} D_{\mathbf{v}} f(\mathbf{x}) &= \nabla f(\mathbf{x}) \cdot (\nabla f(\mathbf{x}) / |\nabla f(\mathbf{x})|) \\ &= |\nabla f(\mathbf{x})|^2 / |\nabla f(\mathbf{x})| = |\nabla f(\mathbf{x})| . \end{aligned} \qquad \blacksquare$$

For a different approach to the proposition, see Problem 7 which follows.

The conclusion of the above proposition is important in many physical models. For example, consider some material which is at various temperatures depending on location. Because it has cool places and hot places, it is expected that the heat will flow from the hot places to the cool places. Consider a small surface having a unit normal $\mathbf{n}$. Thus $\mathbf{n}$ is a normal to this surface and has unit length. If it is desired to find the rate in calories per second at which heat crosses this little surface in the direction of $\mathbf{n}$ it is defined as $\mathbf{J} \cdot \mathbf{n} A$ where A is the area of the surface and $\mathbf{J}$ is called the heat flux. It is reasonable to suppose the rate at which heat flows across this surface will be largest when $\mathbf{n}$ is in the direction of greatest rate of decrease of the temperature. In other words, heat flows most readily in the direction which involves the maximum rate of decrease in temperature. This expectation will be realized by taking $\mathbf{J} = -K \nabla u$ where K is a positive scalar function which can depend on a variety of things. The above relation between the heat flux and ∇u is usually called the Fourier heat conduction law and the constant K is known as the coefficient of thermal conductivity. It is a material property, different for iron than for aluminum. In most applications, K is considered to be a constant but this is wrong. Experiments show that this scalar should depend on temperature. Nevertheless, things get very difficult if this dependence is allowed. The constant can depend on position in the material or even on time.

An identical relationship is usually postulated for the flow of a diffusing species. In this problem, something like a pollutant diffuses. It may be an insecticide in ground water for example. Like heat, it tries to move from areas of high concentration toward areas of low concentration. In this case $\mathbf{J} = -K\nabla c$ where c is the concentration of the diffusing species. When applied to diffusion, this relationship is known as Fick's law. Mathematically, it is indistinguishable from the problem of heat flow.

Note the importance of the gradient in formulating these models.

8.10 The Gradient And Tangent Planes

The gradient has fundamental geometric significance illustrated by the following picture.

In this picture, the surface is a piece of a level surface of a function of three variables $f(x, y, z)$. Thus the surface is defined by $f(x, y, z) = c$ or more completely as $\{(x, y, z) : f(x, y, z) = c\}$. For example, if $f(x, y, z) = x^2 + y^2 + z^2$, this would be a piece of a sphere. There are two smooth curves in this picture which lie in the surface having parameterizations, $\mathbf{x}_1(t) = (x_1(t), y_1(t), z_1(t))$ and $\mathbf{x}_2(s) = (x_2(s), y_2(s), z_2(s))$ which intersect at the point (x_0, y_0, z_0) on this surface[1]. This intersection occurs when $t = t_0$ and $s = s_0$. Since the points $\mathbf{x}_1(t)$ for t in an interval lie in the level surface, it follows

$$f(x_1(t), y_1(t), z_1(t)) = c$$

for all t in some interval. Therefore, taking the derivative of both sides and using the chain rule on the left,

$$\frac{\partial f}{\partial x}(x_1(t), y_1(t), z_1(t)) x_1'(t) +$$

---

[1] Do there exist any smooth curves which lie in the level surface of f and pass through the point (x_0, y_0, z_0)? It turns out there do if $\nabla f(x_0, y_0, z_0) \neq \mathbf{0}$ and if the function f, is C^1. However, this is a consequence of the implicit function theorem, one of the greatest theorems in all mathematics and a topic for an advanced calculus class. See the appendix on the subject found in this book for a relatively elementary presentation of this theorem.

$$\frac{\partial f}{\partial y}\left(x_1\left(t\right), y_1\left(t\right), z_1\left(t\right)\right) y_1'\left(t\right) + \frac{\partial f}{\partial z}\left(x_1\left(t\right), y_1\left(t\right), z_1\left(t\right)\right) z_1'\left(t\right) = 0.$$

In terms of the gradient, this merely states

$$\nabla f\left(x_1\left(t\right), y_1\left(t\right), z_1\left(t\right)\right) \cdot \mathbf{x}_1'\left(t\right) = 0.$$

Similarly,

$$\nabla f\left(x_2\left(s\right), y_2\left(s\right), z_2\left(s\right)\right) \cdot \mathbf{x}_2'\left(s\right) = 0.$$

Letting $s = s_0$ and $t = t_0$, it follows

$$\nabla f\left(x_0, y_0, z_0\right) \cdot \mathbf{x}_1'\left(t_0\right) = 0, \ \nabla f\left(x_0, y_0, z_0\right) \cdot \mathbf{x}_2'\left(s_0\right) = 0.$$

It follows $\nabla f\left(x_0, y_0, z_0\right)$ is perpendicular to both the direction vectors of the two indicated curves shown. Surely if things are as they should be, these two direction vectors would determine a plane which deserves to be called the tangent plane to the level surface of f at the point $\left(x_0, y_0, z_0\right)$ and that $\nabla f\left(x_0, y_0, z_0\right)$ is perpendicular to this tangent plane at the point $\left(x_0, y_0, z_0\right)$.

Example 8.25. Find the equation of the tangent plane to the level surface $f\left(x, y, z\right) = 6$ of the function $f\left(x, y, z\right) = x^2 + 2y^2 + 3z^2$ at the point $\left(1, 1, 1\right)$.

First note that $\left(1, 1, 1\right)$ is a point on this level surface. To find the desired plane it suffices to find the normal vector to the proposed plane. But $\nabla f\left(x, y, z\right) = \left(2x, 4y, 6z\right)$ and so $\nabla f\left(1, 1, 1\right) = \left(2, 4, 6\right)$. Therefore, from this problem, the equation of the plane is $\left(2, 4, 6\right) \cdot \left(x - 1, y - 1, z - 1\right) = 0$ or in other words, $2x - 12 + 4y + 6z = 0$.

Example 8.26. The point $\left(\sqrt{3}, 1, 4\right)$ is on both the surfaces, $z = x^2 + y^2$ and $z = 8 - \left(x^2 + y^2\right)$. Find the cosine of the angle between the two tangent planes at this point.

Recall this is the same as the angle between two normal vectors. Of course there is some ambiguity here because if $\mathbf{n}$ is a normal vector, then so is $-\mathbf{n}$ and replacing $\mathbf{n}$ with $-\mathbf{n}$ in the formula for the cosine of the angle will change the sign. We agree to look for the acute angle and its cosine rather than the obtuse angle. The normals are $\left(2\sqrt{3}, 2, -1\right)$ and $\left(2\sqrt{3}, 2, 1\right)$. Therefore, the cosine of the angle desired is

$$\frac{\left(2\sqrt{3}\right)^2 + 4 - 1}{17} = \frac{15}{17}.$$

Example 8.27. The point $\left(1, \sqrt{3}, 4\right)$ is on the surface $z = x^2 + y^2$. Find the line perpendicular to the surface at this point.

All that is needed is the direction vector of this line. The surface is the level surface $x^2 + y^2 - z = 0$. The normal to this surface is given by the gradient at this point. Thus the desired line is

$$\left(1, \sqrt{3}, 4\right) + t\left(2, 2\sqrt{3}, -1\right).$$

8.11 Exercises

(1) Find the gradient of $f =$

    (a) $x^2 y + z^3$ at $(1, 1, 2)$

    (b) $z \sin\left(x^2 y\right) + 2^{x+y}$ at $(1, 1, 0)$

    (c) $u \ln\left(x + y + z^2 + w\right)$ at $(x, y, z, w, u) = (1, 1, 1, 1, 2)$

    (d) $\sin\left(xy\right) + z^3$ at $(1, \pi, 1)$

    (e) $\ln\left(x + y^2\right) z$

    (f) $z \ln\left(4 + \sin\left(xy\right)\right)$ at the point $(0, \pi, 1)$

(2) Find the directional derivatives of f at the indicated point in the direction $\left(\frac{1}{2}, \frac{1}{2}, \frac{1}{\sqrt{2}}\right)$.

    (a) $x^2 y + z^3$ at $(1, 1, 1)$

    (b) $z \sin\left(x^2 y\right) + 2^{x+y}$ at $(1, 1, 0)$

    (c) $xy + z^2 + 1$ at $(1, 2, 3)$

    (d) $\sin\left(xy\right) + z$ at $(0, 1, 1)$

    (e) $x^y + z$ at $(1, 1, 1)$.

    (f) $\sin\left(\sin\left(x + y\right)\right) + z$ at the point $(1, 0, 1)$.

(3) Find the directional derivatives of the given function at the indicated point in the indicated direction.

    (a) $\sin\left(x^2 + y\right) + z^2$ at $(0, \pi/2, 1)$ in direction of $(1, 1, 2)$.

    (b) $x^{(x+y)} + \sin\left(zx\right)$ at $(1, 0, 0)$ in the direction of $(2, -1, 0)$.

    (c) $z^{\sin(x)} + y$ at $(0, 1, 1)$ in the direction of $(1, 1, 3)$.

(4) Find the tangent plane to the indicated level surface at the indicated point.

    (a) $x^2 y + z^3 = 2$ at $(1, 1, 1)$

    (b) $z \sin\left(x^2 y\right) + 2^{x+y} = 2 \sin 1 + 4$ at $(1, 1, 2)$

    (c) $\cos\left(x\right) + z \sin\left(x + y\right) = 1$ at $\left(-\pi, \frac{3\pi}{2}, 2\right)$

(5) The point $(1, 1, \sqrt{2})$ is a point on the level surface $x^2 + y^2 + z^2 = 4$. Find the line perpendicular to the surface at this point.

(6) The level surfaces $x^2 + y^2 + z^2 = 4$ and $z + x^2 + y^2 = 4$ have the point $\left(\frac{\sqrt{2}}{2}, \frac{\sqrt{2}}{2}, 1\right)$ in the curve formed by the intersection of these surfaces. Find a direction vector for this curve at this point. **Hint:** Recall the gradients of the two surfaces are perpendicular to the corresponding surfaces at this point. A direction vector for the desired curve should be perpendicular to both of these gradients.

(7) For $\mathbf{v}$ a unit vector, recall that $D_{\mathbf{v}} f\left(\mathbf{x}\right) = \nabla f\left(\mathbf{x}\right) \cdot \mathbf{v}$. It was shown above that the largest directional derivative is in the direction of the gradient and the smallest in the direction of $-\nabla f$. Establish the same result using the geometric description of the dot product, the one which says the dot product

is the product of the lengths of the vectors times the cosine of the included angle.

(8) The point $\left(1, 1, \sqrt{2}\right)$ is on the level surface $x^2 + y^2 + z^2 = 4$ and the level surface $y^2 + 2z^2 = 5$. Find an equation for the line tangent to the curve of intersection of these two surfaces at this point.

(9) *In a slightly more general setting, suppose $f_1\left(x, y, z\right) = 0$ and $f_2\left(x, y, z\right) = 0$ are two level surfaces which intersect in a curve which has parametrization, $\left(x\left(t\right), y\left(t\right), z\left(t\right)\right)$. Find a system of differential equations for $(x(t), y(t), z(t))$ where as t varies, the point determined by $(x(t), y(t), z(t))$ moves over the curve.

Chapter 9

Optimization

9.1 Local Extrema

The following definition describes what is meant by a local maximum or local minimum.

Definition 9.1. Suppose $f : D(f) \to \mathbb{R}$ where $D(f) \subseteq \mathbb{R}^n$. A point $\mathbf{x} \in D(f) \subseteq \mathbb{R}^n$ is called a **local minimum** if $f(\mathbf{x}) \leq f(\mathbf{y})$ for all $\mathbf{y} \in D(f)$ sufficiently close to $\mathbf{x}$. A point $\mathbf{x} \in D(f)$ is called a **local maximum** if $f(\mathbf{x}) \geq f(\mathbf{y})$ for all $\mathbf{y} \in D(f)$ sufficiently close to $\mathbf{x}$. A **local extremum** is a point of $D(f)$ which is either a local minimum or a local maximum. The plural for extremum is extrema. The plural for minimum is **minima** and the plural for maximum is **maxima**.

Procedure 9.1. To find candidates for local extrema which are interior points of $D(f)$ where f is a differentiable function, you simply identify those points where ∇f equals the zero vector.

To justify this, note that the graph of f is the level surface $F(\mathbf{x}, z) \equiv z - f(\mathbf{x}) = 0$ and the local extrema at such interior points must have horizontal tangent planes. Therefore, a normal vector at such points must be a multiple of $(0, \cdots, 0, 1)$. Thus ∇F at such points must be a multiple of this vector. That is, if $\mathbf{x}$ is such a point $k(0, \cdots, 0, 1) = (-f_{x_1}(\mathbf{x}), \cdots, -f_{x_n}(\mathbf{x}), 1)$ and so $\nabla f(\mathbf{x}) = \mathbf{0}$.

This is illustrated in the following picture.

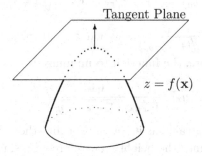

Tangent Plane

$z = f(\mathbf{x})$

A more rigorous explanation is as follows. Let $\mathbf{v}$ be any vector in $\mathbb{R}^n$ and suppose $\mathbf{x}$ is a local maximum (minimum) for $\mathbf{f}$. Then consider the real valued function of one variable, $h(t) \equiv f(\mathbf{x} + t\mathbf{v})$ for small $|t|$. Since $\mathbf{f}$ has a local maximum (minimum), it follows that h is a differentiable function of the single variable t for small t which has a local maximum (minimum) when $t = 0$. Therefore, $h'(0) = 0$. But $h'(t) = Df(\mathbf{x} + t\mathbf{v})\mathbf{v}$ by the chain rule. Therefore,

$$h'(0) = Df(\mathbf{x})\mathbf{v} = 0$$

and since $\mathbf{v}$ is arbitrary, it follows $Df(\mathbf{x}) = 0$. However,

$$Df(\mathbf{x}) = \left(\, f_{x_1}(\mathbf{x}) \quad \cdots \quad f_{x_n}(\mathbf{x}) \, \right)$$

and so $\nabla f(\mathbf{x}) = \mathbf{0}$. This proves the following theorem.

Theorem 9.1. *Suppose U is an open set contained in $D(f)$ such that f is differentiable on U and suppose $\mathbf{x} \in U$ is a local minimum or local maximum for f. Then $\nabla f(\mathbf{x}) = \mathbf{0}$.*

Definition 9.2. A **singular point** for f is a point $\mathbf{x}$ where $\nabla f(\mathbf{x}) = \mathbf{0}$. This is also called a **critical point**. By analogy with the one variable case, a point where the gradient does not exist will also be called a critical point.

Example 9.1. Find the critical points for the function $f(x, y) \equiv xy - x - y$ for $x, y > 0$.

Note that here $D(f)$ is an open set and so every point is an interior point. Where is the gradient equal to zero? $f_x = y - 1 = 0$, $f_y = x - 1 = 0$, and so there is exactly one critical point $(1, 1)$.

Example 9.2. Find the volume of the smallest tetrahedron made up of the coordinate planes in the first octant and a plane which is tangent to the sphere $x^2 + y^2 + z^2 = 4$.

The normal to the sphere at a point (x_0, y_0, z_0) of the sphere is $\left(x_0, y_0, \sqrt{4 - x_0^2 - y_0^2}\right)$ and so the equation of the tangent plane at this point is

$$x_0(x - x_0) + y_0(y - y_0) + \sqrt{4 - x_0^2 - y_0^2}\left(z - \sqrt{4 - x_0^2 - y_0^2}\right) = 0$$

When $x = y = 0$, $z = \dfrac{4}{\sqrt{(4 - x_0^2 - y_0^2)}}$. When $z = 0 = y$, $x = \dfrac{4}{x_0}$, and when $z = x = 0$, $y = \dfrac{4}{y_0}$. Therefore, the function to minimize is

$$f(x, y) = \frac{1}{6} \frac{64}{xy\sqrt{(4 - x^2 - y^2)}}$$

This is because in beginning calculus it was shown that the volume of a pyramid is $1/3$ the area of the base times the height. Therefore, you simply need to find the gradient of this and set it equal to zero. Thus upon taking the partial derivatives, you need to have

$$\frac{-4 + 2x^2 + y^2}{x^2 y \left(-4 + x^2 + y^2\right) \sqrt{(4 - x^2 - y^2)}} = 0,$$

and

$$\frac{-4 + x^2 + 2y^2}{xy^2 \left(-4 + x^2 + y^2\right) \sqrt{(4 - x^2 - y^2)}} = 0.$$

Therefore, $x^2 + 2y^2 = 4$ and $2x^2 + y^2 = 4$. Thus $x = y$ and so $x = y = \frac{2}{\sqrt{3}}$. It follows from the equation for z that $z = \frac{2}{\sqrt{3}}$ also. How do you know this is not the largest tetrahedron?

Example 9.3. An open box is to contain 32 cubic feet. Find the dimensions which will result in the least surface area.

Let the height of the box be z and the length and width be x and y respectively. Then $xyz = 32$ and so $z = 32/xy$. The total area is $xy + 2xz + 2yz$ and so in terms of the two variables x and y, the area is $A = xy + \frac{64}{y} + \frac{64}{x}$. To find best dimensions you note these must result in a local minimum.

$$A_x = \frac{yx^2 - 64}{x^2} = 0, \ A_y = \frac{xy^2 - 64}{y^2}.$$

Therefore, $yx^2 - 64 = 0$ and $xy^2 - 64 = 0$ so $xy^2 = yx^2$. For sure the answer excludes the case where any of the variables equals zero. Therefore, $x = y$ and so $x = 4 = y$. Then $z = 2$ from the requirement that $xyz = 32$. How do you know this gives the least surface area? Why is this not the largest surface area?

9.2 Exercises

(1) Find the points where possible local minima or local maxima occur in the following functions.
   (a) $x^2 - 2x + 5 + y^2 - 4y$
   (b) $-xy + y^2 - y + x$
   (c) $3x^2 - 4xy + 2y^2 - 2y + 2x$
   (d) $\cos(x) + \sin(2y)$
   (e) $x^4 - 4x^3y + 6x^2y^2 - 4xy^3 + y^4 + x^2 - 2x$
   (f) $y^2 x^2 - 2xy^2 + y^2$
(2) Find the volume of the largest box which can be inscribed in a sphere of radius a.
(3) Find in terms of a, b, c the volume of the largest box which can be inscribed in the ellipsoid $\frac{x^2}{a^2} + \frac{y^2}{b^2} + \frac{z^2}{c^2} = 1$.
(4) Find three numbers which add to 36 whose product is as large as possible.

(5) Find three numbers x, y, z such that $x^2 + y^2 + z^2 = 1$ and $x + y + z$ is as large as possible.

(6) Find three numbers x, y, z such that $x^2 + y^2 + z^2 = 4$ and xyz is as large as possible.

(7) A feeding trough in the form of a trapezoid with equal base angles is made from a long rectangular piece of metal of width 24 inches by bending up equal strips along both sides. Find the base angles and the width of these strips which will maximize the volume of the feeding trough.

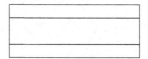

(8) An open box (no top) is to contain 40 cubic feet. The material for the bottom costs twice as much as the material for the sides. Find the dimensions of the box which is cheapest.

(9) The function $f(x, y) = 2x^2 + y^2$ is defined on the disk $x^2 + y^2 \leq 1$. Find its maximum value.

(10) Find the point on the surface $z = x^2 + y + 1$ which is closest to $(0, 0, 0)$.

(11) Let $L_1 = (t, 2t, 3 - t)$ and $L_2 = (2s, s + 2, 4 - s)$ be two lines. Find a pair of points, one on the first line and the other on the second such that these two points are closer together than any other pair of points on the two lines.

(12) *Let

$$
f(x, y) = \begin{cases} -1 \text{ if } y = x^2, x \neq 0 \\ (y - x^2)^2 \text{ if } y \neq x^2 \\ 0 \text{ if } (x, y) = (0, 0) \end{cases}
$$

Show that $\nabla f(0, 0) = \mathbf{0}$. Now show that if (a, b) is any nonzero unit vector, the function $t \to f(ta, tb)$ has a local minimum of 0 when $t = 0$. Thus in every direction, this function has a local minimum at $(0, 0)$ but the function f does not have a local minimum at $(0, 0)$.

9.3   The Second Derivative Test

There is a version of the second derivative test in the case that the function and its first and second partial derivatives are all continuous. The proof of this theorem is dependent on fundamental results in linear algebra which are in an appendix. You can skip the proof if you like.

Definition 9.3. The matrix $H(\mathbf{x})$ whose ij^{th} entry at the point $\mathbf{x}$ is $\frac{\partial^2 f}{\partial x_i \partial x_j}(\mathbf{x})$ is called the **Hessian matrix**. The eigenvalues of $H(\mathbf{x})$ are the solutions λ to the equation $\det(\lambda I - H(\mathbf{x})) = 0$.

The following theorem says that if all the eigenvalues of the Hessian matrix at a critical point are positive, then the critical point is a local minimum. If all the eigenvalues of the Hessian matrix at a critical point are negative, then the critical point is a local maximum. Finally, if some of the eigenvalues of the Hessian matrix at the critical point are positive and some are negative then the critical point is a saddle point. The following picture illustrates the situation.

Theorem 9.2. *Let $f : U \to \mathbb{R}$ for U an open set in $\mathbb{R}^n$ and let f be a C^2 function and suppose that at some $\mathbf{x} \in U$, $\nabla f(\mathbf{x}) = \mathbf{0}$. Also let μ and λ be respectively, the largest and smallest eigenvalues of the matrix $H(\mathbf{x})$. If $\lambda > 0$ then f has a local minimum at $\mathbf{x}$. If $\mu < 0$ then f has a local maximum at $\mathbf{x}$. If either λ or μ equals zero, the test fails. If $\lambda < 0$ and $\mu > 0$ there exists a direction in which when f is evaluated on the line through the critical point having this direction, the resulting function of one variable has a local minimum and there exists a direction in which when f is evaluated on the line through the critical point having this direction, the resulting function of one variable has a local maximum. This last case is called a* **saddle point.**

Here is an example.

Example 9.4. Let $f(x, y) = 10xy + y^2$. Find the critical points and determine whether they are local minima, local maxima or saddle points.

First $\nabla\left(10xy + y^2\right) = (10y, 10x + 2y)$ and so there is one critical point at the point $(0, 0)$. What is it? The Hessian matrix is

$$\begin{pmatrix} 0 & 10 \\ 10 & 2 \end{pmatrix}$$

and the eigenvalues are of different signs. Therefore, the critical point $(0, 0)$ is a saddle point. Here is a graph drawn by Maple.

Here is another example.

Example 9.5. Let $f(x,y) = 2x^4 - 4x^3 + 14x^2 + 12yx^2 - 12yx - 12x + 2y^2 + 4y + 2$. Find the critical points and determine whether they are local minima, local maxima, or saddle points.

$f_x(x,y) = 8x^3 - 12x^2 + 28x + 24yx - 12y - 12$ and $f_y(x,y) = 12x^2 - 12x + 4y + 4$. The points at which both f_x and f_y equal zero are $\left(\frac{1}{2}, -\frac{1}{4}\right), (0,-1)$, and $(1,-1)$.

The Hessian matrix is

$$\begin{pmatrix} 24x^2 + 28 + 24y - 24x & 24x - 12 \\ 24x - 12 & 4 \end{pmatrix}$$

and the thing to determine is the sign of its eigenvalues evaluated at the critical points.

First consider the point $\left(\frac{1}{2}, -\frac{1}{4}\right)$. The Hessian matrix is $\begin{pmatrix} 16 & 0 \\ 0 & 4 \end{pmatrix}$ and its eigenvalues are $16, 4$ showing that this is a local minimum.

Next consider $(0,-1)$ at this point the Hessian matrix is $\begin{pmatrix} 4 & -12 \\ -12 & 4 \end{pmatrix}$ and the eigenvalues are $16, -8$. Therefore, this point is a saddle point. To determine this, find the eigenvalues.

$$\det\left(\lambda \begin{pmatrix} 1 & 0 \\ 0 & 1 \end{pmatrix} - \begin{pmatrix} 4 & -12 \\ -12 & 4 \end{pmatrix}\right) = \lambda^2 - 8\lambda - 128 = (\lambda + 8)(\lambda - 16)$$

so the eigenvalues are -8 and 16 as claimed.

Finally consider the point $(1,-1)$. At this point the Hessian is $\begin{pmatrix} 4 & 12 \\ 12 & 4 \end{pmatrix}$ and the eigenvalues are $16, -8$ so this point is also a saddle point.

Below is a graph of this function which illustrates the behavior near saddle points.

Or course sometimes the second derivative test is inadequate to determine what is going on. This should be no surprise since this was the case even for a function of one variable. For a function of two variables, a nice example is the Monkey saddle.

Example 9.6. Suppose $f(x,y) = 6xy^2 - 2x^3 - 3y^4$. Show that $(0,0)$ is a critical point for which the second derivative test gives no information.

Before doing anything it might be interesting to look at the graph of this function of two variables plotted using Maple.

This picture should indicate why this is called a monkey saddle. It is because the monkey can sit in the saddle and have a place for his tail. Now to see $(0,0)$ is a critical point, note that $f_x(0,0) = f_y(0,0) = 0$ because $f_x(x,y) = 6y^2 - 6x^2$, $f_y(x,y) = 12xy - 12y^3$ and so $(0,0)$ is a critical point. So are $(1,1)$ and $(1,-1)$. Now $f_{xx}(0,0) = 0$ and so are $f_{xy}(0,0)$ and $f_{yy}(0,0)$. Therefore, the Hessian matrix is the zero matrix and clearly has only the zero eigenvalue. Therefore, the second derivative test is totally useless at this point.

However, suppose you took $x = t$ and $y = t$ and evaluated this function on this line. This reduces to $h(t) = f(t,t) = 4t^3 - 3t^4)$, which is strictly increasing near $t = 0$. This shows the critical point $(0,0)$ of f is neither a local max. nor a local min. Next let $x = 0$ and $y = t$. Then $p(t) \equiv f(0,t) = -3t^4$. Therefore, along the line, $(0,t)$, f has a local maximum at $(0,0)$.

Example 9.7. Find the critical points of the following function of three variables and classify them as local minimums, local maximums or saddle points.

$$f(x,y,z) = \frac{5}{6}x^2 + 4x + 16 - \frac{7}{3}xy - 4y - \frac{4}{3}xz + 12z + \frac{5}{6}y^2 - \frac{4}{3}zy + \frac{1}{3}z^2$$

First you need to locate the critical points. This involves taking the gradient.

$$\nabla\left(\frac{5}{6}x^2 + 4x + 16 - \frac{7}{3}xy - 4y - \frac{4}{3}xz + 12z + \frac{5}{6}y^2 - \frac{4}{3}zy + \frac{1}{3}z^2\right)$$
$$= \left(\frac{5}{3}x + 4 - \frac{7}{3}y - \frac{4}{3}z, -\frac{7}{3}x - 4 + \frac{5}{3}y - \frac{4}{3}z, -\frac{4}{3}x + 12 - \frac{4}{3}y + \frac{2}{3}z\right)$$

Next you need to set the gradient equal to zero and solve the equations. This yields $y = 5, x = 3, z = -2$. Now to use the second derivative test, you assemble the Hessian matrix which is

$$\begin{pmatrix} \frac{5}{3} & -\frac{7}{3} & -\frac{4}{3} \\ -\frac{7}{3} & \frac{5}{3} & -\frac{4}{3} \\ -\frac{4}{3} & -\frac{4}{3} & \frac{2}{3} \end{pmatrix}.$$

Note that in this simple example, the Hessian matrix is constant and so all that is left is to consider the eigenvalues. Writing the characteristic equation and solving yields the eigenvalues are $2, -2, 4$. Thus the given point is a saddle point.

9.4   Exercises

(1) Use the second derivative test on the critical points $(1, 1)$, and $(1, -1)$ for Example 9.6. The function is arctan $6xy^2 - 2x^3 - 3x^4$.

(2) If $H = H^T$ and $H\mathbf{x} = \lambda\mathbf{x}$ while $H\mathbf{x} = \mu\mathbf{x}$ for $\lambda \neq \mu$, show that $\mathbf{x} \cdot \mathbf{y} = 0$.

(3) Show that the points $\left(\frac{1}{2}, -\frac{21}{4}\right), (0, -4)$, and $(1, -4)$ are critical points of the following function of two variables and classify them as local minima, local maxima or saddle points.

(4) Show that the points $\left(\frac{1}{2}, -\frac{53}{12}\right), (0, -4)$, and $(1, -4)$ are critical points of the following function of two variables and classify them according to whether they are local minima, local maxima or saddle points.

(5) Show that the points $\left(\frac{1}{2}, \frac{37}{20}\right), (0, 2)$, and $(1, 2)$ are critical points of the following function of two variables and classify them according to whether they are local minima, local maxima or saddle points.

(6) Show that the points $\left(\frac{1}{2}, -\frac{17}{8}\right), (0, -2)$, and $(1, -2)$ are critical points of the following function of two variables and classify them according to whether they are local minima, local maxima or saddle points.

(7) Find the critical points of the following function of three variables and classify them according to whether they are local minima, local maxima or saddle points.

(8) Find the critical points of the following function of three variables and classify them according to whether they are local minima, local maxima or saddle points.

(9) Find the critical points of the following function of three variables and classify them according to whether they are local minima, local maxima or saddle points.

(10) Find the critical points of the following function of three variables and classify them according to whether they are local minima, local maxima or saddle points.

(11) *Show that if f has a critical point and some eigenvalue of the Hessian matrix is positive, then there exists a direction in which when f is evaluated on the line through the critical point having this direction, the resulting function of one variable has a local minimum. State and prove a similar result in the case where some eigenvalue of the Hessian matrix is negative.

(12) Suppose $\mu = 0$ but there are negative eigenvalues of the Hessian at a critical point. Show by giving examples that the second derivative tests fails.

(13) Show that the points $\left(\frac{1}{2}, -\frac{9}{2}\right), (0, -5)$, and $(1, -5)$ are critical points of the following function of two variables and classify them as local minima, local

maxima or saddle points.
$$f(x,y) = 2x^4 - 4x^3 + 42x^2 + 8yx^2 - 8yx - 40x + 2y^2 + 20y + 50.$$

(14) Show that the points $\left(1, -\frac{11}{2}\right), (0, -5)$, and $(2, -5)$ are critical points of the following function of two variables and classify them as local minima, local maxima or saddle points.
$$f(x,y) = 4x^4 - 16x^3 - 4x^2 - 4yx^2 + 8yx + 40x + 4y^2 + 40y + 100.$$

(15) Show that the points $\left(\frac{3}{2}, \frac{27}{20}\right), (0,0)$, and $(3,0)$ are critical points of the following function of two variables and classify them as local minima, local maxima or saddle points.
$$f(x,y) = 5x^4 - 30x^3 + 45x^2 + 6yx^2 - 18yx + 5y^2.$$

(16) Find the critical points of the following function of three variables and classify them as local minima, local maxima or saddle points.
$$f(x,y,z) = \frac{10}{3}x^2 - \frac{44}{3}x + \frac{64}{3} - \frac{10}{3}yx + \frac{16}{3}y + \frac{2}{3}zx - \frac{20}{3}z + \frac{10}{3}y^2 + \frac{2}{3}zy + \frac{4}{3}z^2.$$

(17) Find the critical points of the following function of three variables and classify them as local minima, local maxima or saddle points.
$$f(x,y,z) = -\frac{7}{3}x^2 - \frac{146}{3}x + \frac{83}{3} + \frac{16}{3}yx + \frac{4}{3}y - \frac{14}{3}zx + \frac{94}{3}z - \frac{7}{3}y^2 - \frac{14}{3}zy + \frac{8}{3}z^2.$$

(18) Find the critical points of the following function of three variables and classify them as local minima, local maxima or saddle points.
$$f(x,y,z) = \frac{2}{3}x^2 + 4x + 75 - \frac{14}{3}yx - 38y - \frac{8}{3}zx - 2z + \frac{2}{3}y^2 - \frac{8}{3}zy - \frac{1}{3}z^2.$$

(19) Find the critical points of the following function of three variables and classify them as local minima, local maxima or saddle points.
$$f(x,y,z) = 4x^2 - 30x + 510 - 2yx + 60y - 2zx - 70z + 4y^2 - 2zy + 4z^2.$$

(20) Show that the critical points of the following function are points of the form, $(x,y,z) = (t, 2t^2 - 10t, -t^2 + 5t)$ for $t \in \mathbb{R}$ and classify them as local minima, local maxima or saddle points.
$$f(x,y,z) = -\frac{1}{6}x^4 + \frac{5}{3}x^3 - \frac{25}{6}x^2 + \frac{10}{3}yx^2 - \frac{50}{3}yx + \frac{19}{3}zx^2 - \frac{95}{3}zx - \frac{5}{3}y^2 - \frac{10}{3}zy - \frac{1}{6}z^2.$$

(21) Show that the critical points of the following function are
$$(0,-3,0), (2,-3,0), \text{ and } \left(1,-3,-\frac{1}{3}\right)$$
and classify them as local minima, local maxima or saddle points.
$$f(x,y,z) = -\frac{3}{2}x^4 + 6x^3 - 6x^2 + zx^2 - 2zx - 2y^2 - 12y - 18 - \frac{3}{2}z^2.$$

(22) Show that the critical points of the function $f(x,y,z) = -2yx^2 - 6yx - 4zx^2 - 12zx + y^2 + 2yz$. are points of the form, $(x,y,z) = (t, 2t^2 + 6t, -t^2 - 3t)$ for $t \in \mathbb{R}$ and classify them as local minima, local maxima or saddle points.

(23) Show that the critical points of the function
$$f(x,y,z) = \frac{1}{2}x^4 - 4x^3 + 8x^2 - 3zx^2 + 12zx + 2y^2 + 4y + 2 + \frac{1}{2}z^2.$$
are $(0,-1,0), (4,-1,0)$, and $(2,-1,-12)$ and classify them as local minima, local maxima or saddle points.

(24) Suppose $f(x,y)$, a function of two variables defined on all $\mathbb{R}^n$ has all directional derivatives at $(0,0)$ and they are all equal to 0 there. Suppose also that for $h(t) \equiv f(tu, tv)$ and (u,v) a unit vector, it follows that $h''(0) > 0$. By the

one variable second derivative test, this implies that along every straight line through $(0,0)$ the function restricted to this line has a local minimum at $(0,0)$. Can it be concluded that f has a local minimum at $(0,0)$. In other words, can you conclude a point is a local minimum if it appears to be so along every straight line through the point? **Hint:** Consider $f(x,y) = x^2 + y^2$ for (x,y) not on the curve $y = x^2$ for $x \neq 0$ and on this curve, let $f = -1$.

9.5 Lagrange Multipliers

Lagrange multipliers are used to solve extremum problems for a function defined on a level set of another function. For example, suppose you want to maximize xy given that $x + y = 4$. This is not too hard to do using methods developed earlier. Solve for one of the variables, say y, in the constraint equation $x + y = 4$ to find $y = 4 - x$. Then the function to maximize is $f(x) = x(4 - x)$ and the answer is clearly $x = 2$. Thus the two numbers are $x = y = 2$. This was easy because you could easily solve the constraint equation for one of the variables in terms of the other. Now what if you wanted to maximize $f(x, y, z) = xyz$ subject to the constraint that $x^2 + y^2 + z^2 = 4$? It is still possible to do this using similar techniques. Solve for one of the variables in the constraint equation, say z, substitute it into f, and then find where the partial derivatives equal zero to find candidates for the extremum. However, it seems you might encounter many cases and it does look a little fussy. However, sometimes you can't solve the constraint equation for one variable in terms of the others. Also, what if you had many constraints? What if you wanted to maximize $f(x, y, z)$ subject to the constraints $x^2 + y^2 = 4$ and $z = 2x + 3y^2$. Things are clearly getting more involved and messy. It turns out that at an extremum, there is a simple relationship between the gradient of the function to be maximized and the gradient of the constraint function.

This relation can be seen geometrically in the following picture.

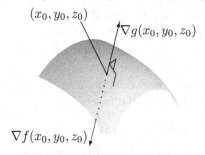

(x_0, y_0, z_0)

$\nabla g(x_0, y_0, z_0)$

$\nabla f(x_0, y_0, z_0)$

In the picture, the surface represents a piece of the level surface of $g(x, y, z) = 0$ and $f(x, y, z)$ is the function of three variables which is being maximized or minimized on the level surface. Suppose the extremum of f occurs at the point (x_0, y_0, z_0). As shown above, $\nabla g(x_0, y_0, z_0)$ is perpendicular to the surface or more precisely to the tangent plane. However, if $\mathbf{x}(t) = (x(t), y(t), z(t))$ is a point on a smooth curve which passes through (x_0, y_0, z_0) when $t = t_0$, then the function $h(t) = f(x(t), y(t), z(t))$ must have either a maximum or a minimum at the point $t = t_0$. Therefore, $h'(t_0) = 0$. But this means

$$0 = h'(t_0) = \nabla f(x(t_0), y(t_0), z(t_0)) \cdot \mathbf{x}'(t_0) = \nabla f(x_0, y_0, z_0) \cdot \mathbf{x}'(t_0)$$

and since this holds for any such smooth curve, $\nabla f(x_0, y_0, z_0)$ is also perpendicular to the surface. This picture represents a situation in three dimensions and you can see that it is intuitively clear that this implies $\nabla f(x_0, y_0, z_0)$ is some scalar multiple of $\nabla g(x_0, y_0, z_0)$. Thus

$$\nabla f(x_0, y_0, z_0) = \lambda \nabla g(x_0, y_0, z_0)$$

This λ is called a **Lagrange multiplier** after Lagrange who considered such problems in the 1700's.

Of course the above argument is at best only heuristic. It does not deal with the question of existence of smooth curves lying in the constraint surface passing through (x_0, y_0, z_0). Nor does it consider all cases, being essentially confined to three dimensions. In addition to this, it fails to consider the situation in which there are many constraints. However, I think it is likely a geometric notion like that presented above which led Lagrange to formulate the method.

Example 9.8. Maximize xyz subject to $x^2 + y^2 + z^2 = 27$.

Here $f(x, y, z) = xyz$ while $g(x, y, z) = x^2 + y^2 + z^2 - 27$. Then $\nabla g(x, y, z) = (2x, 2y, 2z)$ and $\nabla f(x, y, z) = (yz, xz, xy)$. Then at the point which maximizes this function[1], $(yz, xz, xy) = \lambda(2x, 2y, 2z)$. Therefore, each of $2\lambda x^2, 2\lambda y^2, 2\lambda z^2$ equals xyz. It follows that at any point which maximizes xyz, $|x| = |y| = |z|$. Therefore, the only candidates for the point where the maximum occurs are $(3, 3, 3), (-3, -3, 3)(-3, 3, 3)$, etc. The maximum occurs at $(3, 3, 3)$ which can be verified by plugging in to the function which is being maximized.

The method of Lagrange multipliers allows you to consider maximization of functions defined on closed and bounded sets. Recall that any continuous function defined on a closed and bounded set has a maximum and a minimum on the set. Candidates for the extremum on the interior of the set can be located by setting the gradient equal to zero. The consideration of the boundary can then sometimes be handled with the method of Lagrange multipliers.

Example 9.9. Maximize $f(x, y) = xy + y$ subject to the constraint, $x^2 + y^2 \le 1$.

Here I know there is a maximum because the set is the closed disk, a closed and bounded set. Therefore, it is just a matter of finding it. Look for singular points on the interior of the circle. $\nabla f(x, y) = (y, x + 1) = (0, 0)$. There are no points on the interior of the circle where the gradient equals zero. Therefore, the maximum

---

[1]There exists such a point because the sphere is closed and bounded.

occurs on the boundary of the circle. That is, the problem reduces to maximizing $xy + y$ subject to $x^2 + y^2 = 1$. From the above,

$$(y, x + 1) - \lambda (2x, 2y) = 0.$$

Hence $y^2 - 2\lambda xy = 0$ and $x (x + 1) - 2\lambda xy = 0$ so $y^2 = x (x + 1)$. Therefore from the constraint, $x^2 + x (x + 1) = 1$ and the solution is $x = -1, x = \frac{1}{2}$. Then the candidates for a solution are $(-1, 0), \left(\frac{1}{2}, \frac{\sqrt{3}}{2} \right), \left(\frac{1}{2}, \frac{-\sqrt{3}}{2} \right)$. Then

$$f(-1, 0) = 0, f\left(\frac{1}{2}, \frac{\sqrt{3}}{2} \right) = \frac{3\sqrt{3}}{4}, f\left(\frac{1}{2}, -\frac{\sqrt{3}}{2} \right) = -\frac{3\sqrt{3}}{4}.$$

It follows the maximum value of this function is $\frac{3\sqrt{3}}{4}$ and it occurs at $\left(\frac{1}{2}, \frac{\sqrt{3}}{2} \right)$. The minimum value is $-\frac{3\sqrt{3}}{4}$ and it occurs at $\left(\frac{1}{2}, -\frac{\sqrt{3}}{2} \right)$.

Example 9.10. Find candidates for the maximum and minimum values of the function $f(x, y) = xy - x^2$ on the set $\{(x, y) : x^2 + 2xy + y^2 \le 4\}$.

First, the only point where ∇f equals zero is $(x, y) = (0, 0)$ and this is in the desired set. In fact it is an interior point of this set. This takes care of the interior points. What about those on the boundary $x^2 + 2xy + y^2 = 4$? The problem is to maximize $xy - x^2$ subject to the constraint, $x^2 + 2xy + y^2 = 4$. The Lagrangian is $xy - x^2 - \lambda (x^2 + 2xy + y^2 - 4)$ and this yields the following system.

$$y - 2x - \lambda (2x + 2y) = 0$$
$$x - 2\lambda (x + y) = 0$$
$$x^2 + 2xy + y^2 = 4$$

From the first two equations,

$$(2 + 2\lambda) x - (1 - 2\lambda) y = 0, \ (1 - 2\lambda) x - 2\lambda y = 0$$

Since not both x and y equal zero, it follows

$$\det \begin{pmatrix} 2 + 2\lambda & 2\lambda - 1 \\ 1 - 2\lambda & -2\lambda \end{pmatrix} = 0$$

which yields $\lambda = 1/8$. Therefore, $y = 3x$. From the constraint equation $x^2 + 2x (3x) + (3x)^2 = 4$ and so $x = \frac{1}{2}$ or $-\frac{1}{2}$. Now since $y = 3x$, the points of interest on the boundary of this set are

$$\left(\frac{1}{2}, \frac{3}{2} \right), \text{ and } \left(-\frac{1}{2}, -\frac{3}{2} \right). \tag{9.1}$$

$$f\left(\frac{1}{2}, \frac{3}{2} \right) = \left(\frac{1}{2} \right) \left(\frac{3}{2} \right) - \left(\frac{1}{2} \right)^2 = \frac{1}{2}$$

$$f\left(-\frac{1}{2}, -\frac{3}{2} \right) = \left(-\frac{1}{2} \right) \left(-\frac{3}{2} \right) - \left(-\frac{1}{2} \right)^2 = \frac{1}{2}$$

It follows the candidates for maximum and minimum are $\left(\frac{1}{2}, \frac{3}{2}\right), (0,0)$, and $\left(-\frac{1}{2}, -\frac{3}{2}\right)$. Therefore it appears that $(0,0)$ yields a minimum and either $\left(\frac{1}{2}, \frac{3}{2}\right)$ or $\left(-\frac{1}{2}, -\frac{3}{2}\right)$ yields a maximum. However, this is a little misleading. How do you even know a maximum or a minimum exists? The set $x^2 + 2xy + y^2 \leq 4$ is an unbounded set which lies between the two lines $x + y = 2$ and $x + y = -2$. In fact there is no minimum. For example, take $x = 100, y = -98$. Then $xy - x^2 = x(y - x) = 100(-98 - 100)$ which is a large negative number much less than 0, the answer for the point $(0,0)$.

There are no magic bullets here. It was still required to solve a system of nonlinear equations to get the answer. However, it does often help to do it this way.

A nice observation in the case that the function f, which you are trying to maximize, and the function g, which defines the constraint, are functions of two or three variables is the following.

At points of interest,

$$\nabla f \times \nabla g = \mathbf{0}$$

This follows from the above because at these points,

$$\nabla f = \lambda \nabla g$$

so the angle between the two vectors ∇f and ∇g is either 0 or π. Therefore, the sine of this angle equals 0. By the geometric description of the cross product, this implies the cross product equals 0. Here is an example.

Example 9.11. Minimize $f(x, y) = xy - x^2$ on the set

$$\left\{(x, y) : x^2 + 2xy + y^2 = 4\right\}$$

Using the observation about the cross product, and letting $f(x, y, z) = f(x, y)$ with a similar convention for g, $\nabla f = (y - 2x, x, 0)$, $\nabla g = (2x + 2y, 2x + 2y, 0)$ and so

$$(y - 2x, x, 0) \times (2x + 2y, 2x + 2y, 0)$$
$$= (0, 0, (y - 2x)(2x + 2y) - x(2x + 2y)) = 0$$

Thus there are two equations, $x^2 + 2xy + y^2 = 4$ and $4xy - 2y^2 + 6x^2 = 0$. Solving these two yields the points of interest $\left(-\frac{1}{2}, -\frac{3}{2}\right), \left(\frac{1}{2}, \frac{3}{2}\right)$. Now one of these gives the minimum and the other gives the maximum.

The above generalizes to a general procedure which is described in the following major Theorem. All correct proofs of this theorem will involve some appeal to the implicit function theorem or to fundamental existence theorems from differential equations. A complete proof is very fascinating but it will not come cheap. Good advanced calculus books will usually give a correct proof. If you are interested, there is a complete proof in an appendix to this book. First here is a simple definition explaining one of the terms in the statement of this theorem.

Definition 9.4. Let A be an $m \times n$ matrix. A submatrix is any matrix which can be obtained from A by deleting some rows and some columns.

Theorem 9.3. *Let U be an open subset of $\mathbb{R}^n$ and let $f : U \to \mathbb{R}$ be a C^1 function. Then if $\mathbf{x}_0 \in U$ is either a local maximum or local minimum of f subject to the constraints*

$$g_i(\mathbf{x}) = 0, \; i = 1, \cdots, m \tag{9.2}$$

and if some $m \times m$ submatrix of

$$D\mathbf{g}(\mathbf{x}_0) \equiv \begin{pmatrix} g_{1x_1}(\mathbf{x}_0) & g_{1x_2}(\mathbf{x}_0) & \cdots & g_{1x_n}(\mathbf{x}_0) \\ \vdots & \vdots & & \vdots \\ g_{mx_1}(\mathbf{x}_0) & g_{mx_2}(\mathbf{x}_0) & \cdots & g_{mx_n}(\mathbf{x}_0) \end{pmatrix}$$

has nonzero determinant, then there exist scalars, $\lambda_1, \cdots, \lambda_m$ such that

$$\begin{pmatrix} f_{x_1}(\mathbf{x}_0) \\ \vdots \\ f_{x_n}(\mathbf{x}_0) \end{pmatrix} = \lambda_1 \begin{pmatrix} g_{1x_1}(\mathbf{x}_0) \\ \vdots \\ g_{1x_n}(\mathbf{x}_0) \end{pmatrix} + \cdots + \lambda_m \begin{pmatrix} g_{mx_1}(\mathbf{x}_0) \\ \vdots \\ g_{mx_n}(\mathbf{x}_0) \end{pmatrix} \tag{9.3}$$

holds.

To help remember how to use (9.3), do the following. First write the Lagrangian,

$$L = f(\mathbf{x}) - \sum_{i=1}^{m} \lambda_i g_i(\mathbf{x})$$

and then proceed to take derivatives with respect to each of the components of $\mathbf{x}$ and also derivatives with respect to each λ_i and set all of these equations equal to 0. The formula (9.3) is what results from taking the derivatives of L with respect to the components of $\mathbf{x}$. When you take the derivatives with respect to the Lagrange multipliers, and set what results equal to 0, you just pick up the constraint equations. This yields $n + m$ equations for the $n + m$ unknowns $x_1, \cdots, x_n, \lambda_1, \cdots, \lambda_m$. Then you proceed to look for solutions to these equations. Of course these might be impossible to find using methods of algebra, but you just do your best and hope it will work out.

Example 9.12. Minimize xyz subject to the constraints $x^2 + y^2 + z^2 = 4$ and $x - 2y = 0$.

Form the Lagrangian,

$$L = xyz - \lambda(x^2 + y^2 + z^2 - 4) - \mu(x - 2y)$$

and proceed to take derivatives with respect to every possible variable, leading to the following system of equations.

$$\begin{aligned} yz - 2\lambda x - \mu &= 0 \\ xz - 2\lambda y + 2\mu &= 0 \\ xy - 2\lambda z &= 0 \\ x^2 + y^2 + z^2 &= 4 \\ x - 2y &= 0 \end{aligned}$$

Now you have to find the solutions to this system of equations. In general, this could be very hard or even impossible. If $\lambda = 0$, then from the third equation, either x or y must equal 0. Therefore, from the first two equations, $\mu = 0$ also. If $\mu = 0$ and $\lambda \neq 0$, then from the first two equations, $xyz = 2\lambda x^2$ and $xyz = 2\lambda y^2$ and so either $x = y$ or $x = -y$, which requires that both x and y equal zero thanks to the last equation. But then from the fourth equation, $z = \pm 2$ and now this contradicts the third equation. Thus μ and λ are either both equal to zero or neither one is and the expression, xyz equals zero in this case. However, I know this is not the best value for a minimizer because I can take $x = 2\sqrt{\frac{3}{5}}, y = \sqrt{\frac{3}{5}}$, and $z = -1$. This satisfies the constraints and the product of these numbers equals a negative number. Therefore, both μ and λ must be non zero. Now use the last equation eliminate x and write the following system.

$$5y^2 + z^2 = 4$$
$$y^2 - \lambda z = 0$$
$$yz - \lambda y + \mu = 0$$
$$yz - 4\lambda y - \mu = 0$$

From the last equation, $\mu = (yz - 4\lambda y)$. Substitute this into the third and get

$$5y^2 + z^2 = 4$$
$$y^2 - \lambda z = 0$$
$$yz - \lambda y + yz - 4\lambda y = 0$$

$y = 0$ will not yield the minimum value from the above example. Therefore, divide the last equation by y and solve for λ to get $\lambda = (2/5) z$. Now put this in the second equation to conclude

$$5y^2 + z^2 = 4$$
$$y^2 - (2/5) z^2 = 0 \; ,$$

a system which is easy to solve. Thus $y^2 = 8/15$ and $z^2 = 4/3$. Therefore, candidates for minima are $\left(2\sqrt{\frac{8}{15}}, \sqrt{\frac{8}{15}}, \pm\sqrt{\frac{4}{3}} \right)$, and $\left(-2\sqrt{\frac{8}{15}}, -\sqrt{\frac{8}{15}}, \pm\sqrt{\frac{4}{3}} \right)$, a choice of 4 points to check. Clearly the one which gives the smallest value is

$$\left(2\sqrt{\frac{8}{15}}, \sqrt{\frac{8}{15}}, -\sqrt{\frac{4}{3}} \right)$$

or $\left(-2\sqrt{\frac{8}{15}}, -\sqrt{\frac{8}{15}}, -\sqrt{\frac{4}{3}} \right)$ and the minimum value of the function subject to the constraints is $-\frac{2}{5}\sqrt{30} - \frac{2}{3}\sqrt{3}$.

You should rework this problem first solving the second easy constraint for x and then producing a simpler problem involving only the variables y and z.

9.6 Exercises

(1) Maximize $x + y + z$ subject to the constraint $x^2 + y^2 + z^2 = 3$.

(2) Minimize $2x - y + z$ subject to the constraint $2x^2 + y^2 + z^2 = 36$.

(3) Minimize $x + 3y - z$ subject to the constraint $2x^2 + y^2 - 2z^2 = 36$ if possible. Note there is no guaranty this function has either a maximum or a minimum. Determine whether there exists a minimum also.

(4) Find the dimensions of the largest rectangle which can be inscribed in a circle of radius r.

(5) Maximize $2x + y$ subject to the condition that $\frac{x^2}{4} + \frac{y^2}{9} \leq 1$.

(6) Maximize $x + 2y$ subject to the condition that $x^2 + \frac{y^2}{9} \leq 1$.

(7) Maximize $x + y$ subject to the condition that $x^2 + \frac{y^2}{9} + z^2 \leq 1$.

(8) Minimize $x + y + z$ subject to the condition that $x^2 + \frac{y^2}{9} + z^2 \leq 1$.

(9) Find the points on $y^2 x = 16$ which are closest to $(0, 0)$.

(10) Find the points on $\sqrt{2} y^2 x = 1$ which are closest to $(0, 0)$.

(11) Find points on $xy = 4$ farthest from $(0, 0)$ if any exist. If none exist, tell why. What does this say about the method of Lagrange multipliers?

(12) A can is supposed to have a volume of 36π cubic centimeters. Find the dimensions of the can which minimizes the surface area.

(13) A can is supposed to have a volume of 36π cubic centimeters. The top and bottom of the can are made of tin costing 4 cents per square centimeter and the sides of the can are made of aluminum costing 5 cents per square centimeter. Find the dimensions of the can which minimizes the cost.

(14) Minimize and maximize $\sum_{j=1}^{n} x_j$ subject to the constraint $\sum_{j=1}^{n} x_j^2 = a^2$. Your answer should be some function of a which you may assume is a positive number.

(15) Find the point (x, y, z) on the level surface $4x^2 + y^2 - z^2 = 1$ which is closest to $(0, 0, 0)$.

(16) A curve is formed from the intersection of the plane, $2x + y + z = 3$ and the cylinder $x^2 + y^2 = 4$. Find the point on this curve which is closest to $(0, 0, 0)$.

(17) A curve is formed from the intersection of the plane, $2x + 3y + z = 3$ and the sphere $x^2 + y^2 + z^2 = 16$. Find the point on this curve which is closest to $(0, 0, 0)$.

(18) Find the point on the plane, $2x + 3y + z = 4$ which is closest to the point $(1, 2, 3)$.

(19) Let $A = (A_{ij})$ be an $n \times n$ matrix which is symmetric. Thus $A_{ij} = A_{ji}$ and recall $(\mathbf{A}\mathbf{x})_i = A_{ij}x_j$ where as usual, sum over the repeated index. Show that $\frac{\partial}{\partial x_k}(A_{ij}x_j x_i) = 2A_{ij}x_j$. Show that when you use the method of Lagrange multipliers to maximize the function $A_{ij}x_j x_i$ subject to the constraint, $\sum_{j=1}^{n} x_j^2 = 1$, the value of λ which corresponds to the maximum value of this functions is such that $A_{ij}x_j = \lambda x_i$. Thus $\mathbf{A}\mathbf{x} = \lambda\mathbf{x}$. Thus λ is an eigenvalue of the matrix A.

(20) Here are two lines.
$$\mathbf{x} = (1 + 2t, 2 + t, 3 + t)^T$$
and $\mathbf{x} = (2 + s, 1 + 2s, 1 + 3s)^T$. Find points $\mathbf{p_1}$ on the first line and $\mathbf{p_2}$ on

the second with the property that $|\mathbf{p}_1 - \mathbf{p}_2|$ is at least as small as the distance between any other pair of points, one chosen on one line and the other on the other line.

(21) * Find points on the circle of radius r for the largest triangle which can be inscribed in it.

(22) Find the point on the intersection of $z = x^2 + y^2$ and $x + y + z = 1$ which is closest to $(0, 0, 0)$.

(23) Minimize xyz subject to the constraints $x^2 + y^2 + z^2 = r^2$ and $x - y = 0$.

(24) Let n be a positive integer. Find n numbers whose sum is $8n$ and the sum of the squares is as small as possible.

(25) Find the point on the level surface $2x^2 + xy + z^2 = 16$ which is closest to $(0, 0, 0)$.

(26) Find the point on $x^2 + y^2 + z^2 = 1$ closest to the plane $x + y + z = 10$.

(27) Find the point on $\frac{x^2}{4} + \frac{y^2}{9} + z^2 = 1$ closest to the plane $x + y + z = 10$.

(28) Let $x_1, \cdots, x_5$ be 5 positive numbers. Maximize their product subject to the constraint that

$$x_1 + 2x_2 + 3x_3 + 4x_4 + 5x_5 = 300.$$

(29) Let $f(x_1, \cdots, x_n) = x_1^n x_2^{n-1} \cdots x_n^1$. Then f achieves a maximum on the set $S \equiv$

$$\left\{ \mathbf{x} \in \mathbb{R}^n : \sum_{i=1}^{n} i x_i = 1, \text{each } x_i \geq 0 \right\}$$

If $\mathbf{x} \in S$ is the point where this maximum is achieved, find x_1/x_n.

(30) * Let (x, y) be a point on the ellipse, $x^2/a^2 + y^2/b^2 = 1$ which is in the first quadrant. Extend the tangent line through (x, y) till it intersects the x and y axes and let $A(x, y)$ denote the area of the triangle formed by this line and the two coordinate axes. Find the minimum value of the area of this triangle as a function of a and b.

(31) Maximize $\prod_{i=1}^{n} x_i^2$

$$(\equiv x_1^2 \times x_2^2 \times x_3^2 \times \cdots \times x_n^2)$$

subject to the constraint, $\sum_{i=1}^{n} x_i^2 = r^2$. Show that the maximum is $(r^2/n)^n$. Now show from this that

$$\left(\prod_{i=1}^{n} x_i^2 \right)^{1/n} \leq \frac{1}{n} \sum_{i=1}^{n} x_i^2$$

and finally, conclude that if each number $x_i \geq 0$, then

$$\left(\prod_{i=1}^{n} x_i \right)^{1/n} \leq \frac{1}{n} \sum_{i=1}^{n} x_i$$

and there exist values of the x_i for which equality holds. This says the "geometric mean" is always smaller than the arithmetic mean.

(32) Maximize $x^2 y^2$ subject to the constraint

$$\frac{x^{2p}}{p} + \frac{y^{2q}}{q} = r^2$$

where p, q are real numbers larger than 1 which have the property that

$$\frac{1}{p} + \frac{1}{q} = 1$$

show that the maximum is achieved when $x^{2p} = y^{2q}$ and equals r^2. Now conclude that if $x, y > 0$, then

$$xy \leq \frac{x^p}{p} + \frac{y^q}{q}$$

and there are values of x and y where this inequality is an equation.

(33) The area of the ellipse $x^2/a^2 + y^2/b^2 \leq 1$ is πab which is given to equal π. The length of the ellipse is $\int_0^{2\pi} \sqrt{a^2 \sin^2(t) + b^2 \cos^2(t)} dt$. Find a, b such that the ellipse having this volume is as short as possible.

9.7   Proof Of The Second Derivative Test*

Definition 9.5. The matrix $\left(\frac{\partial^2 f}{\partial x_i \partial x_j}(\mathbf{x}) \right)$ is called the Hessian matrix, denoted by $H(\mathbf{x})$.

Now recall the Taylor formula with the Lagrange form of the remainder. See Volume I for a proof.

Theorem 9.4. *Let $h : (-\delta, 1 + \delta) \to \mathbb{R}$ have $m + 1$ derivatives. Then there exists $t \in (0, 1)$ such that*

$$h(1) = h(0) + \sum_{k=1}^{m} \frac{h^{(k)}(0)}{k!} + \frac{h^{(m+1)}(t)}{(m+1)!}.$$

Now let $f : U \to \mathbb{R}$ where U is an open subset of $\mathbb{R}^n$. Suppose $f \in C^2(U)$. Let $\mathbf{x} \in U$ and let $r > 0$ be such that

$$B(\mathbf{x}, r) \subseteq U.$$

Then for $\|\mathbf{v}\| < r$ consider

$$f(\mathbf{x} + t\mathbf{v}) - f(\mathbf{x}) \equiv h(t)$$

for $t \in [0, 1]$. Then from Taylor's theorem for the case where $m = 2$ and the chain rule, using the repeated index summation convention and the chain rule,

$$h'(t) = \frac{\partial f}{\partial x_i}(\mathbf{x} + t\mathbf{v}) v_i, \ h''(t) = \frac{\partial^2 f}{\partial x_j \partial x_i}(\mathbf{x} + t\mathbf{v}) v_i v_j.$$

Thus

$$h''(t) = \mathbf{v}^T H (\mathbf{x} + t\mathbf{v}) \mathbf{v}.$$

From Theorem 9.4 there exists $t \in (0, 1)$ such that

$$f(\mathbf{x} + \mathbf{v}) = f(\mathbf{x}) + \frac{\partial f}{\partial x_i}(\mathbf{x}) v_i + \frac{1}{2} \mathbf{v}^T H (\mathbf{x} + t\mathbf{v}) \mathbf{v}$$

By the continuity of the second partial derivative

$$f(\mathbf{x} + \mathbf{v}) = f(\mathbf{x}) + \nabla f(\mathbf{x}) \cdot \mathbf{v} + \frac{1}{2} \mathbf{v}^T H (\mathbf{x}) \mathbf{v} +$$

$$\frac{1}{2} \left(\mathbf{v}^T (H(\mathbf{x}+t\mathbf{v}) - H(\mathbf{x})) \mathbf{v} \right) \tag{9.4}$$

where the last term satisfies

$$\lim_{|\mathbf{v}| \to 0} \frac{1}{2} \frac{\left(\mathbf{v}^T (H(\mathbf{x}+t\mathbf{v}) - H(\mathbf{x})) \mathbf{v} \right)}{|\mathbf{v}|^2} = 0 \tag{9.5}$$

because of the continuity of the entries of $H(\mathbf{x})$.

Theorem 9.5. *Suppose $\mathbf{x}$ is a critical point for f. That is, suppose $\frac{\partial f}{\partial x_i}(\mathbf{x}) = 0$ for each i. Then if $H(\mathbf{x})$ has all positive eigenvalues, $\mathbf{x}$ is a local minimum. If $H(\mathbf{x})$ has all negative eigenvalues, then $\mathbf{x}$ is a local maximum. If $H(\mathbf{x})$ has a positive eigenvalue, then there exists a direction in which f has a local minimum at $\mathbf{x}$, while if $H(\mathbf{x})$ has a negative eigenvalue, there exists a direction in which f has a local maximum at $\mathbf{x}$.*

Proof: Since $\nabla f(\mathbf{x}) = \mathbf{0}$, formula (9.4) implies

$$f(\mathbf{x} + \mathbf{v}) = f(\mathbf{x}) + \frac{1}{2} \mathbf{v}^T H (\mathbf{x}) \mathbf{v} + \frac{1}{2} \left(\mathbf{v}^T (H(\mathbf{x}+t\mathbf{v}) - H(\mathbf{x})) \mathbf{v} \right) \tag{9.6}$$

and by continuity of the second derivatives, these mixed second derivatives are equal and so $H(\mathbf{x})$ is a symmetric matrix. Thus, by Lemma A.3 on Page 302 in the appendix, $H(\mathbf{x})$ has all real eigenvalues. Suppose first that $H(\mathbf{x})$ has all positive eigenvalues and that all are larger than $\delta^2 > 0$. Then by Theorem A.10 on Page 303 of the appendix,

$$\mathbf{u}^T H (\mathbf{x}) \mathbf{u} \geq \delta^2 |\mathbf{u}|^2$$

By continuity of H, if $\mathbf{v}$ is small enough,

$$f(\mathbf{x} + \mathbf{v}) \geq f(\mathbf{x}) + \frac{1}{2} \delta^2 |\mathbf{v}|^2 - \frac{1}{4} \delta^2 |\mathbf{v}|^2 = f(\mathbf{x}) + \frac{\delta^2}{4} |\mathbf{v}|^2 .$$

This shows the first claim of the theorem. The second claim follows from similar reasoning or applying the above to $-f$.

Suppose $H(\mathbf{x})$ has a positive eigenvalue λ^2. Then let $\mathbf{v}$ be an eigenvector for this eigenvalue. Then from (9.6), replacing $\mathbf{v}$ with $s\mathbf{v}$ and letting t depend on s,

$$f(\mathbf{x}+s\mathbf{v}) = f(\mathbf{x}) + \frac{1}{2} s^2 \mathbf{v}^T H (\mathbf{x}) \mathbf{v} +$$

$$\frac{1}{2}s^2 \left(\mathbf{v}^T \left(H\left(\mathbf{x}+ts\mathbf{v}\right) - H\left(\mathbf{x}\right)\right)\mathbf{v}\right)$$

which implies

$$
\begin{aligned}
f\left(\mathbf{x}+s\mathbf{v}\right) &= f\left(\mathbf{x}\right) + \frac{1}{2}s^2\lambda^2 \left|\mathbf{v}\right|^2 + \frac{1}{2}s^2 \left(\mathbf{v}^T \left(H\left(\mathbf{x}+ts\mathbf{v}\right) - H\left(\mathbf{x}\right)\right)\mathbf{v}\right) \\
&\geq f\left(\mathbf{x}\right) + \frac{1}{4}s^2\lambda^2 \left|\mathbf{v}\right|^2
\end{aligned}
$$

whenever s is small enough. Thus in the direction $\mathbf{v}$ the function has a local minimum at $\mathbf{x}$. The assertion about the local maximum in some direction follows similarly. ∎

Chapter 10

The Riemann Integral On $\mathbb{R}^n$

10.1 Methods For Double Integrals

This chapter is on the Riemann integral for a function of n variables. It begins by introducing the basic concepts and applications of the integral. The proofs of the theorems involved are difficult and are left for an appendix for those who are interested. To begin with, consider the problem of finding the volume under a surface of the form $z = f(x, y)$ where $f(x, y) \geq 0$ and $f(x, y) = 0$ for all (x, y) outside of some bounded set. To solve this problem, consider the following picture.

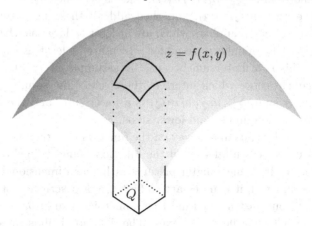

In this picture, the volume of the little prism which lies above the rectangle Q below the graph of the function is between $M_Q(f) v(Q)$ and $m_Q(f) v(Q)$ where

$$M_Q(f) \equiv \sup\{f(\mathbf{x}) : \mathbf{x} \in Q\}, \ m_Q(f) \equiv \inf\{f(\mathbf{x}) : \mathbf{x} \in Q\}, \qquad (10.1)$$

and $v(Q)$ is defined as the area of Q. Now consider the following picture.

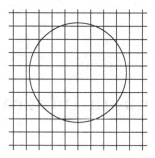

In this picture, it is assumed f equals zero outside the circle and f is a bounded nonnegative function. Then each of those little squares is the base of a prism of the sort in the previous picture, and the sum of the volumes of those prisms should be the volume under the surface $z = f(x, y)$. Therefore, the desired volume must lie between the two numbers,

$$\sum_Q M_Q(f) v(Q) \text{ and } \sum_Q m_Q(f) v(Q)$$

where the notation, $\sum_Q M_Q(f) v(Q)$, means for each Q, take $M_Q(f)$, multiply it by the area of $Q, v(Q)$, and then add all these numbers together. Thus $\sum_Q M_Q(f) v(Q)$ adds numbers which are at least as large as what is desired while in $\sum_Q m_Q(f) v(Q)$ numbers are added which are at least as small as what is desired. Note this is a finite sum because by assumption, $f = 0$ except for finitely many Q, namely those which intersect the circle. The sum, $\sum_Q M_Q(f) v(Q)$ is called an upper sum, $\sum_Q m_Q(f) v(Q)$ is a lower sum, and the desired volume is caught between these upper and lower sums.

None of this depends in any way on the function being nonnegative. It also does not depend in any essential way on the function being defined on $\mathbb{R}^2$, although it is impossible to draw meaningful pictures in higher dimensional cases. To define the Riemann integral, it is necessary to first give a description of something called a **grid**. First you must understand that something like $[a, b] \times [c, d]$ is a rectangle in $\mathbb{R}^2$, having sides parallel to the axes. The situation is illustrated in the following picture.

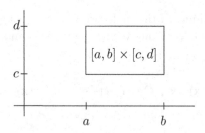

$(x, y) \in [a, b] \times [c, d]$, means $x \in [a, b]$ and also $y \in [c, d]$ and the points which do this comprise the rectangle just as shown in the picture.

Definition 10.1. For $i = 1, 2$, let $\left\{\alpha_k^i\right\}_{k=-\infty}^{\infty}$ be points on $\mathbb{R}$ which satisfy

$$\lim_{k \to \infty} \alpha_k^i = \infty, \quad \lim_{k \to -\infty} \alpha_k^i = -\infty, \quad \alpha_k^i < \alpha_{k+1}^i. \tag{10.2}$$

For such sequences, define a **grid** on $\mathbb{R}^2$ denoted by $\mathcal{G}$ or $\mathcal{F}$ as the collection of rectangles of the form

$$Q = \left[\alpha_k^1, \alpha_{k+1}^1\right] \times \left[\alpha_l^2, \alpha_{l+1}^2\right]. \tag{10.3}$$

For $\mathcal{G}$ a grid, the expression,

$$\sum_{Q \in \mathcal{G}} M_Q\left(f\right) v\left(Q\right)$$

is called the upper sum associated with the grid $\mathcal{G}$ as described above in the discussion of the volume under a surface. Again, this means to take a rectangle from $\mathcal{G}$ multiply $M_Q\left(f\right)$ defined in (10.1) by its area $v\left(Q\right)$ and sum all these products for every $Q \in \mathcal{G}$. The symbol

$$\sum_{Q \in \mathcal{G}} m_Q\left(f\right) v\left(Q\right),$$

called a lower sum, is defined similarly. With this preparation it is time to give a definition of the **Riemann integral** of a function of two variables.

Definition 10.2. Let $f : \mathbb{R}^2 \to \mathbb{R}$ be a bounded function which equals zero for all (x, y) outside some bounded set. Then $\int f \, dV$ is defined to be the unique number which lies between all upper sums and all lower sums. In the case of $\mathbb{R}^2$, it is common to replace the V with A and write this symbol as $\int f \, dA$ where A stands for area.

This definition begs a difficult question. For which functions does there exist a unique number between all the upper and lower sums? This interesting and fundamental question is discussed in most advanced calculus books and may be seen in the appendix on the theory of the Riemann integral. It is a hard problem which was only solved in the first part of the twentieth century. When it was solved, it was also realized that the Riemann integral was not the right integral to use.

Consider the question: How can the Riemann integral be computed? Consider the following picture where f is assumed to be 0 outside the base of the solid which is contained in some rectangle $[a, b] \times [c, d]$.

It depicts a slice taken from the solid defined by $\{(x, y) : 0 \leq y \leq f(x, y)\}$. You see these when you look at a loaf of bread. If you wanted to find the volume of the loaf of bread, and you knew the volume of each slice of bread, you could find the volume of the whole loaf by adding the volumes of individual slices. It is the same here. If you could find the volume of the slice represented in this picture, you could add these up and get the volume of the solid. The slice in the picture corresponds to y and $y + h$ and is assumed to be very thin, having thickness equal to h. Denote the volume of the solid under the graph of $z = f(x, y)$ on $[a, b] \times [c, y]$ by $V(y)$. Then

$$V(y + h) - V(y) \approx h \int_a^b f(x, y) \, dx$$

where the integral is obtained by fixing y and integrating with respect to x and is the area of the cross section corresponding to y. It is hoped that the approximation would be increasingly good as h gets smaller. Thus, dividing by h and taking a limit, it is expected that

$$V'(y) = \int_a^b f(x, y) \, dx, \ V(c) = 0.$$

Therefore, as in the method of cross sections, the volume of the solid under the graph of $z = f(x, y)$ is obtained by doing $\int_c^d$ to both sides,

$$\int_c^d \left(\int_a^b f(x, y) \, dx \right) dy \qquad (10.4)$$

but this volume was also the result of $\int f \, dV$. Therefore, it is expected that this is a way to evaluate $\int f \, dV$.

Note what has been gained here. A hard problem, finding $\int f \, dV$, is reduced to a sequence of easier problems. First do

$$\int_a^b f(x, y) \, dx$$

getting a function of y, say $F(y)$ and then do

$$\int_c^d \left(\int_a^b f(x, y) \, dx \right) dy = \int_c^d F(y) \, dy.$$

Of course there is nothing special about fixing y first. The same thing should be obtained from the integral

$$\int_a^b \left(\int_c^d f(x, y) \, dy \right) dx \qquad (10.5)$$

These expressions in (10.4) and (10.5) are called **iterated integrals**. They are tools for evaluating $\int f \, dV$ which would be hard to find otherwise. In practice, the parenthesis is usually omitted in these expressions. Thus

$$\int_a^b \left(\int_c^d f(x, y) \, dy \right) dx = \int_a^b \int_c^d f(x, y) \, dy \, dx$$

and it is understood that you are to do the inside integral first and then when you have done it, obtaining a function of x, you integrate this function of x. Note that this is nothing more than using an integral to compute the area of a cross section and then using this method to find a volume.

However, there is no difference in the general case where f is not necessarily nonnegative as can be seen by applying the method to the nonnegative functions f^+, f^- given by

$$f^+ \equiv \frac{|f| - f}{2}, \ f^- \equiv \frac{|f| - f}{2}$$

and then noting that $f = f^+ - f^-$ and the integral is linear. Thus

$$\int f dV = \int f^+ - f^- dV = \int f^+ dV - \int f^- dV$$

$$= \int_a^b \int_c^d f^+(x,y) \, dy \, dx - \int_a^b \int_c^d f^-(x,y) \, dy \, dx$$

$$= \int_a^b \int_c^d f(x,y) \, dy \, dx$$

A careful presentation which is not for the faint of heart is in an appendix.

Another aspect of this is the notion of integrating a function which is defined on some set, not on all $\mathbb{R}^2$. For example, suppose f is defined on the set $S \subseteq \mathbb{R}^2$. What is meant by $\int_S f \, dV$?

Definition 10.3. Let $f : S \to \mathbb{R}$ where S is a subset of $\mathbb{R}^2$. Then denote by f_1 the function defined by

$$f_1(x,y) \equiv \begin{cases} f(x,y) \text{ if } (x,y) \in S \\ 0 \text{ if } (x,y) \notin S \end{cases}.$$

Then

$$\int_S f \, dV \equiv \int f_1 \, dV.$$

Example 10.1. Let $f(x,y) = x^2 y + yx$ for $(x,y) \in [0,1] \times [0,2] \equiv R$. Find $\int_R f \, dV$.

This is done using iterated integrals like those defined above. Thus

$$\int_R f \, dV = \int_0^1 \int_0^2 (x^2 y + yx) \, dy \, dx.$$

The inside integral yields

$$\int_0^2 (x^2 y + yx) \, dy = 2x^2 + 2x$$

and now the process is completed by doing $\int_0^1$ to what was just obtained. Thus

$$\int_0^1 \int_0^2 (x^2 y + yx) \, dy \, dx = \int_0^1 (2x^2 + 2x) \, dx = \frac{5}{3}.$$

If the integration is done in the opposite order, the same answer should be obtained.

$$\int_0^2 \int_0^1 \left(x^2 y + yx\right)\, dx\, dy$$

$$\int_0^1 \left(x^2 y + yx\right)\, dx = \frac{5}{6} y$$

Now

$$\int_0^2 \int_0^1 \left(x^2 y + yx\right)\, dx\, dy = \int_0^2 \left(\frac{5}{6} y\right)\, dy = \frac{5}{3}.$$

If a different answer had been obtained it would have been a sign that a mistake had been made.

Example 10.2. Let $f(x, y) = x^2 y + yx$ for $(x, y) \in R$ where R is the triangular region defined to be in the first quadrant, below the line $y = x$ and to the left of the line $x = 4$. Find $\int_R f\, dV$.

From the above discussion,

$$\int_R f\, dV = \int_0^4 \int_0^x \left(x^2 y + yx\right)\, dy\, dx$$

The reason for this is that x goes from 0 to 4 and for each fixed x between 0 and 4, y goes from 0 to the slanted line, $y = x$, the function being defined to be 0 for larger y. Thus y goes from 0 to x. This explains the inside integral. Now $\int_0^x \left(x^2 y + yx\right)\, dy = \frac{1}{2} x^4 + \frac{1}{2} x^3$ and so

$$\int_R f\, dV = \int_0^4 \left(\frac{1}{2} x^4 + \frac{1}{2} x^3\right)\, dx = \frac{672}{5}.$$

What of integration in a different order? Lets put the integral with respect to y on the outside and the integral with respect to x on the inside. Then

$$\int_R f\, dV = \int_0^4 \int_y^4 \left(x^2 y + yx\right)\, dx\, dy$$

For each y between 0 and 4, the variable x, goes from y to 4.

$$\int_y^4 \left(x^2 y + yx\right)\, dx = \frac{88}{3} y - \frac{1}{3} y^4 - \frac{1}{2} y^3$$

Now

$$\int_R f\, dV = \int_0^4 \left(\frac{88}{3} y - \frac{1}{3} y^4 - \frac{1}{2} y^3\right)\, dy = \frac{672}{5}.$$

Here is a similar example.

Example 10.3. Let $f(x, y) = x^2 y$ for $(x, y) \in R$ where R is the triangular region defined to be in the first quadrant, below the line $y = 2x$ and to the left of the line $x = 4$. Find $\int_R f\, dV$.

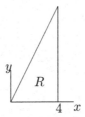

Put the integral with respect to x on the outside first. Then

$$\int_R f\, dV = \int_0^4 \int_0^{2x} \left(x^2 y\right)\, dy\, dx$$

because for each $x \in [0, 4]$, y goes from 0 to $2x$. Then

$$\int_0^{2x} \left(x^2 y\right)\, dy = 2x^4$$

and so

$$\int_R f\, dV = \int_0^4 \left(2x^4\right)\, dx = \frac{2048}{5}$$

Now do the integral in the other order. Here the integral with respect to y will be on the outside. What are the limits of this integral? Look at the triangle and note that x goes from 0 to 4 and so $2x = y$ goes from 0 to 8. Now for fixed y between 0 and 8, where does x go? It goes from the x coordinate on the line $y = 2x$ which corresponds to this y to 4. What is the x coordinate on this line which goes with y? It is $x = y/2$. Therefore, the iterated integral is

$$\int_0^8 \int_{y/2}^4 \left(x^2 y\right)\, dx\, dy.$$

Now

$$\int_{y/2}^4 \left(x^2 y\right)\, dx = \frac{64}{3} y - \frac{1}{24} y^4$$

and so

$$\int_R f\, dV = \int_0^8 \left(\frac{64}{3} y - \frac{1}{24} y^4\right) dy = \frac{2048}{5}$$

the same answer.

A few observations are in order here. In finding $\int_S f\, dV$ there is no problem in setting things up if S is a rectangle. However, if S is not a rectangle, the procedure **always** is agonizing. A good rule of thumb is that if what you do is easy it will be wrong. There are no shortcuts! There are no quick fixes which require no thought! Pain and suffering is inevitable and you must not expect it to be otherwise. Always draw a picture and then begin **agonizing** over the correct limits. Even when you are careful you will make lots of mistakes until you get used to the process.

Sometimes an integral can be evaluated in one order but not in another.

Example 10.4. For R as shown below, find $\int_R \sin\left(y^2\right) dV$.

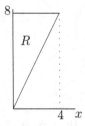

Setting this up to have the integral with respect to y on the inside yields

$$\int_0^4 \int_{2x}^8 \sin\left(y^2\right) \, dy \, dx.$$

Unfortunately, there is no antiderivative in terms of elementary functions for $\sin\left(y^2\right)$ so there is an immediate problem in evaluating the inside integral. It doesn't work out so the next step is to do the integration in another order and see if some progress can be made. This yields

$$\int_0^8 \int_0^{y/2} \sin\left(y^2\right) \, dx \, dy = \int_0^8 \frac{y}{2} \sin\left(y^2\right) \, dy$$

and $\int_0^8 \frac{y}{2} \sin\left(y^2\right) \, dy = -\frac{1}{4}\cos 64 + \frac{1}{4}$ which you can verify by making the substitution, $u = y^2$. Thus

$$\int_R \sin\left(y^2\right) \, dy = -\frac{1}{4}\cos 64 + \frac{1}{4}.$$

This illustrates an important idea. The integral $\int_R \sin\left(y^2\right) \, dV$ is defined as a number. It is the unique number between all the upper sums and all the lower sums. Finding it is another matter. In this case it was possible to find it using one order of integration but not the other. The iterated integral in this other order also is defined as a number but it cannot be found directly without interchanging the order of integration. Of course sometimes nothing you try will work out.

10.1.1   *Density And Mass*

Consider a two dimensional material. Of course there is no such thing but a flat plate might be modeled as one. The density ρ is a function of position and is defined as follows. Consider a small chunk of area dV located at the point whose Cartesian coordinates are (x, y). Then the mass of this small chunk of material is given by $\rho(x, y) \, dV$. Thus if the material occupies a region in two dimensional space U, the total mass of this material would be

$$\int_U \rho \, dV$$

In other words you integrate the density to get the mass. Now by letting ρ depend on position, you can include the case where the material is not homogeneous. Here is an example.

Example 10.5. Let $\rho(x, y)$ denote the density of the plane region determined by the curves $\frac{1}{3}x + y = 2$, $x = 3y^2$, and $x = 9y$. Find the total mass if $\rho(x, y) = y$.

You need to first draw a picture of the region R. A rough sketch follows.

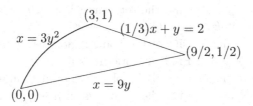

This region is in two pieces, one having the graph of $x = 9y$ on the bottom and the graph of $x = 3y^2$ on the top and another piece having the graph of $x = 9y$ on the bottom and the graph of $\frac{1}{3}x + y = 2$ on the top. Therefore, in setting up the integrals, with the integral with respect to x on the outside, the double integral equals the following sum of iterated integrals.

$$\overbrace{\int_0^3 \int_{x/9}^{\sqrt{x/3}} y \, dy \, dx}^{\text{has } x=3y^2 \text{ on top}} + \overbrace{\int_3^{\frac{9}{2}} \int_{x/9}^{2-\frac{1}{3}x} y \, dy \, dx}^{\text{has } \frac{1}{3}x+y=2 \text{ on top}}$$

You notice it is not necessary to have a perfect picture, just one which is good enough to figure out what the limits should be. The dividing line between the two cases is $x = 3$ and this was shown in the picture. Now it is only a matter of evaluating the iterated integrals which in this case is routine and gives 1.

10.2 Exercises

(1) Evaluate the iterated integral and then write the iterated integral with the order of integration reversed. $\int_0^4 \int_0^{3y} x \, dx \, dy$.

(2) Evaluate the iterated integral and then write the iterated integral with the order of integration reversed. $\int_0^3 \int_0^{3y} y \, dx \, dy$.

(3) Evaluate the iterated integral and then write the iterated integral with the order of integration reversed. $\int_0^2 \int_0^{2y} (x+1) \, dx \, dy$.

(4) Evaluate the iterated integral and then write the iterated integral with the order of integration reversed. $\int_0^3 \int_0^y \sin(x) \, dx \, dy$.

(5) Evaluate the iterated integral and then write the iterated integral with the order of integration reversed. $\int_0^1 \int_0^y \exp(y) \, dx \, dy$.

(6) Let $\rho(x, y)$ denote the density of the plane region closest to $(0, 0)$ which is between the curves $x + 2y = 3$, $x = y^2$, and $x = 0$. Find the total mass if $\rho(x, y) = y$. Set up the integral in terms of $dxdy$ and in terms of $dydx$.

(7) Let $\rho(x, y)$ denote the density of the plane region determined by the curves $x + 2y = 3$, $x = y^2$, and $x = 4y$. Find the total mass if $\rho(x, y) = x$. Set up the

integral in terms of $dx\,dy$ and $dy\,dx$.

(8) Let $\rho(x,y)$ denote the density of the plane region determined by the curves $y = 2x, y = x, x + y = 3$. Find the total mass if $\rho(x,y) = y + 1$. Set up the integrals in terms of $dx\,dy$ and $dy\,dx$.

(9) Let $\rho(x,y)$ denote the density of the plane region determined by the curves $y = 3x, y = x, 2x + y = 4$. Find the total mass if $\rho(x,y) = 1$.

(10) Let $\rho(x,y)$ denote the density of the plane region determined by the curves $y = 3x, y = x, x + y = 2$. Find the total mass if $\rho(x,y) = x + 1$. Set up the integrals in terms of $dx\,dy$ and $dy\,dx$.

(11) Let $\rho(x,y)$ denote the density of the plane region determined by the curves $y = 5x, y = x, 5x + 2y = 10$. Find the total mass if $\rho(x,y) = 1$. Set up the integrals in terms of $dx\,dy$ and $dy\,dx$.

(12) Find $\int_0^4 \int_{y/2}^2 \frac{1}{x} e^{2\frac{y}{x}}\, dx\, dy$. You might need to interchange the order of integration.

(13) Find $\int_0^8 \int_{y/2}^4 \frac{1}{x} e^{3\frac{y}{x}}\, dx\, dy$.

(14) Find $\int_0^{\frac{1}{3}\pi} \int_x^{\frac{1}{3}\pi} \frac{\sin y}{y}\, dy\, dx$.

(15) Find $\int_0^{\frac{1}{2}\pi} \int_x^{\frac{1}{2}\pi} \frac{\sin y}{y}\, dy\, dx$.

(16) Find $\int_0^\pi \int_x^\pi \frac{\sin y}{y}\, dy\, dx$

(17) * Evaluate the iterated integral and then write the iterated integral with the order of integration reversed. $\int_{-3}^3 \int_{-x}^x x^2\, dy\, dx$

Your answer for the iterated integral should be $\int_3^0 \int_{-3}^{-y} x^2\, dx\, dy + \int_0^{-3} \int_{-3}^y x^2\, dx\, dy + \int_0^3 \int_y^3 x^2\, dx\, dy + \int_{-3}^0 \int_{-y}^3 x^2\, dx\, dy$. This is a very interesting example which shows that iterated integrals have a life of their own, not just as a method for evaluating double integrals.

(18) * Evaluate the iterated integral and then write the iterated integral with the order of integration reversed. $\int_{-2}^2 \int_{-x}^x x^2\, dy\, dx$.

10.3 Methods For Triple Integrals

10.3.1 *Definition Of The Integral*

The integral of a function of three variables is similar to the integral of a function of two variables.

Definition 10.4. For $i = 1, 2, 3$ let $\{\alpha_k^i\}_{k=-\infty}^\infty$ be points on $\mathbb{R}$ which satisfy

$$\lim_{k\to\infty} \alpha_k^i = \infty, \quad \lim_{k\to-\infty} \alpha_k^i = -\infty, \quad \alpha_k^i < \alpha_{k+1}^i. \tag{10.6}$$

For such sequences, define a **grid** on $\mathbb{R}^3$ denoted by $\mathcal{G}$ or $\mathcal{F}$ as the collection of boxes of the form

$$Q = \left[\alpha_k^1, \alpha_{k+1}^1\right] \times \left[\alpha_l^2, \alpha_{l+1}^2\right] \times \left[\alpha_p^3, \alpha_{p+1}^3\right]. \tag{10.7}$$

If $\mathcal{G}$ is a grid $\mathcal{F}$ is called a **refinement** of $\mathcal{G}$ if every box of $\mathcal{G}$ is the union of boxes of $\mathcal{F}$.

For $\mathcal{G}$ a grid,

$$\sum_{Q\in\mathcal{G}} M_Q\left(f\right)v\left(Q\right)$$

is the upper sum associated with the grid $\mathcal{G}$ where

$$M_Q\left(f\right) \equiv \sup\left\{f\left(\mathbf{x}\right):\mathbf{x}\in Q\right\}$$

and if $Q = [a,b]\times[c,d]\times[e,f]$, then $v\left(Q\right)$ is the volume of Q given by $(b-a)(d-c)(f-e)$. Letting

$$m_Q\left(f\right) \equiv \inf\left\{f\left(\mathbf{x}\right):\mathbf{x}\in Q\right\}$$

the lower sum associated with this partition is

$$\sum_{Q\in\mathcal{G}} m_Q\left(f\right)v\left(Q\right).$$

With this preparation, it is time to give a definition of the **Riemann integral** of a function of three variables. This definition is just like the one for a function of two variables.

Definition 10.5. Let $f:\mathbb{R}^3\to\mathbb{R}$ be a bounded function which equals zero outside some bounded subset of $\mathbb{R}^3$. $\int f\,dV$ is defined as the unique number between all the upper sums and lower sums.

As in the case of a function of two variables there are all sorts of mathematical questions which are dealt with later.

The way to think of integrals is as follows. Located at a point $\mathbf{x}$, there is an "infinitesimal" chunk of volume dV. The integral involves taking this little chunk of volume dV, multiplying it by $f\left(\mathbf{x}\right)$ and then adding up all such products. Upper sums are too large and lower sums are too small but the unique number between all the lower and upper sums is just right and corresponds to the notion of adding up all the $f\left(\mathbf{x}\right)dV$. Even the notation is suggestive of this concept of sum. It is a long thin S denoting sum. This is the fundamental concept for the integral in any number of dimensions and all the definitions and technicalities are designed to give precision and mathematical respectability to this notion.

Integrals of functions of three variables are also evaluated by using iterated integrals. Imagine a sum of the form $\sum_{ijk}a_{ijk}$ where there are only finitely many choices for i,j, and k and the symbol means you simply add up all the a_{ijk}. By the commutative law of addition, these may be added systematically in the form, $\sum_k\sum_j\sum_i a_{ijk}$. A similar process is used to evaluate triple integrals and since integrals are like sums, you might expect it to be valid. Specifically,

$$\int f\,dV = \int_?^?\int_?^?\int_?^? f\left(x,y,z\right)dx\,dy\,dz.$$

In words, sum with respect to x and then sum what you get with respect to y and finally, with respect to z. Of course this should hold in any other order such as

$$\int f\,dV = \int_?^?\int_?^?\int_?^? f\left(x,y,z\right)dz\,dy\,dx.$$

This is proved in an appendix[1].

Having discussed double and triple integrals, the definition of the integral of a function of n variables is accomplished in the same way.

Definition 10.6. For $i = 1, \cdots, n$, let $\{\alpha_k^i\}_{k=-\infty}^{\infty}$ be points on $\mathbb{R}$ which satisfy

$$\lim_{k \to \infty} \alpha_k^i = \infty, \quad \lim_{k \to -\infty} \alpha_k^i = -\infty, \quad \alpha_k^i < \alpha_{k+1}^i. \tag{10.8}$$

For such sequences, define a grid on $\mathbb{R}^n$ denoted by $\mathcal{G}$ or $\mathcal{F}$ as the collection of boxes of the form

$$Q = \prod_{i=1}^{n} \left[\alpha_{j_i}^i, \alpha_{j_i+1}^i \right]. \tag{10.9}$$

Definition 10.7. Let f be a bounded function which equals zero off a bounded set D, and let $\mathcal{G}$ be a grid. For $Q \in \mathcal{G}$, define

$$M_Q(f) \equiv \sup \{f(\mathbf{x}) : \mathbf{x} \in Q\}, \quad m_Q(f) \equiv \inf \{f(\mathbf{x}) : \mathbf{x} \in Q\}. \tag{10.10}$$

Also define for Q a box, the volume of Q, denoted by $v(Q)$ by

$$v(Q) \equiv \prod_{i=1}^{n} (b_i - a_i), \quad Q \equiv \prod_{i=1}^{n} [a_i, b_i].$$

Now define upper sums, $\mathcal{U}_\mathcal{G}(f)$ and lower sums, $\mathcal{L}_\mathcal{G}(f)$ with respect to the indicated grid, by the formulas

$$\mathcal{U}_\mathcal{G}(f) \equiv \sum_{Q \in \mathcal{G}} M_Q(f) v(Q), \quad \mathcal{L}_\mathcal{G}(f) \equiv \sum_{Q \in \mathcal{G}} m_Q(f) v(Q).$$

Then a function of n variables is Riemann integrable if there is a unique number between all the upper and lower sums. This number is the value of the integral.

In this book most integrals will involve no more than three variables. However, this does not mean an integral of a function of more than three variables is unimportant. Therefore, I will begin to refer to the general case when theorems are stated.

Definition 10.8. For $E \subseteq \mathbb{R}^n$,

$$\mathcal{X}_E(\mathbf{x}) \equiv \begin{cases} 1 \text{ if } \mathbf{x} \in E \\ 0 \text{ if } \mathbf{x} \notin E \end{cases}.$$

Define $\int_E f \, dV \equiv \int \mathcal{X}_E f \, dV$ when $f\mathcal{X}_E \in \mathcal{R}(\mathbb{R}^n)$.

---

[1] All of these fundamental questions about integrals can be considered more easily in the context of the Lebesgue integral. However, this integral is more abstract than the Riemann integral.

10.3.2 *Iterated Integrals*

As before, the integral is often computed by using an iterated integral. In general it is impossible to set up an iterated integral for finding $\int_E f dV$ for arbitrary regions, E but when the region is sufficiently simple, one can make progress. Suppose the region E over which the integral is to be taken is of the form $E = \{(x, y, z) : a(x, y) \le z \le b(x, y)\}$ for $(x, y) \in R$, a two dimensional region. This is illustrated in the following picture in which the bottom surface is the graph of $z = a(x, y)$ and the top is the graph of $z = b(x, y)$.

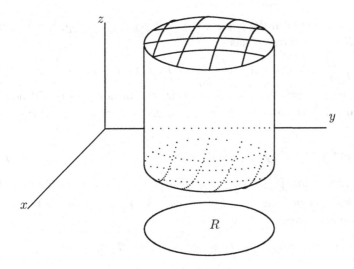

Then

$$\int_E f dV = \int_R \int_{a(x,y)}^{b(x,y)} f(x, y, z) \, dz dA$$

It might be helpful to think of $dV = dz dA$. Now $\int_{a(x,y)}^{b(x,y)} f(x, y, z) \, dz$ is a function of x and y and so you have reduced the triple integral to a double integral over R of this function of x and y. Similar reasoning would apply if the region in $\mathbb{R}^3$ were of the form $\{(x, y, z) : a(y, z) \le x \le b(y, z)\}$ or $\{(x, y, z) : a(x, z) \le y \le b(x, z)\}$.

Example 10.6. Find the volume of the region E in the first octant between $z = 1 - (x + y)$ and $z = 0$.

In this case, R is the region shown.

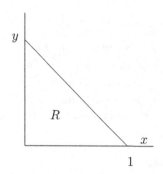

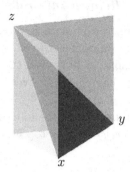

Thus the region E is between the plane $z = 1 - (x+y)$ on the top, $z = 0$ on the bottom, and over R shown above. Thus

$$\int_E 1\,dV = \int_R \int_0^{1-(x+y)} dz\,dA = \int_0^1 \int_0^{1-x} \int_0^{1-(x+y)} dz\,dy\,dx = \frac{1}{6}$$

Of course iterated integrals have a life of their own although this will not be explored here. You can just write them down and go to work on them. Here are some examples.

Example 10.7. Find $\int_2^3 \int_3^x \int_{3y}^x (x-y)\,dz\,dy\,dx$.

The inside integral yields $\int_{3y}^x (x-y)\,dz = x^2 - 4xy + 3y^2$. Next this must be integrated with respect to y to give $\int_3^x \left(x^2 - 4xy + 3y^2\right)\,dy = -3x^2 + 18x - 27$. Finally the third integral gives

$$\int_2^3 \int_3^x \int_{3y}^x (x-y)\,dz\,dy\,dx = \int_2^3 \left(-3x^2 + 18x - 27\right)\,dx = -1.$$

Example 10.8. Find $\int_0^\pi \int_0^{3y} \int_0^{y+z} \cos(x+y)\,dx\,dz\,dy$.

The inside integral is $\int_0^{y+z} \cos(x+y)\,dx = 2\cos z \sin y \cos y + 2 \sin z \cos^2 y - \sin z - \sin y$. Now this has to be integrated.

$$\int_0^{3y} \int_0^{y+z} \cos(x+y)\,dx\,dz$$
$$= \int_0^{3y} \left(2\cos z \sin y \cos y + 2\sin z \cos^2 y - \sin z - \sin y\right)\,dz$$
$$= -1 - 16\cos^5 y + 20 \cos^3 y - 5\cos y - 3(\sin y)y + 2\cos^2 y.$$

Finally, this last expression must be integrated from 0 to π. Thus

$$\int_0^\pi \int_0^{3y} \int_0^{y+z} \cos(x+y)\,dx\,dz\,dy$$
$$= \int_0^\pi \left(-1 - 16\cos^5 y + 20\cos^3 y - 5\cos y - 3(\sin y)y + 2\cos^2 y\right)\,dy = -3\pi$$

Example 10.9. Here is an iterated integral: $\int_0^2 \int_0^{3-\frac{3}{2}x} \int_0^{x^2} dz\,dy\,dx$. Write as an iterated integral in the order $dz\,dx\,dy$.

The inside integral is just a function of x and y. (In fact, only a function of x.) The order of the last two integrals must be interchanged. Thus the iterated integral which needs to be done in a different order is

$$\int_0^2 \int_0^{3-\frac{3}{2}x} f(x,y)\, dy\, dx.$$

As usual, it is important to draw a picture and then go from there.

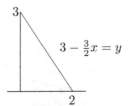

$$3 - \tfrac{3}{2}x = y$$

Thus this double integral equals

$$\int_0^3 \int_0^{\frac{2}{3}(3-y)} f(x,y)\, dx\, dy.$$

Now substituting in for $f(x,y)$,

$$\int_0^3 \int_0^{\frac{2}{3}(3-y)} \int_0^{x^2} dz\, dx\, dy.$$

Example 10.10. Find the volume of the bounded region determined by $3y + 3z = 2, x = 16 - y^2, y = 0, x = 0$.

In the yz plane, the first of the following pictures corresponds to $x = 0$.

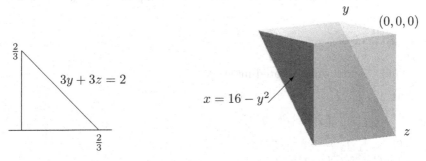

$$3y + 3z = 2$$

Therefore, the outside integrals taken with respect to z and y are of the form $\int_0^{\frac{2}{3}} \int_0^{\frac{2}{3}-y} dz\, dy$, and now for any choice of (y,z) in the above triangular region, x goes from 0 to $16 - y^2$. Therefore, the iterated integral is

$$\int_0^{\frac{2}{3}} \int_0^{\frac{2}{3}-y} \int_0^{16-y^2} dx\, dz\, dy = \frac{860}{243}.$$

Example 10.11. Find the volume of the region determined by the intersection of the two cylinders, $x^2 + y^2 \le 9$ and $y^2 + z^2 \le 9$.

The first listed cylinder intersects the xy plane in the disk, $x^2 + y^2 \le 9$. What is the volume of the three dimensional region which is between this disk and the two surfaces, $z = \sqrt{9 - y^2}$ and $z = -\sqrt{9 - y^2}$? An iterated integral for the volume is

$$\int_{-3}^{3} \int_{-\sqrt{9-y^2}}^{\sqrt{9-y^2}} \int_{-\sqrt{9-y^2}}^{\sqrt{9-y^2}} dz\, dx\, dy = 144.$$

Note I drew no picture of the three dimensional region. If you are interested, here it is.

One of the cylinders is parallel to the z axis, $x^2 + y^2 \le 9$ and the other is parallel to the x axis, $y^2 + z^2 \le 9$. I did not need to be able to draw such a nice picture in order to work this problem. This is the key to doing these. Draw pictures in two dimensions and reason from the two dimensional pictures rather than attempt to wax artistic and consider all three dimensions at once. These problems are hard enough without making them even harder by attempting to be an artist.

10.4   Exercises

(1) Find the following iterated integrals.

 (a) $\int_{-1}^{3} \int_{0}^{2z} \int_{y}^{z+1} (x + y)\, dx\, dy\, dz$

 (b) $\int_{0}^{1} \int_{0}^{z} \int_{y}^{z^2} (y + z)\, dx\, dy\, dz$

 (c) $\int_{0}^{3} \int_{1}^{x} \int_{2}^{3x-y} \sin(x)\, dz\, dy\, dx$

 (d) $\int_{0}^{1} \int_{x}^{2x} \int_{y}^{2y} dz\, dy\, dx$

 (e) $\int_{2}^{4} \int_{2}^{2x} \int_{2y}^{x} dz\, dy\, dx$

 (f) $\int_{0}^{3} \int_{0}^{2-5x} \int_{0}^{2-x-2y} 2x\ dz\, dy\, dx$

 (g) $\int_{0}^{2} \int_{0}^{1-3x} \int_{0}^{3-3x-2y} x\ dz\, dy\, dx$

 (h) $\int_{0}^{\pi} \int_{0}^{3y} \int_{0}^{y+z} \cos(x + y)\ dx\, dz\, dy$

 (i) $\int_{0}^{\pi} \int_{0}^{4y} \int_{0}^{y+z} \sin(x + y)\ dx\, dz\, dy$

(2) Fill in the missing limits.

$$\int_0^1 \int_0^z \int_0^z f(x,y,z)\,dx\,dy\,dz = \int_?^? \int_?^? \int_?^? f(x,y,z)\,dx\,dz\,dy,$$

$$\int_0^1 \int_0^z \int_0^{2z} f(x,y,z)\,dx\,dy\,dz = \int_?^? \int_?^? \int_?^? f(x,y,z)\,dy\,dz\,dx,$$

$$\int_0^1 \int_0^z \int_0^z f(x,y,z)\,dx\,dy\,dz = \int_?^? \int_?^? \int_?^? f(x,y,z)\,dz\,dy\,dx,$$

$$\int_0^1 \int_{z/2}^{\sqrt{z}} \int_0^{y+z} f(x,y,z)\,dx\,dy\,dz = \int_?^? \int_?^? \int_?^? f(x,y,z)\,dx\,dz\,dy,$$

$$\int_4^6 \int_2^6 \int_0^4 f(x,y,z)\,dx\,dy\,dz = \int_?^? \int_?^? \int_?^? f(x,y,z)\,dz\,dy\,dx.$$

(3) Find the volume of R where R is the bounded region formed by the plane $\frac{1}{5}x + y + \frac{1}{4}z = 1$ and the planes $x = 0, y = 0, z = 0$.

(4) Find the volume of R where R is the bounded region formed by the plane $\frac{1}{5}x + \frac{1}{2}y + \frac{1}{4}z = 1$ and the planes $x = 0, y = 0, z = 0$.

(5) Find the volume of R where R is the bounded region formed by the plane $\frac{1}{5}x + \frac{1}{3}y + \frac{1}{4}z = 1$ and the planes $x = 0, y = 0, z = 0$.

(6) Find the volume of the bounded region determined by $3y + z = 3, x = 4 - y^2, y = 0, x = 0$.

(7) Find the volume of the region bounded by $x^2 + y^2 = 16, z = 3x, z = 0$, and $x \geq 0$.

(8) Find the volume of R where R is the bounded region formed by the plane $\frac{1}{4}x + \frac{1}{2}y + \frac{1}{4}z = 1$ and the planes $x = 0, y = 0, z = 0$.

(9) Here is an iterated integral: $\int_0^3 \int_0^{3-x} \int_0^{x^2} dz\,dy\,dx$. Write as an iterated integral in the following orders: $dz\,dx\,dy, \ dx\,dz\,dy, \ dx\,dy\,dz, \ dy\,dx\,dz, \ dy\,dz\,dx$.

(10) Find the volume of the bounded region determined by $2y + z = 3, x = 9 - y^2, y = 0, x = 0, z = 0$.

(11) Find the volume of the bounded region determined by $y + 2z = 3, x = 9 - y^2, y = 0, x = 0$.

(12) Find the volume of the bounded region determined by $y + z = 2, x = 3 - y^2, y = 0, x = 0$.

(13) Find the volume of the region bounded by $x^2 + y^2 = 25, z = x, z = 0$, and $x \geq 0$.

Your answer should be $\frac{250}{3}$.

(14) Find the volume of the region bounded by $x^2 + y^2 = 9, z = 3x, z = 0$, and $x \geq 0$.

10.4.1 Mass And Density

As an example of the use of triple integrals, consider a solid occupying a set of points $U \subseteq \mathbb{R}^3$ having density ρ. Thus ρ is a function of position and the total mass of the solid equals $\int_U \rho\,dV$. This is just like the two dimensional case. The mass of an infinitesimal chunk of the solid located at $\mathbf{x}$ would be $\rho(\mathbf{x})\,dV$ and so the total mass is just the sum of all these, $\int_U \rho(\mathbf{x})\,dV$.

Example 10.12. Find the volume of R where R is the bounded region formed by the plane $\frac{1}{5}x + y + \frac{1}{5}z = 1$ and the planes $x = 0, y = 0, z = 0$.

When $z = 0$, the plane becomes $\frac{1}{5}x + y = 1$. Thus the intersection of this plane with the xy plane is this line shown in the following picture.

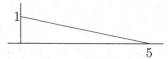

Therefore, the bounded region is between the triangle formed in the above picture by the x axis, the y axis and the above line and the surface given by $\frac{1}{5}x + y + \frac{1}{5}z = 1$ or $z = 5\left(1 - \left(\frac{1}{5}x + y\right)\right) = 5 - x - 5y$. Therefore, an iterated integral which yields the volume is

$$\int_0^5 \int_0^{1 - \frac{1}{5}x} \int_0^{5 - x - 5y} dz\, dy\, dx = \frac{25}{6}.$$

Example 10.13. Find the mass of the bounded region R formed by the plane $\frac{1}{3}x + \frac{1}{3}y + \frac{1}{5}z = 1$ and the planes $x = 0, y = 0, z = 0$ if the density is $\rho(x, y, z) = z$.

This is done just like the previous example except in this case, there is a function to integrate. Thus the answer is

$$\int_0^3 \int_0^{3 - x} \int_0^{5 - \frac{5}{3}x - \frac{5}{3}y} z\, dz\, dy\, dx = \frac{75}{8}.$$

Example 10.14. Find the total mass of the bounded solid determined by $z = 9 - x^2 - y^2$ and $x, y, z \geq 0$ if the mass is given by $\rho(x, y, z) = z$

When $z = 0$ the surface $z = 9 - x^2 - y^2$ intersects the xy plane in a circle of radius 3 centered at $(0, 0)$. Since $x, y \geq 0$, it is only a quarter of a circle of interest, the part where both these variables are nonnegative. For each (x, y) inside this quarter circle, z goes from 0 to $9 - x^2 - y^2$. Therefore, the iterated integral is of the form,

$$\int_0^3 \int_0^{\sqrt{(9 - x^2)}} \int_0^{9 - x^2 - y^2} z\, dz\, dy\, dx = \frac{243}{8}\pi.$$

Example 10.15. Find the volume of the bounded region determined by $x \geq 0, y \geq 0, z \geq 0$, and $\frac{1}{7}x + y + \frac{1}{4}z = 1$, and $x + \frac{1}{7}y + \frac{1}{4}z = 1$.

When $z = 0$, the plane $\frac{1}{7}x + y + \frac{1}{4}z = 1$ intersects the xy plane in the line whose equation is $\frac{1}{7}x + y = 1$, while the plane, $x + \frac{1}{7}y + \frac{1}{4}z = 1$ intersects the xy plane in the line whose equation is $x + \frac{1}{7}y = 1$. Furthermore, the two planes intersect when $x = y$ as can be seen from the equations, $x + \frac{1}{7}y = 1 - \frac{z}{4}$ and $\frac{1}{7}x + y = 1 - \frac{z}{4}$ which imply $x = y$. Thus the two dimensional picture to look at is depicted in the following picture.

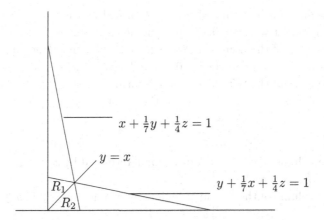

You see in this picture, the base of the region in the xy plane is the union of the two triangles, R_1 and R_2. For $(x, y) \in R_1$, z goes from 0 to what it needs to be to be on the plane, $\frac{1}{7}x + y + \frac{1}{4}z = 1$. Thus z goes from 0 to $4\left(1 - \frac{1}{7}x - y\right)$. Similarly, on R_2, z goes from 0 to $4\left(1 - \frac{1}{7}y - x\right)$. Therefore, the integral needed is

$$\int_{R_1} \int_0^{4\left(1 - \frac{1}{7}x - y\right)} dz \, dV + \int_{R_2} \int_0^{4\left(1 - \frac{1}{7}y - x\right)} dz \, dV$$

and now it only remains to consider $\int_{R_1} dV$ and $\int_{R_2} dV$. The point of intersection of these lines shown in the above picture is $\left(\frac{7}{8}, \frac{7}{8}\right)$ and so an iterated integral is

$$\int_0^{7/8} \int_x^{1 - \frac{x}{7}} \int_0^{4\left(1 - \frac{1}{7}x - y\right)} dz \, dy \, dx + \int_0^{7/8} \int_y^{1 - \frac{y}{7}} \int_0^{4\left(1 - \frac{1}{7}y - x\right)} dz \, dx \, dy = \frac{7}{6}$$

10.5 Exercises

(1) Find the volume of the region determined by the intersection of the two cylinders, $x^2 + y^2 \leq 16$ and $y^2 + z^2 \leq 16$.

(2) Find the volume of the region determined by the intersection of the two cylinders, $x^2 + y^2 \leq 9$ and $y^2 + z^2 \leq 9$.

(3) Find the volume of the region bounded by $x^2 + y^2 = 4$, $z = 0$, $z = 5 - y$

(4) Find $\int_0^2 \int_0^{6-2z} \int_{\frac{1}{2}x}^{3-z} (3 - z) \cos\left(y^2\right) dy \, dx \, dz$.

(5) Find $\int_0^1 \int_0^{18-3z} \int_{\frac{1}{3}x}^{6-z} (6 - z) \exp\left(y^2\right) dy \, dx \, dz$.

(6) Find $\int_0^2 \int_0^{24-4z} \int_{\frac{1}{4}y}^{6-z} (6 - z) \exp\left(x^2\right) dx \, dy \, dz$.

(7) Find $\int_0^1 \int_0^{10-2z} \int_{\frac{1}{2}y}^{5-z} \frac{\sin x}{x} dx \, dy \, dz$.

Hint: Interchange order of integration.

(8) Find the mass of the bounded region R formed by the plane $\frac{1}{4}x + \frac{1}{2}y + \frac{1}{3}z = 1$ and the planes $x = 0, y = 0, z = 0$ if the density is $\rho(x, y, z) = y$

(9) Find the mass of the bounded region R formed by the plane $\frac{1}{2}x + \frac{1}{2}y + \frac{1}{4}z = 1$ and the planes $x = 0, y = 0, z = 0$ if the density is $\rho(x, y, z) = z^2$

(10) Find the mass of the bounded region R formed by the plane $\frac{1}{4}x + \frac{1}{2}y + \frac{1}{4}z = 1$ and the planes $x = 0, y = 0, z = 0$ if the density is $\rho(x, y, z) = y + z$

(11) Find the mass of the bounded region R formed by the plane $\frac{1}{4}x + \frac{1}{2}y + \frac{1}{5}z = 1$ and the planes $x = 0, y = 0, z = 0$ if the density is $\rho(x, y, z) = y$

(12) Find $\int_0^1 \int_0^{12-4z} \int_{\frac{1}{4}y}^{3-z} \frac{\sin x}{x} \, dx \, dy \, dz$.

(13) Find $\int_0^{20} \int_0^2 \int_{\frac{1}{5}y}^{6-z} \frac{\sin x}{x} \, dx \, dz \, dy + \int_{20}^{30} \int_0^{6-\frac{1}{5}y} \int_{\frac{1}{5}y}^{6-z} \frac{\sin x}{x} \, dx \, dz \, dy$.

(14) Find the volume of the bounded region determined by $x \geq 0, y \geq 0, z \geq 0$, and $\frac{1}{2}x + y + \frac{1}{2}z = 1$, and $x + \frac{1}{2}y + \frac{1}{2}z = 1$.

(15) Find the volume of the bounded region determined by $x \geq 0, y \geq 0, z \geq 0$, and $\frac{1}{7}x + y + \frac{1}{3}z = 1$, and $x + \frac{1}{7}y + \frac{1}{3}z = 1$.

(16) Find an iterated integral for the volume of the region between the graphs of $z = x^2 + y^2$ and $z = 2(x + y)$.

(17) Find the volume of the region which lies between $z = x^2 + y^2$ and the plane $z = 4$.

(18) The base of a solid is the region in the xy plane between the curves $y = x^2$ and $y = 1$. The top of the solid is the plane $z = 2 - x$. Find the volume of the solid.

(19) The base of a solid is in the xy plane and is bounded by the lines $y = x, y = 1 - x$, and $y = 0$. The top of the solid is $z = 3 - y$. Find its volume.

(20) The base of a solid is in the xy plane and is bounded by the lines $x = 0, x = \pi, y = 0$, and $y = \sin x$. The top of this solid is $z = x$. Find the volume of this solid.

Chapter 11

The Integral In Other Coordinates

11.1 Polar Coordinates

Recall the relation between the rectangular coordinates and polar coordinates is

$$\mathbf{x}(r, \theta) \equiv \begin{pmatrix} x \\ y \end{pmatrix} = \begin{pmatrix} r\cos(\theta) \\ r\sin(\theta) \end{pmatrix}, \ r \geq 0, \ \theta \in [0, 2\pi)$$

Now consider the part of grid obtained by fixing θ at various values and varying r and then by fixing r at various values and varying θ.

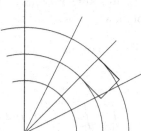

The idea is that these lines obtained by fixing one or the other coordinate are very close together, much closer than drawn and so we would expect the area of one of the little curvy quadrilaterals to be close to the area of the parallelogram shown. Consider this parallelogram. The two sides originating at the intersection of two of the grid lines as shown are approximately equal to

$$\mathbf{x}_r(r, \theta) \, dr, \ \mathbf{x}_\theta(r, \theta) \, d\theta$$

where dr and $d\theta$ are the respective small changes in the variables r and θ. Thus the area of one of those little curvy shapes should be approximately equal to

$$|\mathbf{x}_r(r, \theta) \, dr \times \mathbf{x}_\theta(r, \theta) \, d\theta|$$

by the geometric description of the cross product. These vectors are extended as 0 in the third component in order to take the cross product. This reduces to

$$dA = \left| \det \begin{pmatrix} \cos(\theta) & -r\sin(\theta) \\ \sin(\theta) & r\cos(\theta) \end{pmatrix} \right| dr d\theta = r dr d\theta$$

which is the increment of area in polar coordinates, taking the place of $dxdy$. The integral is really about taking the value of the function integrated multiplied by dA and adding these products. Here is an example.

Example 11.1. Find the area of a circle of radius a.

The variable r goes from 0 to a and the angle θ goes from 0 to 2π. Therefore, the area is

$$\int_D dA = \int_0^{2\pi} \int_0^a r\,dr\,d\theta = \pi a^2$$

Example 11.2. The density equals r. Find the total mass of a disk of radius a.

This is easy to do in polar coordinates. The disk involved has θ going from 0 to 2π and r from 0 to 2. Therefore, the integral to work is just

$$\int_0^{2\pi} \int_0^a \overbrace{rr\,dr\,d\theta}^{dA} = \frac{2}{3}\pi a^3$$

Notice how in these examples the circular disk is really a rectangle $[0, 2\pi] \times [0, a]$. This is why polar coordinates are so useful. The next example was worked earlier from a different point of view.

Example 11.3. Find the area of the inside of the cardioid $r = 1 + \cos\theta$, $\theta \in [0, 2\pi]$.

Here the integral is

$$\int_0^{2\pi} \int_0^{1+\cos(\theta)} r\,dr\,d\theta = \frac{3}{2}\pi$$

To see how impossible this problem is in rectangular coordinates, draw the graph of the cardioid.

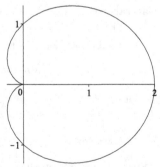

How would you go about setting this up in rectangular coordinates? It would be very hard if not impossible, but is easy in polar coordinates. This is because in polar coordinates the region integrated over is the region below the curve in the following picture.

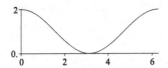

Example 11.4. Let R denote the inside of the cardioid $r = 1 + \cos\theta$ for $\theta \in [0, 2\pi]$. Find

$$\int_R x\,dA$$

Here the convenient increment of area is $r\,dr\,d\theta$ and so the integral is

$$\int_0^{2\pi} \int_0^{1+\cos(\theta)} x\,r\,dr\,d\theta$$

Now you need to change x to the right coordinates. Thus the integral equals

$$\int_0^{2\pi} \int_0^{1+\cos(\theta)} (r\cos(\theta))\,r\,dr\,d\theta = \frac{5}{4}\pi$$

A case where this sort of problem occurs is when you find the mass of a plate given the density.

Definition 11.1. Suppose a material occupies a region of the plane R. The density λ is a nonnegative function of position with the property that if $B \subseteq R$, then the mass of B is given by $\int_B \lambda\,dA$. In particular, this is true of $B = R$.

Example 11.5. Let R denote the inside of the polar curve $r = 2 + \sin\theta$. Let $\lambda = 3 + x$. Find the total mass of R.

As above, this is

$$\int_0^{2\pi} \int_0^{2+\sin(\theta)} (3 + r\cos(\theta))\,r\,dr\,d\theta = \frac{27}{2}\pi$$

11.2 Exercises

(1) Sketch a graph in polar coordinates of $r = 2 + \sin(\theta)$ and find the area of the enclosed region.

(2) Sketch a graph in polar coordinates of $r = \sin(4\theta)$ and find the area of the region enclosed. **Hint:** In this case, you need to worry and fuss about $r < 0$.

(3) Suppose the density is $\lambda(x, y) = 2 - x$ and the region is the interior of the cardioid $r = 1 + \cos\theta$. Find the total mass.

(4) Suppose the density is $\lambda = 4 - x - y$ and find the mass of the plate which is between the concentric circles $r = 1$ and $r = 2$.

(5) Suppose the density is $\lambda = 4 - x - y$ and find the mass of the plate which is inside the polar graph of $r = 1 + \sin(\theta)$.

(6) Suppose the density is $2 + x$. Find the mass of the plate which is the inside of the polar curve $r = \sin(2\theta)$. **Hint:** This is one of those fussy things with negative radius.

(7) The area density of a plate is given by $\lambda = 1 + x$ and the plate occupies the inside of the cardioid $r = 1 + \cos\theta$. Find its mass.

(8) The moment about the x axis of a plate with density λ occupying the region R is defined as $m_y = \int_R y\lambda dA$. The moment about the y axis of the same plate is $m_x = \int_R x\lambda dA$. If $\lambda = 2 - x$, find the moments about the x and y axes of the plate inside $r = 2 + \sin(\theta)$.

(9) Using the above problem, find the moments about the x and y axes of a plate having density $1+x$ for the plate which is the inside of the cardioid $r = 1+\cos\theta$.

(10) Use the same plate as the above but this time, let the density be $(2 + x + y)$. Find the moments.

(11) Let $D = \{(x,y) : x^2 + y^2 \le 25\}$. Find $\int_D e^{25x^2 + 25y^2}\, dx\, dy$. **Hint:** This is an integral of the form $\int_D f(x,y)\, dA$. Write in polar coordinates and it will be fairly easy.

(12) Let $D = \{(x,y) : x^2 + y^2 \le 16\}$. Find $\int_D \cos\left(9x^2 + 9y^2\right)\, dx\, dy$.**Hint:** This is an integral of the form $\int_D f(x,y)\, dA$. Write in polar coordinates and it will be fairly easy.

(13) Derive a formula for area between two polar graphs using the increment of area of polar coordinates.

(14) Use polar coordinates to evaluate the following integral. Here S is given in terms of the polar coordinates. $\int_S \sin\left(2x^2 + 2y^2\right)\, dV$ where $r \le 2$ and $0 \le \theta \le \frac{3}{2}\pi$.

(15) Find $\int_S e^{2x^2 + 2y^2}\, dV$ where S is given in terms of the polar coordinates $r \le 2$ and $0 \le \theta \le \pi$.

(16) Find $\int_S \frac{y}{x}\, dV$ where S is described in polar coordinates as $1 \le r \le 2$ and $0 \le \theta \le \pi/4$.

(17) Find $\int_S \left(\left(\frac{y}{x}\right)^2 + 1\right)\, dV$ where S is given in polar coordinates as $1 \le r \le 2$ and $0 \le \theta \le \frac{1}{6}\pi$.

(18) A right circular cone has a base of radius 2 and a height equal to 2. Use polar coordinates to find its volume.

(19) Now suppose in the above problem, it is not really a cone but instead $z = 2 - \frac{1}{2}r^2$. Find its volume.

11.3   Cylindrical And Spherical Coordinates

Cylindrical coordinates are defined as follows.

$$
\mathbf{x}(r, \theta, z) \equiv \begin{pmatrix} x \\ y \\ z \end{pmatrix} = \begin{pmatrix} r\cos(\theta) \\ r\sin(\theta) \\ z \end{pmatrix},
$$

$$
r \geq 0, \theta \in [0, 2\pi), z \in \mathbb{R}
$$

Spherical coordinates are a little harder. Recall these are given by

$$
\mathbf{x}(\rho, \theta, \phi) \equiv \begin{pmatrix} x \\ y \\ z \end{pmatrix} = \begin{pmatrix} \rho\sin(\phi)\cos(\theta) \\ \rho\sin(\phi)\sin(\theta) \\ \rho\cos(\phi) \end{pmatrix},
$$

$$
\rho \geq 0, \theta \in [0, 2\pi), \phi \in [0, \pi]
$$

The following picture relates the various coordinates.

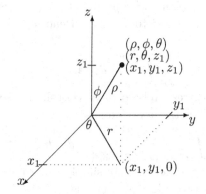

In this picture, ρ is the distance between the origin, the point whose Cartesian coordinates are $(0, 0, 0)$ and the point indicated by a dot and labelled as (x_1, y_1, z_1), (r, θ, z_1), and (ρ, ϕ, θ). The angle between the positive z axis and the line between the origin and the point indicated by a dot is denoted by ϕ, and θ is the angle between the positive x axis and the line joining the origin to the point $(x_1, y_1, 0)$ as shown, while r is the length of this line. Thus $r = \rho\sin(\phi)$ and is the usual polar coordinate while θ is the other polar coordinate. Letting z_1 denote the usual z coordinate of a point in three dimensions, like the one shown as a dot, (r, θ, z_1) are the cylindrical coordinates of the dotted point. The spherical coordinates are determined by (ρ, ϕ, θ). When ρ is specified, this indicates that the point of interest is on some sphere of radius ρ which is centered at the origin. Then when ϕ is given, the location of the point is narrowed down to a circle of "latitude" and finally, θ determines which point is on this circle by specifying a circle of "longitude". Let $\phi \in [0, \pi], \theta \in [0, 2\pi)$, and $\rho \in [0, \infty)$. The picture shows how to relate these new coordinate systems to Cartesian coordinates. Note that θ **is the same** in the two coordinate systems and that $\rho\sin\phi = r$.

The increment of three dimensional volume in cylindrical coordinates is $dV = rdrd\theta dz$. It is just a chunk of two dimensional area $rdrd\theta$ times the height dz which gives three dimensional volume. Here is an example.

Example 11.6. Find the volume of the three dimensional region between the graphs of $z = 4 - 2y^2$ and $z = 4x^2 + 2y^2$.

Where do the two surfaces intersect? This happens when $4x^2 + 2y^2 = 4 - 2y^2$ which is the curve in the xy plane, $x^2 + y^2 = 1$. Thus (x, y) is on the inside of this circle while z goes from $4x^2 + 2y^2$ to $4 - 2y^2$. Denoting the unit disk by D, the desired integral is

$$\int_D \int_{4x^2+2y^2}^{4-2y^2} dz dA$$

I will use the dA which corresponds to polar coordinates so this will then be in cylindrical coordinates. Thus the above equals

$$\int_0^{2\pi} \int_0^1 \int_{4(r^2 \cos^2(\theta))+2(r^2 \sin^2(\theta))}^{4-2(r^2 \sin^2(\theta))} dz dr d\theta = 2\pi$$

Note this is really not much different than simply using polar coordinates to integrate the difference of the two values of z This is

$$\int_D 4 - 2y^2 - (4x^2 + 2y^2) \, dA = \int_D (4 - 4r^2) \, dA$$

$$= \int_0^{2\pi} \int_0^1 (4 - 4r^2) \, r dr d\theta = 2\pi$$

Here is another example.

Example 11.7. Find the volume of the three dimensional region between the graphs of $z = 0, z = \sqrt{x^2 + y^2}$, and the cylinder $(x - 1)^2 + y^2 = 1$.

Consider the cylinder. It reduces to $r^2 = 2r\cos\theta$ or more simply $r = 2\cos\theta$. This is the graph of a circle having radius 1 and centered at $(1,0)$. Therefore, $\theta \in [-\pi/2, \pi/2]$. It follows that the cylindrical coordinate description of this volume is

$$\int_{-\pi/2}^{\pi/2} \int_0^{2\cos\theta} \int_0^r dz r \, dr \, d\theta = \frac{32}{9}$$

What is the increment of volume in spherical coordinates? There are two ways to see what this is, through art and through a systematic procedure. First consider art. Here is a picture.

In the picture there are two concentric spheres formed by making ρ two different constants and surfaces which correspond to θ assuming two different constants and ϕ assuming two different constants. These intersecting surfaces form the little box in the picture. What is the volume of this little box? Length $\approx \rho d\phi$, width $\approx \rho \sin(\phi) \, d\theta$, height $\approx d\rho$ and so the volume increment for spherical coordinates is the product of these

$$dV = \rho^2 \sin(\phi) \, d\rho d\theta d\phi$$

Now what is really going on? Consider the lower left corner of the little box. Fixing θ and ϕ at their values at this point and differentiating with respect to ρ leads to a little vector of the form

$$\begin{pmatrix} \sin(\phi)\cos(\theta) \\ \sin(\phi)\sin(\theta) \\ \cos(\phi) \end{pmatrix} d\rho$$

which points out from the surface of the sphere. Next keeping ρ and θ constant and differentiating only with respect to ϕ leads to an infinitesimal vector in the direction of a line of longitude,

$$\begin{pmatrix} \rho\cos(\phi)\cos(\theta) \\ \rho\cos(\phi)\sin(\theta) \\ -\rho\sin(\phi) \end{pmatrix} d\phi$$

and finally keeping ρ and ϕ constant and differentiating with respect to θ leads to the third infinitesimal vector which points in the direction of a line of latitude.

$$\begin{pmatrix} -\rho\sin(\phi)\sin(\theta) \\ \rho\sin(\phi)\cos(\theta) \\ 0 \end{pmatrix} d\theta$$

To find the increment of volume, we just need to take the absolute value of the determinant which has these vectors as columns, (Remember this is the absolute value of the box product.) exactly as was the case for polar coordinates. This will also yield

$$dV = \rho^2 \sin(\phi)\, d\rho d\theta d\phi.$$

However, in contrast to the drawing of pictures, this procedure is completely general and will handle all curvilinear coordinate systems in any dimension. This is discussed more later.

Example 11.8. Find the volume of a ball, B_R of radius R.

In this case, $U = (0, R] \times [0, \pi] \times [0, 2\pi)$ and use spherical coordinates. Then (11.4) yields a set in $\mathbb{R}^3$ which clearly differs from the ball of radius R only by a set having volume equal to zero. It leaves out the point at the origin is all. Therefore, the volume of the ball is

$$\int_{B_R} 1\, dV = \int_U \rho^2 \sin\phi\, dV = \int_0^R \int_0^\pi \int_0^{2\pi} \rho^2 \sin\phi\, d\theta\, d\phi\, d\rho = \frac{4}{3} R^3 \pi.$$

The reason this was effortless, is that the ball, B_R is realized as a box in terms of the spherical coordinates. Remember what was pointed out earlier about setting up iterated integrals over boxes.

Example 11.9. A cone is cut out of a ball of radius R as shown in the following picture, the diagram on the left being a side view. The angle of the cone is $\pi/3$. Find the volume of what is left.

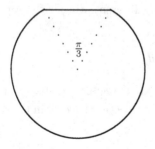 

Use spherical coordinates. This volume is then

$$\int_{\pi/6}^\pi \int_0^{2\pi} \int_0^R \rho^2 \sin(\phi)\, d\rho d\theta d\phi = \frac{2}{3}\pi R^3 + \frac{1}{3}\sqrt{3}\pi R^3$$

Now change the example a little by cutting out a cone at the bottom which has an angle of $\pi/2$ as shown. What is the volume of what is left?

This time you would have the volume equals

$$\int_{\pi/6}^{3\pi/4} \int_0^{2\pi} \int_0^R \rho^2 \sin(\phi)\, d\rho d\theta d\phi = \frac{1}{3}\sqrt{2}\pi R^3 + \frac{1}{3}\sqrt{3}\pi R^3$$

Example 11.10. Next suppose the ball of radius R is a sort of an orange and you remove a slice as shown in the picture. What is the volume of what is left? Assume the slice is formed by the two half planes $\theta = 0$ and $\theta = \pi/4$.

Using spherical coordinates, this gives for the volume

$$\int_0^\pi \int_{\pi/4}^{2\pi} \int_0^R \rho^2 \sin(\phi)\, d\rho d\theta d\phi = \frac{7}{6}\pi R^3$$

Example 11.11. Now remove the same two cones as in the above examples along with the same slice and find the volume of what is left.

This time you need

$$\int_{\pi/6}^{3\pi/4} \int_{\pi/4}^{2\pi} \int_0^R \rho^2 \sin(\phi)\, d\rho d\theta d\phi = \frac{7}{24}\sqrt{2}\pi R^3 + \frac{7}{24}\sqrt{3}\pi R^3$$

Example 11.12. Set up the integrals to find the volume of the cone $0 \le z \le 4$, $z = \sqrt{x^2 + y^2}$.

It would be easier to use cylindrical coordinates for this problem but it is a good exercise to use spherical coordinates. Here is a side view.

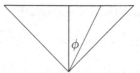

You need to figure out what ρ is as a function of ϕ which goes from 0 to $\pi/4$. You should get

$$\int_0^{2\pi} \int_0^{\pi/4} \int_0^{4\sec(\phi)} \rho^2 \sin(\phi)\, d\rho d\phi d\theta = \frac{64}{3}\pi$$

Example 11.13. Find the volume element for cylindrical coordinates.

In cylindrical coordinates,

$$\begin{pmatrix} x \\ y \\ z \end{pmatrix} = \begin{pmatrix} r\cos\theta \\ r\sin\theta \\ z \end{pmatrix}$$

Therefore, the Jacobian determinant is

$$\det \begin{pmatrix} \cos\theta & -r\sin\theta & 0 \\ \sin\theta & r\cos\theta & 0 \\ 0 & 0 & 1 \end{pmatrix} = r.$$

It follows the volume element in cylindrical coordinates is $r\, d\theta\, dr\, dz$.

Example 11.14. In the cone of Example 11.12 set up the integrals for finding the volume in cylindrical coordinates.

This is a better coordinate system for this example than spherical coordinates. This time you should get

$$\int_0^{2\pi} \int_0^4 \int_r^4 r\, dz dr d\theta = \frac{64}{3}\pi$$

Example 11.15. This example uses spherical coordinates to verify an important conclusion about gravitational force. Let the hollow sphere, H be defined by $a^2 < x^2 + y^2 + z^2 < b^2$

and suppose this hollow sphere has constant density taken to equal 1. Now place a unit mass at the point $(0, 0, z_0)$ where $|z_0| \notin [a, b]$. Show that the force of gravity acting on this unit mass is $\left(\alpha G \int_H \frac{(z-z_0)}{[x^2+y^2+(z-z_0)^2]^{3/2}} \, dV \right) \mathbf{k}$ and then show that if $|z_0| > b$ then the force of gravity acting on this point mass is the same as if the entire mass of the hollow sphere were placed at the origin, while if $|z_0| < a$, the total force acting on the point mass from gravity equals zero. Here G is the gravitation constant and α is the density. In particular, this shows that the force a planet exerts on an object is as though the entire mass of the planet were situated at its center[1].

Without loss of generality, assume $z_0 > 0$. Let dV be a little chunk of material located at the point (x, y, z) of H the hollow sphere. Then according to Newton's law of gravity, the force this small chunk of material exerts on the given point mass equals

$$\frac{x\mathbf{i} + y\mathbf{j} + (z - z_0)\mathbf{k}}{|x\mathbf{i} + y\mathbf{j} + (z - z_0)\mathbf{k}|} \frac{1}{\left(x^2 + y^2 + (z - z_0)^2\right)} G\alpha \, dV =$$

$$(x\mathbf{i} + y\mathbf{j} + (z - z_0)\mathbf{k}) \frac{1}{\left(x^2 + y^2 + (z - z_0)^2\right)^{3/2}} G\alpha \, dV$$

Therefore, the total force is

$$\int_H (x\mathbf{i} + y\mathbf{j} + (z - z_0)\mathbf{k}) \frac{1}{\left(x^2 + y^2 + (z - z_0)^2\right)^{3/2}} G\alpha \, dV.$$

By the symmetry of the sphere, the $\mathbf{i}$ and $\mathbf{j}$ components will cancel out when the integral is taken. This is because there is the same amount of stuff for negative x and y as there is for positive x and y. Hence what remains is

$$\alpha G \mathbf{k} \int_H \frac{(z - z_0)}{\left[x^2 + y^2 + (z - z_0)^2\right]^{3/2}} \, dV$$

as claimed. Now for the interesting part, the integral is evaluated. In spherical coordinates this integral is.

$$\int_0^{2\pi} \int_a^b \int_0^\pi \frac{(\rho \cos \phi - z_0) \rho^2 \sin \phi}{(\rho^2 + z_0^2 - 2\rho z_0 \cos \phi)^{3/2}} \, d\phi \, d\rho \, d\theta. \tag{11.1}$$

---

[1]This was shown by Newton in 1685 and allowed him to assert his law of gravitation applied to the planets as though they were point masses. It was a major accomplishment.

Rewrite the inside integral and use integration by parts to obtain this inside integral equals

$$\frac{1}{2z_0} \int_0^\pi (\rho^2 \cos\phi - \rho z_0) \, \frac{(2z_0\rho \sin\phi)}{(\rho^2 + z_0^2 - 2\rho z_0 \cos\phi)^{3/2}} \, d\phi =$$

$$\frac{1}{2z_0} \left(-2\frac{-\rho^2 - \rho z_0}{\sqrt{(\rho^2 + z_0^2 + 2\rho z_0)}} + 2\frac{\rho^2 - \rho z_0}{\sqrt{(\rho^2 + z_0^2 - 2\rho z_0)}} \right.$$

$$\left. - \int_0^\pi 2\rho^2 \frac{\sin\phi}{\sqrt{(\rho^2 + z_0^2 - 2\rho z_0 \cos\phi)}} \, d\phi \right). \tag{11.2}$$

There are some cases to consider here.

First suppose $z_0 < a$ so the point is on the inside of the hollow sphere and it is always the case that $\rho > z_0$. Then in this case, the two first terms reduce to

$$\frac{2\rho(\rho + z_0)}{\sqrt{(\rho + z_0)^2}} + \frac{2\rho(\rho - z_0)}{\sqrt{(\rho - z_0)^2}} = \frac{2\rho(\rho + z_0)}{(\rho + z_0)} + \frac{2\rho(\rho - z_0)}{\rho - z_0} = 4\rho$$

and so the expression in (11.2) equals

$$\frac{1}{2z_0} \left(4\rho - \int_0^\pi 2\rho^2 \frac{\sin\phi}{\sqrt{(\rho^2 + z_0^2 - 2\rho z_0 \cos\phi)}} \, d\phi \right)$$

$$= \frac{1}{2z_0} \left(4\rho - \frac{1}{z_0} \int_0^\pi \rho \frac{2\rho z_0 \sin\phi}{\sqrt{(\rho^2 + z_0^2 - 2\rho z_0 \cos\phi)}} \, d\phi \right)$$

$$= \frac{1}{2z_0} \left(4\rho - \frac{2\rho}{z_0} (\rho^2 + z_0^2 - 2\rho z_0 \cos\phi)^{1/2} \Big|_0^\pi \right)$$

$$= \frac{1}{2z_0} \left(4\rho - \frac{2\rho}{z_0} [(\rho + z_0) - (\rho - z_0)] \right) = 0.$$

Therefore, in this case the inner integral of (11.1) equals zero and so the original integral will also be zero.

The other case is when $z_0 > b$ and so it is always the case that $z_0 > \rho$. In this case the first two terms of (11.2) are

$$\frac{2\rho(\rho + z_0)}{\sqrt{(\rho + z_0)^2}} + \frac{2\rho(\rho - z_0)}{\sqrt{(\rho - z_0)^2}} = \frac{2\rho(\rho + z_0)}{(\rho + z_0)} + \frac{2\rho(\rho - z_0)}{z_0 - \rho} = 0.$$

Therefore in this case, (11.2) equals

$$\frac{1}{2z_0} \left(-\int_0^\pi 2\rho^2 \frac{\sin\phi}{\sqrt{(\rho^2 + z_0^2 - 2\rho z_0 \cos\phi)}} \, d\phi \right)$$

$$= \frac{-\rho}{2z_0^2} \left(\int_0^\pi \frac{2\rho z_0 \sin\phi}{\sqrt{(\rho^2 + z_0^2 - 2\rho z_0 \cos\phi)}} \, d\phi \right)$$

which equals

$$\frac{-\rho}{z_0^2}\left(\left(\rho^2 + z_0^2 - 2\rho z_0 \cos\phi\right)^{1/2}\Big|_0^\pi\right) = \frac{-\rho}{z_0^2}\left[(\rho + z_0) - (z_0 - \rho)\right] = -\frac{2\rho^2}{z_0^2}.$$

Thus the inner integral of (11.1) reduces to the above simple expression. Therefore, (11.1) equals

$$\int_0^{2\pi}\int_a^b \left(-\frac{2}{z_0^2}\rho^2\right)\, d\rho\, d\theta = -\frac{4}{3}\pi\frac{b^3 - a^3}{z_0^2}$$

and so

$$\alpha Gk \int_H \frac{(z - z_0)}{\left[x^2 + y^2 + (z - z_0)^2\right]^{3/2}}\, dV$$

$$= \alpha Gk\left(-\frac{4}{3}\pi\frac{b^3 - a^3}{z_0^2}\right) = -kG\frac{\text{total mass}}{z_0^2}.$$

11.4 Exercises

(1) Find the volume of the region bounded by $z = 0, x^2 + (y - 2)^2 = 4$, and $z = \sqrt{x^2 + y^2}$.

(2) Find the volume of the region $z \geq 0, x^2 + y^2 \leq 4$, and $z \leq 4 - \sqrt{x^2 + y^2}$.
(3) Find the volume of the region which is between the surfaces $z = 5y^2 + 9x^2$ and $z = 9 - 4y^2$.

(4) Find the volume of the region which is between $z = x^2 + y^2$ and $z = 5 - 4x$.
   Hint: You might want to change variables at some point.
(5) The ice cream in a sugar cone is described in spherical coordinates by $\rho \in [0, 10], \phi \in [0, \frac{1}{3}\pi], \theta \in [0, 2\pi]$. If the units are in centimeters, find the total volume in cubic centimeters of this ice cream.

(6) Find the volume between $z = 3 - x^2 - y^2$ and $z = 2\sqrt{(x^2 + y^2)}$.

(7) A ball of radius 3 is placed in a drill press and a hole of radius 2 is drilled out with the center of the hole a diameter of the ball. What is the volume of the material which remains?

(8) Find the volume of the cone defined by $z \in [0, 4]$ having angle $\pi/2$. Use spherical coordinates.

(9) A ball of radius 9 has density equal to $\sqrt{x^2 + y^2 + z^2}$ in rectangular coordinates. The top of this ball is sliced off by a plane of the form $z = 2$. Write integrals for the mass of what is left. In spherical coordinates and in cylindrical coordinates.

(10) A ball of radius 4 has a cone taken out of the top which has an angle of $\pi/2$ and then a cone taken out of the bottom which has an angle of $\pi/3$. Then a slice, $\theta \in [0, \pi/4]$ is removed. What is the volume of what is left?

(11) In Example 11.15 on Page 212 check out all the details by working the integrals to be sure the steps are right.

(12) What if the hollow sphere in Example 11.15 were in two dimensions and everything, including Newton's law still held? Would similar conclusions hold? Explain.

(13) Convert the following integrals into integrals involving cylindrical coordinates and then evaluate them.

(a) $\int_{-2}^{2} \int_{0}^{\sqrt{4-x^2}} \int_{0}^{x} xy \, dz \, dy \, dx$

(b) $\int_{-1}^{1} \int_{-\sqrt{1-y^2}}^{\sqrt{1-y^2}} \int_{0}^{x+y} dz \, dx \, dy$

(c) $\int_{0}^{1} \int_{0}^{\sqrt{1-x^2}} \int_{x}^{1} dz \, dy \, dx$

(d) For $a > 0$, $\int_{-a}^{a} \int_{-\sqrt{a^2-x^2}}^{\sqrt{a^2-x^2}} \int_{-\sqrt{a^2-x^2-y^2}}^{\sqrt{a^2-x^2-y^2}} dz \, dy \, dx$

(e) $\int_{-1}^{1} \int_{-\sqrt{1-x^2}}^{\sqrt{1-x^2}} \int_{-\sqrt{4-x^2-y^2}}^{\sqrt{4-x^2-y^2}} dz \, dy \, dx$

(14) Convert the following integrals into integrals involving spherical coordinates and then evaluate them.

(a) $\int_{-a}^{a} \int_{-\sqrt{a^2-x^2}}^{\sqrt{a^2-x^2}} \int_{-\sqrt{a^2-x^2-y^2}}^{\sqrt{a^2-x^2-y^2}} dz \, dy \, dx$

(b) $\int_{-1}^{1} \int_{0}^{\sqrt{1-x^2}} \int_{-\sqrt{1-x^2-y^2}}^{\sqrt{1-x^2-y^2}} dz \, dy \, dx$

(c) $\int_{-\sqrt{2}}^{\sqrt{2}} \int_{-\sqrt{2-x^2}}^{\sqrt{2-x^2}} \int_{\sqrt{x^2+y^2}}^{\sqrt{4-x^2-y^2}} dz \, dy \, dx$

(d) $\int_{-\sqrt{3}}^{\sqrt{3}} \int_{-\sqrt{3-x^2}}^{\sqrt{3-x^2}} \int_{1}^{\sqrt{4-x^2-y^2}} dz \, dy \, dx$

(e) $\int_{-1}^{1} \int_{-\sqrt{1-x^2}}^{\sqrt{1-x^2}} \int_{-\sqrt{4-x^2-y^2}}^{\sqrt{4-x^2-y^2}} dz \, dy \, dx$

11.5 The General Procedure

As mentioned above, the fundamental concept of an integral is a sum of things of the form $f(\mathbf{x})\, dV$ where dV is an "infinitesimal" chunk of volume located at the point $\mathbf{x}$. Up to now, this infinitesimal chunk of volume has had the form of a box with sides $dx_1, \cdots, dx_n$ so $dV = dx_1\, dx_2 \cdots dx_n$ but its form is not important. It could just as well be an infinitesimal parallelepiped for example. In what follows, this is what it will be.

First recall the definition of a parallelepiped.

Definition 11.2. Let $\mathbf{u}_1, \cdots, \mathbf{u}_p$ be vectors in $\mathbb{R}^k$. The parallelepiped determined by these vectors will be denoted by $P(\mathbf{u}_1, \cdots, \mathbf{u}_p)$ and it is defined as

$$P(\mathbf{u}_1, \cdots, \mathbf{u}_p) \equiv \left\{ \sum_{j=1}^{p} s_j \mathbf{u}_j : s_j \in [0,1] \right\}.$$

Now define the volume of this parallelepiped.

$$\text{volume of } P(\mathbf{u}_1, \cdots, \mathbf{u}_p) \equiv \left(\det \left(\mathbf{u}_i \cdot \mathbf{u}_j \right) \right)^{1/2}.$$

The dot product is used to determine this volume of a parallelepiped spanned by the given vectors and you should note that it is only the dot product that matters. Let

$$x = f_1(u_1, u_2, u_3),\ y = f_2(u_1, u_2, u_3),\ z = f_3(u_1, u_2, u_3) \tag{11.3}$$

where $\mathbf{u} \in U$ an open set in $\mathbb{R}^3$ and corresponding to such a $\mathbf{u} \in U$ there exists a unique point $(x, y, z) \in V$ as above. Suppose at the point $\mathbf{u}_0 \in U$, there is an infinitesimal box having sides du_1, du_2, du_3. Then this little box would correspond to something in V. What? Consider the mapping from U to V defined by

$$\mathbf{x} = \begin{pmatrix} x \\ y \\ z \end{pmatrix} = \begin{pmatrix} f_1(u_1, u_2, u_3) \\ f_2(u_1, u_2, u_3) \\ f_3(u_1, u_2, u_3) \end{pmatrix} = \mathbf{f}(\mathbf{u}) \tag{11.4}$$

which takes a point $\mathbf{u}$ in U and sends it to the point in V which is identified as $(x, y, z)^T \equiv \mathbf{x}$. What happens to a point of the infinitesimal box? Such a point is of the form

$$(u_{01} + s_1 du_1, u_{02} + s_2\, du_2, u_{03} + s_3 du_3),$$

where $s_i \geq 0$ and $\sum_i s_i \leq 1$. Also, from the definition of the derivative,

$$\mathbf{f}(u_{10} + s_1 du_1, u_{20} + s_2\, du_2, u_{30} + s_3 du_3) - \mathbf{f}(u_{01}, u_{02}, u_{03}) =$$

$$D\mathbf{f}(u_{10}, u_{20}, u_{30}) \begin{pmatrix} s_1 du_1 \\ s_2 du_2 \\ s_3 du_3 \end{pmatrix} + \mathbf{o} \begin{pmatrix} s_1 du_1 \\ s_2 du_2 \\ s_3 du_3 \end{pmatrix}$$

where the last term may be taken equal to $\mathbf{0}$ because the vector $(s_1 du_1, s_2 du_2, s_3 du_3)^T$ is infinitesimal, meaning nothing precise, but conveying the idea that it is surpassingly small. Therefore, a point of this infinitesimal box is sent to the vector

$$\overbrace{\left(\frac{\partial \mathbf{x}\,(\mathbf{u}_0)}{\partial u_1}, \frac{\partial \mathbf{x}\,(\mathbf{u}_0)}{\partial u_2}, \frac{\partial \mathbf{x}\,(\mathbf{u}_0)}{\partial u_3} \right)}^{=D\mathbf{f}(u_{10},u_{20},u_{30})} \begin{pmatrix} s_1 du_1 \\ s_2 du_2 \\ s_3 du_3 \end{pmatrix} =$$

$$s_1 \frac{\partial \mathbf{x}\,(\mathbf{u}_0)}{\partial u_1} du_1 + s_2 \frac{\partial \mathbf{x}\,(\mathbf{u}_0)}{\partial u_2} du_2 + s_3 \frac{\partial \mathbf{x}\,(\mathbf{u}_0)}{\partial u_3} du_3,$$

a point of the infinitesimal parallelepiped determined by the vectors

$$\left\{ \frac{\partial \mathbf{x}\,(u_{10}, u_{20}, u_{30})}{\partial u_1} du_1, \frac{\partial \mathbf{x}\,(u_{10}, u_{20}, u_{30})}{\partial u_2} du_2, \frac{\partial \mathbf{x}\,(u_{10}, u_{20}, u_{30})}{\partial u_3} du_3 \right\}.$$

The situation is no different for general coordinate systems in any dimension. In general, $\mathbf{x} = \mathbf{f}\,(\mathbf{u})$ where $\mathbf{u} \in U$, a subset of $\mathbb{R}^n$ and $\mathbf{x}$ is a point in V, a subset of n dimensional space. Thus, letting the Cartesian coordinates of $\mathbf{x}$ be given by $\mathbf{x} = (x_1, \cdots, x_n)^T$, each x_i being a function of $\mathbf{u}$, an infinitesimal box located at $\mathbf{u}_0$ corresponds to an infinitesimal parallelepiped located at $\mathbf{f}\,(\mathbf{u}_0)$ which is determined by the n vectors $\left\{ \frac{\partial \mathbf{x}(\mathbf{u}_0)}{\partial u_i} du_i \right\}_{i=1}^{n}$. From Definition 11.2, the volume of this infinitesimal parallelepiped located at $\mathbf{f}\,(\mathbf{u}_0)$ is given by

$$\left(\det \left(\frac{\partial \mathbf{x}\,(\mathbf{u}_0)}{\partial u_i} du_i \cdot \frac{\partial \mathbf{x}\,(\mathbf{u}_0)}{\partial u_j} du_j \right) \right)^{1/2} \tag{11.5}$$

in which there is no sum on the repeated index. Now in general, if there are n vectors in $\mathbb{R}^n$, $\{\mathbf{v}_1, \cdots, \mathbf{v}_n\}$,

$$\det (\mathbf{v}_i \cdot \mathbf{v}_j)^{1/2} = |\det (\mathbf{v}_1, \cdots, \mathbf{v}_n)| \tag{11.6}$$

where this last matrix is the $n \times n$ matrix which has the ith column equal to $\mathbf{v}_i$. The reason for this is that the matrix whose ij^{th} entry is $\mathbf{v}_i \cdot \mathbf{v}_j$ is just the product of the two matrices,

$$\begin{pmatrix} \mathbf{v}_1^T \\ \vdots \\ \mathbf{v}_n^T \end{pmatrix} (\mathbf{v}_1, \cdots, \mathbf{v}_n)$$

where the first on the left is the matrix having the ith row equal to $\mathbf{v}_i^T$ while the matrix on the right is just the matrix having the ith column equal to $\mathbf{v}_i$. Therefore, since the determinant of a matrix equals the determinant of its transpose,

$$\det (\mathbf{v}_i \cdot \mathbf{v}_j) = \det \left(\begin{pmatrix} \mathbf{v}_1^T \\ \vdots \\ \mathbf{v}_n^T \end{pmatrix} (\mathbf{v}_1, \cdots, \mathbf{v}_n) \right) = \det (\mathbf{v}_1, \cdots, \mathbf{v}_n)^2$$

and so taking square roots yields (11.6). Therefore, from the properties of determinants, (11.5) equals

$$\left| \det\left(\frac{\partial \mathbf{x}(\mathbf{u}_0)}{\partial u_1} du_1, \cdots, \frac{\partial \mathbf{x}(\mathbf{u}_0)}{\partial u_n} du_n \right) \right| = \left| \det\left(\frac{\partial \mathbf{x}(\mathbf{u}_0)}{\partial u_1}, \cdots, \frac{\partial \mathbf{x}(\mathbf{u}_0)}{\partial u_n} \right) \right| du_1 \cdots du_n$$

This is the infinitesimal chunk of volume corresponding to the point $\mathbf{f}(\mathbf{u}_0)$ in V.

Definition 11.3. Let $\mathbf{x} = \mathbf{f}(\mathbf{u})$ be as described above. Then the symbol

$$\frac{\partial (x_1, \cdots x_n)}{\partial (u_1, \cdots, u_n)},$$

called the Jacobian determinant, is defined by

$$\det\left(\frac{\partial \mathbf{x}(\mathbf{u}_0)}{\partial u_1}, \cdots, \frac{\partial \mathbf{x}(\mathbf{u}_0)}{\partial u_n} \right) \equiv \frac{\partial (x_1, \cdots x_n)}{\partial (u_1, \cdots, u_n)}.$$

Also, the symbol $\left| \frac{\partial (x_1, \cdots x_n)}{\partial (u_1, \cdots, u_n)} \right| du_1 \cdots du_n$ is called the volume element or increment of volume, or increment of area.

This has given motivation for the following fundamental procedure often called the **change of variables formula** which holds under fairly general conditions.

Procedure 11.1. Suppose U is an open subset of $\mathbb{R}^n$ for $n > 0$ and suppose $\mathbf{f} : U \to \mathbf{f}(U)$ is a C^1 function which is one to one, $\mathbf{x} = \mathbf{f}(\mathbf{u})$. [2]Then if $h : \mathbf{f}(U) \to \mathbb{R}$,

$$\int_U h(\mathbf{f}(\mathbf{u})) \left| \frac{\partial (x_1, \cdots, x_n)}{\partial (u_1, \cdots, u_n)} \right| dV = \int_{\mathbf{f}(U)} h(\mathbf{x}) \, dV.$$

Example 11.16. Find the area of the region in $\mathbb{R}^2$ which is determined by the lines $y = 2x, y = (1/2)x, x + y = 1, x + y = 3$.

You might sketch this region. You will find it is an ugly quadrilateral. Let $u = x + y$ and $v = \frac{y}{x}$. The reason for this is that the given region corresponds to $(u, v) \in [1, 3] \times \left[\frac{1}{2}, 2 \right]$, a nice rectangle. Now we need to solve for x, y to obtain the Jacobian. A little computation shows that

$$x = \frac{u}{v+1}, \quad y = \frac{uv}{v+1}$$

Therefore, $\frac{\partial (x,y)}{\partial (u,v)}$ is

$$\det\begin{pmatrix} \frac{1}{v+1} & -\frac{u}{(v+1)^2} \\ \frac{v}{v+1} & \frac{u}{(v+1)^2} \end{pmatrix} = \frac{u}{(v+1)^2}.$$

Therefore, the area of this quadrilateral is

$$\int_{1/2}^{2} \int_{1}^{3} \frac{u}{(v+1)^2} du\, dv = \frac{4}{3}.$$

---

[2]This will cause non overlapping infinitesimal boxes in U to be mapped to non overlapping infinitesimal parallelepipeds in V.

Also, in the context of the Riemann integral we should say more about the set U in any case the function h. These conditions are mainly technical however, and since a mathematically respectable treatment will not be attempted for this theorem in this part of the book, I think it best to give a memorable version of it which is essentially correct in all examples of interest.

11.6   Exercises

(1) Verify the three dimensional volume increment in spherical coordinates is

$$\rho^2 \sin(\phi)\, d\rho d\phi d\theta.$$

(2) Find the area of the bounded region R, determined by $5x + y = 1, 5x + y = 9, y = 2x$, and $y = 5x$.

(3) Find the area of the bounded region R, determined by $y + 2x = 6, y + 2x = 10, y = 3x$, and $y = 4x$.

(4) A solid, R is determined by $3x + y = 2, 3x + y = 4, y = x$, and $y = 2x$ and the density is $\rho = x$. Find the total mass of R.

(5) A solid, R is determined by $4x + 2y = 1, 4x + 2y = 9, y = x$, and $y = 6x$ and the density is $\rho = y$. Find the total mass of R.

(6) A solid, R is determined by $3x + y = 3, 3x + y = 10, y = 3x$, and $y = 5x$ and the density is $\rho = y^{-1}$. Find the total mass of R.

(7) Find a 2×2 matrix A which maps the equilateral triangle having vertices at

$$(0,0), (1,0), \text{and } \left(1/2, \sqrt{3}/2\right)$$

to the triangle having vertices at $(0,0), (a,b)$, and (c,d) where (c,d) is not a multiple of (a,b). Find the area of this last triangle by using the cross product. Next find the area of this triangle using the change of variables formula and the fact that the area of the equilateral triangle is $\frac{\sqrt{3}}{4}$.

(8) Find the volume of the region E, bounded by the ellipsoid, $\frac{1}{4}x^2 + y^2 + z^2 = 1$.

(9) Here are three vectors. $(4,1,2)^T, (5,0,2)^T$, and $(3,1,3)^T$. These vectors determine a parallelepiped, R, which is occupied by a solid having density $\rho = x$. Find the mass of this solid.

(10) Here are three vectors. $(5,1,6)^T, (6,0,6)^T$, and $(4,1,7)^T$. These vectors determine a parallelepiped, R, which is occupied by a solid having density $\rho = y$. Find the mass of this solid.

(11) Here are three vectors. $(5,2,9)^T, (6,1,9)^T$, and $(4,2,10)^T$. These vectors determine a parallelepiped, R, which is occupied by a solid having density $\rho = y + x$. Find the mass of this solid.

(12) Compute the volume of a sphere of radius R using cylindrical coordinates.

(13) Fill in all details for the following argument that

$$\int_0^\infty e^{-x^2}\, dx = \frac{1}{2}\sqrt{\pi}.$$

Let $I = \int_0^\infty e^{-x^2}\, dx$. Then

$$I^2 = \int_0^\infty \int_0^\infty e^{-(x^2+y^2)}\, dx\, dy = \int_0^{\pi/2} \int_0^\infty re^{-r^2}\, dr\, d\theta = \frac{1}{4}\pi$$

from which the result follows.

(14) Show that $\int_{-\infty}^{\infty} \frac{1}{\sqrt{2\pi}\sigma} e^{-\frac{(x-\mu)^2}{2\sigma^2}} dx = 1$. Here σ is a positive number called the standard deviation and μ is a number called the mean.

(15) Show using Problem 13 $\Gamma\left(\frac{1}{2}\right) = \sqrt{\pi}$,. Recall $\Gamma(\alpha) \equiv \int_0^\infty e^{-t} t^{\alpha-1} dt$.

(16) Let $p, q > 0$ and define $B(p, q) = \int_0^1 x^{p-1}(1-x)^{q-1}$. Show that

$$\Gamma(p)\Gamma(q) = B(p, q)\Gamma(p+q).$$

Hint: It is fairly routine if you start with the left side and proceed to change variables.

11.7   The Moment Of Inertia And Center Of Mass

The methods used to evaluate multiple integrals make possible the determination of centers of mass and moments of inertia for solids. This leads to the following definition.

Definition 11.4. Let a solid occupy a region R such that its density is $\rho(\mathbf{x})$ for $\mathbf{x}$ a point in R and let L be a line. For $\mathbf{x} \in R$, let $l(\mathbf{x})$ be the distance from the point $\mathbf{x}$ to the line L. The moment of inertia of the solid is defined as

$$I = \int_R l(\mathbf{x})^2 \rho(\mathbf{x}) \, dV.$$

Letting $(\overline{x}, \overline{y}, \overline{z})$ denote the Cartesian coordinates of the center of mass,

$$\overline{x} = \frac{\int_R x\rho(\mathbf{x}) \, dV}{\int_R \rho(\mathbf{x}) \, dV}, \quad \overline{y} = \frac{\int_R y\rho(\mathbf{x}) \, dV}{\int_R \rho(\mathbf{x}) \, dV}, \quad \overline{z} = \frac{\int_R z\rho(\mathbf{x}) \, dV}{\int_R \rho(\mathbf{x}) \, dV}$$

where x, y, z are the Cartesian coordinates of the point at $\mathbf{x}$.

The reason the moment of inertia is of interest has to do with the total kinetic energy of a solid occupying the region R which is rotating about the line L. Suppose its angular velocity is ω. Then the kinetic energy of an infinitesimal chunk of volume located at point $\mathbf{x}$ is $\frac{1}{2}\rho(\mathbf{x})(l(\mathbf{x})\omega)^2 dV$. Then using an integral to add these up, it follows the total kinetic energy is

$$\frac{1}{2}\int_R \rho(\mathbf{x})l(\mathbf{x})^2 \, dV\omega^2 = \frac{1}{2}I\omega^2$$

Thus in the consideration of a rotating body, the moment of inertia takes the place of mass when angular velocity takes the place of speed.

As to the center of mass, its significance is that it gives the point at which the mass will balance. See Volume 1 to see this explained with point masses. The only difference is that here the sums need to be replaced with integrals.

Example 11.17. Let a solid occupy the three dimensional region R and suppose the density is ρ. What is the moment of inertia of this solid about the z axis? What is the center of mass?

Here the little masses would be of the form $\rho(\mathbf{x}) \, dV$ where $\mathbf{x}$ is a point of R. Therefore, the contribution of this mass to the moment of inertia would be $(x^2 + y^2) \rho(\mathbf{x}) \, dV$ where the Cartesian coordinates of the point $\mathbf{x}$ are (x, y, z). Then summing these up as an integral, yields the following for the moment of inertia.

$$\int_R (x^2 + y^2) \rho(\mathbf{x}) \, dV. \tag{11.7}$$

To find the center of mass, sum up $\mathbf{r}\rho \, dV$ for the points in R and divide by the total mass. In Cartesian coordinates, where $\mathbf{r} = (x, y, z)$, this means to sum up vectors of the form $(x\rho \, dV, y\rho \, dV, z\rho \, dV)$ and divide by the total mass. Thus the Cartesian coordinates of the center of mass are

$$\left(\frac{\int_R x\rho \, dV}{\int_R \rho \, dV}, \frac{\int_R y\rho \, dV}{\int_R \rho \, dV}, \frac{\int_R z\rho \, dV}{\int_R \rho \, dV} \right) \equiv \frac{\int_R \mathbf{r}\rho \, dV}{\int_R \rho \, dV}.$$

Here is a specific example.

Example 11.18. Find the moment of inertia about the z axis and center of mass of the solid which occupies the region R defined by $9 - (x^2 + y^2) \geq z \geq 0$ if the density is $\rho(x, y, z) = \sqrt{x^2 + y^2}$.

This moment of inertia is $\int_R (x^2 + y^2) \sqrt{x^2 + y^2} \, dV$ and the easiest way to find this integral is to use cylindrical coordinates. Thus the answer is

$$\int_0^{2\pi} \int_0^3 \int_0^{9-r^2} r^3 r \, dz \, dr \, d\theta = \frac{8748}{35}\pi.$$

To find the center of mass, note the x and y coordinates of the center of mass,

$$\frac{\int_R x\rho \, dV}{\int_R \rho \, dV}, \quad \frac{\int_R y\rho \, dV}{\int_R \rho \, dV}$$

both equal zero because the above shape is symmetric about the z axis and ρ is also symmetric in its values. Thus $x\rho \, dV$ will cancel with $-x\rho \, dV$ and a similar conclusion will hold for the y coordinate. It only remains to find the z coordinate of the center of mass, $\bar{z}$. In polar coordinates, $\rho = r$ and so,

$$\bar{z} = \frac{\int_R z\rho \, dV}{\int_R \rho \, dV} = \frac{\int_0^{2\pi} \int_0^3 \int_0^{9-r^2} zr^2 \, dz \, dr \, d\theta}{\int_0^{2\pi} \int_0^3 \int_0^{9-r^2} r^2 \, dz \, dr \, d\theta} = \frac{18}{7}.$$

Thus the center of mass will be $\left(0, 0, \frac{18}{7}\right)$.

11.8 Exercises

(1) Let R denote the finite region bounded by $z = 4 - x^2 - y^2$ and the xy plane. Find z_c, the z coordinate of the center of mass if the density σ is a constant.

(2) Let R denote the finite region bounded by $z = 4 - x^2 - y^2$ and the xy plane. Find z_c, the z coordinate of the center of mass if the density σ is equals $\sigma(x, y, z) = z$.

(3) Find the mass and center of mass of the region between the surfaces $z = -y^2 + 8$ and $z = 2x^2 + y^2$ if the density equals $\sigma = 1$.

(4) Find the mass and center of mass of the region between the surfaces $z = -y^2 + 8$ and $z = 2x^2 + y^2$ if the density equals $\sigma(x, y, z) = x^2$.

(5) The two cylinders, $x^2 + y^2 = 4$ and $y^2 + z^2 = 4$ intersect in a region R. Find the mass and center of mass if the density σ, is given by $\sigma(x, y, z) = z^2$.

(6) The two cylinders, $x^2 + y^2 = 4$ and $y^2 + z^2 = 4$ intersect in a region R. Find the mass and center of mass if the density σ, is given by $\sigma(x, y, z) = 4 + z$.

(7) Find the mass and center of mass of the set (x, y, z) such that $\frac{x^2}{4} + \frac{y^2}{9} + z^2 \leq 1$ if the density is $\sigma(x, y, z) = 4 + y + z$.

(8) Let R denote the finite region bounded by $z = 9 - x^2 - y^2$ and the xy plane. Find the moment of inertia of this shape about the z axis given the density equals 1.

(9) Let R denote the finite region bounded by $z = 9 - x^2 - y^2$ and the xy plane. Find the moment of inertia of this shape about the x axis given the density equals 1.

(10) Let B be a solid ball of constant density and radius R. Find the moment of inertia about a line through a diameter of the ball. You should get $\frac{2}{5} R^2 M$ where M is the mass..

(11) Let B be a solid ball of density $\sigma = \rho$ where ρ is the distance to the center of the ball which has radius R. Find the moment of inertia about a line through a diameter of the ball. Write your answer in terms of the total mass and the radius as was done in the constant density case.

(12) Let C be a solid cylinder of constant density and radius R. Find the moment of inertia about the axis of the cylinder
You should get $\frac{1}{2} R^2 M$ where M is the mass.

(13) Let C be a solid cylinder of constant density and radius R and mass M and let B be a solid ball of radius R and mass M. The cylinder and the ball are placed on the top of an inclined plane and allowed to roll to the bottom. Which one will arrive first and why?

(14) A ball of radius 4 has a cone taken out of the top which has an angle of $\pi/2$ and then a cone taken out of the bottom which has an angle of $\pi/3$. If the density is $\lambda = \rho$, find the z component of the center of mass.

(15) A ball of radius 4 has a cone taken out of the top which has an angle of $\pi/2$ and then a cone taken out of the bottom which has an angle of $\pi/3$. If the density is $\lambda = \rho$, find the moment of inertia about the z axis.

(16) Suppose a solid of mass M occupying the region B has moment of inertia, I_l about a line, l which passes through the center of mass of M and let l_1 be another line parallel to l and at a distance of a from l. Then the parallel axis

theorem states $I_{l_1} = I_l + a^2 M$. Prove the parallel axis theorem. **Hint:** Choose axes such that the z axis is l and l_1 passes through the point $(a, 0)$ in the xy plane.

(17) * Using the parallel axis theorem find the moment of inertia of a solid ball of radius R and mass M about an axis located at a distance of a from the center of the ball. Your answer should be $Ma^2 + \frac{2}{5} MR^2$.

(18) Consider all axes in computing the moment of inertia of a solid. Will the smallest possible moment of inertia always result from using an axis which goes through the center of mass?

(19) Find the moment of inertia of a solid thin rod of length l, mass M, and constant density about an axis through the center of the rod perpendicular to the axis of the rod. You should get $\frac{1}{12} l^2 M$.

(20) Using the parallel axis theorem, find the moment of inertia of a solid thin rod of length l, mass M, and constant density about an axis through an end of the rod perpendicular to the axis of the rod. You should get $\frac{1}{3} l^2 M$.

(21) Let the angle between the z axis and the sides of a right circular cone be α. Also assume the height of this cone is h. Find the z coordinate of the center of mass of this cone in terms of α and h assuming the density is constant.

(22) Let the angle between the z axis and the sides of a right circular cone be α. Also assume the height of this cone is h. Assuming the density is $\sigma = 1$, find the moment of inertia about the z axis in terms of α and h.

(23) Let R denote the part of the solid ball, $x^2 + y^2 + z^2 \leq R^2$ which lies in the first octant. That is $x, y, z \geq 0$. Find the coordinates of the center of mass if the density is constant. Your answer for one of the coordinates for the center of mass should be $(3/8) R$.

(24) Show that in general for $\mathbf{L}$ angular momentum,

$$\frac{d\mathbf{L}}{dt} = \mathbf{\Gamma}$$

where $\mathbf{\Gamma}$ is the total torque,

$$\mathbf{\Gamma} \equiv \sum \mathbf{r}_i \times \mathbf{F}_i$$

where $\mathbf{F}_i$ is the force on the ith point mass.

Chapter 12

The Integral On Two Dimensional Surfaces In $\mathbb{R}^3$

12.1 The Two Dimensional Area In $\mathbb{R}^3$

Consider the boundary of some three dimensional region such that a function f is defined on this boundary. Imagine taking the value of this function at a point, multiplying this value by the area of an infinitesimal chunk of area located at this point and then adding these up. This is just the notion of the integral presented earlier only now there is a difference because this infinitesimal chunk of area should be considered as two dimensional even though it is in three dimensions. However, it is not really all that different from what was done earlier. It all depends on the following fundamental definition which is just a review of the fact presented earlier that the area of a parallelogram determined by two vectors in $\mathbb{R}^3$ is the norm of the cross product of the two vectors.

Definition 12.1. Let $\mathbf{u}_1, \mathbf{u}_2$ be vectors in $\mathbb{R}^3$. The 2 dimensional parallelogram determined by these vectors will be denoted by $P(\mathbf{u}_1, \mathbf{u}_2)$ and it is defined as

$$P(\mathbf{u}_1, \mathbf{u}_2) \equiv \left\{ \sum_{j=1}^{2} s_j \mathbf{u}_j : s_j \in [0, 1] \right\}.$$

Then the area of this parallelogram is

$$\text{area } P(\mathbf{u}_1, \mathbf{u}_2) \equiv |\mathbf{u}_1 \times \mathbf{u}_2|.$$

Suppose then that $\mathbf{x} = \mathbf{f}(\mathbf{u})$ where $\mathbf{u} \in U$, a subset of $\mathbb{R}^2$ and $\mathbf{x}$ is a point in V, a subset of 3 dimensional space. Thus, letting the Cartesian coordinates of $\mathbf{x}$ be given by $\mathbf{x} = (x_1, x_2, x_3)^T$, each x_i being a function of $\mathbf{u}$, an infinitesimal rectangle located at $\mathbf{u}_0$ corresponds to an infinitesimal parallelogram located at $\mathbf{f}(\mathbf{u}_0)$ which is determined by the 2 vectors $\left\{ \frac{\partial \mathbf{f}(\mathbf{u}_0)}{\partial u_i} du_i \right\}_{i=1}^{2}$, each of which is tangent to the surface defined by $\mathbf{x} = \mathbf{f}(\mathbf{u})$. (No sum on the repeated index.)

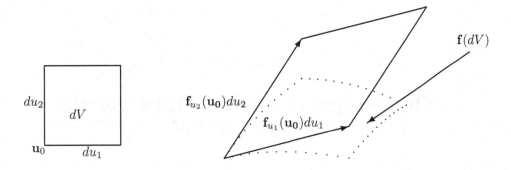

From Definition (12.1), the two dimensional volume of this infinitesimal parallelepiped located at $\mathbf{f}(\mathbf{u_0})$ is given by

$$\left| \frac{\partial \mathbf{f}(\mathbf{u_0})}{\partial u_1} du_1 \times \frac{\partial \mathbf{f}(\mathbf{u_0})}{\partial u_2} du_2 \right| = \left| \frac{\partial \mathbf{f}(\mathbf{u_0})}{\partial u_1} \times \frac{\partial \mathbf{f}(\mathbf{u_0})}{\partial u_2} \right| du_1 du_2 \qquad (12.1)$$

$$= \left| \mathbf{f}_{u_1} \times \mathbf{f}_{u_2} \right| du_1 du_2 \qquad (12.2)$$

It might help to think of a lizard. The infinitesimal parallelepiped is like a very small scale on a lizard. This is the essence of the idea. To define the area of the lizard sum up areas of individual scales[1]. If the scales are small enough, their sum would serve as a good approximation to the area of the lizard.

This motivates the following fundamental procedure which I hope is extremely familiar from the earlier material.

Procedure 12.1. Suppose U is a subset of $\mathbb{R}^2$ and suppose $\mathbf{f} : U \to \mathbf{f}(U) \subseteq \mathbb{R}^3$ is a one to one and C^1 function. Then if $h : \mathbf{f}(U) \to \mathbb{R}$, define the

---

[1]This beautiful lizard is a *Sceloporus magister*. It was photographed by C. Riley Nelson who is in the Zoology department at Brigham Young University © 2004 in Kane Co. Utah. The lizard is a little less than one foot in length.

2 dimensional surface integral $\int_{\mathbf{f}(U)} h(\mathbf{x})\, dA$ according to the following formula.

$$\int_{\mathbf{f}(U)} h(\mathbf{x})\, dA \equiv \int_U h\left(\mathbf{f}(\mathbf{u})\right) \left|\mathbf{f}_{u_1}(\mathbf{u}) \times \mathbf{f}_{u_2}(\mathbf{u})\right| du_1 du_2.$$

Definition 12.2. It is customary to write $\left|\mathbf{f}_{u_1}(\mathbf{u}) \times \mathbf{f}_{u_2}(\mathbf{u})\right| = \frac{\partial(x_1, x_2, x_3)}{\partial(u_1, u_2)}$ because this new notation generalizes to far more general situations for which the cross product is not defined. For example, one can consider three dimensional surfaces in $\mathbb{R}^8$.

Example 12.1. Find the area of the region labeled A in the following picture. The two circles are of radius 1, one has center $(0,0)$ and the other has center $(1,0)$.

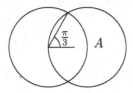

The circles bounding these disks are $x^2 + y^2 = 1$ and $(x-1)^2 + y^2 = x^2 + y^2 - 2x + 1 = 1$. Therefore, in polar coordinates these are of the form $r = 1$ and $r = 2\cos\theta$.

The set A corresponds to the set U, in the (θ, r) plane determined by $\theta \in \left[-\frac{\pi}{3}, \frac{\pi}{3}\right]$ and for each value of θ in this interval, r goes from 1 up to $2\cos\theta$. Therefore, the area of this region is of the form,

$$\int_U 1\, dV = \int_{-\pi/3}^{\pi/3} \int_1^{2\cos\theta} \frac{\partial(x_1, x_2, x_3)}{\partial(\theta, r)}\, dr\, d\theta.$$

The mapping $\mathbf{f} : U \to \mathbb{R}^2$ takes the form $\mathbf{f}(\theta, r) = (r\cos\theta, r\sin\theta)^T$. Here $x_3 = 0$ and so

$$\frac{\partial(x_1, x_2, x_3)}{\partial(u_1, u_2)} = \left\|\begin{array}{ccc} \mathbf{i} & \mathbf{j} & \mathbf{k} \\ \frac{\partial x_1}{\partial \theta} & \frac{\partial x_2}{\partial \theta} & \frac{\partial x_3}{\partial \theta} \\ \frac{\partial x_1}{\partial r} & \frac{\partial x_2}{\partial r} & \frac{\partial x_3}{\partial r} \end{array}\right\| = \left\|\begin{array}{ccc} \mathbf{i} & \mathbf{j} & \mathbf{k} \\ -r\sin\theta & r\cos\theta & 0 \\ \cos\theta & \sin\theta & 0 \end{array}\right\| = r$$

Therefore, the area element is $r\, dr\, d\theta$. It follows the desired area is

$$\int_{-\pi/3}^{\pi/3} \int_1^{2\cos\theta} r\, dr\, d\theta = \frac{1}{2}\sqrt{3} + \frac{1}{3}\pi.$$

Example 12.2. Consider the surface given by $z = x^2$ for $(x, y) \in [0,1] \times [0,1] = U$. Find the surface area of this surface.

The first step in using the above is to write this surface in the form $\mathbf{x} = \mathbf{f}(\mathbf{u})$. This is easy to do if you let $\mathbf{u} = (x, y)$. Then $\mathbf{f}(x, y) = (x, y, x^2)$. If you like, let

$x = u_1$ and $y = u_2$. What is $\frac{\partial(x_1, x_2, x_3)}{\partial(x,y)} = |\mathbf{f}_x \times \mathbf{f}_y|$?

$$\mathbf{f}_x = \begin{pmatrix} 1 \\ 0 \\ 2x \end{pmatrix}, \ \mathbf{f}_y = \begin{pmatrix} 0 \\ 1 \\ 0 \end{pmatrix}$$

and so

$$|\mathbf{f}_x \times \mathbf{f}_y| = \left| \begin{pmatrix} 1 \\ 0 \\ 2x \end{pmatrix} \times \begin{pmatrix} 0 \\ 1 \\ 0 \end{pmatrix} \right| = \sqrt{1 + 4x^2}$$

and so the area element is $\sqrt{1 + 4x^2}\, dx\, dy$ and the surface area is obtained by integrating the function $h(\mathbf{x}) \equiv 1$. Therefore, this area is

$$\int_{\mathbf{f}(U)} dA = \int_0^1 \int_0^1 \sqrt{1 + 4x^2}\, dx\, dy = \frac{1}{2}\sqrt{5} - \frac{1}{4}\ln\left(-2 + \sqrt{5}\right)$$

which can be obtained by using the trig. substitution, $2x = \tan\theta$ on the inside integral.

Note this all depends on being able to write the surface in the form, $\mathbf{x} = \mathbf{f}(\mathbf{u})$ for $\mathbf{u} \in U \subseteq \mathbb{R}^p$. Surfaces obtained in this form are called parametrically defined surfaces. These are best but sometimes you have some other description of a surface and in these cases things can get pretty intractable. For example, you might have a level surface of the form $3x^2 + 4y^4 + z^6 = 10$. In this case, you could solve for z using methods of algebra. Thus $z = \sqrt[6]{10 - 3x^2 - 4y^4}$ and a parametric description of part of this level surface is $\left(x, y, \sqrt[6]{10 - 3x^2 - 4y^4}\right)$ for $(x, y) \in U$ where $U = \{(x, y) : 3x^2 + 4y^4 \leq 10\}$. But what if the level surface was something like

$$\sin\left(x^2 + \ln\left(7 + y^2 \sin x\right)\right) + \sin(zx)\, e^z = 11 \sin(xyz)?$$

I really do not see how to use methods of algebra to solve for some variable in terms of the others. It is not even clear to me whether there are any points $(x, y, z) \in \mathbb{R}^3$ satisfying this particular relation. However, if a point satisfying this relation can be identified, the implicit function theorem from advanced calculus can usually be used to assert one of the variables is a function of the others, proving the existence of a parametrization at least locally. The problem is, this theorem does not give the answer in terms of known functions so this is not much help. Finding a parametric description of a surface is a hard problem and there are no easy answers. This is a good example which illustrates the gulf between theory and practice.

Example 12.3. Let $U = [0, 12] \times [0, 2\pi]$ and let $\mathbf{f} : U \to \mathbb{R}^3$ be given by $\mathbf{f}(t, s) \equiv (2 \cos t + \cos s, 2 \sin t + \sin s, t)^T$. Find a double integral for the surface area. A graph of this surface is drawn below.

Then

$$\mathbf{f}_t = \begin{pmatrix} -2\sin t \\ 2\cos t \\ 1 \end{pmatrix}, \mathbf{f}_s = \begin{pmatrix} -\sin s \\ \cos s \\ 0 \end{pmatrix}$$

and

$$\mathbf{f}_t \times \mathbf{f}_s = \begin{pmatrix} -\cos s \\ -\sin s \\ -2\sin t \cos s + 2\cos t \sin s \end{pmatrix}$$

and so $\frac{\partial(x_1,x_2,x_3)}{\partial(t,s)} =$

$$|\mathbf{f}_t \times \mathbf{f}_s| = \sqrt{5 - 4\sin^2 t \sin^2 s - 8\sin t \sin s \cos t \cos s - 4\cos^2 t \cos^2 s}.$$

Therefore, the desired integral giving the area is

$$\int_0^{2\pi} \int_0^{12} \sqrt{5 - 4\sin^2 t \sin^2 s - 8\sin t \sin s \cos t \cos s - 4\cos^2 t \cos^2 s}\, dt\, ds.$$

If you really needed to find the number this equals, how would you go about finding it? This is an interesting question and there is no single right answer. You should think about this. Here is an example for which you will be able to find the integrals.

Example 12.4. Let $U = [0, 2\pi] \times [0, 2\pi]$ and for $(t, s) \in U$, let

$$\mathbf{f}(t, s) = (2\cos t + \cos t \cos s, -2\sin t - \sin t \cos s, \sin s)^T.$$

Find the area of $\mathbf{f}(U)$. This is the surface of a donut shown below. The fancy name for this shape is a torus.

To find its area,

$$\mathbf{f}_t = \begin{pmatrix} -2\sin t - \sin t \cos s \\ -2\cos t - \cos t \cos s \\ 0 \end{pmatrix}, \mathbf{f}_s = \begin{pmatrix} -\cos t \sin s \\ \sin t \sin s \\ \cos s \end{pmatrix}$$

and so $|\mathbf{f}_t \times \mathbf{f}_s| = (\cos s + 2)$ so the area element is $(\cos s + 2)\, ds\, dt$ and the area is

$$\int_0^{2\pi} \int_0^{2\pi} (\cos s + 2)\, ds\, dt = 8\pi^2$$

Example 12.5. Let $U = [0, 2\pi] \times [0, 2\pi]$ and for $(t, s) \in U$, let

$$\mathbf{f}(t, s) = (2\cos t + \cos t \cos s, -2\sin t - \sin t \cos s, \sin s)^T.$$

Find $\int_{\mathbf{f}(U)} h\, dV$ where $h(x, y, z) = x^2$.

Everything is the same as the preceding example except this time it is an integral of a function. The area element is $(\cos s + 2)\, ds\, dt$ and so the integral called for is

$$\int_{\mathbf{f}(U)} h\, dA = \int_0^{2\pi} \int_0^{2\pi} \left(\overbrace{2\cos t + \cos t \cos s}^{x \text{ on the surface}} \right)^2 (\cos s + 2)\, ds\, dt = 22\pi^2$$

12.1.1   *Surfaces Of The Form $z = f(x, y)$*

The special case where a surface is in the form $z = f(x, y)$, $(x, y) \in U$, yields a simple formula which is used most often in this situation. You write the surface parametrically in the form $\mathbf{f}(x, y) = (x, y, f(x, y))^T$ such that $(x, y) \in U$. Then

$$\mathbf{f}_x = \begin{pmatrix} 1 \\ 0 \\ f_x \end{pmatrix}, \quad \mathbf{f}_y = \begin{pmatrix} 0 \\ 1 \\ f_y \end{pmatrix}$$

and

$$|\mathbf{f}_x \times \mathbf{f}_y| = \sqrt{1 + f_y^2 + f_x^2}$$

so the area element is

$$\sqrt{1 + f_y^2 + f_x^2}\, dx\, dy.$$

When the surface of interest comes in this simple form, people generally use this area element directly rather than worrying about a parametrization and taking cross products.

In the case where the surface is of the form $x = f(y, z)$ for $(y, z) \in U$, the area element is obtained similarly and is $\sqrt{1 + f_y^2 + f_z^2}\, dy\, dz$. I think you can guess what the area element is if $y = f(x, z)$.

There is also a simple geometric description of these area elements. Consider the surface $z = f(x, y)$. This is a level surface of the function of three variables $z - f(x, y)$. In fact the surface is simply $z - f(x, y) = 0$. Now consider the gradient of this function of three variables. The gradient is perpendicular to the surface and the third component is positive in this case. This gradient is $(-f_x, -f_y, 1)$ and so the unit upward normal is just $\frac{1}{\sqrt{1 + f_x^2 + f_y^2}}(-f_x, -f_y, 1)$. Now consider the following picture.

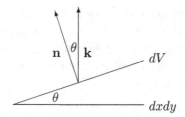

In this picture, you are looking at a chunk of area on the surface seen on edge and so it seems reasonable to expect to have $dx\,dy = dV\cos\theta$. But it is easy to find $\cos\theta$ from the picture and the properties of the dot product.

$$\cos\theta = \frac{\mathbf{n}\cdot\mathbf{k}}{|\mathbf{n}|\,|\mathbf{k}|} = \frac{1}{\sqrt{1+f_x^2+f_y^2}}.$$

Therefore, $dA = \sqrt{1+f_x^2+f_y^2}\,dx\,dy$ as claimed.

Example 12.6. Let $z = \sqrt{x^2+y^2}$ where $(x,y)\in U$ for $U = \{(x,y): x^2+y^2 \le 4\}$ Find $\int_S h\,dS$ where $h(x,y,z) = x+z$ and S is the surface described as $\left(x,y,\sqrt{x^2+y^2}\right)$ for $(x,y)\in U$.

Here you can see directly the angle in the above picture is $\frac{\pi}{4}$ and so $dV = \sqrt{2}\,dx\,dy$. If you do not see this or if it is unclear, simply compute $\sqrt{1+f_x^2+f_y^2}$ and you will find it is $\sqrt{2}$. Therefore, using polar coordinates,

$$\int_S h\,dS = \int_U \left(x+\sqrt{x^2+y^2}\right)\sqrt{2}\,dA$$
$$= \sqrt{2}\int_0^{2\pi}\int_0^2 (r\cos\theta + r)\,r\,dr\,d\theta = \frac{16}{3}\sqrt{2}\pi.$$

One other issue is worth mentioning. Suppose $\mathbf{f}_i : U_i \to \mathbb{R}^3$ where U_i are sets in $\mathbb{R}^2$ and suppose $\mathbf{f}_1(U_1)$ intersects $\mathbf{f}_2(U_2)$ along C where $C = \mathbf{h}(V)$ for $V \subseteq \mathbb{R}^1$. Then define integrals and areas over $\mathbf{f}_1(U_1)\cup\mathbf{f}_2(U_2)$ as follows.

$$\int_{\mathbf{f}_1(U_1)\cup\mathbf{f}_2(U_2)} g\,dA \equiv \int_{\mathbf{f}_1(U_1)} g\,dA + \int_{\mathbf{f}_2(U_2)} g\,dA.$$

Admittedly, the set C gets added in twice but this does not matter because its 2 dimensional volume equals zero and therefore, the integrals over this set will also be zero.

I have been purposely vague about precise mathematical conditions necessary for the above procedures. This is because the precise mathematical conditions which are usually cited are very technical and at the same time far too restrictive. The most general conditions under which these sorts of procedures are valid include things like Lipschitz functions defined on very general sets. These are functions satisfying a Lipschitz condition of the form $|\mathbf{f}(\mathbf{x}) - \mathbf{f}(\mathbf{y})| \le K|\mathbf{x}-\mathbf{y}|$. For example, $y = |x|$ is

Lipschitz continuous. This function does not have a derivative at every point. So it is with Lipschitz functions. However, it turns out these functions have derivatives at enough points to push everything through but this requires considerations involving the Lebesgue integral. Lipschitz functions are also not the most general kind of function for which the above is valid.

12.2   Flux Integrals

These will be important in the next chapter. The idea is this. You have a surface S and a field of unit normal vectors $\mathbf{n}$ on S. That is, for each point of S there exists a unit normal. There is also a vector field $\mathbf{F}$ and you want to find $\int_S \mathbf{F} \cdot \mathbf{n} dS$. There is really nothing new here. You just need to compute the function $\mathbf{F} \cdot \mathbf{n}$ and then integrate it over the surface. Here is an example.

Example 12.7. Let $\mathbf{F}(x, y, z) = (x, x + z, y)$ and let S be the hemisphere $x^2 + y^2 + z^2 = 4, z \geq 0$. Let $\mathbf{n}$ be the unit normal to S which has nonnegative z component. Find $\int_S \mathbf{F} \cdot \mathbf{n} dS$.

First find the function

$$\mathbf{F} \cdot \mathbf{n} \equiv (x, x + z, y) \cdot \overbrace{(x, y, z)\frac{1}{2}}^{=\mathbf{n}} = \frac{1}{2}x^2 + \frac{1}{2}(x + z)y + \frac{1}{2}yz$$

This follows because the normal is of the form $(2x, 2y, 2z)$ and then when you divide by its length using the fact that $x^2 + y^2 + z^2 = 4$, you obtain that $\mathbf{n} = (x, y, z)\frac{1}{2}$ as claimed. Next it remains to choose a coordinate system for the surface and then to compute the integral. A parametrization is

$$x = 2 \sin \phi \cos \theta, \ y = 2 \sin \phi \sin \theta, \ z = 2 \cos \phi$$

and the increment of surface area is then

$$\left| \begin{pmatrix} -2\sin\phi\sin\theta \\ 2\sin\phi\cos\theta \\ 0 \end{pmatrix} \times \begin{pmatrix} 2\cos\phi\cos\theta \\ 2\cos\phi\sin\theta \\ -2\sin\phi \end{pmatrix} \right| d\theta d\phi$$

$$= \left| \begin{pmatrix} -4\sin^2\phi\cos\theta \\ -4\sin^2\phi\sin\theta \\ -4\sin\phi\cos\phi \end{pmatrix} \right| d\theta d\phi = 4\sin\phi d\theta d\phi$$

Therefore, since the hemisphere corresponds to $\theta \in [0, 2\pi]$ and $\phi \in [0, \pi/2]$, the integral to work is

$$\int_0^{2\pi} \int_0^{\pi/2} \left[\frac{1}{2}(2\sin\phi\cos\theta)^2 + \left(\frac{1}{2}(2\sin\phi\cos\theta + 2\cos\phi)\right) \cdot \right.$$

$$\left. (2\sin\phi\sin\theta) + \frac{1}{2}(2\sin\phi\sin\theta)2\cos\phi \right] 4\sin(\phi)\, d\phi d\theta$$

Doing the integration, this reduces to $\frac{16}{3}\pi$.

The important thing to notice is that there is no new mathematics here. That which is new is the significance of a flux integral which will be discussed more in the next chapter. In short, this integral often has the interpretation of a measure of how fast something is crossing a surface.

12.3 Exercises

(1) Find a parametrization for the intersection of the planes $4x + 2y + 4z = 3$ and $6x - 2y = -1$.
(2) Find a parametrization for the intersection of the plane $3x + y + z = 1$ and the circular cylinder $x^2 + y^2 = 1$.
(3) Find a parametrization for the intersection of the plane $3x + 2y + 4z = 4$ and the elliptic cylinder $x^2 + 4z^2 = 16$.
(4) Find a parametrization for the straight line joining $(1, 3, 1)$ and $(-2, 5, 3)$.
(5) Find a parametrization for the intersection of the surfaces $4y + 3z = 3x^2 + 2$ and $3y + 2z = -x + 3$.
(6) Find the area of S if S is the part of the circular cylinder $x^2 + y^2 = 4$ which lies between $z = 0$ and $z = 2 + y$.
(7) Find the area of S if S is the part of the cone $x^2 + y^2 = 16z^2$ between $z = 0$ and $z = h$.
(8) Parametrizing the cylinder $x^2 + y^2 = a^2$ by $x = a\cos v, y = a\sin v, z = u$, show that the area element is $dA = a\, du\, dv$
(9) Find the area enclosed by the limacon $r = 2 + \cos\theta$.
(10) Find the surface area of the paraboloid $z = h\left(1 - x^2 - y^2\right)$ between $z = 0$ and $z = h$. Take a limit of this area as h decreases to 0.
(11) Evaluate $\int_S (1 + x)\, dA$ where S is the part of the plane $4x + y + 3z = 12$ which is in the first octant.
(12) Evaluate $\int_S (1 + x)\, dA$ where S is the part of the cylinder $x^2 + y^2 = 9$ between $z = 0$ and $z = h$.
(13) Evaluate $\int_S (1 + x)\, dA$ where S is the hemisphere $x^2 + y^2 + z^2 = 4$ between $x = 0$ and $x = 2$.
(14) For $(\theta, \alpha) \in [0, 2\pi] \times [0, 2\pi]$, let
$$\mathbf{f}(\theta, \alpha) \equiv (\cos\theta\,(4 + \cos\alpha),\, -\sin\theta\,(4 + \cos\alpha),\, \sin\alpha)^T.$$
Find the area of $\mathbf{f}\left([0, 2\pi] \times [0, 2\pi]\right)$. **Hint:** Check whether $\mathbf{f}_\theta \cdot \mathbf{f}_\alpha = 0$. This might make the computations reasonable.
(15) For $(\theta, \alpha) \in [0, 2\pi] \times [0, 2\pi]$, let
$$\mathbf{f}(\theta, \alpha) \equiv (\cos\theta\,(3 + 2\cos\alpha),\, -\sin\theta\,(3 + 2\cos\alpha),\, 2\sin\alpha)^T,\ h(\mathbf{x}) = \cos\alpha,$$
where α is such that $\mathbf{x} = (\cos\theta\,(3 + 2\cos\alpha),\, -\sin\theta\,(3 + 2\cos\alpha),\, 2\sin\alpha)^T$.
Find $\int_{\mathbf{f}([0,2\pi] \times [0,2\pi])} h\, dA$. **Hint:** Check whether $\mathbf{f}_\theta \cdot \mathbf{f}_\alpha = 0$. This might make the computations reasonable.

(16) For $(\theta, \alpha) \in [0, 2\pi] \times [0, 2\pi]$, let

$$\mathbf{f}(\theta, \alpha) \equiv \left(\cos\theta\,(4 + 3\cos\alpha), -\sin\theta\,(4 + 3\cos\alpha), 3\sin\alpha\right)^T, \quad h(\mathbf{x}) = \cos^2\theta,$$

where θ is such that $\mathbf{x} = \left(\cos\theta\,(4 + 3\cos\alpha), -\sin\theta\,(4 + 3\cos\alpha), 3\sin\alpha\right)^T$. Find $\int_{\mathbf{f}([0,2\pi] \times [0,2\pi])} h\, dA$. **Hint:** Check whether $\mathbf{f}_\theta \cdot \mathbf{f}_\alpha = 0$. This might make the computations reasonable.

(17) In spherical coordinates, $\phi = c, \rho \in [0, R]$ determines a cone. Find the area of this cone.

(18) Let $\mathbf{F} = (x, y, z)$ and let S be the surface which comes from the intersection of the plane $z = x$ with the paraboloid $z = x^2 + y^2$. Find an iterated integral for the flux integral $\int_S \mathbf{F} \cdot \mathbf{n}\, dS$ where $\mathbf{n}$ is the field of unit normals which has negative z component.

(19) Let $\mathbf{F} = (x, 0, 0)$ and let S denote the surface which consists of the part of the sphere $x^2 + y^2 + z^2 = 9$ which lies between the planes $z = 1$ and $z = 2$. Find $\int_S \mathbf{F} \cdot \mathbf{n}\, dS$ where $\mathbf{n}$ is the unit normal to this surface which has positive z component.

(20) In the situation of the above problem change the vector field to $\mathbf{F} = (0, 0, z)$ and do the same problem.

(21) Show that for a sphere of radius a parameterized with spherical coordinates so that

$$x = a \sin\phi \cos\theta, \ y = a \sin\phi \sin\theta, \ z = a \cos\phi$$

the increment of surface area is $a^2 \sin\phi\, d\theta\, d\phi$. Use to show that the area of a sphere of radius a is $4\pi a^2$.

Chapter 13

Calculus Of Vector Fields

13.1 Some Algebraic Preliminaries

This is a review of material in Volume 1. There are two special symbols, δ_{ij} and ε_{ijk} which are very useful in dealing with vector identities. To begin with, here is the definition of these symbols.

Definition 13.1. The symbol δ_{ij}, called the Kroneker delta symbol is defined as follows.

$$\delta_{ij} \equiv \begin{cases} 1 \text{ if } i = j \\ 0 \text{ if } i \neq j \end{cases}.$$

With the Kroneker symbol, i and j can equal any integer in $\{1, 2, \cdots, n\}$ for any $n \in \mathbb{N}$.

Definition 13.2. For i, j, and k integers in the set, $\{1, 2, 3\}$, ε_{ijk} is defined as follows.

$$\varepsilon_{ijk} \equiv \begin{cases} 1 \text{ if } (i, j, k) = (1, 2, 3), (2, 3, 1), \text{ or } (3, 1, 2) \\ -1 \text{ if } (i, j, k) = (2, 1, 3), (1, 3, 2), \text{ or } (3, 2, 1) \\ 0 \text{ if there are any repeated integers} \end{cases}.$$

The subscripts ijk and ij in the above are called indices. A single one is called an index. This symbol ε_{ijk} is also called the permutation symbol.

The way to think of ε_{ijk} is that $\varepsilon_{123} = 1$ and if you switch any two of the numbers in the list i, j, k, it changes the sign. Thus $\varepsilon_{ijk} = -\varepsilon_{jik}$ and $\varepsilon_{ijk} = -\varepsilon_{kji}$ etc. You should check that this rule reduces to the above definition. For example, it immediately implies that if there is a repeated index, the answer is zero. This follows because $\varepsilon_{iij} = -\varepsilon_{iij}$ and so $\varepsilon_{iij} = 0$.

It is useful to use the Einstein summation convention when dealing with these symbols. Simply stated, the convention is that you sum over the repeated index. Thus $a_i b_i$ means $\sum_i a_i b_i$. Also, $\delta_{ij} x_j$ means $\sum_j \delta_{ij} x_j = x_i$. When you use this convention, there is one very important thing to never forget. It is this: Never have an index be repeated more than once. Thus $a_i b_i$ is all right but $a_{ii} b_i$ is not. The reason for this is that you end up getting confused about what is meant. If you want to write $\sum_i a_i b_i c_i$ it is best to simply use the summation notation. There is a very important reduction identity connecting these two symbols.

Lemma 13.1. *The following holds.*

$$\varepsilon_{ijk}\varepsilon_{irs} = (\delta_{jr}\delta_{ks} - \delta_{kr}\delta_{js}).$$

Proof: If $\{j,k\} \neq \{r,s\}$ then every term in the sum on the left must have either ε_{ijk} or ε_{irs} contains a repeated index. Therefore, the left side equals zero. The right side also equals zero in this case. To see this, note that if the two sets are not equal, then there is one of the indices in one of the sets which is not in the other set. For example, it could be that j is not equal to either r or s. Then the right side equals zero.

Therefore, it can be assumed $\{j,k\} = \{r,s\}$. If $i = r$ and $j = s$ for $s \neq r$, then there is exactly one term in the sum on the left and it equals 1. The right also reduces to 1 in this case. If $i = s$ and $j = r$, there is exactly one term in the sum on the left which is nonzero and it must equal -1. The right side also reduces to -1 in this case. If there is a repeated index in $\{j,k\}$, then every term in the sum on the left equals zero. The right also reduces to zero in this case because then $j = k = r = s$ and so the right side becomes $(1)(1) - (-1)(-1) = 0$. ∎

You should verify the following proposition from Volume 1 using this notation.

Proposition 13.1. Let $\mathbf{u}, \mathbf{v}$ be vectors in $\mathbb{R}^n$ where the Cartesian coordinates of $\mathbf{u}$ are $(u_1, \cdots, u_n)$ and the Cartesian coordinates of $\mathbf{v}$ are $(v_1, \cdots, v_n)$. Then $\mathbf{u} \cdot \mathbf{v} = u_i v_i$. If $\mathbf{u}, \mathbf{v}$ are vectors in $\mathbb{R}^3$, then

$$(\mathbf{u} \times \mathbf{v})_i = \varepsilon_{ijk} u_j v_k.$$

Also, $\delta_{ik} a_k = a_i$.

With this notation, you can easily discover vector identities and simplify expressions which involve the cross product. For example, you should use this notation to verify the following proposition from Volume 1.

Example 13.1. Show $(\mathbf{u} \times \mathbf{v}) \times \mathbf{w} = ((\mathbf{u} \cdot \mathbf{w})\mathbf{v} - (\mathbf{v} \cdot \mathbf{w})\mathbf{u})$.

13.2   Divergence And Curl Of A Vector Field

Here the important concepts of divergence and curl are defined.

Definition 13.3. Let $\mathbf{f} : U \to \mathbb{R}^p$ for $U \subseteq \mathbb{R}^p$ denote a vector field. A scalar valued function is called a **scalar field**. The function $\mathbf{f}$ is called a C^k **vector field** if the function $\mathbf{f}$ is a C^k function. For a C^1 vector field, as just described $\nabla \cdot \mathbf{f}(\mathbf{x}) \equiv \operatorname{div} \mathbf{f}(\mathbf{x})$ known as the **divergence**, is defined as

$$\nabla \cdot \mathbf{f}(\mathbf{x}) \equiv \operatorname{div} \mathbf{f}(\mathbf{x}) \equiv \sum_{i=1}^p \frac{\partial f_i}{\partial x_i}(\mathbf{x}).$$

Using the repeated summation convention, this is often written as

$$f_{i,i}\left(\mathbf{x}\right) \equiv \partial_i f_i\left(\mathbf{x}\right)$$

where the comma indicates a partial derivative is being taken with respect to the ith variable and ∂_i denotes differentiation with respect to the ith variable. In words, the divergence is the sum of the ith derivative of the ith component function of $\mathbf{f}$ for all values of i. If $p = 3$, the **curl** of the vector field yields another vector field and it is defined as follows.

$$\left(\text{curl}\left(\mathbf{f}\right)\left(\mathbf{x}\right)\right)_i \equiv \left(\nabla \times \mathbf{f}\left(\mathbf{x}\right)\right)_i \equiv \varepsilon_{ijk}\partial_j f_k\left(\mathbf{x}\right)$$

where here ∂_j means the partial derivative with respect to x_j and the subscript of i in $\left(\text{curl}\left(\mathbf{f}\right)\left(\mathbf{x}\right)\right)_i$ means the ith Cartesian component of the vector $\text{curl}\left(\mathbf{f}\right)\left(\mathbf{x}\right)$. Thus the curl is evaluated by expanding the following determinant along the top row.

$$\begin{vmatrix} \mathbf{i} & \mathbf{j} & \mathbf{k} \\ \frac{\partial}{\partial x} & \frac{\partial}{\partial y} & \frac{\partial}{\partial z} \\ f_1\left(x,y,z\right) & f_2\left(x,y,z\right) & f_3\left(x,y,z\right) \end{vmatrix}.$$

Note the similarity with the cross product. Sometimes the curl is called rot. (Short for rotation not decay.) Also

$$\nabla^2 f \equiv \nabla \cdot \left(\nabla f\right).$$

This last symbol is important enough that it is given a name, the **Laplacian**. It is also denoted by Δ. Thus $\nabla^2 f = \Delta f$. In addition for $\mathbf{f}$ a vector field, the symbol $\mathbf{f} \cdot \nabla$ is defined as a "differential operator" in the following way.

$$\mathbf{f} \cdot \nabla\left(\mathbf{g}\right) \equiv f_1\left(\mathbf{x}\right)\frac{\partial \mathbf{g}\left(\mathbf{x}\right)}{\partial x_1} + f_2\left(\mathbf{x}\right)\frac{\partial \mathbf{g}\left(\mathbf{x}\right)}{\partial x_2} + \cdots + f_p\left(\mathbf{x}\right)\frac{\partial \mathbf{g}\left(\mathbf{x}\right)}{\partial x_p}.$$

Thus $\mathbf{f} \cdot \nabla$ takes vector fields and makes them into new vector fields.

This definition is in terms of a given coordinate system but later coordinate free definitions of the curl and div are presented. For now, everything is defined in terms of a given Cartesian coordinate system. The divergence and curl have profound physical significance and this will be discussed later. For now it is important to understand their definition in terms of coordinates. Be sure you understand that for $\mathbf{f}$ a vector field, $\text{div}\,\mathbf{f}$ is a scalar field meaning it is a scalar valued function of three variables. For a scalar field f, ∇f is a vector field described earlier. For $\mathbf{f}$ a vector field having values in $\mathbb{R}^3$, $\text{curl}\,\mathbf{f}$ is another vector field.

Example 13.2. Let $\mathbf{f}\left(\mathbf{x}\right) = xy\mathbf{i} + \left(z - y\right)\mathbf{j} + \left(\sin\left(x\right) + z\right)\mathbf{k}$. Find $\text{div}\,\mathbf{f}$ and $\text{curl}\,\mathbf{f}$.

First the divergence of $\mathbf{f}$ is

$$\frac{\partial\left(xy\right)}{\partial x} + \frac{\partial\left(z - y\right)}{\partial y} + \frac{\partial\left(\sin\left(x\right) + z\right)}{\partial z} = y + \left(-1\right) + 1 = y.$$

Now curl **f** is obtained by evaluating

$$\begin{vmatrix} \mathbf{i} & \mathbf{j} & \mathbf{k} \\ \frac{\partial}{\partial x} & \frac{\partial}{\partial y} & \frac{\partial}{\partial z} \\ xy & z-y & \sin(x)+z \end{vmatrix} =$$

$$\mathbf{i}\left(\frac{\partial}{\partial y}\left(\sin(x)+z\right)-\frac{\partial}{\partial z}\left(z-y\right)\right)-\mathbf{j}\left(\frac{\partial}{\partial x}\left(\sin(x)+z\right)-\frac{\partial}{\partial z}\left(xy\right)\right)+$$

$$\mathbf{k}\left(\frac{\partial}{\partial x}\left(z-y\right)-\frac{\partial}{\partial y}\left(xy\right)\right)=-\mathbf{i}-\cos(x)\mathbf{j}-x\mathbf{k}.$$

13.2.1   *Vector Identities*

There are many interesting identities which relate the gradient, divergence and curl.

Theorem 13.1. *Assuming* **f**, **g** *are a* C^2 *vector fields whenever necessary, the following identities are valid.*

(1) $\nabla \cdot (\nabla \times \mathbf{f}) = 0$
(2) $\nabla \times \nabla \phi = \mathbf{0}$
(3) $\nabla \times (\nabla \times \mathbf{f}) = \nabla (\nabla \cdot \mathbf{f}) - \nabla^2 \mathbf{f}$ *where* $\nabla^2 \mathbf{f}$ *is a vector field whose* i^{th} *component is* $\nabla^2 f_i$.
(4) $\nabla \cdot (\mathbf{f} \times \mathbf{g}) = \mathbf{g} \cdot (\nabla \times \mathbf{f}) - \mathbf{f} \cdot (\nabla \times \mathbf{g})$
(5) $\nabla \times (\mathbf{f} \times \mathbf{g}) = (\nabla \cdot \mathbf{g}) \mathbf{f} - (\nabla \cdot \mathbf{f}) \mathbf{g} + (\mathbf{g} \cdot \nabla) \mathbf{f} - (\mathbf{f} \cdot \nabla) \mathbf{g}$

   Proof: These are all easy to establish if you use the repeated index summation convention and the reduction identities.

$$\begin{aligned} \nabla \cdot (\nabla \times \mathbf{f}) &= \partial_i (\nabla \times \mathbf{f})_i = \partial_i (\varepsilon_{ijk} \partial_j f_k) = \varepsilon_{ijk} \partial_i (\partial_j f_k) \\ &= \varepsilon_{jik} \partial_j (\partial_i f_k) = -\varepsilon_{ijk} \partial_j (\partial_i f_k) = -\varepsilon_{ijk} \partial_i (\partial_j f_k) \\ &= -\nabla \cdot (\nabla \times \mathbf{f}). \end{aligned}$$

This establishes the first formula. The second formula is done similarly. Now consider the third.

$$\begin{aligned} (\nabla \times (\nabla \times \mathbf{f}))_i &= \varepsilon_{ijk} \partial_j (\nabla \times \mathbf{f})_k = \varepsilon_{ijk} \partial_j (\varepsilon_{krs} \partial_r f_s) \\ &\overset{=\varepsilon_{ijk}}{} \\ &= \overbrace{\varepsilon_{kij}} \varepsilon_{krs} \partial_j (\partial_r f_s) = (\delta_{ir} \delta_{js} - \delta_{is} \delta_{jr}) \partial_j (\partial_r f_s) \\ &= \partial_j (\partial_i f_j) - \partial_j (\partial_j f_i) = \partial_i (\partial_j f_j) - \partial_j (\partial_j f_i) \\ &= \left(\nabla (\nabla \cdot \mathbf{f}) - \nabla^2 \mathbf{f}\right)_i \end{aligned}$$

This establishes the third identity.
   Consider the fourth identity.

$$\begin{aligned} \nabla \cdot (\mathbf{f} \times \mathbf{g}) &= \partial_i (\mathbf{f} \times \mathbf{g})_i = \partial_i \varepsilon_{ijk} f_j g_k \\ &= \varepsilon_{ijk} (\partial_i f_j) g_k + \varepsilon_{ijk} f_j (\partial_i g_k) \\ &= (\varepsilon_{kij} \partial_i f_j) g_k - (\varepsilon_{jik} \partial_i g_k) f_k \\ &= \nabla \times \mathbf{f} \cdot \mathbf{g} - \nabla \times \mathbf{g} \cdot \mathbf{f}. \end{aligned}$$

This proves the fourth identity.

Consider the fifth.

$$
\begin{aligned}
(\nabla \times (\mathbf{f} \times \mathbf{g}))_i &= \varepsilon_{ijk} \partial_j (\mathbf{f} \times \mathbf{g})_k = \varepsilon_{ijk} \partial_j \varepsilon_{krs} f_r g_s \\
&= \varepsilon_{kij} \varepsilon_{krs} \partial_j (f_r g_s) = (\delta_{ir} \delta_{js} - \delta_{is} \delta_{jr}) \partial_j (f_r g_s) \\
&= \partial_j (f_i g_j) - \partial_j (f_j g_i) \\
&= (\partial_j g_j) f_i + g_j \partial_j f_i - (\partial_j f_j) g_i - f_j (\partial_j g_i) \\
&= ((\nabla \cdot \mathbf{g}) \mathbf{f} + (\mathbf{g} \cdot \nabla) (\mathbf{f}) - (\nabla \cdot \mathbf{f}) \mathbf{g} - (\mathbf{f} \cdot \nabla) (\mathbf{g}))_i
\end{aligned}
$$

and this establishes the fifth identity. ∎

13.2.2 Vector Potentials

One of the above identities says $\nabla \cdot (\nabla \times \mathbf{f}) = 0$. Suppose now $\nabla \cdot \mathbf{g} = 0$. Does it follow that there exists $\mathbf{f}$ such that $\mathbf{g} = \nabla \times \mathbf{f}$? It turns out that this is usually the case and when such an $\mathbf{f}$ exists, it is called a **vector potential**. Here is one way to do it, assuming everything is defined so the following formulas make sense.

$$
\mathbf{f}(x, y, z) = \left(\int_0^z g_2(x, y, t) \, dt, - \int_0^z g_1(x, y, t) \, dt + \int_0^x g_3(t, y, 0) \, dt, 0 \right)^T. \quad (13.1)
$$

In verifying this you need to use the following manipulation which will generally hold under reasonable conditions but which has not been carefully shown yet.

$$
\frac{\partial}{\partial x} \int_a^b h(x, t) \, dt = \int_a^b \frac{\partial h}{\partial x}(x, t) \, dt. \quad (13.2)
$$

The above formula seems plausible because the integral is a sort of a sum and the derivative of a sum is the sum of the derivatives. However, this sort of sloppy reasoning will get you into all sorts of trouble. The formula involves the interchange of two limit operations, the integral and the limit of a difference quotient. Such an interchange can only be accomplished through a theorem. The following gives the necessary result.

Lemma 13.2. *Suppose h and $\frac{\partial h}{\partial x}$ are continuous on the rectangle $R = [c, d] \times [a, b]$. Then (13.2) holds.*

Proof: Let Δx be such that $x, x + \Delta x$ are both in $[c, d]$. By Theorem 4.17 on Page 82 there exists $\delta > 0$ such that if $|(x, t) - (x_1, t_1)| < \delta$, then

$$
\left| \frac{\partial h}{\partial x}(x, t) - \frac{\partial h}{\partial x}(x_1, t_1) \right| < \frac{\varepsilon}{b - a}.
$$

Let $|\Delta x| < \delta$. Then

$$
\left| \int_a^b \frac{h(x + \Delta x, t) - h(x, t)}{\Delta x} \, dt - \int_a^b \frac{\partial h}{\partial x}(x, t) \, dt \right|
$$

$$
\leq \int_a^b \left| \frac{h(x + \Delta x, t) - h(x, t)}{\Delta x} - \frac{\partial h}{\partial x}(x, t) \right| \, dt
$$

$$
= \int_a^b \left| \frac{\partial h(x + \theta_t \Delta x)}{\partial x} - \frac{\partial h}{\partial x}(x, t) \right| \, dt < \int_a^b \frac{\varepsilon}{b - a} \, dt = \varepsilon.
$$

Here θ_t is a number between 0 and 1 and going from the second to the third line is an application of the mean value theorem. ∎

The second formula of Theorem 13.1 states $\nabla \times \nabla \phi = \mathbf{0}$. This suggests the following question: Suppose $\nabla \times \mathbf{f} = \mathbf{0}$, does it follow there exists ϕ, a scalar field such that $\nabla \phi = \mathbf{f}$? The answer to this is often yes and a theorem will be given and proved after the presentation of Stoke's theorem. This scalar field ϕ, is called a **scalar potential** for $\mathbf{f}$.

13.2.3 The Weak Maximum Principle

There is also a fundamental result having great significance which involves ∇^2 called the maximum principle. This principle says that if $\nabla^2 u \geq 0$ on a bounded open set U, then u achieves its maximum value on the boundary of U.

Theorem 13.2. *Let U be a bounded open set in $\mathbb{R}^n$ and suppose*

$$u \in C^2 (U) \cap C (\overline{U})$$

such that $\nabla^2 u \geq 0$ in U. Then letting $\partial U = \overline{U} \setminus U$, it follows that

$$\max \{u (\mathbf{x}) : \mathbf{x} \in \overline{U}\} = \max \{u (\mathbf{x}) : x \in \partial U\}.$$

Proof: If this is not so, there exists $\mathbf{x}_0 \in U$ such that $u (\mathbf{x}_0) > \max \{u (\mathbf{x}) : x \in \partial U\} \equiv M$. Since U is bounded, there exists $\varepsilon > 0$ such that

$$u (\mathbf{x}_0) > \max \left\{u (\mathbf{x}) + \varepsilon |\mathbf{x}|^2 : \mathbf{x} \in \partial U\right\}.$$

Therefore, $u (\mathbf{x}) + \varepsilon |\mathbf{x}|^2$ also has its maximum in U because for ε small enough,

$$u (\mathbf{x}_0) + \varepsilon |\mathbf{x}_0|^2 > u (\mathbf{x}_0) > \max \left\{u (\mathbf{x}) + \varepsilon |\mathbf{x}|^2 : \mathbf{x} \in \partial U\right\}$$

for all $\mathbf{x} \in \partial U$.

Now let $\mathbf{x}_1$ be the point in U at which $u (\mathbf{x}) + \varepsilon |\mathbf{x}|^2$ achieves its maximum. As an exercise you should show that $\nabla^2 (f + g) = \nabla^2 f + \nabla^2 g$ and therefore, $\nabla^2 \left(u (\mathbf{x}) + \varepsilon |\mathbf{x}|^2\right) = \nabla^2 u (\mathbf{x}) + 2n\varepsilon$. (Why?) Therefore,

$$0 \geq \nabla^2 u (\mathbf{x}_1) + 2n\varepsilon \geq 2n\varepsilon,$$

a contradiction. ∎

13.3 Exercises

(1) Find $\operatorname{div} \mathbf{f}$ and $\operatorname{curl} \mathbf{f}$ where $\mathbf{f}$ is

    (a) $\left(xyz, x^2 + \ln (xy), \sin x^2 + z\right)^T$

    (b) $(\sin x, \sin y, \sin z)^T$

    (c) $(f (x), g (y), h (z))^T$

(d) $(x - 2, y - 3, z - 6)^T$

(e) $\left(y^2, 2xy, \cos z\right)^T$

(f) $\left(f\left(y, z\right), g\left(x, z\right), h\left(y, z\right)\right)^T$

(2) Prove formula (2) of Theorem 13.1.

(3) Show that if u and v are C^2 functions, then $\operatorname{curl}\left(u\nabla v\right) = \nabla u \times \nabla v$.

(4) Simplify the expression $\mathbf{f} \times \left(\nabla \times \mathbf{g}\right) + \mathbf{g} \times \left(\nabla \times \mathbf{f}\right) + \left(\mathbf{f} \cdot \nabla\right)\mathbf{g} + \left(\mathbf{g} \cdot \nabla\right)\mathbf{f}$.

(5) Simplify $\nabla \times \left(\mathbf{v} \times \mathbf{r}\right)$ where $\mathbf{r} = \left(x, y, z\right)^T = x\mathbf{i} + y\mathbf{j} + z\mathbf{k}$ and $\mathbf{v}$ is a constant vector.

(6) Discover a formula which simplifies $\nabla \cdot \left(v\nabla u\right)$.

(7) Verify that $\nabla \cdot \left(u\nabla v\right) - \nabla \cdot \left(v\nabla u\right) = u\nabla^2 v - v\nabla^2 u$.

(8) Verify that $\nabla^2 \left(uv\right) = v\nabla^2 u + 2\left(\nabla u \cdot \nabla v\right) + u\nabla^2 v$.

(9) Functions u, which satisfy $\nabla^2 u = 0$ are called harmonic functions. Show that the following functions are harmonic where ever they are defined.

(a) $2xy$

(b) $x^2 - y^2$

(c) $\sin x \cosh y$

(d) $\ln\left(x^2 + y^2\right)$

(e) $1/\sqrt{x^2 + y^2 + z^2}$

(10) Verify the formula given in (13.1) is a vector potential for $\mathbf{g}$ assuming that $\operatorname{div} \mathbf{g} = 0$.

(11) Show that if $\nabla^2 u_k = 0$ for each $k = 1, 2, \cdots, m$, and c_k is a constant, then

$$\nabla^2 \left(\sum_{k=1}^m c_k u_k\right) = 0$$

also.

(12) In Theorem 13.2, why is $\nabla^2 \left(\varepsilon \left|\mathbf{x}\right|^2\right) = 2n\varepsilon$?

(13) Using Theorem 13.2, prove the following: Let $f \in C\left(\partial U\right)$ (f is continuous on ∂U.) where U is a bounded open set. Then there exists at most one solution $u \in C^2\left(U\right) \cap C\left(\overline{U}\right)$ and $\nabla^2 u = 0$ in U with $u = f$ on ∂U. **Hint:** Suppose there are two solutions u_i, $i = 1, 2$ and let $w = u_1 - u_2$. Then use the maximum principle.

(14) Suppose $\mathbf{B}$ is a vector field and $\nabla \times \mathbf{A} = \mathbf{B}$. Thus $\mathbf{A}$ is a vector potential for $\mathbf{B}$. Show that $\mathbf{A} + \nabla \phi$ is also a vector potential for $\mathbf{B}$. Here ϕ is just a C^2 scalar field. Thus the vector potential is not unique.

13.4 The Divergence Theorem

The divergence theorem relates an integral over a set to one on the boundary of the set. It is also called Gauss's theorem.

Definition 13.4. A subset V of $\mathbb{R}^3$ is called cylindrical in the x direction if it is of

the form
$$V = \{(x, y, z) : \phi(y, z) \leq x \leq \psi(y, z) \text{ for } (y, z) \in D\}$$
where D is a subset of the yz plane. V is cylindrical in the z direction if
$$V = \{(x, y, z) : \phi(x, y) \leq z \leq \psi(x, y) \text{ for } (x, y) \in D\}$$
where D is a subset of the xy plane, and V is cylindrical in the y direction if
$$V = \{(x, y, z) : \phi(x, z) \leq y \leq \psi(x, z) \text{ for } (x, z) \in D\}$$
where D is a subset of the xz plane. If V is cylindrical in the z direction, denote by ∂V the boundary of V defined to be the points of the form $(x, y, \phi(x, y)), (x, y, \psi(x, y))$ for $(x, y) \in D$, along with points of the form (x, y, z) where $(x, y) \in \partial D$ and $\phi(x, y) \leq z \leq \psi(x, y)$. Points on ∂D are defined to be those for which every open ball contains points which are in D as well as points which are not in D. A similar definition holds for ∂V in the case that V is cylindrical in one of the other directions.

The following picture illustrates the above definition in the case of V cylindrical in the z direction. Also labeled are the z components of the respective outer unit normals on the sides and top and bottom.

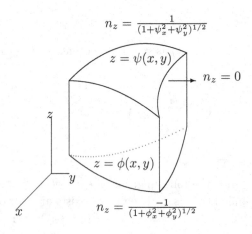

Of course, many three dimensional sets are cylindrical in each of the coordinate directions. For example, a ball or a rectangle or a tetrahedron are all cylindrical in each direction. The following lemma allows the exchange of the volume integral of a partial derivative for an area integral in which the derivative is replaced with multiplication by an appropriate component of the unit exterior normal.

Lemma 13.3. *Suppose V is cylindrical in the z direction and that ϕ and ψ are the functions in the above definition. Assume ϕ and ψ are C^1 functions and suppose F is a C^1 function defined on V. Also, let $\mathbf{n} = (n_x, n_y, n_z)$ be the unit exterior normal to ∂V. Then*
$$\int_V \frac{\partial F}{\partial z}(x, y, z)\, dV = \int_{\partial V} F n_z\, dA.$$

Proof: From the fundamental theorem of calculus,

$$\int_V \frac{\partial F}{\partial z}(x,y,z)\,dV \;=\; \int_D \int_{\phi(x,y)}^{\psi(x,y)} \frac{\partial F}{\partial z}(x,y,z)\,dz\,dx\,dy \tag{13.3}$$

$$= \int_D [F(x,y,\psi(x,y)) - F(x,y,\phi(x,y))]\,dx\,dy$$

Now the unit exterior normal on the top of V, the surface $(x,y,\psi(x,y))$ is

$$\frac{1}{\sqrt{\psi_x^2 + \psi_y^2 + 1}}\,(-\psi_x, -\psi_y, 1).$$

This follows from the observation that the top surface is the level surface $z - \psi(x,y) = 0$ and so the gradient of this function of three variables is perpendicular to the level surface. It points in the correct direction because the z component is positive. Therefore, on the top surface

$$n_z = \frac{1}{\sqrt{\psi_x^2 + \psi_y^2 + 1}}$$

Similarly, the unit normal to the surface on the bottom is

$$\frac{1}{\sqrt{\phi_x^2 + \phi_y^2 + 1}}\,(\phi_x, \phi_y, -1)$$

and so on the bottom surface,

$$n_z = \frac{-1}{\sqrt{\phi_x^2 + \phi_y^2 + 1}}$$

Note that here the z component is negative because since it is the outer normal it must point down. On the lateral surface, the one where $(x,y) \in \partial D$ and $z \in [\phi(x,y), \psi(x,y)]$, $n_z = 0$.

The area element on the top surface is $dA = \sqrt{\psi_x^2 + \psi_y^2 + 1}\,dx\,dy$ while the area element on the bottom surface is $\sqrt{\phi_x^2 + \phi_y^2 + 1}\,dx\,dy$. Therefore, the last expression in (13.3) is of the form,

$$\int_D F(x,y,\psi(x,y)) \overbrace{\frac{1}{\sqrt{\psi_x^2 + \psi_y^2 + 1}}}^{n_z} \overbrace{\sqrt{\psi_x^2 + \psi_y^2 + 1}\,dx\,dy}^{dA} +$$

$$\int_D F(x,y,\phi(x,y)) \left(\underbrace{\frac{-1}{\sqrt{\phi_x^2 + \phi_y^2 + 1}}}_{n_z} \right) \overbrace{\sqrt{\phi_x^2 + \phi_y^2 + 1}\,dx\,dy}^{dA}$$

$$+ \int_{\text{Lateral surface}} F n_z\,dA,$$

the last term equaling zero because on the lateral surface, $n_z = 0$. Therefore, this reduces to $\int_{\partial V} F n_z \, dA$ as claimed. ∎

The following corollary is entirely similar to the above.

Corollary 13.1. *If V is cylindrical in the y direction, then*

$$\int_V \frac{\partial F}{\partial y} \, dV = \int_{\partial V} F n_y \, dA$$

and if V is cylindrical in the x direction, then

$$\int_V \frac{\partial F}{\partial x} \, dV = \int_{\partial V} F n_x \, dA$$

With this corollary, here is a proof of the divergence theorem.

Theorem 13.3. *Let V be cylindrical in each of the coordinate directions and let $\mathbf{F}$ be a C^1 vector field defined on V. Then*

$$\int_V \nabla \cdot \mathbf{F} \, dV = \int_{\partial V} \mathbf{F} \cdot \mathbf{n} \, dA.$$

Proof: From the above lemma and corollary,

$$
\begin{aligned}
\int_V \nabla \cdot \mathbf{F} \, dV &= \int_V \frac{\partial F_1}{\partial x} + \frac{\partial F_2}{\partial y} + \frac{\partial F_3}{\partial y} \, dV \\
&= \int_{\partial V} (F_1 n_x + F_2 n_y + F_3 n_z) \, dA \\
&= \int_{\partial V} \mathbf{F} \cdot \mathbf{n} \, dA. \qquad\qquad ∎
\end{aligned}
$$

The divergence theorem holds for much more general regions than this. Suppose for example you have a complicated region which is the union of finitely many disjoint regions of the sort just described which are cylindrical in each of the coordinate directions. Then the volume integral over the union of these would equal the sum of the integrals over the disjoint regions. If the boundaries of two of these regions intersect, then the area integrals will cancel out on the intersection because the unit exterior normals will point in opposite directions. Therefore, the sum of the integrals over the boundaries of these disjoint regions will reduce to an integral over the boundary of the union of these. Hence the divergence theorem will continue to hold. For example, consider the following picture. If the divergence theorem holds for each V_i in the following picture, then it holds for the union of these two.

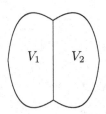

General formulations of the divergence theorem involve Hausdorff measures and the Lebesgue integral, a better integral than the old fashioned Riemann integral which has been obsolete now for almost 100 years. When all is said and done, one finds that the conclusion of the divergence theorem is usually true and the theorem can be used with confidence.

Example 13.3. Let $V = [0,1] \times [0,1] \times [0,1]$. That is, V is the cube in the first octant having the lower left corner at $(0,0,0)$ and the sides of length 1. Let $\mathbf{F}(x,y,z) = x\mathbf{i} + y\mathbf{j} + z\mathbf{k}$. Find the flux integral in which $\mathbf{n}$ is the unit exterior normal.

$$\int_{\partial V} \mathbf{F} \cdot \mathbf{n} \, dS$$

You can certainly inflict much suffering on yourself by breaking the surface up into 6 pieces corresponding to the 6 sides of the cube, finding a parametrization for each face and adding up the appropriate flux integrals. For example, $\mathbf{n} = \mathbf{k}$ on the top face and $\mathbf{n} = -\mathbf{k}$ on the bottom face. On the top face, a parametrization is $(x,y,1) : (x,y) \in [0,1] \times [0,1]$. The area element is just $dxdy$. It is not really all that hard to do it this way but it is much easier to use the divergence theorem. The above integral equals

$$\int_V \operatorname{div}(\mathbf{F}) \, dV = \int_V 3 dV = 3.$$

Example 13.4. This time, let V be the unit ball, $\{(x,y,z) : x^2 + y^2 + z^2 \le 1\}$ and let $\mathbf{F}(x,y,z) = x^2\mathbf{i} + y\mathbf{j} + (z-1)\mathbf{k}$. Find

$$\int_{\partial V} \mathbf{F} \cdot \mathbf{n} \, dS.$$

As in the above you could do this by brute force. A parametrization of the ∂V is obtained as

$$x = \sin\phi\cos\theta, \; y = \sin\phi\sin\theta, \; z = \cos\phi$$

where $(\phi, \theta) \in (0,\pi) \times (0, 2\pi]$. Now this does not include all the ball but it includes all but the point at the top and at the bottom. As far as the flux integral is concerned these points contribute nothing to the integral so you can neglect them. Then you can grind away and get the flux integral which is desired. However, it is so much easier to use the divergence theorem! Using spherical coordinates,

$$\int_{\partial V} \mathbf{F} \cdot \mathbf{n} \, dS = \int_V \operatorname{div}(\mathbf{F}) \, dV = \int_V (2x + 1 + 1) \, dV$$

$$= \int_0^\pi \int_0^{2\pi} \int_0^1 (2 + 2\rho\sin(\phi)\cos\theta) \, \rho^2 \sin(\phi) \, d\rho d\theta d\phi = \frac{8}{3}\pi$$

Example 13.5. Suppose V is an open set in $\mathbb{R}^3$ for which the divergence theorem holds. Let $\mathbf{F}(x,y,z) = x\mathbf{i} + y\mathbf{j} + z\mathbf{k}$. Then show that

$$\int_{\partial V} \mathbf{F} \cdot \mathbf{n} \, dS = 3 \times \operatorname{volume}(V).$$

This follows from the divergence theorem.

$$\int_{\partial V} \mathbf{F} \cdot \mathbf{n} dS = \int_V \operatorname{div}(\mathbf{F}) \, dV = 3 \int_V dV = 3 \times \operatorname{volume}(V).$$

The message of the divergence theorem is the relation between the volume integral and an area integral. This is the exciting thing about this marvelous theorem. It is not its utility as a method for evaluations of boring problems. This will be shown in the examples of its use which follow.

13.4.1   *Coordinate Free Concept Of Divergence*

The divergence theorem also makes possible a coordinate free definition of the divergence.

Theorem 13.4. *Let $B(\mathbf{x}, \delta)$ be the ball centered at $\mathbf{x}$ having radius δ and let $\mathbf{F}$ be a C^1 vector field. Then letting $v(B(\mathbf{x}, \delta))$ denote the volume of $B(\mathbf{x}, \delta)$ given by*

$$\int_{B(\mathbf{x}, \delta)} dV,$$

it follows

$$\operatorname{div} \mathbf{F}(\mathbf{x}) = \lim_{\delta \to 0+} \frac{1}{v(B(\mathbf{x}, \delta))} \int_{\partial B(\mathbf{x}, \delta)} \mathbf{F} \cdot \mathbf{n} \, dA. \qquad (13.4)$$

Proof: The divergence theorem holds for balls because they are cylindrical in every direction. Therefore,

$$\frac{1}{v(B(\mathbf{x}, \delta))} \int_{\partial B(\mathbf{x}, \delta)} \mathbf{F} \cdot \mathbf{n} \, dA = \frac{1}{v(B(\mathbf{x}, \delta))} \int_{B(\mathbf{x}, \delta)} \operatorname{div} \mathbf{F}(\mathbf{y}) \, dV.$$

Therefore, since $\operatorname{div} \mathbf{F}(\mathbf{x})$ is a constant,

$$\left| \operatorname{div} \mathbf{F}(\mathbf{x}) - \frac{1}{v(B(\mathbf{x}, \delta))} \int_{\partial B(\mathbf{x}, \delta)} \mathbf{F} \cdot \mathbf{n} \, dA \right|$$

$$= \left| \operatorname{div} \mathbf{F}(\mathbf{x}) - \frac{1}{v(B(\mathbf{x}, \delta))} \int_{B(\mathbf{x}, \delta)} \operatorname{div} \mathbf{F}(\mathbf{y}) \, dV \right|$$

$$= \left| \frac{1}{v(B(\mathbf{x}, \delta))} \int_{B(\mathbf{x}, \delta)} (\operatorname{div} \mathbf{F}(\mathbf{x}) - \operatorname{div} \mathbf{F}(\mathbf{y})) \, dV \right|$$

$$\leq \frac{1}{v(B(\mathbf{x}, \delta))} \int_{B(\mathbf{x}, \delta)} |\operatorname{div} \mathbf{F}(\mathbf{x}) - \operatorname{div} \mathbf{F}(\mathbf{y})| \, dV$$

$$\leq \frac{1}{v(B(\mathbf{x}, \delta))} \int_{B(\mathbf{x}, \delta)} \frac{\varepsilon}{2} \, dV < \varepsilon$$

whenever ε is small enough, due to the continuity of $\operatorname{div} \mathbf{F}$. Since ε is arbitrary, this shows (13.4). ∎

How is this definition independent of coordinates? It only involves geometrical notions of volume and dot product. This is why. Imagine rotating the coordinate axes, keeping all distances the same and expressing everything in terms of the new coordinates. The divergence would still have the same value because of this theorem.

13.5 Some Applications Of The Divergence Theorem

13.5.1 *Hydrostatic Pressure*

Imagine a fluid which does not move which is acted on by an acceleration **g**. Of course the acceleration is usually the acceleration of gravity. Also let the density of the fluid be ρ, a function of position. What can be said about the pressure p in the fluid? Let $B(\mathbf{x}, \varepsilon)$ be a small ball centered at the point $\mathbf{x}$. Then the force the fluid exerts on this ball would equal

$$- \int_{\partial B(\mathbf{x}, \varepsilon)} p\mathbf{n} \, dA.$$

Here **n** is the unit exterior normal at a small piece of $\partial B(\mathbf{x}, \varepsilon)$ having area dA. By the divergence theorem, (see Problem 1 on Page 263) this integral equals

$$- \int_{B(\mathbf{x}, \varepsilon)} \nabla p \, dV.$$

Also the force acting on this small ball of fluid is

$$\int_{B(\mathbf{x}, \varepsilon)} \rho \mathbf{g} \, dV.$$

Since it is given that the fluid does not move, the sum of these forces must equal zero. Thus

$$\int_{B(\mathbf{x}, \varepsilon)} \rho \mathbf{g} \, dV = \int_{B(\mathbf{x}, \varepsilon)} \nabla p \, dV.$$

Since this must hold for any ball in the fluid of any radius, it must be that

$$\nabla p = \rho \mathbf{g}. \tag{13.5}$$

It turns out that the pressure in a lake at depth z is equal to $62.5z$. This is easy to see from (13.5). In this case, $\mathbf{g} = g\mathbf{k}$ where $g = 32$ feet/sec^2. The weight of a cubic foot of water is 62.5 pounds. Therefore, the mass in slugs of this water is $62.5/32$. Since it is a cubic foot, this is also the density of the water in slugs per cubic foot. Also, it is normally assumed that water is incompressible[1]. Therefore, this is the mass of water at any depth. Therefore,

$$\frac{\partial p}{\partial x}\mathbf{i} + \frac{\partial p}{\partial y}\mathbf{j} + \frac{\partial p}{\partial z}\mathbf{k} = \frac{62.5}{32} \times 32\mathbf{k}.$$

and so p does not depend on x and y and is only a function of z. It follows $p(0) = 0$, and $p'(z) = 62.5$. Therefore, $p(x, y, z) = 62.5z$. This establishes the claim. This is interesting but (13.5) is more interesting because it does not require ρ to be constant.

---

[1] There is no such thing as an incompressible fluid but this does not stop people from making this assumption.

13.5.2   *Archimedes Law Of Buoyancy*

Archimedes principle states that when a solid body is immersed in a fluid, the net force acting on the body by the fluid is directly up and equals the total weight of the fluid displaced.

Denote the set of points in three dimensions occupied by the body as V. Then for dA an increment of area on the surface of this body, the force acting on this increment of area would equal $-p\, dA\mathbf{n}$ where $\mathbf{n}$ is the exterior unit normal. Therefore, since the fluid does not move,

$$\int_{\partial V} -p\mathbf{n}\, dA = \int_V -\nabla p\, dV = \int_V \rho g\, dV\mathbf{k}$$

which equals the total weight of the displaced fluid and you note the force is directed upward as claimed. Here ρ is the density and (13.5) is being used. There is an interesting point in the above explanation. Why does the second equation hold? Imagine that V were filled with fluid. Then the equation follows from (13.5) because in this equation $\mathbf{g} = -g\mathbf{k}$.

13.5.3   *Equations Of Heat And Diffusion*

Let $\mathbf{x}$ be a point in three dimensional space and let (x_1, x_2, x_3) be Cartesian coordinates of this point. Let there be a three dimensional body having density $\rho = \rho(\mathbf{x}, t)$.

The heat flux $\mathbf{J}$, in the body is defined as a vector which has the following property.

$$\text{Rate at which heat crosses } S = \int_S \mathbf{J} \cdot \mathbf{n}\, dA$$

where $\mathbf{n}$ is the unit normal in the desired direction. Thus if V is a three dimensional body,

$$\text{Rate at which heat leaves } V = \int_{\partial V} \mathbf{J} \cdot \mathbf{n}\, dA$$

where $\mathbf{n}$ is the unit exterior normal.

Fourier's law of heat conduction states that the heat flux $\mathbf{J}$ satisfies $\mathbf{J} = -k\nabla(u)$ where u is the temperature and $k = k(u, \mathbf{x}, t)$ is called the coefficient of thermal conductivity. This changes depending on the material. It also can be shown by experiment to change with temperature. This equation for the heat flux states that the heat flows from hot places toward colder places in the direction of greatest rate of decrease in temperature. Let $c(\mathbf{x}, t)$ denote the specific heat of the material in the body. This means the amount of heat within V is given by the formula $\int_V \rho(\mathbf{x}, t)c(\mathbf{x}, t)u(\mathbf{x}, t)\, dV$. Suppose also there are sources for the heat within the material given by $f(\mathbf{x}, u, t)$. If f is positive, the heat is increasing while if f is negative the heat is decreasing. For example such sources could result from a chemical reaction taking place. Then the divergence theorem can be used to verify the following equation for u. Such an equation is called a reaction diffusion equation.

$$\frac{\partial}{\partial t} \left(\rho\left(\mathbf{x}, t\right) c\left(\mathbf{x}, t\right) u\left(\mathbf{x}, t\right) \right) = \nabla \cdot \left(k\left(u, \mathbf{x}, t\right) \nabla u\left(\mathbf{x}, t\right) \right) + f\left(\mathbf{x}, u, t\right). \qquad (13.6)$$

Take an arbitrary V for which the divergence theorem holds. Then the time rate of change of the heat in V is

$$\frac{d}{dt} \int_V \rho\left(\mathbf{x}, t\right) c\left(\mathbf{x}, t\right) u\left(\mathbf{x}, t\right) \, dV = \int_V \frac{\partial \left(\rho\left(\mathbf{x}, t\right) c\left(\mathbf{x}, t\right) u\left(\mathbf{x}, t\right) \right)}{\partial t} \, dV$$

where, as in the preceding example, this is a physical derivation so the consideration of hard mathematics is not necessary. Therefore, from the Fourier law of heat conduction, $\frac{d}{dt} \int_V \rho\left(\mathbf{x}, t\right) c\left(\mathbf{x}, t\right) u\left(\mathbf{x}, t\right) \, dV =$

$$\int_V \frac{\partial \left(\rho\left(\mathbf{x}, t\right) c\left(\mathbf{x}, t\right) u\left(\mathbf{x}, t\right) \right)}{\partial t} \, dV = \overbrace{\int_{\partial V} -\mathbf{J} \cdot \mathbf{n} \, dA}^{\text{rate at which heat enters}} + \int_V f\left(\mathbf{x}, u, t\right) \, dV$$

$$= \int_{\partial V} k\nabla\left(u\right) \cdot \mathbf{n} \, dA + \int_V f\left(\mathbf{x}, u, t\right) \, dV = \int_V \left(\nabla \cdot \left(k\nabla\left(u\right) \right) + f \right) \, dV.$$

Since this holds for every sample volume V it must be the case that the above reaction diffusion equation (13.6) holds. Note that more interesting equations can be obtained by letting more of the quantities in the equation depend on temperature. However, the above is a fairly hard equation and people usually assume the coefficient of thermal conductivity depends only on $\mathbf{x}$ and that the reaction term f depends only on $\mathbf{x}$ and t and that ρ and c are constant. Then it reduces to the much easier equation

$$\frac{\partial}{\partial t} u\left(\mathbf{x}, t\right) = \frac{1}{\rho c} \nabla \cdot \left(k\left(\mathbf{x}\right) \nabla u\left(\mathbf{x}, t\right) \right) + f\left(\mathbf{x}, t\right). \qquad (13.7)$$

This is often referred to as the heat equation. Sometimes there are modifications of this in which k is not just a scalar but a matrix to account for different heat flow properties in different directions. However, they are not much harder than the above. The major mathematical difficulties result from allowing k to depend on temperature.

It is known that the heat equation is not correct even if the thermal conductivity did not depend on u because it implies infinite speed of propagation of heat. However, this does not prevent people from using it.

13.5.4 Balance Of Mass

Let $\mathbf{y}$ be a point in three dimensional space and let (y_1, y_2, y_3) be Cartesian coordinates of this point. Let V be a region in three dimensional space and suppose a fluid having density $\rho\left(\mathbf{y}, t\right)$ and velocity, $\mathbf{v}\left(\mathbf{y}, t\right)$ is flowing through this region. Then the mass of fluid leaving V per unit time is given by the area integral $\int_{\partial V} \rho(\mathbf{y}, t) \mathbf{v}(\mathbf{y}, t) \cdot \mathbf{n} \, dA$ while the total mass of the fluid enclosed in V at a given time is $\int_V \rho(\mathbf{y}, t) \, dV$. Also suppose mass originates at the rate $f(\mathbf{y}, t)$ per cubic unit per unit time within this fluid. Then the conclusion which can be drawn through the use of the divergence theorem is the following fundamental equation known as the mass balance equation.

$$\frac{\partial \rho}{\partial t} + \nabla \cdot (\rho \mathbf{v}) = f(\mathbf{y}, t) \tag{13.8}$$

To see this is so, take an arbitrary V for which the divergence theorem holds. Then the time rate of change of the mass in V is

$$\frac{\partial}{\partial t} \int_V \rho(\mathbf{y}, t) \, dV = \int_V \frac{\partial \rho(\mathbf{y}, t)}{\partial t} \, dV$$

where the derivative was taken under the integral sign with respect to t. (This is a physical derivation and therefore, it is not necessary to fuss with the hard mathematics related to the change of limit operations. You should expect this to be true under fairly general conditions because the integral is a sort of sum and the derivative of a sum is the sum of the derivatives.) Therefore, the rate of change of mass $\frac{\partial}{\partial t} \int_V \rho(\mathbf{y}, t) \, dV$, equals

$$\int_V \frac{\partial \rho(\mathbf{y}, t)}{\partial t} \, dV = \overbrace{-\int_{\partial V} \rho(\mathbf{y}, t) \mathbf{v}(\mathbf{y}, t) \cdot \mathbf{n} \, dA}^{\text{rate at which mass enters}} + \int_V f(\mathbf{y}, t) \, dV$$

$$= -\int_V \left(\nabla \cdot (\rho(\mathbf{y}, t) \mathbf{v}(\mathbf{y}, t)) + f(\mathbf{y}, t) \right) \, dV.$$

Since this holds for every sample volume V it must be the case that the equation of continuity holds. Again, there are interesting mathematical questions here which can be explored but since it is a physical derivation, it is not necessary to dwell too much on them. If all the functions involved are continuous, it is certainly true but it is true under far more general conditions than that.

Also note this equation applies to many situations and f might depend on more than just $\mathbf{y}$ and t. In particular, f might depend also on temperature and the density ρ. This would be the case for example if you were considering the mass of some chemical and f represented a chemical reaction. Mass balance is a general sort of equation valid in many contexts.

13.5.5 Balance Of Momentum

This example is a little more substantial than the above. It concerns the balance of momentum for a continuum. To see a full description of all the physics involved, you should consult a book on continuum mechanics. The situation is of a material in three dimensions and it deforms and moves about in three dimensions. This means this material is not a rigid body. Let B_0 denote an open set identifying a chunk of this material at time $t = 0$ and let B_t be an open set which identifies the same chunk of material at time $t > 0$.

Let $\mathbf{y}(t, \mathbf{x}) = (y_1(t, \mathbf{x}), y_2(t, \mathbf{x}), y_3(t, \mathbf{x}))$ denote the position with respect to Cartesian coordinates at time t of the point whose position at time $t = 0$ is $\mathbf{x} = (x_1, x_2, x_3)$. The coordinates $\mathbf{x}$ are sometimes called the reference coordinates and sometimes the material coordinates and sometimes the Lagrangian coordinates. The coordinates $\mathbf{y}$ are called the Eulerian coordinates or sometimes the spacial coordinates and the function $(t, \mathbf{x}) \to \mathbf{y}(t, \mathbf{x})$ is called the motion. Thus

$$\mathbf{y}(0, \mathbf{x}) = \mathbf{x}. \tag{13.9}$$

The derivative,

$$D_2 \mathbf{y}(t, \mathbf{x}) \equiv D_{\mathbf{x}} \mathbf{y}(t, \mathbf{x})$$

is called the deformation gradient. Recall the notation means you fix t and consider the function $\mathbf{x} \to \mathbf{y}(t, \mathbf{x})$, taking its derivative. Since it is a linear transformation, it is represented by the usual matrix, whose ij^{th} entry is given by

$$F_{ij}(\mathbf{x}) = \frac{\partial y_i(t, \mathbf{x})}{\partial x_j}.$$

Let $\rho(t, \mathbf{y})$ denote the density of the material at time t at the point $\mathbf{y}$ and let $\rho_0(\mathbf{x})$ denote the density of the material at the point $\mathbf{x}$. Thus $\rho_0(\mathbf{x}) = \rho(0, \mathbf{x}) = \rho(0, \mathbf{y}(0, \mathbf{x}))$. The first task is to consider the relationship between $\rho(t, \mathbf{y})$ and $\rho_0(\mathbf{x})$. The following picture is useful to illustrate the ideas.

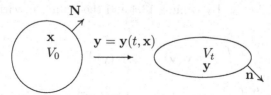

Lemma 13.4. $\rho_0(\mathbf{x}) = \rho(t, \mathbf{y}(t, \mathbf{x})) \det(F)$ *and in any reasonable physical motion* $\det(F) > 0$.

Proof: Let V_0 represent a small chunk of material at $t = 0$ and let V_t represent the same chunk of material at time t. I will be a little sloppy and refer to V_0 as the small chunk of material at time $t = 0$ and V_t as the chunk of material at time t rather than an open set representing the chunk of material. Then by the change of variables formula for multiple integrals,

$$\int_{V_t} dV = \int_{V_0} |\det(F)| \, dV.$$

If $\det(F) = 0$ for some t the above formula shows that the chunk of material went from positive volume to zero volume and this is not physically possible. Therefore, it is impossible that $\det(F)$ can equal zero. However, at $t = 0$, $F = I$, the identity because of (13.9). Therefore, $\det(F) = 1$ at $t = 0$ and if it is assumed $t \to \det(F)$ is continuous it follows by the intermediate value theorem that $\det(F) > 0$ for all t.

Of course it is not known for sure that this function is continuous but the above shows why it is at least reasonable to expect $\det(F) > 0$.

Now using the change of variables formula

$$\text{mass of } V_t = \int_{V_t} \rho(t, \mathbf{y}) \, dV = \int_{V_0} \rho(t, \mathbf{y}(t, \mathbf{x})) \det(F) \, dV$$

$$= \text{mass of } V_0 = \int_{V_0} \rho_0(\mathbf{x}) \, dV.$$

Since V_0 is arbitrary, it follows

$$\rho_0(\mathbf{x}) = \rho(t, \mathbf{y}(t, \mathbf{x})) \det(F)$$

as claimed. Note this shows that $\det(F)$ is a magnification factor for the density.

Now consider a small chunk of material, V_t at time t which corresponds to V_0 at time $t = 0$. The total linear momentum of this material at time t is

$$\int_{V_t} \rho(t, \mathbf{y}) \mathbf{v}(t, \mathbf{y}) \, dV$$

where $\mathbf{v}$ is the velocity. By Newton's second law, the time rate of change of this linear momentum should equal the total force acting on the chunk of material. In the following derivation, $dV(\mathbf{y})$ will indicate the integration is taking place with respect to the variable, $\mathbf{y}$. By Lemma 13.4 and the change of variables formula for multiple integrals

$$\frac{d}{dt}\left(\int_{V_t} \rho(t, \mathbf{y}) \mathbf{v}(t, \mathbf{y}) \, dV(\mathbf{y}) \right)$$

$$= \frac{d}{dt}\left(\int_{V_0} \rho(t, \mathbf{y}(t, \mathbf{x})) \mathbf{v}(t, \mathbf{y}(t, \mathbf{x})) \det(F) \, dV(\mathbf{x}) \right)$$

$$= \frac{d}{dt}\left(\int_{V_0} \rho_0(\mathbf{x}) \mathbf{v}(t, \mathbf{y}(t, \mathbf{x})) \, dV(\mathbf{x}) \right) dV(\mathbf{x}) = \int_{V_0} \rho_0(\mathbf{x}) \left[\frac{\partial \mathbf{v}}{\partial t} + \frac{\partial \mathbf{v}}{\partial y_i} \frac{\partial y_i}{\partial t} \right]$$

$$= \int_{V_t} \overbrace{\rho(t, \mathbf{y}) \det(F)}^{\rho_0(\mathbf{x})} \left[\frac{\partial \mathbf{v}}{\partial t} + \frac{\partial \mathbf{v}}{\partial y_i} \frac{\partial y_i}{\partial t} \right] \frac{1}{\det(F)} \, dV(\mathbf{y})$$

$$= \int_{V_t} \rho(t, \mathbf{y}) \left[\frac{\partial \mathbf{v}}{\partial t} + \frac{\partial \mathbf{v}}{\partial y_i} \frac{\partial y_i}{\partial t} \right] dV(\mathbf{y}).$$

Having taken the derivative of the total momentum, it is time to consider the total force acting on the chunk of material.

The force comes from two sources, a body force $\mathbf{b}$ and a force which acts on the boundary of the chunk of material called a traction force. Typically, the body force is something like gravity in which case, $\mathbf{b} = -g\rho\mathbf{k}$, assuming the Cartesian coordinate system has been chosen in the usual manner. The traction force is of the form

$$\int_{\partial V_t} \mathbf{s}\,(t,\mathbf{y},\mathbf{n})\ dA$$

where $\mathbf{n}$ is the unit exterior normal. Thus the traction force depends on position, time, and the orientation of the boundary of V_t. Cauchy showed the existence of a linear transformation $T\,(t,\mathbf{y})$ such that $T\,(t,\mathbf{y})\,\mathbf{n} = \mathbf{s}\,(t,\mathbf{y},\mathbf{n})$. It follows there is a matrix $T_{ij}\,(t,\mathbf{y})$ such that the ith component of $\mathbf{s}$ is given by $s_i\,(t,\mathbf{y},\mathbf{n}) = T_{ij}\,(t,\mathbf{y})\,n_j$. Cauchy also showed this matrix is symmetric, $T_{ij} = T_{ji}$. It is called the Cauchy stress. Using Newton's second law to equate the time derivative of the total linear momentum with the applied forces and using the usual repeated index summation convention,

$$\int_{V_t} \rho\,(t,\mathbf{y}) \left[\frac{\partial \mathbf{v}}{\partial t} + \frac{\partial \mathbf{v}}{\partial y_i}\frac{\partial y_i}{\partial t} \right] dV\,(\mathbf{y}) = \int_{V_t} \mathbf{b}\,(t,\mathbf{y})\ dV\,(\mathbf{y}) + \int_{\partial B_t} \mathbf{e}_i T_{ij}\,(t,\mathbf{y})\,n_j\ dA,$$

the sum taken over repeated indices. Here is where the divergence theorem is used. In the last integral, the multiplication by n_j is exchanged for the jth partial derivative and an integral over V_t. Thus

$$\int_{V_t} \rho\,(t,\mathbf{y}) \left[\frac{\partial \mathbf{v}}{\partial t} + \frac{\partial \mathbf{v}}{\partial y_i}\frac{\partial y_i}{\partial t} \right] dV\,(\mathbf{y}) = \int_{V_t} \mathbf{b}\,(t,\mathbf{y})\ dV\,(\mathbf{y}) + \int_{V_t} \frac{\mathbf{e}_i \partial\,(T_{ij}\,(t,\mathbf{y}))}{\partial y_j}\ dV\,(\mathbf{y}),$$

the sum taken over repeated indices. Since V_t was arbitrary, it follows

$$\rho\,(t,\mathbf{y}) \left[\frac{\partial \mathbf{v}}{\partial t} + \frac{\partial \mathbf{v}}{\partial y_i}\frac{\partial y_i}{\partial t} \right] = \mathbf{b}\,(t,\mathbf{y}) + \mathbf{e}_i \frac{\partial\,(T_{ij}\,(t,\mathbf{y}))}{\partial y_j}$$
$$\equiv \mathbf{b}\,(t,\mathbf{y}) + \operatorname{div}\,(T)$$

where here $\operatorname{div} T$ is a vector whose i^{th} component is given by

$$(\operatorname{div} T)_i = \frac{\partial T_{ij}}{\partial y_j}.$$

The term $\frac{\partial \mathbf{v}}{\partial t} + \frac{\partial \mathbf{v}}{\partial y_i}\frac{\partial y_i}{\partial t}$, is the total derivative with respect to t of the velocity $\mathbf{v}$. Thus you might see this written as

$$\rho\dot{\mathbf{v}} = \mathbf{b} + \operatorname{div}\,(T).$$

The above formulation of the balance of momentum involves the spatial coordinates $\mathbf{y}$ but people also like to formulate momentum balance in terms of the material coordinates $\mathbf{x}$. Of course this changes everything.

The momentum in terms of the material coordinates is

$$\int_{V_0} \rho_0\,(\mathbf{x})\,\mathbf{v}\,(t,\mathbf{x})\ dV$$

and so, since $\mathbf{x}$ does not depend on t,

$$\frac{d}{dt}\left(\int_{V_0} \rho_0\,(\mathbf{x})\,\mathbf{v}\,(t,\mathbf{x})\ dV \right) = \int_{V_0} \rho_0\,(\mathbf{x})\,\mathbf{v}_t\,(t,\mathbf{x})\ dV.$$

As indicated earlier, this is a physical derivation, so the mathematical questions related to interchange of limit operations are ignored. This must equal the total applied force. Thus using the repeated index summation convention,

$$\int_{V_0} \rho_0\left(\mathbf{x}\right) \mathbf{v}_t\left(t, \mathbf{x}\right) dV = \int_{V_0} \mathbf{b}_0\left(t, \mathbf{x}\right) dV + \int_{\partial V_t} \mathbf{e}_i T_{ij} n_j dA, \tag{13.10}$$

the first term on the right being the contribution of the body force given per unit volume in the material coordinates and the last term being the traction force discussed earlier. The task is to write this last integral as one over ∂V_0. For $\mathbf{y} \in \partial V_t$ there is a unit outer normal $\mathbf{n}$. Here $\mathbf{y} = \mathbf{y}\left(t, \mathbf{x}\right)$ for $\mathbf{x} \in \partial V_0$. Then define $\mathbf{N}$ to be the unit outer normal to V_0 at the point $\mathbf{x}$. Near the point $\mathbf{y} \in \partial V_t$ the surface ∂V_t is given parametrically in the form $\mathbf{y} = \mathbf{y}\left(s, t\right)$ for $\left(s, t\right) \in D \subseteq \mathbb{R}^2$ and it can be assumed the unit normal to ∂V_t near this point is

$$\mathbf{n} = \frac{\mathbf{y}_s\left(s, t\right) \times \mathbf{y}_t\left(s, t\right)}{\left|\mathbf{y}_s\left(s, t\right) \times \mathbf{y}_t\left(s, t\right)\right|}$$

with the area element given by $\left|\mathbf{y}_s\left(s, t\right) \times \mathbf{y}_t\left(s, t\right)\right| ds\, dt$. This is true for $\mathbf{y} \in P_t \subseteq \partial V_t$, a small piece of ∂V_t. Therefore, the last integral in (13.10) is the sum of integrals over small pieces of the form

$$\int_{P_t} T_{ij} n_j dA \tag{13.11}$$

where P_t is parameterized by $\mathbf{y}\left(s, t\right)$, $\left(s, t\right) \in D$. Thus the integral in (13.11) is of the form

$$\int_D T_{ij}\left(\mathbf{y}\left(s, t\right)\right)\left(\mathbf{y}_s\left(s, t\right) \times \mathbf{y}_t\left(s, t\right)\right)_j ds\, dt.$$

By the chain rule this equals

$$\int_D T_{ij}\left(\mathbf{y}\left(s, t\right)\right)\left(\frac{\partial \mathbf{y}}{\partial x_\alpha} \frac{\partial x_\alpha}{\partial s} \times \frac{\partial \mathbf{y}}{\partial x_\beta} \frac{\partial x_\beta}{\partial t}\right)_j ds\, dt.$$

Summation over repeated indices is used. Remember $\mathbf{y} = \mathbf{y}\left(t, \mathbf{x}\right)$ and it is always assumed the mapping $\mathbf{x} \to \mathbf{y}\left(t, \mathbf{x}\right)$ is one to one and so, since on the surface ∂V_t near $\mathbf{y}$, the points are functions of $\left(s, t\right)$, it follows $\mathbf{x}$ is also a function of $\left(s, t\right)$. Now by the properties of the cross product, this last integral equals

$$\int_D T_{ij}\left(\mathbf{x}\left(s, t\right)\right) \frac{\partial x_\alpha}{\partial s} \frac{\partial x_\beta}{\partial t}\left(\frac{\partial \mathbf{y}}{\partial x_\alpha} \times \frac{\partial \mathbf{y}}{\partial x_\beta}\right)_j ds\, dt \tag{13.12}$$

where here $\mathbf{x}\left(s, t\right)$ is the point of ∂V_0 which corresponds with $\mathbf{y}\left(s, t\right) \in \partial V_t$. Thus

$$T_{ij}\left(\mathbf{x}\left(s, t\right)\right) = T_{ij}\left(\mathbf{y}\left(s, t\right)\right).$$

(Perhaps this is a slight abuse of notation because T_{ij} is defined on ∂V_t, not on ∂V_0, but it avoids introducing extra symbols.) Next (13.12) equals

$$\int_D T_{ij}\left(\mathbf{x}\left(s, t\right)\right) \frac{\partial x_\alpha}{\partial s} \frac{\partial x_\beta}{\partial t} \varepsilon_{jab} \frac{\partial y_a}{\partial x_\alpha} \frac{\partial y_b}{\partial x_\beta} ds\, dt$$

$$= \int_D T_{ij}\left(\mathbf{x}\left(s,t\right)\right) \frac{\partial x_\alpha}{\partial s} \frac{\partial x_\beta}{\partial t} \varepsilon_{cab}\delta_{jc} \frac{\partial y_a}{\partial x_\alpha} \frac{\partial y_b}{\partial x_\beta}\, ds\, dt$$

$$= \int_D T_{ij}\left(\mathbf{x}\left(s,t\right)\right) \frac{\partial x_\alpha}{\partial s} \frac{\partial x_\beta}{\partial t} \varepsilon_{cab} \overbrace{\frac{\partial y_c}{\partial x_p} \frac{\partial x_p}{\partial y_j}}^{=\delta_{jc}} \frac{\partial y_a}{\partial x_\alpha} \frac{\partial y_b}{\partial x_\beta}\, ds\, dt$$

$$= \int_D T_{ij}\left(\mathbf{x}\left(s,t\right)\right) \frac{\partial x_\alpha}{\partial s} \frac{\partial x_\beta}{\partial t} \frac{\partial x_p}{\partial y_j} \varepsilon_{cab} \overbrace{\frac{\partial y_c}{\partial x_p} \frac{\partial y_a}{\partial x_\alpha} \frac{\partial y_b}{\partial x_\beta}}^{=\varepsilon_{p\alpha\beta}\det(F)}\, ds\, dt$$

$$= \int_D \left(\det F\right) T_{ij}\left(\mathbf{x}\left(s,t\right)\right) \varepsilon_{p\alpha\beta} \frac{\partial x_\alpha}{\partial s} \frac{\partial x_\beta}{\partial t} \frac{\partial x_p}{\partial y_j}\, ds\, dt.$$

Now $\frac{\partial x_p}{\partial y_j} = F_{pj}^{-1}$ and also

$$\varepsilon_{p\alpha\beta} \frac{\partial x_\alpha}{\partial s} \frac{\partial x_\beta}{\partial t} = \left(\mathbf{x}_s \times \mathbf{x}_t\right)_p$$

so the result just obtained is of the form

$$\int_D \left(\det F\right) F_{pj}^{-1} T_{ij}\left(\mathbf{x}\left(s,t\right)\right) \left(\mathbf{x}_s \times \mathbf{x}_t\right)_p\, ds\, dt =$$

$$\int_D \left(\det F\right) T_{ij}\left(\mathbf{x}\left(s,t\right)\right) \left(F^{-T}\right)_{jp} \left(\mathbf{x}_s \times \mathbf{x}_t\right)_p\, ds\, dt.$$

This has transformed the integral over P_t to one over P_0, the part of ∂V_0 which corresponds with P_t. Thus the last integral is of the form

$$\int_{P_0} \det\left(F\right) \left(TF^{-T}\right)_{ip} N_p dA$$

Summing these up over the pieces of ∂V_t and ∂V_0, yields the last integral in (13.10) equals

$$\int_{\partial V_0} \det\left(F\right) \left(TF^{-T}\right)_{ip} N_p dA$$

and so the balance of momentum in terms of the material coordinates becomes

$$\int_{V_0} \rho_0\left(\mathbf{x}\right) \mathbf{v}_t\left(t,\mathbf{x}\right) dV = \int_{V_0} \mathbf{b}_0\left(t,\mathbf{x}\right) dV + \int_{\partial V_0} \mathbf{e}_i \det\left(F\right) \left(TF^{-T}\right)_{ip} N_p dA$$

The matrix $\det\left(F\right) \left(TF^{-T}\right)_{ip}$ is called the Piola Kirchhoff stress S. An application of the divergence theorem yields

$$\int_{V_0} \rho_0\left(\mathbf{x}\right) \mathbf{v}_t\left(t,\mathbf{x}\right) dV = \int_{V_0} \mathbf{b}_0\left(t,\mathbf{x}\right) dV + \int_{V_0} \mathbf{e}_i \frac{\partial\left(\det\left(F\right)\left(TF^{-T}\right)_{ip}\right)}{\partial x_p} dV.$$

Since V_0 is arbitrary, a balance law for momentum in terms of the material coordinates is obtained

$$\rho_0 \left(\mathbf{x} \right) \mathbf{v}_t \left(t, \mathbf{x} \right) \;\; = \;\; \mathbf{b}_0 \left(t, \mathbf{x} \right) + \mathbf{e}_i \, \frac{\partial \left(\det \left(F \right) \left(T F^{-T} \right)_{ip} \right)}{\partial x_p}$$

$$= \;\; \mathbf{b}_0 \left(t, \mathbf{x} \right) + \operatorname{div} \left(\det \left(F \right) \left(T F^{-T} \right) \right)$$

$$= \;\; \mathbf{b}_0 \left(t, \mathbf{x} \right) + \operatorname{div} S. \tag{13.13}$$

As just shown, the relation between the Cauchy stress and the Piola Kirchhoff stress is

$$S = \det \left(F \right) \left(T F^{-T} \right), \tag{13.14}$$

perhaps not the first thing you would think of.

The main purpose of this presentation is to show how the divergence theorem is used in a significant way to obtain balance laws and to indicate a very interesting direction for further study. To continue, one needs to specify T or S as an appropriate function of things related to the motion $\mathbf{y}$. Often the thing related to the motion is something called the strain and such relationships are known as constitutive laws.

13.5.6   *Frame Indifference*

The proper formulation of constitutive laws involves more physical considerations such as frame indifference in which it is required that the response of the system cannot depend on the manner in which the Cartesian coordinate system for the spacial coordinates was chosen.

For $Q \left(t \right)$ an orthogonal transformation, (see Problem 17 on Page 39) and

$$\mathbf{y}' = \mathbf{q} \left(t \right) + Q \left(t \right) \mathbf{y}, \; \mathbf{n}' = Q \mathbf{n},$$

the new spacial coordinates are denoted by $\mathbf{y}'$. Recall an orthogonal transformation is just one which satisfies

$$Q \left(t \right)^T Q \left(t \right) = Q \left(t \right) Q \left(t \right)^T = I.$$

The stress has to do with the traction force area density produced by internal changes in the body and has nothing to do with the way the body is observed. Therefore, it is required that

$$T' \mathbf{n}' = Q T \mathbf{n}$$

Thus

$$T' Q \mathbf{n} = Q T \mathbf{n}$$

Since this is true for any $\mathbf{n}$ normal to the boundary of any piece of the material considered, it must be the case that

$$T' Q = Q T$$

and so

$$T' = QTQ^T.$$

This is called frame indifference.

By (13.14), the Piola Kirchhoff stress S is related to T by

$$S = \det(F) TF^{-T}, \ F \equiv D_{\mathbf{x}}\mathbf{y}.$$

This stress involves the use of the material coordinates and a normal $\mathbf{N}$ to a piece of the body in reference configuration. Thus $S\mathbf{N}$ gives the force on a part of ∂V_t per unit area on ∂V_0. Then for a different choice of spacial coordinates, $\mathbf{y}' = \mathbf{q}(t) + Q(t)\mathbf{y}$,

$$S' = \det(F') T' (F')^{-T}$$

but

$$F' = D_{\mathbf{x}}\mathbf{y}' = Q(t) D_{\mathbf{x}}\mathbf{y} = QF$$

and so frame indifference in terms of S is

$$S' = \det(F) QTQ^T (QF)^{-T} = \det(F) QTQ^T QF^{-T} = QS$$

This principle of frame indifference is sometimes ignored and there are certainly interesting mathematical models which have resulted from doing this, but such things cannot be considered physically acceptable.

There are also many other physical properties which can be included, which require a certain form for the constitutive equations. These considerations are outside the scope of this book and require a considerable amount of linear algebra.

There are also balance laws for energy which you may study later but these are more problematic than the balance laws for mass and momentum. However, the divergence theorem is used in these also.

13.5.7 *Bernoulli's Principle*

Consider a possibly moving fluid with constant density ρ and let P denote the pressure in this fluid. If B is a part of this fluid the force exerted on B by the rest of the fluid is $\int_{\partial B} -P\mathbf{n}dA$ where $\mathbf{n}$ is the outer normal from B. Assume this is the only force which matters so for example there is no viscosity in the fluid. Thus the Cauchy stress in rectangular coordinates should be

$$T = \begin{pmatrix} -P & 0 & 0 \\ 0 & -P & 0 \\ 0 & 0 & -P \end{pmatrix}.$$

Then $\text{div}\, T = -\nabla P$. Also suppose the only body force is from gravity, a force of the form $-\rho g\mathbf{k}$, so from the balance of momentum

$$\rho\dot{\mathbf{v}} = -\rho g\mathbf{k} - \nabla P(\mathbf{x}). \tag{13.15}$$

Now in all this, the coordinates are the spacial coordinates, and it is assumed they are rectangular. Thus $\mathbf{x} = (x, y, z)^T$ and $\mathbf{v}$ is the velocity while $\dot{\mathbf{v}}$ is the total derivative of $\mathbf{v} = (v_1, v_2, v_3)^T$ given by $\mathbf{v}_t + v_i \mathbf{v}_{,i}$. Take the dot product of both sides of (13.15) with $\mathbf{v}$. This yields

$$(\rho/2) \frac{d}{dt} |\mathbf{v}|^2 = -\rho g \frac{dz}{dt} - \frac{d}{dt} P(\mathbf{x}).$$

Therefore,

$$\frac{d}{dt} \left(\frac{\rho |\mathbf{v}|^2}{2} + \rho g z + P(\mathbf{x}) \right) = 0,$$

so there is a constant C' such that

$$\frac{\rho |\mathbf{v}|^2}{2} + \rho g z + P(\mathbf{x}) = C'$$

For convenience define γ to be the weight density of this fluid. Thus $\gamma = \rho g$. Divide by γ. Then

$$\frac{|\mathbf{v}|^2}{2g} + z + \frac{P(\mathbf{x})}{\gamma} = C.$$

This is Bernoulli's[2] principle. Note how, if you keep the height the same, then if you raise $|\mathbf{v}|$, it follows the pressure drops.

This is often used to explain the lift of an airplane wing. The top surface is curved, which forces the air to go faster over the top of the wing, causing a drop in pressure which creates lift. It is also used to explain the concept of a venturi tube in which the air loses pressure due to being pinched which causes it to flow faster. In many of these applications, the assumptions used in which ρ is constant, and there is no other contribution to the traction force on ∂B than pressure, so in particular, there is no viscosity, are not correct. However, it is hoped that the effects of these deviations from the ideal situation are small enough that the conclusions are still roughly true. You can see how using balance of momentum can be used to consider more difficult situations. For example, you might have a body force which is more involved than gravity.

13.5.8   *The Wave Equation*

As an example of how the balance law of momentum is used to obtain an important equation of mathematical physics, suppose $S = kF$ where k is a constant and F is the deformation gradient and let $\mathbf{u} \equiv \mathbf{y} - \mathbf{x}$. Thus $\mathbf{u}$ is the displacement. Then from (13.13) you can verify the following holds.

$$\rho_0(\mathbf{x}) \mathbf{u}_{tt}(t, \mathbf{x}) = \mathbf{b}_0(t, \mathbf{x}) + k \Delta \mathbf{u}(t, \mathbf{x}) \tag{13.16}$$

---

[2]There were many Bernoullis. This is Daniel Bernoulli. He seems to have been nicer than some of the others. Daniel was actually a doctor who was interested in mathematics. He lived from 1700-1782.

In the case where ρ_0 is a constant and $\mathbf{b}_0 = 0$, this yields

$$\mathbf{u}_{tt} - c\Delta\mathbf{u} = \mathbf{0}.$$

The wave equation is $u_{tt} - c\Delta u = 0$ and so the above gives three wave equations, one for each component.

13.5.9 A Negative Observation

Many of the above applications of the divergence theorem are based on the assumption that matter is continuously distributed in a way that the above arguments are correct. In other words, a continuum. However, there is no such thing as a continuum. It has been known for some time now that matter is composed of atoms. It is not continuously distributed through some region of space as it is in the above. Apologists for this contradiction with reality sometimes say to consider enough of the material in question that it is reasonable to think of it as a continuum. This mystical reasoning is then violated as soon as they go from the integral form of the balance laws to the differential equations expressing the traditional formulation of these laws. See Problem 10 below, for example. However, these laws continue to be used and seem to lead to useful physical models which have value in predicting the behavior of physical systems. This is what justifies their use, not any fundamental truth.

13.5.10 Volumes Of Balls In $\mathbb{R}^n$

Recall, $B(\mathbf{x}, r)$ denotes the set of all $\mathbf{y} \in \mathbb{R}^n$ such that $|\mathbf{y} - \mathbf{x}| < r$. By the change of variables formula for multiple integrals or simple geometric reasoning, all balls of radius r have the same volume. Furthermore, simple reasoning or change of variables formula will show that the volume of the ball of radius r equals $\alpha_n r^n$ where α_n will denote the volume of the unit ball in $\mathbb{R}^n$. With the divergence theorem, it is now easy to give a simple relationship between the surface area of the ball of radius r and the volume. By the divergence theorem,

$$\int_{B(\mathbf{0},r)} \operatorname{div}\mathbf{x}\, dx = \int_{\partial B(\mathbf{0},r)} \mathbf{x}\cdot\frac{\mathbf{x}}{|\mathbf{x}|}\, dA$$

because the unit outward normal on $\partial B(\mathbf{0},r)$ is $\frac{\mathbf{x}}{|\mathbf{x}|}$. Therefore, denoting $A(\partial B)$ as the area of ∂B,

$$n\alpha_n r^n = rA(\partial B(\mathbf{0},r))$$

and so

$$A(\partial B(\mathbf{0},r)) = n\alpha_n r^{n-1}.$$

You recall the surface area of $S^2 \equiv \{\mathbf{x} \in \mathbb{R}^3 : |\mathbf{x}| = r\}$ is given by $4\pi r^2$ while the volume of the ball, $B(\mathbf{0},r)$ is $\frac{4}{3}\pi r^3$. This follows the above pattern. You just take the derivative with respect to the radius of the volume of the ball of radius

r to get the area of the surface of this ball. Let ω_n denote the area of the sphere $S^{n-1} = \{\mathbf{x} \in \mathbb{R}^n : |\mathbf{x}| = 1\}$. I just showed that $\omega_n = n\alpha_n$.

I want to find α_n now and also to get a relationship between ω_n and ω_{n-1}. Consider the following picture of the ball of radius ρ seen on the side.

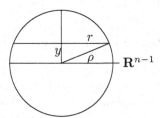

Taking slices at height y as shown and using that these slices have $n-1$ dimensional area equal to $\alpha_{n-1} r^{n-1}$, it follows

$$\alpha_n \rho^n = 2 \int_0^\rho \alpha_{n-1} \left(\rho^2 - y^2\right)^{(n-1)/2} dy$$

In the integral, change variables, letting $y = \rho \cos \theta$. Then

$$\alpha_n \rho^n = 2\rho^n \alpha_{n-1} \int_0^{\pi/2} \sin^n (\theta)\, d\theta.$$

It follows that

$$\alpha_n = 2\alpha_{n-1} \int_0^{\pi/2} \sin^n (\theta)\, d\theta. \tag{13.17}$$

Consequently,

$$\omega_n = \frac{2n\omega_{n-1}}{n-1} \int_0^{\pi/2} \sin^n (\theta)\, d\theta. \tag{13.18}$$

This is a little messier than I would like.

$$\begin{aligned}
\int_0^{\pi/2} \sin^n (\theta)\, d\theta &= \left. -\cos\theta \sin^{n-1}\theta \right|_0^{\pi/2} + (n-1)\int_0^{\pi/2} \cos^2\theta \sin^{n-2}\theta \\
&= (n-1)\int_0^{\pi/2} \left(1 - \sin^2\theta\right) \sin^{n-2}(\theta)\, d\theta \\
&= (n-1)\int_0^{\pi/2} \sin^{n-2}(\theta)\, d\theta - (n-1)\int_0^{\pi/2} \sin^n(\theta)\, d\theta
\end{aligned}$$

Hence

$$n \int_0^{\pi/2} \sin^n (\theta)\, d\theta = (n-1) \int_0^{\pi/2} \sin^{n-2}(\theta)\, d\theta \tag{13.19}$$

and so (13.18) is of the form

$$\omega_n = 2\omega_{n-1} \int_0^{\pi/2} \sin^{n-2}(\theta)\, d\theta. \tag{13.20}$$

So what is α_n explicitly? Clearly $\alpha_1 = 2$ and $\alpha_2 = \pi$.

Theorem 13.5. $\alpha_n = \dfrac{\pi^{n/2}}{\Gamma(\frac{n}{2}+1)}$ *where Γ denotes the gamma function, defined for $\alpha > 0$ by*

$$\Gamma(\alpha) \equiv \int_0^\infty e^{-t} t^{\alpha-1} dt.$$

Proof: Recall that $\Gamma(\alpha + 1) = \alpha \Gamma(\alpha)$. Now note the given formula holds if $n = 1$ because

$$\Gamma\left(\frac{1}{2}+1\right) = \frac{1}{2}\Gamma\left(\frac{1}{2}\right) = \frac{\sqrt{\pi}}{2}.$$

(I leave it as an exercise for you to verify that $\Gamma\left(\frac{1}{2}\right) = \sqrt{\pi}$. This is also outlined in an exercise in Volume 1.) Thus $\alpha_1 = 2 = \frac{\sqrt{\pi}}{\sqrt{\pi}/2}$ satisfying the formula. Now suppose this formula holds for $k \leq n$. Then from the induction hypothesis, (13.20), (13.19), (13.17) and (13.18),

$$\alpha_{n+1} = 2\alpha_n \int_0^{\pi/2} \sin^{n+1}(\theta)\, d\theta = 2\alpha_n \frac{n}{n+1} \int_0^{\pi/2} \sin^{n-1}(\theta)\, d\theta$$

$$= 2\alpha_n \frac{n}{n+1} \frac{\alpha_{n-1}}{2\alpha_{n-2}} = \frac{\pi^{n/2}}{\Gamma\left(\frac{n}{2}+1\right)} \frac{n}{n+1} \pi^{1/2} \frac{\Gamma\left(\frac{n-2}{2}+1\right)}{\Gamma\left(\frac{n-1}{2}+1\right)}$$

$$= \frac{\pi^{n/2}}{\Gamma\left(\frac{n-2}{2}+1\right)\left(\frac{n}{2}\right)} \frac{n}{n+1} \pi^{1/2} \frac{\Gamma\left(\frac{n-2}{2}+1\right)}{\Gamma\left(\frac{n-1}{2}+1\right)}$$

$$= 2\pi^{(n+1)/2} \frac{1}{n+1} \frac{1}{\Gamma\left(\frac{n-1}{2}+1\right)} = \pi^{(n+1)/2} \frac{1}{\left(\frac{n+1}{2}\right)} \frac{1}{\Gamma\left(\frac{n-1}{2}+1\right)}$$

$$= \pi^{(n+1)/2} \frac{1}{\left(\frac{n+1}{2}\right)\Gamma\left(\frac{n+1}{2}\right)} = \frac{\pi^{(n+1)/2}}{\Gamma\left(\frac{n+1}{2}+1\right)}. \qquad \blacksquare$$

13.5.11 Electrostatics

Coloumb's law says that the electric field intensity at $\mathbf{x}$ of a charge q located at point $\mathbf{x}_0$ is given by

$$\mathbf{E} = k \frac{q(\mathbf{x} - \mathbf{x}_0)}{|\mathbf{x} - \mathbf{x}_0|^3}$$

where the electric field intensity is defined to be the force experienced by a unit positive charge placed at the point $\mathbf{x}$. Note that this is a vector and that its direction depends on the sign of q. It points away from $\mathbf{x}_0$ if q is positive and points toward $\mathbf{x}_0$ if q is negative. The constant k is a physical constant like the gravitation constant. It has been computed through careful experiments similar to those used with the calculation of the gravitation constant.

The interesting thing about Coloumb's law is that $\mathbf{E}$ is the gradient of a function. In fact,

$$\mathbf{E} = \nabla \left(qk \frac{1}{|\mathbf{x} - \mathbf{x}_0|} \right).$$

The other thing which is significant about this is that in three dimensions and for $\mathbf{x} \neq \mathbf{x}_0$,

$$\nabla \cdot \nabla \left(qk \frac{1}{|\mathbf{x} - \mathbf{x}_0|} \right) = \nabla \cdot \mathbf{E} = 0. \tag{13.21}$$

This is left as an exercise for you to verify.

These observations will be used to derive a very important formula for the integral

$$\int_{\partial U} \mathbf{E} \cdot \mathbf{n} dS$$

where $\mathbf{E}$ is the electric field intensity due to a charge, q located at the point $\mathbf{x}_0 \in U$, a bounded open set for which the divergence theorem holds.

Let U_ε denote the open set obtained by removing the open ball centered at $\mathbf{x}_0$ which has radius ε where ε is small enough that the following picture is a correct representation of the situation.

Then on the boundary of B_ε the unit outer normal to U_ε is $-\frac{\mathbf{x} - \mathbf{x}_0}{|\mathbf{x} - \mathbf{x}_0|}$. Therefore,

$$
\begin{aligned}
\int_{\partial B_\varepsilon} \mathbf{E} \cdot \mathbf{n} dS &= -\int_{\partial B_\varepsilon} k \frac{q(\mathbf{x} - \mathbf{x}_0)}{|\mathbf{x} - \mathbf{x}_0|^3} \cdot \frac{\mathbf{x} - \mathbf{x}_0}{|\mathbf{x} - \mathbf{x}_0|} dS \\
&= -kq \int_{\partial B_\varepsilon} \frac{1}{|\mathbf{x} - \mathbf{x}_0|^2} dS = \frac{-kq}{\varepsilon^2} \int_{\partial B_\varepsilon} dS \\
&= \frac{-kq}{\varepsilon^2} 4\pi\varepsilon^2 = -4\pi kq.
\end{aligned}
$$

Therefore, from the divergence theorem and observation (13.21),

$$-4\pi kq + \int_{\partial U} \mathbf{E} \cdot \mathbf{n} dS = \int_{\partial U_\varepsilon} \mathbf{E} \cdot \mathbf{n} dS = \int_{U_\varepsilon} \nabla \cdot \mathbf{E} dV = 0.$$

It follows that $4\pi kq = \int_{\partial U} \mathbf{E} \cdot \mathbf{n} dS$. If there are several charges located inside U, say $q_1, q_2, \cdots, q_n$, then letting $\mathbf{E}_i$ denote the electric field intensity of the ith charge and $\mathbf{E}$ denoting the total resulting electric field intensity due to all these charges,

$$\int_{\partial U} \mathbf{E} \cdot \mathbf{n} dS = \sum_{i=1}^n \int_{\partial U} \mathbf{E}_i \cdot \mathbf{n} dS = \sum_{i=1}^n 4\pi kq_i = 4\pi k \sum_{i=1}^n q_i.$$

This is known as Gauss's law and it is the fundamental result in electrostatics.

13.6 Exercises

(1) To prove the divergence theorem, it was shown first that the spacial partial derivative in the volume integral could be exchanged for multiplication by an appropriate component of the exterior normal. This problem starts with the divergence theorem and goes the other direction. Assuming the divergence theorem, holds for a region V, show that $\int_{\partial V} n u \, dA = \int_V \nabla u \, dV$. Note this implies $\int_V \frac{\partial u}{\partial x} \, dV = \int_{\partial V} n_1 u \, dA$.

(2) Fick's law for diffusion states the flux of a diffusing species, $\mathbf{J}$ is proportional to the gradient of the concentration, c. Write this law getting the sign right for the constant of proportionality and derive an equation similar to the heat equation for the concentration, c. Typically, c is the concentration of some sort of pollutant or a chemical.

(3) Sometimes people consider diffusion in materials which are not homogeneous. This means that $\mathbf{J} = -K\nabla c$ where K is a 3×3 matrix. Thus in terms of components, $J_i = -\sum_j K_{ij} \frac{\partial c}{\partial x_j}$. Here c is the concentration which means the amount of pollutant or whatever is diffusing in a volume is obtained by integrating c over the volume. Derive a formula for a nonhomogeneous model of diffusion based on the above.

(4) Let V be such that the divergence theorem holds. Show that $\int_V \nabla \cdot (u\nabla v) \, dV = \int_{\partial V} u \frac{\partial v}{\partial n} \, dA$ where $\mathbf{n}$ is the exterior normal and $\frac{\partial v}{\partial n}$ denotes the directional derivative of v in the direction $\mathbf{n}$.

(5) Let V be such that the divergence theorem holds. Show that

$$\int_V \left(v\nabla^2 u - u\nabla^2 v \right) dV = \int_{\partial V} \left(v \frac{\partial u}{\partial n} - u \frac{\partial v}{\partial n} \right) dA$$

where $\mathbf{n}$ is the exterior normal and $\frac{\partial u}{\partial n}$ is defined in Problem 4.

(6) Let V be a ball and suppose $\nabla^2 u = f$ in V while $u = g$ on ∂V. Show that there is at most one solution to this boundary value problem which is C^2 in V and continuous on V with its boundary. **Hint:** You might consider $w = u - v$ where u and v are solutions to the problem. Then use the result of Problem 4 and the identity $w\nabla^2 w = \nabla \cdot (w\nabla w) - \nabla w \cdot \nabla w$ to conclude $\nabla w = 0$. Then show this implies w must be a constant by considering $h(t) = w(t\mathbf{x} + (1-t)\mathbf{y})$ and showing h is a constant. Alternatively, you might consider the maximum principle.

(7) Show that $\int_{\partial V} \nabla \times \mathbf{v} \cdot \mathbf{n} \, dA = 0$ where V is a region for which the divergence theorem holds and $\mathbf{v}$ is a C^2 vector field.

(8) Let $\mathbf{F}(x, y, z) = (x, y, z)$ be a vector field in $\mathbb{R}^3$ and let V be a three dimensional shape and let $\mathbf{n} = (n_1, n_2, n_3)$. Show that $\int_{\partial V} (xn_1 + yn_2 + zn_3) \, dA = 3\times$ volume of V.

(9) Let $\mathbf{F} = x\mathbf{i} + y\mathbf{j} + z\mathbf{k}$ and let V denote the tetrahedron formed by the planes, $x = 0, y = 0, z = 0$, and $\frac{1}{3}x + \frac{1}{3}y + \frac{1}{5}z = 1$. Verify the divergence theorem for this example.

(10) Suppose $f : U \to \mathbb{R}$ is continuous where U is some open set and for all $B \subseteq U$ where B is a ball, $\int_B f(\mathbf{x}) \, dV = 0$. Show that this implies $f(\mathbf{x}) = 0$ for all $\mathbf{x} \in U$.

(11) Let U denote the box centered at $(0,0,0)$ with sides parallel to the coordinate planes which has width 4, length 2 and height 3. Find the flux integral $\int_{\partial U} \mathbf{F} \cdot \mathbf{n} \, dS$ where $\mathbf{F} = (x+3, 2y, 3z)$. **Hint:** If you like, you might want to use the divergence theorem.

(12) Find the flux out of the cylinder whose base is $x^2 + y^2 \leq 1$ which has height 2 of the vector field $\mathbf{F} = (xy, zy, z^2 + x)$.

(13) Find the flux out of the ball of radius 4 centered at $\mathbf{0}$ of the vector field $\mathbf{F} = (x, zy, z + x)$.

(14) Verify (13.16) from (13.13) and the assumption that $S = kF$.

(15) Show that if $u_k, k = 1, 2, \cdots, n$ each satisfies (13.7) with $f = 0$ then for any choice of constants $c_1, \cdots, c_n$, so does $\sum_{k=1}^{n} c_k u_k$.

(16) Suppose $k(\mathbf{x}) = k$, a constant and $f = 0$. Then in one dimension, the heat equation is of the form $u_t = \alpha u_{xx}$. Show that $u(x,t) = e^{-\alpha n^2 t} \sin(nx)$ satisfies the heat equation[3].

(17) Let U be a three dimensional region for which the divergence theorem holds. Show that $\int_U \nabla \times \mathbf{F} \, dx = \int_{\partial U} \mathbf{n} \times \mathbf{F} \, dS$ where $\mathbf{n}$ is the unit outer normal.

(18) In a linear, viscous, incompressible fluid, the Cauchy stress is of the form

$$T_{ij}(t, \mathbf{y}) = \lambda \left(\frac{v_{i,j}(t, \mathbf{y}) + v_{j,i}(t, \mathbf{y})}{2} \right) - p \delta_{ij}$$

where p is the pressure, δ_{ij} equals 0 if $i \neq j$ and 1 if $i = j$, and the comma followed by an index indicates the partial derivative with respect to that variable and $\mathbf{v}$ is the velocity. Thus $v_{i,j} = \frac{\partial v_i}{\partial y_j}$. Also, p denotes the pressure. Show, using the balance of mass equation that incompressible implies div $\mathbf{v} = 0$. Next show that the balance of momentum equation requires

$$\rho \dot{\mathbf{v}} - \frac{\lambda}{2} \Delta \mathbf{v} = \rho \left[\frac{\partial \mathbf{v}}{\partial t} + \frac{\partial \mathbf{v}}{\partial y_i} v_i \right] - \frac{\lambda}{2} \Delta \mathbf{v} = \mathbf{b} - \nabla p.$$

This is the famous Navier Stokes equation for incompressible viscous linear fluids. There are still open questions related to this equation, one of which is worth $1,000,000 at this time.

---

[3]Fourier, an officer in Napoleon's army studied solutions to the heat equation back in 1813. He was interested in heat flow in cannons. He sought to find solutions by adding up infinitely many solutions of this form. Actually, it was a little more complicated because cannons are not one dimensional but it was the beginning of the study of Fourier series, a topic which fascinated mathematicians for the next 150 years and motivated the development of analysis.

Chapter 14

Stokes And Green's Theorems

14.1 Green's Theorem

Green's theorem is an important theorem which relates line integrals to integrals over a surface in the plane. It can be used to establish the seemingly more general Stoke's theorem but is interesting for it's own sake. Historically, theorems like it were important in the development of complex analysis. I will first establish Green's theorem for regions of a particular sort and then show that the theorem holds for many other regions also. Suppose a region is of the form indicated in the following picture in which

$$U = \{(x,y) : x \in (a,b) \text{ and } y \in (b(x), t(x))\}$$
$$= \{(x,y) : y \in (c,d) \text{ and } x \in (l(y), r(y))\}.$$

I will refer to such a region as being convex in both the x and y directions.

Lemma 14.1. *Let* $\mathbf{F}(x,y) \equiv (P(x,y), Q(x,y))$ *be a* C^1 *vector field defined near* U *where* U *is a region of the sort indicated in the above picture which is convex in both the* x *and* y *directions. Suppose also that the functions* $r, l, t,$ *and* b *in the above picture are all* C^1 *functions and denote by* ∂U *the boundary of* U *oriented such that the direction of motion is counter clockwise. (As you walk around* U *on* ∂U, *the points of* U *are on your left.) Then*

$$\int_{\partial U} P dx + Q dy \equiv \int_{\partial U} \mathbf{F} \cdot d\mathbf{R} = \int_U \left(\frac{\partial Q}{\partial x} - \frac{\partial P}{\partial y} \right) dA. \tag{14.1}$$

Proof: First consider the right side of (14.1).

$$\int_U \left(\frac{\partial Q}{\partial x} - \frac{\partial P}{\partial y} \right) dA = \int_c^d \int_{l(y)}^{r(y)} \frac{\partial Q}{\partial x} dx dy - \int_a^b \int_{b(x)}^{t(x)} \frac{\partial P}{\partial y} dy dx$$

$$= \int_c^d \left(Q\left(r\left(y\right) , y\right) - Q\left(l\left(y\right) , y\right) \right) dy$$

$$+ \int_a^b \left(P\left(x, b\left(x\right) \right) \right) - P\left(x, t\left(x\right) \right) dx. \tag{14.2}$$

Now consider the left side of (14.1). Denote by V the vertical parts of ∂U and by H the horizontal parts.

$$\int_{\partial U} \mathbf{F} \cdot d\mathbf{R} = \int_{\partial U} \left(\left(0, Q\right) + \left(P, 0\right) \right) \cdot d\mathbf{R}$$

$$= \int_c^d \left(0, Q\left(r\left(s\right) , s\right) \right) \cdot \left(r'\left(s\right) , 1\right) ds + \int_H \left(0, Q\left(r\left(s\right) , s\right) \right) \cdot \left(\pm 1, 0\right) ds$$

$$- \int_c^d \left(0, Q\left(l\left(s\right) , s\right) \right) \cdot \left(l'\left(s\right) , 1\right) ds + \int_a^b \left(P\left(s, b\left(s\right) \right) , 0\right) \cdot \left(1, b'\left(s\right) \right) ds$$

$$+ \int_V \left(P\left(s, b\left(s\right) \right) , 0\right) \cdot \left(0, \pm 1\right) ds - \int_a^b \left(P\left(s, t\left(s\right) \right) , 0\right) \cdot \left(1, t'\left(s\right) \right) ds$$

$$= \int_c^d Q\left(r\left(s\right) , s\right) ds - \int_c^d Q\left(l\left(s\right) , s\right) ds + \int_a^b P\left(s, b\left(s\right) \right) ds - \int_a^b P\left(s, t\left(s\right) \right) ds$$

which coincides with (14.2). ∎

Corollary 14.1. *Let everything be the same as in Lemma 14.1 but only assume the functions $r, l, t,$ and b are continuous and piecewise C^1 functions. Then the conclusion this lemma is still valid.*

Proof: The details are left for you. All you have to do is to break up the various line integrals into the sum of integrals over sub intervals on which the function of interest is C^1. ∎

From this corollary, it follows (14.1) is valid for any triangle for example.

Now suppose (14.1) holds for $U_1, U_2, \cdots, U_m$ and the open sets U_k have the property that no two have nonempty intersection and their boundaries intersect only in a finite number of piecewise smooth curves. Then (14.1) must hold for $U \equiv \cup_{i=1}^m U_i$, the union of these sets. This is because

$$\int_U \left(\frac{\partial Q}{\partial x} - \frac{\partial P}{\partial y} \right) dA$$

$$= \sum_{k=1}^m \int_{U_k} \left(\frac{\partial Q}{\partial x} - \frac{\partial P}{\partial y} \right) dA$$

$$= \sum_{k=1}^m \int_{\partial U_k} \mathbf{F} \cdot d\mathbf{R} = \int_{\partial U} \mathbf{F} \cdot d\mathbf{R}$$

because if $\Gamma = \partial U_k \cap \partial U_j$, then its orientation as a part of ∂U_k is opposite to its orientation as a part of ∂U_j and consequently the line integrals over Γ will cancel, points of Γ also not being in ∂U. As an illustration, consider the following picture for two such U_k.

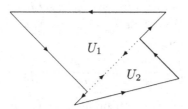

Similarly, if $U \subseteq V$ and if also $\partial U \subseteq V$ and both U and V are open sets for which (14.1) holds, then the open set $V \setminus (U \cup \partial U)$ consisting of what is left in V after deleting U along with its boundary also satisfies (14.1). Roughly speaking, you can drill holes in a region for which (14.1) holds and get another region for which this continues to hold provided (14.1) holds for the holes. To see why this is so, consider the following picture which typifies the situation just described.

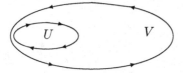

Then

$$\int_{\partial V} \mathbf{F} \cdot d\mathbf{R} = \int_V \left(\frac{\partial Q}{\partial x} - \frac{\partial P}{\partial y} \right) dA$$

$$= \int_U \left(\frac{\partial Q}{\partial x} - \frac{\partial P}{\partial y} \right) dA + \int_{V \setminus U} \left(\frac{\partial Q}{\partial x} - \frac{\partial P}{\partial y} \right) dA$$

$$= \int_{\partial U} \mathbf{F} \cdot d\mathbf{R} + \int_{V \setminus U} \left(\frac{\partial Q}{\partial x} - \frac{\partial P}{\partial y} \right) dA$$

and so

$$\int_{V \setminus U} \left(\frac{\partial Q}{\partial x} - \frac{\partial P}{\partial y} \right) dA = \int_{\partial V} \mathbf{F} \cdot d\mathbf{R} - \int_{\partial U} \mathbf{F} \cdot d\mathbf{R}$$

which equals

$$\int_{\partial (V \setminus U)} \mathbf{F} \cdot d\mathbf{R}$$

where ∂V is oriented as shown in the picture. (If you walk around the region $V \setminus U$ with the area on the left, you get the indicated orientation for this curve.)

You can see that (14.1) is valid quite generally. This verifies the following theorem.

Theorem 14.1. *(Green's Theorem) Let U be an open set in the plane and let ∂U be piecewise smooth and let $\mathbf{F}(x,y) = (P(x,y), Q(x,y))$ be a C^1 vector field defined near U. Then it is often[1] the case that*

$$\int_{\partial U} \mathbf{F} \cdot d\mathbf{R} = \int_U \left(\frac{\partial Q}{\partial x}(x,y) - \frac{\partial P}{\partial y}(x,y) \right) dA.$$

Here is an alternate proof of Green's theorem from the divergence theorem.

Theorem 14.2. *(Green's Theorem) Let U be an open set in the plane and let ∂U be piecewise smooth and let $\mathbf{F}(x,y) = (P(x,y), Q(x,y))$ be a C^1 vector field defined near U. Then it is often the case that*

$$\int_{\partial U} \mathbf{F} \cdot d\mathbf{R} = \int_U \left(\frac{\partial Q}{\partial x}(x,y) - \frac{\partial P}{\partial y}(x,y) \right) dA.$$

Proof: Suppose the divergence theorem holds for U. Consider the following picture.

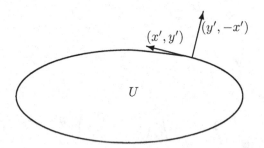

Since it is assumed that motion around U is counter clockwise, the tangent vector (x', y') is as shown. The unit **exterior normal** is a multiple of

$$(x', y', 0) \times (0, 0, 1) = (y', -x', 0).$$

Use your right hand and the geometric description of the cross product to verify this. This would be the case at all the points where the unit exterior normal exists.

Now let $\mathbf{F}(x,y) = (Q(x,y), -P(x,y))$. Also note the area (length) element on the bounding curve ∂U is $\sqrt{(x')^2 + (y')^2}\,dt$. Suppose the boundary of U consists of m smooth curves, the ith of which is parameterized by (x_i, y_i) with the parameter $t \in [a_i, b_i]$. Then by the divergence theorem,

$$\int_U (Q_x - P_y)\, dA = \int_U \text{div}(\mathbf{F})\, dA = \int_{\partial U} \mathbf{F} \cdot \mathbf{n}\, dS$$

---

[1] For a general version see the advanced calculus book by Apostol. The general versions involve the concept of a rectifiable (finite length) Jordan curve.

$$= \sum_{i=1}^{m} \int_{a_i}^{b_i} \left(Q\left(x_i\left(t\right), y_i\left(t\right)\right), -P\left(x_i\left(t\right), y_i\left(t\right)\right)\right)$$

$$\cdot \frac{1}{\sqrt{\left(x_i'\right)^2 + \left(y_i'\right)^2}} \left(y_i', -x_i'\right) \overbrace{\sqrt{\left(x_i'\right)^2 + \left(y_i'\right)^2}dt}^{dS}$$

$$= \sum_{i=1}^{m} \int_{a_i}^{b_i} \left(Q\left(x_i\left(t\right), y_i\left(t\right)\right), -P\left(x_i\left(t\right), y_i\left(t\right)\right)\right) \cdot \left(y_i', -x_i'\right) dt$$

$$= \sum_{i=1}^{m} \int_{a_i}^{b_i} Q\left(x_i\left(t\right), y_i\left(t\right)\right) y_i'\left(t\right) + P\left(x_i\left(t\right), y_i\left(t\right)\right) x_i'\left(t\right) dt \equiv \int_{\partial U} Pdx + Qdy$$

This proves Green's theorem from the divergence theorem. ∎

Proposition 14.1. Let U be an open set in $\mathbb{R}^2$ for which Green's theorem holds. Then

$$\text{Area of } U = \int_{\partial U} \mathbf{F} \cdot d\mathbf{R}$$

where $\mathbf{F}\left(x, y\right) = \frac{1}{2}\left(-y, x\right), \left(0, x\right),$ or $\left(-y, 0\right)$.

Proof: This follows immediately from Green's theorem. ∎

Example 14.1. Use Proposition 14.1 to find the area of the ellipse

$$\frac{x^2}{a^2} + \frac{y^2}{b^2} \leq 1.$$

You can parameterize the boundary of this ellipse as

$$x = a\cos t, \ y = b\sin t, \ t \in \left[0, 2\pi\right].$$

Then from Proposition 14.1,

$$\begin{aligned}
\text{Area equals} \ &= \ \frac{1}{2}\int_{0}^{2\pi} \left(-b\sin t, a\cos t\right) \cdot \left(-a\sin t, b\cos t\right) dt \\
&= \ \frac{1}{2}\int_{0}^{2\pi} \left(ab\right) dt = \pi ab.
\end{aligned}$$

Example 14.2. Find $\int_{\partial U} \mathbf{F} \cdot d\mathbf{R}$ where U is the set

$$\left\{ \left(x, y\right) : x^2 + 3y^2 \leq 9\right\}$$

and $\mathbf{F}\left(x, y\right) = \left(y, -x\right)$.

One way to do this is to parameterize the boundary of U and then compute the line integral directly. It is easier to use Green's theorem. The desired line integral equals

$$\int_{U} \left(\left(-1\right) - 1\right) dA = -2 \int_{U} dA.$$

Now U is an ellipse having area equal to $3\sqrt{3}$ and so the answer is $-6\sqrt{3}$.

Example 14.3. Find $\int_{\partial U} \mathbf{F} \cdot d\mathbf{R}$ where U is the set $\{(x, y) : 2 \le x \le 4, 0 \le y \le 3\}$ and $\mathbf{F}(x, y) = (x \sin y, y^3 \cos x)$.

From Green's theorem this line integral equals

$$\int_2^4 \int_0^3 \left(-y^3 \sin x - x \cos y\right) dy dx = \frac{81}{4} \cos 4 - 6 \sin 3 - \frac{81}{4} \cos 2.$$

This is much easier than computing the line integral because you don't have to break the boundary in pieces and consider each separately.

Example 14.4. Find $\int_{\partial U} \mathbf{F} \cdot d\mathbf{R}$ where U is the set

$$\{(x, y) : 2 \le x \le 4, x \le y \le 3\}$$

and $\mathbf{F}(x, y) = (x \sin y, y \sin x)$.

From Green's theorem, this line integral equals

$$\int_2^4 \int_x^3 (y \cos x - x \cos y) dy dx = -\frac{3}{2} \sin 4 - 6 \sin 3 - 8 \cos 4 - \frac{9}{2} \sin 2 + 4 \cos 2.$$

14.2 Exercises

(1) Find $\int_S x dS$ where S is the surface which results from the intersection of the cone $z = 2 - \sqrt{x^2 + y^2}$ with the cylinder $x^2 + y^2 - 2x = 0$.

(2) Now let $\mathbf{n}$ be the unit normal to the above surface which has positive z component and let $\mathbf{F}(x, y, z) = (x, y, z)$. Find the flux integral $\int_S \mathbf{F} \cdot \mathbf{n} dS$.

(3) Find $\int_S z dS$ where S is the surface which results from the intersection of the hemisphere $z = \sqrt{4 - x^2 - y^2}$ with the cylinder $x^2 + y^2 - 2x = 0$.

(4) In the situation of the above problem, find the flux integral $\int_S \mathbf{F} \cdot \mathbf{n} dS$ where $\mathbf{n}$ is the unit normal to the surface which has positive z component and $\mathbf{F} = (x, y, z)$.

(5) Let $x^2/a^2 + y^2/b^2 = 1$ be an ellipse. Show using Green's theorem that its area is $\pi a b$.

(6) A spherical storage tank having radius a is filled with water which weights 62.5 pounds per cubic foot. It is shown later that this implies that the pressure of the water at depth z equals $62.5z$. Find the total force acting on this storage tank.

(7) Let $\mathbf{n}$ be the unit normal to the cone $z = \sqrt{x^2 + y^2}$ which has negative z component and let $\mathbf{F} = (x, 0, z)$ be a vector field. Let S be the part of this cone which lies between the planes $z = 1$ and $z = 2$.

Find $\int_S \mathbf{F} \cdot \mathbf{n} dS$.

(8) Let S be the surface $z = 9 - x^2 - y^2$ for $x^2 + y^2 \leq 9$. Let $\mathbf{n}$ be the unit normal to S which points up. Let $\mathbf{F} = (y, -x, z)$ and find $\int_S \mathbf{F} \cdot \mathbf{n} dS$.

(9) Let S be the surface $3z = 9 - x^2 - y^2$ for $x^2 + y^2 \leq 9$. Let $\mathbf{n}$ be the unit normal to S which points up. Let $\mathbf{F} = (y, -x, z)$ and find $\int_S \mathbf{F} \cdot \mathbf{n} dS$.

(10) For $\mathbf{F} = (x, y, z)$, S is the part of the cylinder $x^2 + y^2 = 1$ between the planes $z = 1$ and $z = 3$. Letting $\mathbf{n}$ be the unit normal which points away from the z axis, find $\int_S \mathbf{F} \cdot \mathbf{n} dS$.

(11) Let S be the part of the sphere of radius a which lies between the two cones $\phi = \frac{\pi}{4}$ and $\phi = \frac{\pi}{6}$. Let $\mathbf{F} = (z, y, 0)$. Find the flux integral $\int_S \mathbf{F} \cdot \mathbf{n} dS$.

(12) Let S be the part of a sphere of radius a above the plane $z = \frac{a}{2}$, $\mathbf{F} = (2x, 1, 1)$ and let $\mathbf{n}$ be the unit upward normal on S. Find $\int_S \mathbf{F} \cdot \mathbf{n} dS$.

(13) In the above, problem, let C be the boundary of S oriented counter clockwise as viewed from high on the z axis. Find $\int_C 2x dx + dy + dz$.

(14) Let S be the top half of a sphere of radius a centered at $\mathbf{0}$ and let $\mathbf{n}$ be the unit outward normal. Let $\mathbf{F} = (0, 0, z)$. Find $\int_S \mathbf{F} \cdot \mathbf{n} dS$.

(15) Let D be a circle in the plane which has radius 1 and let C be its counter clockwise boundary. Find $\int_C y dx + x dy$.

(16) Let D be a circle in the plane which has radius 1 and let C be its counter clockwise boundary. Find $\int_C y dx - x dy$.

(17) Find $\int_C (x + y) dx$ where C is the square curve which goes from $(0, 0) \to (1, 0) \to (1, 1) \to (0, 1) \to (0, 0)$.

(18) Find the line integral $\int_C (\sin x + y) dx + y^2 dy$ where C is the oriented square
$$(0, 0) \to (1, 0) \to (1, 1) \to (0, 1) \to (0, 0).$$

(19) Let $P(x,y) = \frac{-y}{x^2+y^2}, Q(x,y) = \frac{x}{x^2+y^2}$. Show $Q_x - P_y = 0$. Let D be the unit disk. Compute directly $\int_C P dx + Q dy$ where C is the counter clockwise circle of radius 1 which bounds the unit disk. Why don't you get 0 for the line integral?

(20) Let $\mathbf{F} = \left(2y, \ln\left(1+y^2\right) + x\right)$. Find $\int_C \mathbf{F} \cdot d\mathbf{R}$ where C is the curve consisting of line segments,

$$(0,0) \to (1,0) \to (1,1) \to (0,0).$$

14.3   Stoke's Theorem From Green's Theorem

Stoke's theorem is a generalization of Green's theorem which relates the integral over a surface to the integral around the boundary of the surface. These terms are a little different from what occurs in $\mathbb{R}^2$. To describe this, consider a sock. The surface is the sock and its boundary will be the edge of the opening of the sock in which you place your foot. Another way to think of this is to imagine a region in $\mathbb{R}^2$ of the sort discussed above for Green's theorem. Suppose it is on a sheet of rubber and the sheet of rubber is stretched in three dimensions. The boundary of the resulting surface is the result of the stretching applied to the boundary of the original region in $\mathbb{R}^2$. Here is a picture describing the situation.

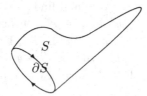

Recall the following definition of the curl of a vector field.

Definition 14.1. Let

$$\mathbf{F}(x,y,z) = (F_1(x,y,z), F_2(x,y,z), F_3(x,y,z))$$

be a C^1 vector field defined on an open set V in $\mathbb{R}^3$. Then

$$\nabla \times \mathbf{F} \equiv \begin{vmatrix} \mathbf{i} & \mathbf{j} & \mathbf{k} \\ \frac{\partial}{\partial x} & \frac{\partial}{\partial y} & \frac{\partial}{\partial z} \\ F_1 & F_2 & F_3 \end{vmatrix} \equiv \left(\frac{\partial F_3}{\partial y} - \frac{\partial F_2}{\partial z}\right)\mathbf{i} + \left(\frac{\partial F_1}{\partial z} - \frac{\partial F_3}{\partial x}\right)\mathbf{j} + \left(\frac{\partial F_2}{\partial x} - \frac{\partial F_1}{\partial y}\right)\mathbf{k}.$$

This is also called curl $(\mathbf{F})$ and written as indicated, $\nabla \times \mathbf{F}$.

The following lemma gives the fundamental identity which will be used in the proof of Stoke's theorem.

Lemma 14.2. *Let* $\mathbf{R}: U \to V \subseteq \mathbb{R}^3$ *where* U *is an open subset of* $\mathbb{R}^2$ *and* V *is an open subset of* $\mathbb{R}^3$. *Suppose* $\mathbf{R}$ *is* C^2 *and let* $\mathbf{F}$ *be a* C^1 *vector field defined in* V.

$$(\mathbf{R}_u \times \mathbf{R}_v) \cdot (\nabla \times \mathbf{F})(\mathbf{R}(u,v)) = ((\mathbf{F} \circ \mathbf{R})_u \cdot \mathbf{R}_v - (\mathbf{F} \circ \mathbf{R})_v \cdot \mathbf{R}_u)(u,v). \quad (14.3)$$

Proof: Start with the left side and let $x_i = R_i(u, v)$ for short.

$$
\begin{aligned}
(\mathbf{R}_u \times \mathbf{R}_v) \cdot (\nabla \times \mathbf{F})(\mathbf{R}(u, v)) &= \varepsilon_{ijk} x_{ju} x_{kv} \varepsilon_{irs} \frac{\partial F_s}{\partial x_r} \\
&= (\delta_{jr}\delta_{ks} - \delta_{js}\delta_{kr}) x_{ju} x_{kv} \frac{\partial F_s}{\partial x_r} \\
&= x_{ju} x_{kv} \frac{\partial F_k}{\partial x_j} - x_{ju} x_{kv} \frac{\partial F_j}{\partial x_k} \\
&= \mathbf{R}_v \cdot \frac{\partial (\mathbf{F} \circ \mathbf{R})}{\partial u} - \mathbf{R}_u \cdot \frac{\partial (\mathbf{F} \circ \mathbf{R})}{\partial v}
\end{aligned}
$$

which proves (14.3). ∎

The proof of Stoke's theorem given next follows [7]. First, it is convenient to give a definition.

Definition 14.2. A vector valued function $\mathbf{R} : U \subseteq \mathbb{R}^m \to \mathbb{R}^n$ is said to be in $C^k\left(\overline{U}, \mathbb{R}^n\right)$ if it is the restriction to $\overline{U}$ of a vector valued function which is defined on $\mathbb{R}^m$ and is C^k. That is, this function has continuous partial derivatives up to order k.

Theorem 14.3. *(Stoke's Theorem) Let U be any region in $\mathbb{R}^2$ for which the conclusion of Green's theorem holds and let $\mathbf{R} \in C^2\left(\overline{U}, \mathbb{R}^3\right)$ be a one to one function satisfying $|(\mathbf{R}_u \times \mathbf{R}_v)(u, v)| \neq 0$ for all $(u, v) \in U$ and let S denote the surface*

$$
\begin{aligned}
S &\equiv \{\mathbf{R}(u, v) : (u, v) \in U\}, \\
\partial S &\equiv \{\mathbf{R}(u, v) : (u, v) \in \partial U\}
\end{aligned}
$$

where the orientation on ∂S is consistent with the counter clockwise orientation on ∂U (U is on the left as you walk around ∂U). Then for $\mathbf{F}$ a C^1 vector field defined near S,

$$
\int_{\partial S} \mathbf{F} \cdot d\mathbf{R} = \int_S \operatorname{curl}(\mathbf{F}) \cdot \mathbf{n}\, dS
$$

where $\mathbf{n}$ is the normal to S defined by

$$
\mathbf{n} \equiv \frac{\mathbf{R}_u \times \mathbf{R}_v}{|\mathbf{R}_u \times \mathbf{R}_v|}.
$$

Proof: Letting C be an oriented part of ∂U having parametrization,

$$
\mathbf{r}(t) \equiv (u(t), v(t))
$$

for $t \in [\alpha, \beta]$ and letting $\mathbf{R}(C)$ denote the oriented part of ∂S corresponding to C,

$$
\int_{\mathbf{R}(C)} \mathbf{F} \cdot d\mathbf{R}
$$

$$= \int_{\alpha}^{\beta} \mathbf{F}\left(\mathbf{R}\left(u\left(t\right),v\left(t\right)\right)\right) \cdot \left(\mathbf{R}_u u'\left(t\right) + \mathbf{R}_v v'\left(t\right)\right) dt$$

$$= \int_{\alpha}^{\beta} \mathbf{F}\left(\mathbf{R}\left(u\left(t\right),v\left(t\right)\right)\right) \mathbf{R}_u \left(u\left(t\right),v\left(t\right)\right) u'\left(t\right) dt$$

$$+ \int_{\alpha}^{\beta} \mathbf{F}\left(\mathbf{R}\left(u\left(t\right),v\left(t\right)\right)\right) \mathbf{R}_v \left(u\left(t\right),v\left(t\right)\right) v'\left(t\right) dt$$

$$= \int_{C} \left(\left(\mathbf{F} \circ \mathbf{R}\right) \cdot \mathbf{R}_u, \left(\mathbf{F} \circ \mathbf{R}\right) \cdot \mathbf{R}_v\right) \cdot d\mathbf{r}.$$

Since this holds for each such piece of ∂U, it follows

$$\int_{\partial S} \mathbf{F} \cdot d\mathbf{R} = \int_{\partial U} \left(\left(\mathbf{F} \circ \mathbf{R}\right) \cdot \mathbf{R}_u, \left(\mathbf{F} \circ \mathbf{R}\right) \cdot \mathbf{R}_v\right) \cdot d\mathbf{r}.$$

By the assumption that the conclusion of Green's theorem holds for U, this equals

$$\int_{U} \left[\left(\left(\mathbf{F} \circ \mathbf{R}\right) \cdot \mathbf{R}_v\right)_u - \left(\left(\mathbf{F} \circ \mathbf{R}\right) \cdot \mathbf{R}_u\right)_v\right] dA$$

$$= \int_{U} \left[\left(\mathbf{F} \circ \mathbf{R}\right)_u \cdot \mathbf{R}_v + \left(\mathbf{F} \circ \mathbf{R}\right) \cdot \mathbf{R}_{vu} - \left(\mathbf{F} \circ \mathbf{R}\right) \cdot \mathbf{R}_{uv} - \left(\mathbf{F} \circ \mathbf{R}\right)_v \cdot \mathbf{R}_u\right] dA$$

$$= \int_{U} \left[\left(\mathbf{F} \circ \mathbf{R}\right)_u \cdot \mathbf{R}_v - \left(\mathbf{F} \circ \mathbf{R}\right)_v \cdot \mathbf{R}_u\right] dA$$

the last step holding by equality of mixed partial derivatives, a result of the assumption that $\mathbf{R}$ is C^2. Now by Lemma 14.2, this equals

$$\int_{U} \left(\mathbf{R}_u \times \mathbf{R}_v\right) \cdot \left(\nabla \times \mathbf{F}\right) dA$$

$$= \int_{U} \nabla \times \mathbf{F} \cdot \left(\mathbf{R}_u \times \mathbf{R}_v\right) dA$$

$$= \int_{S} \nabla \times \mathbf{F} \cdot \mathbf{n} dS$$

because $dS = \left|\left(\mathbf{R}_u \times \mathbf{R}_v\right)\right| dA$ and $\mathbf{n} = \frac{\left(\mathbf{R}_u \times \mathbf{R}_v\right)}{\left|\left(\mathbf{R}_u \times \mathbf{R}_v\right)\right|}$. Thus

$$\left(\mathbf{R}_u \times \mathbf{R}_v\right) dA = \frac{\left(\mathbf{R}_u \times \mathbf{R}_v\right)}{\left|\left(\mathbf{R}_u \times \mathbf{R}_v\right)\right|} \left|\left(\mathbf{R}_u \times \mathbf{R}_v\right)\right| dA$$

$$= \mathbf{n} dS.$$

This proves Stoke's theorem. ∎

Note that there is no mention made in the final result that $\mathbf{R}$ is C^2. Therefore, it is not surprising that versions of this theorem are valid in which this assumption is not present. It is possible to obtain extremely general versions of Stoke's theorem if you use the Lebesgue integral.

14.3.1 *The Normal And The Orientation*

Stoke's theorem as just presented needs no apology. However, it is helpful in applications to have some additional geometric insight.

To begin with, suppose the surface S of interest is a parallelogram in $\mathbb{R}^3$ determined by the two vectors $\mathbf{a}, \mathbf{b}$. Thus $S = \mathbf{R}(Q)$ where $Q = [0, 1] \times [0, 1]$ is the unit square and for $(u, v) \in Q$,

$$\mathbf{R}(u, v) \equiv u\mathbf{a} + v\mathbf{b} + \mathbf{p},$$

the point $\mathbf{p}$ being a corner of the parallelogram S. Then orient ∂S consistent with the counter clockwise orientation on ∂Q. Thus, following this orientation on S you go from $\mathbf{p}$ to $\mathbf{p} + \mathbf{a}$ to $\mathbf{p} + \mathbf{a} + \mathbf{b}$ to $\mathbf{p} + \mathbf{b}$ to $\mathbf{p}$. Then Stoke's theorem implies that with this orientation on ∂S,

$$\int_{\partial S} \mathbf{F} \cdot d\mathbf{R} = \int_S \nabla \times \mathbf{F} \cdot \mathbf{n} \, ds$$

where

$$\mathbf{n} = \mathbf{R}_u \times \mathbf{R}_v / |\mathbf{R}_u \times \mathbf{R}_v| = \mathbf{a} \times \mathbf{b} / |\mathbf{a} \times \mathbf{b}|.$$

Now recall $\mathbf{a}, \mathbf{b}, \mathbf{a} \times \mathbf{b}$ forms a right hand system.

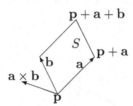

Thus, if you were walking around ∂S in the direction of the orientation with your left hand over the surface S, the normal vector $\mathbf{a} \times \mathbf{b}$ would be pointing in the direction of your head.

More generally, if S is a surface which is not necessarily a parallelogram but is instead as described in Theorem 14.3, you could consider a **small** rectangle Q contained in U and orient the boundary of $\mathbf{R}(Q)$ consistent with the counter clockwise orientation on ∂Q. Then if Q is small enough, as you walk around $\partial \mathbf{R}(Q)$ in the direction of the described orientation with your left hand over $\mathbf{R}(Q)$, your head points roughly in the direction of $\mathbf{R}_u \times \mathbf{R}_v$.

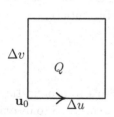
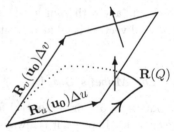

As explained above, this is true of the tangent parallelogram, and by continuity of $\mathbf{R}_v, \mathbf{R}_u$, the normals to the surface $\mathbf{R}(Q)$ $\mathbf{R}_u \times \mathbf{R}_v(\mathbf{u})$ for $\mathbf{u} \in Q$ will still point roughly in the same direction as your head if you walk in the indicated direction over $\partial \mathbf{R}(Q)$, meaning the angle between the vector from your feet to your head and the vector $\mathbf{R}_u \times \mathbf{R}_v(\mathbf{u})$ is less than $\pi/2$.

You can imagine filling U with such non-overlapping regions Q_i. Then orienting $\partial \mathbf{R}(Q_i)$ consistent with the counter clockwise orientation on Q_i, and adding the resulting line integrals, the line integrals over the common sides cancel as indicated in the following picture and the result is the line integral over ∂S.

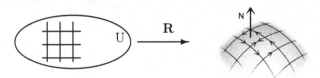

Thus there is a simple relation between the field of normal vectors on S and the orientation of ∂S. It is simply this. If you walk along ∂S in the direction mandated by the orientation, with your left hand over the surface, the nearby normal vectors in Stoke's theorem will point roughly in the direction of your head.

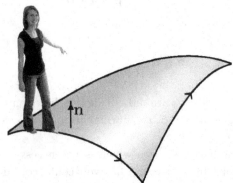

This also illustrates that you can **define** an orientation for ∂S by specifying a field of unit normal vectors for the surface, which varies continuously over the surface, and require that the motion over the boundary of the surface is such that your head points roughly in the direction of nearby normal vectors as you walk along the boundary with your left hand over S. The existence of such a continuous field of normal vectors is what constitutes an **orientable** surface.

14.3.2 *The Mobeus Band*

It turns out there are more general formulations of Stoke's theorem than what is presented above. However, it is always necessary for the surface S to be **orientable**.

This means it is possible to obtain a vector field of unit normals to the surface which is a continuous function of position on S.

An example of a surface which is not orientable is the famous Mobeus band, obtained by taking a long rectangular piece of paper and gluing the ends together after putting a twist in it. Here is a picture of one.

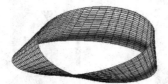

There is something quite interesting about this Mobeus band and this is that it can be written parametrically with a simple parameter domain. The picture above is a maple graph of the parametrically defined surface

$$\mathbf{R}\left(\theta, v\right) \equiv \begin{cases} x = 4\cos\theta + v\cos\frac{\theta}{2} \\ y = 4\sin\theta + v\cos\frac{\theta}{2}, \\ z = v\sin\frac{\theta}{2} \end{cases} \quad \theta \in [0, 2\pi], v \in [-1, 1].$$

An obvious question is why the normal vector $\mathbf{R}_{,\theta} \times \mathbf{R}_{,v}/\left|\mathbf{R}_{,\theta} \times \mathbf{R}_{,v}\right|$ is not a continuous function of position on S. You can see easily that it is a continuous function of both θ and v. However, the map, $\mathbf{R}$ is not one to one. In fact, $\mathbf{R}\left(0,0\right) = \mathbf{R}\left(2\pi, 0\right)$. Therefore, near this point on S, there are two different values for the above normal vector. In fact, a tedious computation will show that this normal vector is

$$\frac{\left(4\sin\frac{1}{2}\theta\cos\theta - \frac{1}{2}v, 4\sin\frac{1}{2}\theta\sin\theta + \frac{1}{2}v, -8\cos^2\frac{1}{2}\theta\sin\frac{1}{2}\theta - 8\cos^3\frac{1}{2}\theta + 4\cos\frac{1}{2}\theta\right)}{D}$$

where

$$D = \left(16\sin^2\left(\frac{\theta}{2}\right) + \frac{v^2}{2} + 4\sin\left(\frac{\theta}{2}\right)v\left(\sin\theta - \cos\theta\right)\right.$$
$$\left. + 4^3\cos^2\left(\frac{\theta}{2}\right)\left(\cos\left(\frac{1}{2}\theta\right)\sin\left(\frac{1}{2}\theta\right) + \cos^2\left(\frac{1}{2}\theta\right) - \frac{1}{2}\right)^2\right)$$

and you can verify that the denominator will not vanish. Letting $v = 0$ and $\theta = 0$ and 2π yields the two vectors $(0, 0, -1), (0, 0, 1)$ so there is a discontinuity. This is why I was careful to say in the statement of Stoke's theorem given above that $\mathbf{R}$ is one to one.

The Mobeus band has some usefulness. In old machine shops the equipment was run by a belt which was given a twist to spread the surface wear on the belt over twice the area.

The above explanation shows that $\mathbf{R}_{,\theta} \times \mathbf{R}_{,v}/\left|\mathbf{R}_{,\theta} \times \mathbf{R}_{,v}\right|$ fails to deliver an orientation for the Mobeus band. However, this does not answer the question whether there is some orientation for it other than this one. In fact there is none. You can see this by looking at the first of the two pictures below or by making one and tracing it with a pencil. There is only one side to the Mobeus band. An oriented surface must have two sides, one side identified by the given unit normal which varies continuously over the surface and the other side identified by the negative of this normal. The second picture below was taken by Ouyang when he was at meetings in Paris and saw it at a museum.

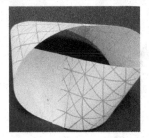 

14.3.3   Conservative Vector Fields

Definition 14.3. A vector field $\mathbf{F}$ defined in a three dimensional region is said to be **conservative**[2] if for every piecewise smooth closed curve C, it follows $\int_C \mathbf{F} \cdot d\mathbf{R} = 0$.

Definition 14.4. Let $(\mathbf{x}, \mathbf{p}_1, \cdots, \mathbf{p}_n, \mathbf{y})$ be an ordered list of points in $\mathbb{R}^p$. Let

$$\mathbf{p}(\mathbf{x}, \mathbf{p}_1, \cdots, \mathbf{p}_n, \mathbf{y})$$

denote the piecewise smooth curve consisting of a straight line segment from $\mathbf{x}$ to $\mathbf{p}_1$ and then the straight line segment from $\mathbf{p}_1$ to $\mathbf{p}_2 \cdots$ and finally the straight line segment from $\mathbf{p}_n$ to $\mathbf{y}$. This is called a **polygonal curve**. An open set in $\mathbb{R}^p$, U, is said to be a **region** if it has the property that for any two points $\mathbf{x}, \mathbf{y} \in U$, there exists a polygonal curve joining the two points.

Conservative vector fields are important because of the following theorem, sometimes called the fundamental theorem for line integrals.

Theorem 14.4. *Let U be a region in $\mathbb{R}^p$ and let $\mathbf{F} : U \to \mathbb{R}^p$ be a continuous vector field. Then $\mathbf{F}$ is conservative if and only if there exists a scalar valued function of p variables ϕ such that $\mathbf{F} = \nabla \phi$. Furthermore, if C is an oriented curve which goes from $\mathbf{x}$ to $\mathbf{y}$ in U, then*

$$\int_C \mathbf{F} \cdot d\mathbf{R} = \phi(\mathbf{y}) - \phi(\mathbf{x}). \tag{14.4}$$

Thus the line integral is path independent in this case. This function ϕ is called a ***scalar potential*** *for $\mathbf{F}$.*

Proof: To save space and fussing over things which are unimportant, denote by $\mathbf{p}(\mathbf{x}_0, \mathbf{x})$ a polygonal curve from $\mathbf{x}_0$ to $\mathbf{x}$. Thus the orientation is such that it

---

[2]There is no such thing as a liberal vector field.

goes from $\mathbf{x}_0$ to $\mathbf{x}$. The curve $\mathbf{p}(\mathbf{x}, \mathbf{x}_0)$ denotes the same set of points but in the opposite order. Suppose first $\mathbf{F}$ is conservative. Fix $\mathbf{x}_0 \in U$ and let

$$\phi(\mathbf{x}) \equiv \int_{\mathbf{p}(\mathbf{x}_0, \mathbf{x})} \mathbf{F} \cdot d\mathbf{R}.$$

This is well defined because if $\mathbf{q}(\mathbf{x}_0, \mathbf{x})$ is another polygonal curve joining $\mathbf{x}_0$ to $\mathbf{x}$, Then the curve obtained by following $\mathbf{p}(\mathbf{x}_0, \mathbf{x})$ from $\mathbf{x}_0$ to $\mathbf{x}$ and then from $\mathbf{x}$ to $\mathbf{x}_0$ along $\mathbf{q}(\mathbf{x}, \mathbf{x}_0)$ is a closed piecewise smooth curve and so by assumption, the line integral along this closed curve equals 0. However, this integral is just

$$\int_{\mathbf{p}(\mathbf{x}_0, \mathbf{x})} \mathbf{F} \cdot d\mathbf{R} + \int_{\mathbf{q}(\mathbf{x}, \mathbf{x}_0)} \mathbf{F} \cdot d\mathbf{R} = \int_{\mathbf{p}(\mathbf{x}_0, \mathbf{x})} \mathbf{F} \cdot d\mathbf{R} - \int_{\mathbf{q}(\mathbf{x}_0, \mathbf{x})} \mathbf{F} \cdot d\mathbf{R}$$

which shows

$$\int_{\mathbf{p}(\mathbf{x}_0, \mathbf{x})} \mathbf{F} \cdot d\mathbf{R} = \int_{\mathbf{q}(\mathbf{x}_0, \mathbf{x})} \mathbf{F} \cdot d\mathbf{R}$$

and that ϕ is well defined. For small t,

$$\frac{\phi(\mathbf{x} + t\mathbf{e}_i) - \phi(\mathbf{x})}{t} = \frac{\int_{\mathbf{p}(\mathbf{x}_0, \mathbf{x} + t\mathbf{e}_i)} \mathbf{F} \cdot d\mathbf{R} - \int_{\mathbf{p}(\mathbf{x}_0, \mathbf{x})} \mathbf{F} \cdot d\mathbf{R}}{t}$$

$$= \frac{\int_{\mathbf{p}(\mathbf{x}_0, \mathbf{x})} \mathbf{F} \cdot d\mathbf{R} + \int_{\mathbf{p}(\mathbf{x}, \mathbf{x} + t\mathbf{e}_i)} \mathbf{F} \cdot d\mathbf{R} - \int_{\mathbf{p}(\mathbf{x}_0, \mathbf{x})} \mathbf{F} \cdot d\mathbf{R}}{t}.$$

Since U is open, for small t, the ball of radius $|t|$ centered at $\mathbf{x}$ is contained in U. Therefore, the line segment from $\mathbf{x}$ to $\mathbf{x} + t\mathbf{e}_i$ is also contained in U and so one can take $\mathbf{p}(\mathbf{x}, \mathbf{x} + t\mathbf{e}_i)(s) = \mathbf{x} + s(t\mathbf{e}_i)$ for $s \in [0, 1]$. Therefore, the above difference quotient reduces to

$$\frac{1}{t} \int_0^1 \mathbf{F}(\mathbf{x} + s(t\mathbf{e}_i)) \cdot t\mathbf{e}_i \, ds = \int_0^1 F_i(\mathbf{x} + s(t\mathbf{e}_i)) \, ds$$

$$= F_i(\mathbf{x} + s_t(t\mathbf{e}_i))$$

by the mean value theorem for integrals. Here s_t is some number between 0 and 1. By continuity of $\mathbf{F}$, this converges to $F_i(\mathbf{x})$ as $t \to 0$. Therefore, $\nabla\phi = \mathbf{F}$ as claimed.

Conversely, if $\nabla\phi = \mathbf{F}$, then if $\mathbf{R} : [a, b] \to \mathbb{R}^p$ is any C^1 curve joining $\mathbf{x}$ to $\mathbf{y}$,

$$\int_a^b \mathbf{F}(\mathbf{R}(t)) \cdot \mathbf{R}'(t) \, dt = \int_a^b \nabla\phi(\mathbf{R}(t)) \cdot \mathbf{R}'(t) \, dt$$

$$= \int_a^b \frac{d}{dt} (\phi(\mathbf{R}(t))) \, dt$$

$$= \phi(\mathbf{R}(b)) - \phi(\mathbf{R}(a))$$

$$= \phi(\mathbf{y}) - \phi(\mathbf{x})$$

and this verifies (14.4) in the case where the curve joining the two points is smooth. The general case follows immediately from this by using this result on each of the pieces of the piecewise smooth curve. For example if the curve goes from $\mathbf{x}$ to $\mathbf{p}$

and then from **p** to **y**, the above would imply the integral over the curve from **x** to **p** is $\phi(\mathbf{p}) - \phi(\mathbf{x})$ while from **p** to **y** the integral would yield $\phi(\mathbf{y}) - \phi(\mathbf{p})$. Adding these gives $\phi(\mathbf{y}) - \phi(\mathbf{x})$. The formula (14.4) implies the line integral over any closed curve equals zero because the starting and ending points of such a curve are the same. ■

Example 14.5. Let $\mathbf{F}(x, y, z) = (\cos x - yz \sin(xz), \cos(xz), -yx \sin(xz))$. Let C be a piecewise smooth curve which goes from $(\pi, 1, 1)$ to $\left(\frac{\pi}{2}, 3, 2\right)$. Find $\int_C \mathbf{F} \cdot d\mathbf{R}$.

The specifics of the curve are not given so the problem is nonsense unless the vector field is conservative. Therefore, it is reasonable to look for the function ϕ satisfying $\nabla \phi = \mathbf{F}$. Such a function satisfies

$$\phi_x = \cos x - y(\sin xz)z$$

and so, assuming ϕ exists,

$$\phi(x, y, z) = \sin x + y \cos(xz) + \psi(y, z).$$

I have to add in the most general thing possible, $\psi(y, z)$ to ensure possible solutions are not being thrown out. It wouldn't be good at this point to only add in a constant since the answer could involve a function of either or both of the other variables. Now from what was just obtained,

$$\phi_y = \cos(xz) + \psi_y = \cos xz$$

and so it is possible to take $\psi_y = 0$. Consequently, ϕ, if it exists is of the form

$$\phi(x, y, z) = \sin x + y \cos(xz) + \psi(z).$$

Now differentiating this with respect to z gives

$$\phi_z = -yx \sin(xz) + \psi_z = -yx \sin(xz)$$

and this shows ψ does not depend on z either. Therefore, it suffices to take $\psi = 0$ and

$$\phi(x, y, z) = \sin(x) + y \cos(xz).$$

Therefore, the desired line integral equals

$$\sin\left(\frac{\pi}{2}\right) + 3 \cos(\pi) - (\sin(\pi) + \cos(\pi)) = -1.$$

The above process for finding ϕ will not lead you astray in the case where there does not exist a scalar potential. As an example, consider the following.

Example 14.6. Let $\mathbf{F}(x, y, z) = (x, y^2 x, z)$. Find a scalar potential for **F** if it exists.

If ϕ exists, then $\phi_x = x$ and so $\phi = \frac{x^2}{2} + \psi(y, z)$. Then $\phi_y = \psi_y(y, z) = xy^2$ but this is impossible because the left side depends only on y and z while the right side depends also on x. Therefore, this vector field is not conservative and there does not exist a scalar potential.

Definition 14.5. A set of points in three dimensional space V is simply connected if every piecewise smooth closed curve C is the edge of a surface S which is contained entirely within V in such a way that Stokes theorem holds for the surface S and its edge, C.

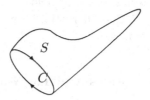

This is like a sock. The surface is the sock and the curve C goes around the opening of the sock.

As an application of Stoke's theorem, here is a useful theorem which gives a way to check whether a vector field is conservative.

Theorem 14.5. *For a three dimensional simply connected open set V and $\mathbf{F}$ a C^1 vector field defined in V, $\mathbf{F}$ is conservative if $\nabla \times \mathbf{F} = \mathbf{0}$ in V.*

Proof: If $\nabla \times \mathbf{F} = \mathbf{0}$ then taking an arbitrary closed curve C, and letting S be a surface bounded by C which is contained in V, Stoke's theorem implies

$$0 = \int_S \nabla \times \mathbf{F} \cdot \mathbf{n}\, dA = \int_C \mathbf{F} \cdot d\mathbf{R}.$$

Thus $\mathbf{F}$ is conservative. ∎

Example 14.7. Determine whether the vector field

$$\left(4x^3 + 2\left(\cos\left(x^2 + z^2\right)\right)x, 1, 2\left(\cos\left(x^2 + z^2\right)\right)z\right)$$

is conservative.

Since this vector field is defined on all of $\mathbb{R}^3$, it only remains to take its curl and see if it is the zero vector.

$$\begin{vmatrix} \mathbf{i} & \mathbf{j} & \mathbf{k} \\ \partial_x & \partial_y & \partial_z \\ 4x^3 + 2\left(\cos\left(x^2 + z^2\right)\right)x & 1 & 2\left(\cos\left(x^2 + z^2\right)\right)z \end{vmatrix}.$$

This is obviously equal to zero. Therefore, the given vector field is conservative. Can you find a potential function for it? Let ϕ be the potential function. Then $\phi_z = 2\left(\cos\left(x^2 + z^2\right)\right)z$ and so $\phi(x, y, z) = \sin\left(x^2 + z^2\right) + g(x, y)$. Now taking the derivative of ϕ with respect to y, you see $g_y = 1$ so $g(x, y) = y + h(x)$. Hence $\phi(x, y, z) = y + g(x) + \sin(x^2 + z^2)$. Taking the derivative with respect to x, you get $4x^3 + 2(\cos(x^2 + z^2)) x = g'(x) + 2x\cos(x^2 + z^2)$ and so it suffices to take $g(x) = x^4$. Hence $\phi(x, y, z) = y + x^4 + \sin(x^2 + z^2)$.

14.3.4 *Some Terminology*

If $\mathbf{F} = (P, Q, R)$ is a vector field. Then the statement that $\mathbf{F}$ is conservative is the same as saying the differential form $P dx + Q dy + R dz$ is exact. Some people like to say things in terms of vector fields and some say it in terms of differential forms. In Example 14.7, the differential form $\left(4x^3 + 2\left(\cos\left(x^2 + z^2\right)\right) x\right) dx + dy + \left(2\left(\cos\left(x^2 + z^2\right)\right) z\right) dz$ is exact.

14.4 Maxwell's Equations And The Wave Equation

Many of the ideas presented above are useful in analyzing Maxwell's equations. These equations are derived in advanced physics courses. They are

$$\nabla \times \mathbf{E} + \frac{1}{c}\frac{\partial \mathbf{B}}{\partial t} = 0 \tag{14.5}$$

$$\nabla \cdot \mathbf{E} = 4\pi\rho \tag{14.6}$$

$$\nabla \times \mathbf{B} - \frac{1}{c}\frac{\partial \mathbf{E}}{\partial t} = \frac{4\pi}{c}\mathbf{f} \tag{14.7}$$

$$\nabla \cdot \mathbf{B} = 0 \tag{14.8}$$

and it is assumed these hold on all of $\mathbb{R}^3$ to eliminate technical considerations having to do with whether something is simply connected.

In these equations, $\mathbf{E}$ is the electrostatic field and $\mathbf{B}$ is the magnetic field while ρ and $\mathbf{f}$ are sources. By (14.8) $\mathbf{B}$ has a vector potential, $\mathbf{A}_1$ such that $\mathbf{B} = \nabla \times \mathbf{A}_1$. Now go to (14.5) and write

$$\nabla \times \mathbf{E} + \frac{1}{c}\nabla \times \frac{\partial \mathbf{A}_1}{\partial t} = 0$$

showing that

$$\nabla \times \left(\mathbf{E} + \frac{1}{c}\frac{\partial \mathbf{A}_1}{\partial t}\right) = 0$$

It follows $\mathbf{E} + \frac{1}{c}\frac{\partial \mathbf{A}_1}{\partial t}$ has a scalar potential, ψ_1 satisfying

$$\nabla\psi_1 = \mathbf{E} + \frac{1}{c}\frac{\partial \mathbf{A}_1}{\partial t}. \tag{14.9}$$

Now suppose ϕ is a time dependent scalar field satisfying

$$\nabla^2\phi - \frac{1}{c^2}\frac{\partial^2\phi}{\partial t^2} = \frac{1}{c}\frac{\partial \psi_1}{\partial t} - \nabla \cdot \mathbf{A}_1. \tag{14.10}$$

Next define

$$\mathbf{A} \equiv \mathbf{A}_1 + \nabla\phi, \quad \psi \equiv \psi_1 + \frac{1}{c}\frac{\partial\phi}{\partial t}. \tag{14.11}$$

Therefore, in terms of the new variables, (14.10) becomes

$$\nabla^2\phi - \frac{1}{c^2}\frac{\partial^2\phi}{\partial t^2} = \frac{1}{c}\left(\frac{\partial\psi}{\partial t} - \frac{1}{c}\frac{\partial^2\phi}{\partial t^2}\right) - \nabla\cdot\mathbf{A} + \nabla^2\phi$$

which yields

$$0 = \frac{\partial\psi}{\partial t} - c\nabla\cdot A. \tag{14.12}$$

Then it follows from Theorem 13.1 on Page 238 that $\mathbf{A}$ is also a vector potential for $\mathbf{B}$. That is

$$\nabla\times\mathbf{A} = \mathbf{B}. \tag{14.13}$$

From (14.9)

$$\nabla\left(\psi - \frac{1}{c}\frac{\partial\phi}{\partial t}\right) = \mathbf{E} + \frac{1}{c}\left(\frac{\partial A}{\partial t} - \nabla\frac{\partial\phi}{\partial t}\right)$$

and so

$$\nabla\psi = \mathbf{E} + \frac{1}{c}\frac{\partial\mathbf{A}}{\partial t}. \tag{14.14}$$

Using (14.7) and (14.14),

$$\nabla\times(\nabla\times\mathbf{A}) - \frac{1}{c}\frac{\partial}{\partial t}\left(\nabla\psi - \frac{1}{c}\frac{\partial\mathbf{A}}{\partial t}\right) = \frac{4\pi}{c}\mathbf{f}. \tag{14.15}$$

Now from Theorem 13.1 on Page 238 this implies

$$\nabla(\nabla\cdot\mathbf{A}) - \nabla^2\mathbf{A} - \nabla\left(\frac{1}{c}\frac{\partial\psi}{\partial t}\right) + \frac{1}{c^2}\frac{\partial^2\mathbf{A}}{\partial t^2} = \frac{4\pi}{c}\mathbf{f}$$

and using (14.12), this gives

$$\frac{1}{c^2}\frac{\partial^2\mathbf{A}}{\partial t^2} - \nabla^2\mathbf{A} = \frac{4\pi}{c}\mathbf{f}. \tag{14.16}$$

Also from (14.14), (14.6), and (14.12),

$$\begin{aligned}\nabla^2\psi &= \nabla\cdot\mathbf{E} + \frac{1}{c}\frac{\partial}{\partial t}(\nabla\cdot\mathbf{A}) \\ &= 4\pi\rho + \frac{1}{c^2}\frac{\partial^2\psi}{\partial t^2}\end{aligned}$$

and so

$$\frac{1}{c^2}\frac{\partial^2\psi}{\partial t^2} - \nabla^2\psi = -4\pi\rho. \tag{14.17}$$

This is very interesting. If a solution to the wave equations, (14.17), and (14.16) can be found along with a solution to (14.12), then letting the magnetic field be given by (14.13) and letting $\mathbf{E}$ be given by (14.14) the result is a solution to Maxwell's equations. This is significant because wave equations are easier to think of than Maxwell's equations. Note the above argument also showed that it is always possible, by solving another wave equation, to get (14.12) to hold.

14.5   Exercises

(1) Determine whether the vector field

$$\left(2xy^3 \sin z^4, 3x^2y^2 \sin z^4 + 1, 4x^2y^3 \left(\cos z^4\right) z^3 + 1\right)$$

is conservative. If it is conservative, find a potential function.

(2) Determine whether the vector field

$$\left(2xy^3 \sin z + y^2 + z, 3x^2y^2 \sin z + 2xy, x^2y^3 \cos z + x\right)$$

is conservative. If it is conservative, find a potential function.

(3) Determine whether the vector field

$$\left(2xy^3 \sin z + z, 3x^2y^2 \sin z + 2xy, x^2y^3 \cos z + x\right)$$

is conservative. If it is conservative, find a potential function.

(4) Find scalar potentials for the following vector fields if it is possible to do so. If it is not possible to do so, explain why.

(a) $\left(y^2, 2xy + \sin z, 2z + y \cos z\right)$

(b) $\left(2z \left(\cos \left(x^2 + y^2\right)\right) x, 2z \left(\cos \left(x^2 + y^2\right)\right) y, \sin \left(x^2 + y^2\right) + 2z\right)$

(c) $\left(f\left(x\right), g\left(y\right), h\left(z\right)\right)$

(d) $\left(xy, z^2, y^3\right)$

(e) $\left(z + 2\frac{x}{x^2+y^2+1}, 2\frac{y}{x^2+y^2+1}, x + 3z^2\right)$

(5) If a vector field is not conservative on the set U, is it possible the same vector field could be conservative on some subset of U? Explain and give examples if it is possible. If it is not possible also explain why.

(6) Prove that if a vector field $\mathbf{F}$ has a scalar potential, then it has infinitely many scalar potentials.

(7) Here is a vector field: $\mathbf{F} \equiv \left(2xy, x^2 - 5y^4, 3z^2\right)$. Find $\int_C \mathbf{F} \cdot d\mathbf{R}$ where C is a curve which goes from $(1, 2, 3)$ to $(4, -2, 1)$.

(8) Here is a vector field: $\mathbf{F} \equiv \left(2xy, x^2 - 5y^4, 3 \left(\cos z^3\right) z^2\right)$. Find $\int_C \mathbf{F} \cdot d\mathbf{R}$ where C is a curve which goes from $(1, 0, 1)$ to $(-4, -2, 1)$.

(9) Find $\int_{\partial U} \mathbf{F} \cdot d\mathbf{R}$ where U is the set $\{(x, y) : 2 \leq x \leq 4, 0 \leq y \leq x\}$ and $\mathbf{F}(x, y) = (x \sin y, y \sin x)$.

(10) Find $\int_{\partial U} \mathbf{F} \cdot d\mathbf{R}$ where U is the set $\{(x, y) : 2 \leq x \leq 3, 0 \leq y \leq x^2\}$ and $\mathbf{F}(x, y) = (x \cos y, y + x)$.

(11) Find $\int_{\partial U} \mathbf{F} \cdot d\mathbf{R}$ where U is the set $\{(x, y) : 1 \leq x \leq 2, x \leq y \leq 3\}$ and $\mathbf{F}(x, y) = (x \sin y, y \sin x)$.

(12) Find $\int_{\partial U} \mathbf{F} \cdot d\mathbf{R}$ where U is the set $\{(x, y) : x^2 + y^2 \leq 2\}$ and $\mathbf{F}(x, y) = (-y^3, x^3)$.

(13) Show that for many open sets in $\mathbb{R}^2$, Area of $U = \int_{\partial U} x dy$, and Area of $U = \int_{\partial U} -y dx$ and Area of $U = \frac{1}{2} \int_{\partial U} -y dx + x dy$. **Hint:** Use Green's theorem.

(14) Two smooth oriented surfaces, S_1 and S_2 intersect in a piecewise smooth oriented closed curve C. Let $\mathbf{F}$ be a C^1 vector field defined on $\mathbb{R}^3$. Explain why $\int_{S_1} \operatorname{curl}(\mathbf{F}) \cdot \mathbf{n}\, dS = \int_{S_2} \operatorname{curl}(\mathbf{F}) \cdot \mathbf{n}\, dS$. Here $\mathbf{n}$ is the normal to the surface which corresponds to the given orientation of the curve C.

(15) Show that $\operatorname{curl}(\psi \nabla \phi) = \nabla \psi \times \nabla \phi$ and explain why $\int_S \nabla \psi \times \nabla \phi \cdot \mathbf{n}\, dS = \int_{\partial S} (\psi \nabla \phi) \cdot d\mathbf{r}$.

(16) Find a simple formula for $\operatorname{div}(\nabla(u^\alpha))$ where $\alpha \in \mathbb{R}$.

(17) Parametric equations for one arch of a cycloid are given by $x = a(t - \sin t)$ and $y = a(1 - \cos t)$ where here $t \in [0, 2\pi]$. Sketch a rough graph of this arch of a cycloid and then find the area between this arch and the x axis. **Hint:** This is very easy using Green's theorem and the vector field $\mathbf{F} = (-y, x)$.

(18) Let $\mathbf{r}(t) = \left(\cos^3(t), \sin^3(t)\right)$ where $t \in [0, 2\pi]$. Sketch this curve and find the area enclosed by it using Green's theorem.

(19) Verify that Green's theorem can be considered a special case of Stoke's theorem.

(20) Consider the vector field $\left(\frac{-y}{(x^2+y^2)}, \frac{x}{(x^2+y^2)}, 0\right) = \mathbf{F}$. Show that $\nabla \times \mathbf{F} = \mathbf{0}$ but that for the closed curve, whose parametrization is $\mathbf{R}(t) = (\cos t, \sin t, 0)$ for $t \in [0, 2\pi]$, $\int_C \mathbf{F} \cdot d\mathbf{R} \neq 0$. Therefore, the vector field is not conservative. Does this contradict Theorem 14.5? Explain.

(21) Let $\mathbf{x}$ be a point of $\mathbb{R}^3$ and let $\mathbf{n}$ be a unit vector. Let D_r be the circular disk of radius r containing $\mathbf{x}$ which is perpendicular to $\mathbf{n}$. Placing the tail of $\mathbf{n}$ at $\mathbf{x}$ and viewing D_r from the point of $\mathbf{n}$, orient ∂D_r in the counter clockwise direction. Now suppose $\mathbf{F}$ is a vector field defined near $\mathbf{x}$. Show that $\operatorname{curl}(\mathbf{F}) \cdot \mathbf{n} = \lim_{r \to 0} \frac{1}{\pi r^2} \int_{\partial D_r} \mathbf{F} \cdot d\mathbf{R}$. This last integral is sometimes called the circulation density of $\mathbf{F}$. Explain how this shows that $\operatorname{curl}(\mathbf{F}) \cdot \mathbf{n}$ measures the tendency for the vector field to "curl" around the point, the vector $\mathbf{n}$ at the point $\mathbf{x}$.

(22) The cylinder $x^2 + y^2 = 4$ is intersected with the plane $x + y + z = 2$. This yields a closed curve C. Orient this curve in the counter clockwise direction when viewed from a point high on the z axis. Let $\mathbf{F} = \left(x^2 y, z + y, x^2\right)$. Find $\int_C \mathbf{F} \cdot d\mathbf{R}$.

(23) The cylinder $x^2 + 4y^2 = 4$ is intersected with the plane $x + 3y + 2z = 1$. This yields a closed curve C. Orient this curve in the counter clockwise direction when viewed from a point high on the z axis. Let $\mathbf{F} = \left(y, z + y, x^2\right)$. Find $\int_C \mathbf{F} \cdot d\mathbf{R}$.

(24) The cylinder $x^2 + y^2 = 4$ is intersected with the plane $x + 3y + 2z = 1$. This yields a closed curve C. Orient this curve in the clockwise direction when viewed from a point high on the z axis. Let $\mathbf{F} = (y, z + y, x)$. Find $\int_C \mathbf{F} \cdot d\mathbf{R}$.

(25) Let $\mathbf{F} = \left(xz, z^2(y + \sin x), z^3 y\right)$. Find the surface integral $\int_S \operatorname{curl}(\mathbf{F}) \cdot \mathbf{n} dA$ where S is the surface $z = 4 - \left(x^2 + y^2\right)$, $z \geq 0$.

(26) Let $\mathbf{F} = \left(xz, \left(y^3 + x\right), z^3 y\right)$. Find the surface integral $\int_S \operatorname{curl}(\mathbf{F}) \cdot \mathbf{n} dA$ where S is the surface $z = 16 - \left(x^2 + y^2\right)$, $z \geq 0$.

(27) The cylinder $z = y^2$ intersects the surface $z = 8 - x^2 - 4y^2$ in a curve C which is oriented in the counter clockwise direction when viewed high on the z axis. Find $\int_C \mathbf{F} \cdot d\mathbf{R}$ if $\mathbf{F} = \left(\frac{z^2}{2}, xy, xz\right)$.

(28) Suppose solutions have been found to (14.17), (14.16), and (14.12). Then define $\mathbf{E}$ and $\mathbf{B}$ using (14.14) and (14.13). Verify Maxwell's equations hold for $\mathbf{E}$ and $\mathbf{B}$.

(29) Suppose now you have found solutions to (14.17) and (14.16), ψ_1 and A_1. Then go show again that if ϕ satisfies (14.10) and $\psi \equiv \psi_1 + \frac{1}{c}\frac{\partial\phi}{\partial t}$, while $\mathbf{A} \equiv \mathbf{A}_1 + \nabla\phi$, then (14.12) holds for $\mathbf{A}$ and ψ.

(30) Why consider Maxwell's equations? Why not just consider (14.17), (14.16), and (14.12)?

(31) Tell which open sets are simply connected. The inside of a car radiator, A donut., The solid part of a cannon ball which contains a void on the interior. The inside of a donut which has had a **large** bite taken out of it, All of $\mathbb{R}^3$ except the z axis, All of $\mathbb{R}^3$ except the xy plane.

(32) Let P be a polygon with vertices $(x_1, y_1), (x_2, y_2), \cdots, (x_n, y_n), (x_1, y_1)$ encountered as you move over the boundary of the polygon which is assumed a simple closed curve in the counter clockwise direction. Using Problem 13, find a nice formula for the area of the polygon in terms of the vertices.

(33) Here is a picture of two regions in the plane, U_1 and U_2. Suppose Green's theorem holds for each of these regions. Explain why Green's theorem must also hold for the region which lies between them if the boundary is oriented as shown in the picture.

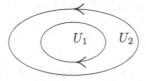

(34) Here is a picture of a surface which has two bounding curves oriented as shown. Explain why Stoke's theorem will hold for such a surface and sketch a region in the plane which could serve as a parameter domain for this surface.

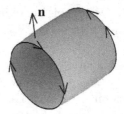

Appendix A

The Mathematical Theory Of Determinants*

A.1 The Function sgn_n

It is easiest to give a different definition of the determinant which is clearly well defined and then prove the earlier one in terms of Laplace expansion. Let $(i_1, \cdots, i_n)$ be an ordered list of numbers from $\{1, \cdots, n\}$. This means the order is important so $(1, 2, 3)$ and $(2, 1, 3)$ are different. There will be some repetition between this section and the earlier section on determinants. The main purpose is to give all the missing proofs. Two books which give a good introduction to determinants are Apostol [2] and Rudin [23]. A recent book which also has a good introduction is Baker [4].

The following Lemma will be essential in the definition of the determinant.

Lemma A.1. *There exists a unique function sgn_n which maps each list of numbers from $\{1, \cdots, n\}$ to one of the three numbers, $0, 1,$ or -1 which also has the following properties.*

$$\text{sgn}_n (1, \cdots, n) = 1 \tag{A.1}$$

$$\text{sgn}_n (i_1, \cdots, p, \cdots, q, \cdots, i_n) = -\text{sgn}_n (i_1, \cdots, q, \cdots, p, \cdots, i_n) \tag{A.2}$$

In words, the second property states that if two of the numbers are switched, the value of the function is multiplied by -1. Also, in the case where $n > 1$ and $\{i_1, \cdots, i_n\} = \{1, \cdots, n\}$ so that every number from $\{1, \cdots, n\}$ appears in the ordered list $(i_1, \cdots, i_n)$,

$$\text{sgn}_n (i_1, \cdots, i_{\theta-1}, n, i_{\theta+1}, \cdots, i_n) \equiv$$

$$(-1)^{n-\theta} \operatorname{sgn}_{n-1}(i_1, \cdots, i_{\theta-1}, i_{\theta+1}, \cdots, i_n) \qquad (\text{A.3})$$

where $n = i_\theta$ *in the ordered list* $(i_1, \cdots, i_n)$.

Proof: To begin with, it is necessary to show the existence of such a function. This is clearly true if $n = 1$. Define $\operatorname{sgn}_1(1) \equiv 1$ and observe that it works. No switching is possible. In the case where $n = 2$, it is also clearly true. Let $\operatorname{sgn}_2(1,2) = 1$ and $\operatorname{sgn}_2(2,1) = -1$ while $\operatorname{sgn}_2(2,2) = \operatorname{sgn}_2(1,1) = 0$ and verify it works. Assuming such a function exists for n, sgn_{n+1} will be defined in terms of sgn_n. If there are any repeated numbers in $(i_1, \cdots, i_{n+1})$, $\operatorname{sgn}_{n+1}(i_1, \cdots, i_{n+1}) \equiv 0$. If there are no repeats, then $n+1$ appears somewhere in the ordered list. Let θ be the position of the number $n+1$ in the list. Thus, the list is of the form $(i_1, \cdots, i_{\theta-1}, n+1, i_{\theta+1}, \cdots, i_{n+1})$. From (A.3) it must be that

$$\operatorname{sgn}_{n+1}(i_1, \cdots, i_{\theta-1}, n+1, i_{\theta+1}, \cdots, i_{n+1}) \equiv$$

$$(-1)^{n+1-\theta} \operatorname{sgn}_n(i_1, \cdots, i_{\theta-1}, i_{\theta+1}, \cdots, i_{n+1}).$$

It is necessary to verify this satisfies (A.1) and (A.2) with n replaced with $n+1$. The first of these is obviously true because

$$\operatorname{sgn}_{n+1}(1, \cdots, n, n+1) \equiv (-1)^{n+1-(n+1)} \operatorname{sgn}_n(1, \cdots, n) = 1.$$

If there are repeated numbers in $(i_1, \cdots, i_{n+1})$, then it is obvious (A.2) holds because both sides would equal zero from the above definition. It remains to verify (A.2) in the case where there are no numbers repeated in $(i_1, \cdots, i_{n+1})$. Consider

$$\operatorname{sgn}_{n+1}\left(i_1, \cdots, \overset{r}{p}, \cdots, \overset{s}{q}, \cdots, i_{n+1}\right),$$

where the r above the p indicates the number p is in the rth position and the s above the q indicates that the number q is in the sth position. Suppose first that $r < \theta < s$. Then

$$\operatorname{sgn}_{n+1}\left(i_1, \cdots, \overset{r}{p}, \cdots, \overset{\theta}{n+1}, \cdots, \overset{s}{q}, \cdots, i_{n+1}\right) \equiv$$

$$(-1)^{n+1-\theta} \operatorname{sgn}_n\left(i_1, \cdots, \overset{r}{p}, \cdots, \overset{s-1}{q}, \cdots, i_{n+1}\right)$$

while

$$\operatorname{sgn}_{n+1}\left(i_1, \cdots, \overset{r}{q}, \cdots, \overset{\theta}{n+1}, \cdots, \overset{s}{p}, \cdots, i_{n+1}\right) =$$

$$(-1)^{n+1-\theta} \operatorname{sgn}_n\left(i_1, \cdots, \overset{r}{q}, \cdots, \overset{s-1}{p}, \cdots, i_{n+1}\right)$$

and so, by induction, a switch of p and q introduces a minus sign in the result. Similarly, if $\theta > s$ or if $\theta < r$, it also follows that (A.2) holds. The interesting case is when $\theta = r$ or $\theta = s$. Consider the case where $\theta = r$ and note the other case is entirely similar.

$$\operatorname{sgn}_{n+1}\left(i_1, \cdots, \overset{r}{n+1}, \cdots, \overset{s}{q}, \cdots, i_{n+1}\right) =$$

$$(-1)^{n+1-r} \operatorname{sgn}_n \left(i_1, \cdots, \overset{s-1}{q}, \cdots, i_{n+1} \right) \tag{A.4}$$

while

$$\operatorname{sgn}_{n+1} \left(i_1, \cdots, \overset{r}{q}, \cdots, n \overset{s}{+} 1, \cdots, i_{n+1} \right) =$$
$$(-1)^{n+1-s} \operatorname{sgn}_n \left(i_1, \cdots, \overset{r}{q}, \cdots, i_{n+1} \right). \tag{A.5}$$

By making $s - 1 - r$ switches, move the q which is in the $s - 1^{th}$ position in (A.4) to the rth position in (A.5). By induction, each of these switches introduces a factor of -1, so

$$\operatorname{sgn}_n \left(i_1, \cdots, \overset{s-1}{q}, \cdots, i_{n+1} \right) = (-1)^{s-1-r} \operatorname{sgn}_n \left(i_1, \cdots, \overset{r}{q}, \cdots, i_{n+1} \right).$$

Therefore,

$$\operatorname{sgn}_{n+1} \left(i_1, \cdots, n \overset{r}{+} 1, \cdots, \overset{s}{q}, \cdots, i_{n+1} \right) = (-1)^{n+1-r} \operatorname{sgn}_n \left(i_1, \cdots, \overset{s-1}{q}, \cdots, i_{n+1} \right)$$

$$= (-1)^{n+1-r} (-1)^{s-1-r} \operatorname{sgn}_n \left(i_1, \cdots, \overset{r}{q}, \cdots, i_{n+1} \right)$$

$$= (-1)^{n+s} \operatorname{sgn}_n \left(i_1, \cdots, \overset{r}{q}, \cdots, i_{n+1} \right)$$

$$= (-1)^{2s-1} (-1)^{n+1-s} \operatorname{sgn}_n \left(i_1, \cdots, \overset{r}{q}, \cdots, i_{n+1} \right)$$

$$= - \operatorname{sgn}_{n+1} \left(i_1, \cdots, \overset{r}{q}, \cdots, n \overset{s}{+} 1, \cdots, i_{n+1} \right).$$

This proves the existence of the desired function.

To see this function is unique, note that you can obtain any ordered list of distinct numbers from a sequence of switches. If there exist two functions f and g both satisfying (A.1) and (A.2), you could start with $f(1, \cdots, n) = g(1, \cdots, n)$ and applying the same sequence of switches, eventually arrive at $f(i_1, \cdots, i_n) = g(i_1, \cdots, i_n)$. If any numbers are repeated, then (A.2) gives both functions are equal to zero for that ordered list. ∎

A.2   The Determinant

A.2.1   *The Definition*

In what follows, sgn will often be used rather than sgn_n because the context supplies the appropriate n.

Definition A.1. Let f be a real valued function which has the set of ordered lists of numbers from $\{1, \cdots, n\}$ as its domain. Define

$$\sum_{(k_1, \cdots, k_n)} f(k_1 \cdots k_n)$$

to be the sum of all the $f(k_1 \cdots k_n)$ for all possible choices of ordered lists $(k_1, \cdots, k_n)$ of numbers of $\{1, \cdots, n\}$. For example,

$$\sum_{(k_1, k_2)} f(k_1, k_2) = f(1, 2) + f(2, 1) + f(1, 1) + f(2, 2).$$

Definition A.2. Let $(a_{ij}) = A$ denote an $n \times n$ matrix. The determinant of A, denoted by $\det(A)$, is defined by

$$\det(A) \equiv \sum_{(k_1, \cdots, k_n)} \text{sgn}(k_1, \cdots, k_n)\, a_{1k_1} \cdots a_{nk_n}$$

where the sum is taken over all ordered lists of numbers from $\{1, \cdots, n\}$. Note it suffices to take the sum over only those ordered lists in which there are no repeats because if there are, $\text{sgn}(k_1, \cdots, k_n) = 0$ and so that term contributes 0 to the sum.

Let A be an $n \times n$ matrix, $A = (a_{ij})$ and let $(r_1, \cdots, r_n)$ denote an ordered list of n numbers from $\{1, \cdots, n\}$. Let $A(r_1, \cdots, r_n)$ denote the matrix whose k^{th} row is the r_k row of the matrix A. Thus

$$\det(A(r_1, \cdots, r_n)) = \sum_{(k_1, \cdots, k_n)} \text{sgn}(k_1, \cdots, k_n)\, a_{r_1 k_1} \cdots a_{r_n k_n} \tag{A.6}$$

and

$$A(1, \cdots, n) = A.$$

A.2.2   *Permuting Rows Or Columns*

Proposition A.1. Let

$$(r_1, \cdots, r_n)$$

be an ordered list of numbers from $\{1, \cdots, n\}$. Then

$$\text{sgn}(r_1, \cdots, r_n) \det(A)$$

$$= \sum_{(k_1, \cdots, k_n)} \text{sgn}(k_1, \cdots, k_n)\, a_{r_1 k_1} \cdots a_{r_n k_n} \tag{A.7}$$

$$= \det(A(r_1, \cdots, r_n)). \tag{A.8}$$

Proof: Let $(1, \cdots, n) = (1, \cdots, r, \cdots s, \cdots, n)$ so $r < s$.

$$\det(A(1, \cdots, r, \cdots, s, \cdots, n)) = \tag{A.9}$$

$$\sum_{(k_1, \cdots, k_n)} \text{sgn}(k_1, \cdots, k_r, \cdots, k_s, \cdots, k_n)\, a_{1k_1} \cdots a_{rk_r} \cdots a_{sk_s} \cdots a_{nk_n},$$

and renaming the variables, calling k_s, k_r and k_r, k_s, this equals

$$= \sum_{(k_1, \cdots, k_n)} \text{sgn}(k_1, \cdots, k_s, \cdots, k_r, \cdots, k_n)\, a_{1k_1} \cdots a_{rk_s} \cdots a_{sk_r} \cdots a_{nk_n}$$

$$= \sum_{(k_1, \cdots, k_n)} - \text{sgn}\left(k_1, \cdots, \overbrace{k_r, \cdots, k_s}^{\text{These got switched}}, \cdots, k_n \right) a_{1k_1} \cdots a_{sk_r} \cdots a_{rk_s} \cdots a_{nk_n}$$

$$= -\det\left(A\left(1,\cdots,s,\cdots,r,\cdots,n\right)\right). \tag{A.10}$$

Consequently,

$$\det\left(A\left(1,\cdots,s,\cdots,r,\cdots,n\right)\right) =$$

$$-\det\left(A\left(1,\cdots,r,\cdots,s,\cdots,n\right)\right) = -\det\left(A\right)$$

Now letting $A\left(1,\cdots,s,\cdots,r,\cdots,n\right)$ play the role of A, and continuing in this way, switching pairs of numbers,

$$\det\left(A\left(r_1,\cdots,r_n\right)\right) = (-1)^p \det\left(A\right)$$

where it took p switches to obtain $(r_1,\cdots,r_n)$ from $(1,\cdots,n)$. By Lemma A.1, this implies

$$\det\left(A\left(r_1,\cdots,r_n\right)\right) = (-1)^p \det\left(A\right) = \mathrm{sgn}\left(r_1,\cdots,r_n\right)\det\left(A\right)$$

and proves the proposition in the case when there are no repeated numbers in the ordered list $(r_1,\cdots,r_n)$. However, if there is a repeat, say the rth row equals the sth row, then the reasoning of (A.9) -(A.10) shows that $A\left(r_1,\cdots,r_n\right) = 0$ and also $\mathrm{sgn}\left(r_1,\cdots,r_n\right) = 0$, so the formula holds in this case also. ∎

Observation A.1. There are $n!$ ordered lists of distinct numbers from $\{1,\cdots,n\}$.

To see this, consider n slots placed in order. There are n choices for the first slot. For each of these choices, there are $n-1$ choices for the second. Thus there are $n(n-1)$ ways to fill the first two slots. Then for each of these ways there are $n-2$ choices left for the third slot. Continuing this way, there are $n!$ ordered lists of distinct numbers from $\{1,\cdots,n\}$ as stated in the observation.

A.2.3 A Symmetric Definition

With the above, it is possible to give a more symmetric description of the determinant from which it will follow that $\det\left(A\right) = \det\left(A^T\right)$.

Corollary A.1. *The following formula for $\det\left(A\right)$ is valid.*

$$\det\left(A\right) = \frac{1}{n!}\cdot$$

$$\sum_{(r_1,\cdots,r_n)}\sum_{(k_1,\cdots,k_n)}\mathrm{sgn}\left(r_1,\cdots,r_n\right)\mathrm{sgn}\left(k_1,\cdots,k_n\right)a_{r_1 k_1}\cdots a_{r_n k_n}. \tag{A.11}$$

And also $\det\left(A^T\right) = \det\left(A\right)$ where A^T is the transpose of A. (Recall that for $A^T = \left(a_{ij}^T\right),\ a_{ij}^T = a_{ji}$.)

Proof: From Proposition A.1, if the r_i are distinct,

$$\det(A) = \sum_{(k_1, \cdots, k_n)} \text{sgn}(r_1, \cdots, r_n) \, \text{sgn}(k_1, \cdots, k_n) \, a_{r_1 k_1} \cdots a_{r_n k_n}.$$

Summing over all ordered lists $(r_1, \cdots, r_n)$ where the r_i are distinct, (If the r_i are not distinct, $\text{sgn}(r_1, \cdots, r_n) = 0$, so there is no contribution to the sum.)

$$n! \det(A) =$$

$$\sum_{(r_1, \cdots, r_n)} \sum_{(k_1, \cdots, k_n)} \text{sgn}(r_1, \cdots, r_n) \, \text{sgn}(k_1, \cdots, k_n) \, a_{r_1 k_1} \cdots a_{r_n k_n}.$$

This proves the corollary since the formula gives the same number for A as it does for A^T. ∎

A.2.4   *The Alternating Property Of The Determinant*

Corollary A.2. *If two rows or two columns in an $n \times n$ matrix A, are switched, the determinant of the resulting matrix equals (-1) times the determinant of the original matrix. If A is an $n \times n$ matrix in which two rows are equal or two columns are equal then $\det(A) = 0$. Suppose the ith row of A equals $(xa_1 + yb_1, \cdots, xa_n + yb_n)$. Then*

$$\det(A) = x \det(A_1) + y \det(A_2)$$

where the ith row of A_1 is $(a_1, \cdots, a_n)$ and the ith row of A_2 is $(b_1, \cdots, b_n)$, all other rows of A_1 and A_2 coinciding with those of A. In other words, $\det$ is a linear function of each row A. The same is true with the word "row" replaced with the word "column".

Proof: By Proposition A.1 when two rows are switched, the determinant of the resulting matrix is (-1) times the determinant of the original matrix. By Corollary A.1, the same holds for columns because the columns of the matrix equal the rows of the transposed matrix. Thus if A_1 is the matrix obtained from A by switching two columns,

$$\det(A) = \det(A^T) = -\det(A_1^T) = -\det(A_1).$$

If A has two equal columns or two equal rows, then switching them results in the same matrix. Therefore, $\det(A) = -\det(A)$ and so $\det(A) = 0$.

It remains to verify the last assertion.

$$\det(A) \equiv \sum_{(k_1, \cdots, k_n)} \text{sgn}(k_1, \cdots, k_n) \, a_{1k_1} \cdots (xa_{k_i} + yb_{k_i}) \cdots a_{nk_n}$$

$$= x \sum_{(k_1, \cdots, k_n)} \text{sgn}(k_1, \cdots, k_n) \, a_{1k_1} \cdots a_{k_i} \cdots a_{nk_n}$$

$$+ y \sum_{(k_1, \cdots, k_n)} \text{sgn}(k_1, \cdots, k_n) \, a_{1k_1} \cdots b_{k_i} \cdots a_{nk_n} \equiv x \det(A_1) + y \det(A_2).$$

The same is true of columns because $\det(A^T) = \det(A)$ and the rows of A^T are the columns of A. ∎

A.2.5   Linear Combinations And Determinants

Definition A.3. A vector $\mathbf{w}$, is a linear combination of the vectors $\{\mathbf{v}_1, \cdots, \mathbf{v}_r\}$ if there exists scalars, $c_1, \cdots c_r$ such that $\mathbf{w} = \sum_{k=1}^{r} c_k \mathbf{v}_k$. This is the same as saying $\mathbf{w} \in \text{span}\{\mathbf{v}_1, \cdots, \mathbf{v}_r\}$.

The following corollary is also of great use.

Corollary A.3. *Suppose A is an $n \times n$ matrix and some column (row) is a linear combination of r other columns (rows). Then $\det(A) = 0$.*

Proof: Let $A = (\begin{array}{ccc} \mathbf{a}_1 & \cdots & \mathbf{a}_n \end{array})$ be the columns of A and suppose the condition that one column is a linear combination of r of the others is satisfied. Then by using Corollary A.2 you may rearrange the columns to have the nth column a linear combination of the first r columns. Thus $\mathbf{a}_n = \sum_{k=1}^{r} c_k \mathbf{a}_k$ and so

$$\det(A) = \det(\begin{array}{ccccc} \mathbf{a}_1 & \cdots & \mathbf{a}_r & \cdots & \mathbf{a}_{n-1} & \sum_{k=1}^{r} c_k \mathbf{a}_k \end{array}).$$

By Corollary A.2

$$\det(A) = \sum_{k=1}^{r} c_k \det(\begin{array}{ccccc} \mathbf{a}_1 & \cdots & \mathbf{a}_r & \cdots & \mathbf{a}_{n-1} & \mathbf{a}_k \end{array}) = 0.$$

The case for rows follows from the fact that $\det(A) = \det(A^T)$. ∎

A.2.6   The Determinant Of A Product

Recall the following definition of matrix multiplication.

Definition A.4. If A and B are $n \times n$ matrices, $A = (a_{ij})$ and $B = (b_{ij})$, $AB = (c_{ij})$ where $c_{ij} \equiv \sum_{k=1}^{n} a_{ik} b_{kj}$.

One of the most important rules about determinants is that the determinant of a product equals the product of the determinants.

Theorem A.1. *Let A and B be $n \times n$ matrices. Then*

$$\det(AB) = \det(A)\det(B).$$

Proof: Let c_{ij} be the ij^{th} entry of AB. Then by Proposition A.1, $\det(AB) =$

$$\sum_{(k_1, \cdots, k_n)} \text{sgn}(k_1, \cdots, k_n) c_{1k_1} \cdots c_{nk_n}$$

$$= \sum_{(k_1, \cdots, k_n)} \text{sgn}(k_1, \cdots, k_n) \left(\sum_{r_1} a_{1r_1} b_{r_1 k_1} \right) \cdots \left(\sum_{r_n} a_{nr_n} b_{r_n k_n} \right)$$

$$= \sum_{(r_1 \cdots, r_n)} \sum_{(k_1, \cdots, k_n)} \text{sgn}(k_1, \cdots, k_n) b_{r_1 k_1} \cdots b_{r_n k_n} (a_{1r_1} \cdots a_{nr_n})$$

$$= \sum_{(r_1 \cdots, r_n)} \text{sgn}(r_1 \cdots r_n) a_{1r_1} \cdots a_{nr_n} \det(B) = \det(A)\det(B). \qquad \blacksquare$$

A.2.7 Cofactor Expansions

Lemma A.2. *Suppose a matrix is of the form*

$$M = \begin{pmatrix} A & * \\ \mathbf{0} & a \end{pmatrix} \tag{A.12}$$

or

$$M = \begin{pmatrix} A & \mathbf{0} \\ * & a \end{pmatrix} \tag{A.13}$$

where a is a number and A is an $(n-1) \times (n-1)$ matrix and $*$ denotes either a column or a row having length $n-1$ and the $\mathbf{0}$ denotes either a column or a row of length $n-1$ consisting entirely of zeros. Then

$$\det(M) = a \det(A).$$

Proof: Denote M by (m_{ij}). Thus in the first case, $m_{nn} = a$ and $m_{ni} = 0$ if $i \neq n$ while in the second case, $m_{nn} = a$ and $m_{in} = 0$ if $i \neq n$. From the definition of the determinant,

$$\det(M) \equiv \sum_{(k_1,\cdots,k_n)} \operatorname{sgn}_n(k_1,\cdots,k_n) m_{1k_1} \cdots m_{nk_n}$$

Letting θ denote the position of n in the ordered list $(k_1,\cdots,k_n)$ then using the earlier conventions used to prove Lemma A.1, $\det(M)$ equals

$$\sum_{(k_1,\cdots,k_n)} (-1)^{n-\theta} \operatorname{sgn}_{n-1}\left(k_1,\cdots,k_{\theta-1},\overset{\theta}{k_{\theta+1}},\cdots,\overset{n-1}{k_n}\right) m_{1k_1}\cdots m_{nk_n}$$

Now suppose (A.13). Then if $k_n \neq n$, the term involving m_{nk_n} in the above expression equals zero. Therefore, the only terms which survive are those for which $\theta = n$ or in other words, those for which $k_n = n$. Therefore, the above expression reduces to

$$a \sum_{(k_1,\cdots,k_{n-1})} \operatorname{sgn}_{n-1}(k_1,\cdots k_{n-1}) m_{1k_1}\cdots m_{(n-1)k_{n-1}} = a \det(A).$$

To get the assertion in the situation of (A.12) use Corollary A.1 and (A.13) to write

$$\det(M) = \det(M^T) = \det\left(\begin{pmatrix} A^T & \mathbf{0} \\ * & a \end{pmatrix}\right) = a\det(A^T) = a\det(A). \qquad \blacksquare$$

In terms of the theory of determinants, arguably the most important idea is that of Laplace expansion along a row or a column. This will follow from the above definition of a determinant.

Definition A.5. Let $A = (a_{ij})$ be an $n \times n$ matrix. Then a new matrix called the cofactor matrix, $\operatorname{cof}(A)$ is defined by $\operatorname{cof}(A) = (c_{ij})$ where, to obtain c_{ij}, delete the ith row and the jth column of A, take the determinant of the $(n-1)\times(n-1)$ matrix which results, (This is called the ij^{th} minor of A.) and then multiply this number by $(-1)^{i+j}$. To make the formulas easier to remember, $\operatorname{cof}(A)_{ij}$ will denote the ij^{th} entry of the cofactor matrix.

The following is the main result. Earlier this was given as a definition and the outrageous totally unjustified assertion was made that the same number would be obtained by expanding the determinant along any row or column. The following theorem proves this assertion.

Theorem A.2. *Let A be an $n \times n$ matrix where $n \geq 2$. Then*

$$\det (A) = \sum_{j=1}^{n} a_{ij} \operatorname{cof} (A)_{ij} = \sum_{i=1}^{n} a_{ij} \operatorname{cof} (A)_{ij} \, .$$

The first formula consists of expanding the determinant along the ith row and the second expands the determinant along the jth column.

Proof: Let $(a_{i1}, \cdots, a_{in})$ be the ith row of A. Let B_j be the matrix obtained from A by leaving every row the same except the ith row which in B_j equals $(0, \cdots, 0, a_{ij}, 0, \cdots, 0)$. Then by Corollary A.2,

$$\det (A) = \sum_{j=1}^{n} \det (B_j)$$

Denote by A^{ij} the $(n-1) \times (n-1)$ matrix obtained by deleting the ith row and the jth column of A. Thus $\operatorname{cof} (A)_{ij} \equiv (-1)^{i+j} \det (A^{ij})$. At this point, recall that from Proposition A.1, when two rows or two columns in a matrix M, are switched, this results in multiplying the determinant of the old matrix by -1 to get the determinant of the new matrix. Therefore, by Lemma A.2,

$$\det (B_j) = (-1)^{n-j} (-1)^{n-i} \det \left(\begin{pmatrix} A^{ij} & * \\ \mathbf{0} & a_{ij} \end{pmatrix} \right)$$

$$= (-1)^{i+j} \det \left(\begin{pmatrix} A^{ij} & * \\ \mathbf{0} & a_{ij} \end{pmatrix} \right) = a_{ij} \operatorname{cof} (A)_{ij} \, .$$

Therefore, $\det (A) = \sum_{j=1}^{n} a_{ij} \operatorname{cof} (A)_{ij}$ which is the formula for expanding $\det (A)$ along the ith row. Also,

$$\det (A) = \det (A^T) = \sum_{j=1}^{n} a_{ij}^T \operatorname{cof} (A^T)_{ij} = \sum_{j=1}^{n} a_{ji} \operatorname{cof} (A)_{ji}$$

which is the formula for expanding $\det (A)$ along the ith column. ∎

Note that this gives an easy way to write a formula for the inverse of an $n \times n$ matrix.

A.2.8   *Formula For The Inverse*

Theorem A.3. A^{-1} *exists if and only if* $\det(A) \neq 0$. *If* $\det(A) \neq 0$, *then* $A^{-1} = (a_{ij}^{-1})$ *where*

$$a_{ij}^{-1} = \det(A)^{-1} \operatorname{cof} (A)_{ji}$$

for $\operatorname{cof} (A)_{ij}$ *the* ij^{th} *cofactor of* A.

Proof: By Theorem A.2, and letting $(a_{ir}) = A$, if $\det(A) \neq 0$,

$$\sum_{i=1}^{n} a_{ir} \operatorname{cof}(A)_{ir} \det(A)^{-1} = \det(A) \det(A)^{-1} = 1.$$

Now consider

$$\sum_{i=1}^{n} a_{ir} \operatorname{cof}(A)_{ik} \det(A)^{-1}$$

when $k \neq r$. Replace the kth column with the rth column to obtain a matrix B_k whose determinant equals zero by Corollary A.2. However, expanding this matrix along the kth column yields

$$0 = \det(B_k) \det(A)^{-1} = \sum_{i=1}^{n} a_{ir} \operatorname{cof}(A)_{ik} \det(A)^{-1}$$

Summarizing,

$$\sum_{i=1}^{n} a_{ir} \operatorname{cof}(A)_{ik} \det(A)^{-1} = \delta_{rk}.$$

Using the other formula in Theorem A.2, and similar reasoning,

$$\sum_{j=1}^{n} a_{rj} \operatorname{cof}(A)_{kj} \det(A)^{-1} = \delta_{rk}$$

This proves that if $\det(A) \neq 0$, then A^{-1} exists with $A^{-1} = \left(a_{ij}^{-1}\right)$, where

$$a_{ij}^{-1} = \operatorname{cof}(A)_{ji} \det(A)^{-1}.$$

Now suppose A^{-1} exists. Then by Theorem A.1,

$$1 = \det(I) = \det\left(AA^{-1}\right) = \det(A) \det\left(A^{-1}\right)$$

so $\det(A) \neq 0$. ∎

The next corollary points out that if an $n \times n$ matrix A has a right or a left inverse, then it has an inverse.

Corollary A.4. *Let A be an $n \times n$ matrix and suppose there exists an $n \times n$ matrix B such that $BA = I$. Then A^{-1} exists and $A^{-1} = B$. Also, if there exists C an $n \times n$ matrix such that $AC = I$, then A^{-1} exists and $A^{-1} = C$.*

Proof: Since $BA = I$, Theorem A.1 implies

$$\det B \det A = 1$$

and so $\det A \neq 0$. Therefore from Theorem A.3, A^{-1} exists. Therefore,

$$A^{-1} = (BA) A^{-1} = B\left(AA^{-1}\right) = BI = B.$$

The case where $CA = I$ is handled similarly. ∎

The conclusion of this corollary is that left inverses, right inverses and inverses are all the same in the context of $n \times n$ matrices.

Theorem A.3 says that to find the inverse, take the transpose of the cofactor matrix and divide by the determinant. The transpose of the cofactor matrix is called the adjugate or sometimes the classical adjoint of the matrix A. In words, A^{-1} is equal to one over the determinant of A times the adjugate matrix of A.

A.2.9   Cramer's Rule

In case you are solving a system of equations, $A\mathbf{x} = \mathbf{y}$ for $\mathbf{x}$, it follows that if A^{-1} exists,

$$\mathbf{x} = \left(A^{-1}A\right)\mathbf{x} = A^{-1}\left(A\mathbf{x}\right) = A^{-1}\mathbf{y}$$

thus solving the system. Now in the case that A^{-1} exists, there is a formula for A^{-1} given above. Using this formula,

$$x_i = \sum_{j=1}^{n} a_{ij}^{-1} y_j = \sum_{j=1}^{n} \frac{1}{\det\left(A\right)} \operatorname{cof}\left(A\right)_{ji} y_j.$$

By the formula for the expansion of a determinant along a column,

$$x_i = \frac{1}{\det\left(A\right)} \det \begin{pmatrix} * & \cdots & y_1 & \cdots & * \\ \vdots & & \vdots & & \vdots \\ * & \cdots & y_n & \cdots & * \end{pmatrix},$$

where here the ith column of A is replaced with the column vector $(y_1 \cdots, y_n)^T$, and the determinant of this modified matrix is taken and divided by $\det\left(A\right)$. This formula is known as Cramer's rule.

A.2.10   Upper Triangular Matrices

Definition A.6. A matrix M, is upper triangular if $M_{ij} = 0$ whenever $i > j$. Thus such a matrix equals zero below the main diagonal, the entries of the form M_{ii} as shown.

$$\begin{pmatrix} * & * & \cdots & * \\ 0 & * & \ddots & \vdots \\ \vdots & \ddots & \ddots & * \\ 0 & \cdots & 0 & * \end{pmatrix}$$

A lower triangular matrix is defined similarly as a matrix for which all entries above the main diagonal are equal to zero.

With this definition, here is a simple corollary of Theorem A.2.

Corollary A.5. *Let M be an upper (lower) triangular matrix. Then $\det\left(M\right)$ is obtained by taking the product of the entries on the main diagonal.*

A.2.11   The Determinant Rank

Definition A.7. A submatrix of a matrix A is the rectangular array of numbers obtained by deleting some rows and columns of A. Let A be an $m \times n$ matrix. The **determinant rank** of the matrix equals r where r is the largest number such that some $r \times r$ submatrix of A has a non zero determinant.

Theorem A.4. *If A, an $m \times n$ matrix has determinant rank r, then there exist r rows (columns) of the matrix such that every other row (column) is a linear combination of these r rows (columns).*

Proof: Suppose the determinant rank of $A = (a_{ij})$ equals r. Thus some $r \times r$ submatrix has non zero determinant and there is no larger square submatrix which has non zero determinant. Suppose such a submatrix is determined by the r columns whose indices are

$$j_1 < \cdots < j_r$$

and the r rows whose indices are

$$i_1 < \cdots < i_r$$

I want to show that every row is a linear combination of these rows. Consider the lth row and let p be an index between 1 and n. Form the following $(r+1) \times (r+1)$ matrix

$$\begin{pmatrix} a_{i_1 j_1} & \cdots & a_{i_1 j_r} & a_{i_1 p} \\ \vdots & & \vdots & \vdots \\ a_{i_r j_1} & \cdots & a_{i_r j_r} & a_{i_r p} \\ a_{l j_1} & \cdots & a_{l j_r} & a_{l p} \end{pmatrix}$$

Of course you can assume $l \notin \{i_1, \cdots, i_r\}$ because there is nothing to prove if the lth row is one of the chosen ones. The above matrix has determinant 0. This is because if $p \notin \{j_1, \cdots, j_r\}$ then the above would be a submatrix of A which is too large to have non zero determinant. On the other hand, if $p \in \{j_1, \cdots, j_r\}$ then the above matrix has two columns which are equal so its determinant is still 0.

Expand the determinant of the above matrix along the last column. Let C_k denote the cofactor associated with the entry $a_{i_k p}$. This is not dependent on the choice of p. Remember, you delete the column and the row the entry is in and take the determinant of what is left and multiply by -1 raised to an appropriate power. Let C denote the cofactor associated with a_{lp}. This is given to be nonzero, it being the determinant of the matrix

$$\begin{pmatrix} a_{i_1 j_1} & \cdots & a_{i_1 j_r} \\ \vdots & & \vdots \\ a_{i_r j_1} & \cdots & a_{i_r j_r} \end{pmatrix}$$

Thus

$$0 = a_{lp}C + \sum_{k=1}^{r} C_k a_{i_k p}$$

which implies

$$a_{lp} = \sum_{k=1}^{r} \frac{-C_k}{C} a_{i_k p} \equiv \sum_{k=1}^{r} m_k a_{i_k p}$$

Since this is true for every p and since m_k does not depend on p, this has shown the lth row is a linear combination of the $i_1, i_2, \cdots, i_r$ rows. The determinant rank does not change when you replace A with A^T. Therefore, the same conclusion holds for the columns. ∎

A.2.12 *Determining Whether A Is One To One Or Onto*

The following theorem is of fundamental importance and ties together many of the ideas presented above.

Theorem A.5. *Let A be an $n \times n$ matrix. Then the following are equivalent.*

(1) $\det(A) = 0$.
(2) A, A^T *are not one to one.*
(3) A *is not onto.*

Proof: Suppose $\det(A) = 0$. Then the determinant rank of $A = r < n$. Therefore, there exist r columns such that every other column is a linear combination of these columns by Theorem A.4. In particular, it follows that for some m, the mth column is a linear combination of all the others. Thus letting $A = (\ \mathbf{a}_1\ \cdots\ \mathbf{a}_m\ \cdots\ \mathbf{a}_n\ )$ where the columns are denoted by $\mathbf{a}_i$, there exists scalars, α_i such that

$$\mathbf{a}_m = \sum_{k \neq m} \alpha_k \mathbf{a}_k.$$

Now consider the column vector $\mathbf{x} \equiv (\ \alpha_1\ \cdots\ -1\ \cdots\ \alpha_n\ )^T$. Then

$$A\mathbf{x} = -\mathbf{a}_m + \sum_{k \neq m} \alpha_k \mathbf{a}_k = \mathbf{0}.$$

Since also $A\mathbf{0} = \mathbf{0}$, it follows A is not one to one. Similarly, A^T is not one to one by the same argument applied to A^T. This verifies that (1) implies (2).

Now suppose (2). Then since A^T is not one to one, it follows there exists $\mathbf{x} \neq \mathbf{0}$ such that $A^T\mathbf{x} = \mathbf{0}$. Taking the transpose of both sides yields $\mathbf{x}^T A = \mathbf{0}^T$ where the $\mathbf{0}^T$ is a $1 \times n$ matrix or row vector. Now if $A\mathbf{y} = \mathbf{x}$, then

$$|\mathbf{x}|^2 = \mathbf{x}^T(A\mathbf{y}) = (\mathbf{x}^T A)\mathbf{y} = \mathbf{0}^T\mathbf{y} = 0$$

contrary to $\mathbf{x} \neq \mathbf{0}$. Consequently there can be no $\mathbf{y}$ such that $A\mathbf{y} = \mathbf{x}$, so A is not onto. This shows that (2) implies (3).

Finally, suppose (3). If (1) does not hold, then $\det(A) \neq 0$ but then from Theorem A.3 A^{-1} exists and so for every $\mathbf{y} \in \mathbb{F}^n$ there exists a unique $\mathbf{x} \in \mathbb{F}^n$ such that $A\mathbf{x} = \mathbf{y}$. In fact $\mathbf{x} = A^{-1}\mathbf{y}$. Thus A would be onto contrary to (3). This shows (3) implies (1). ∎

Corollary A.6. *Let A be an $n \times n$ matrix. Then the following are equivalent.*

(1) $\det(A) \neq 0$.
(2) A *and* A^T *are one to one.*
(3) A *is onto.*

Proof: This follows immediately from the above theorem. ∎

A.2.13   *Schur's Theorem*

Consider the following system of equations for $x_1, x_2, \cdots, x_n$

$$\sum_{j=1}^{n} a_{ij} x_j = 0, \ i = 1, 2, \cdots, m \qquad (A.14)$$

where $m < n$. Then the following theorem is a fundamental observation.

Theorem A.6. *Let the system of equations be as just described in (A.14) where $m < n$. Then letting*

$$\mathbf{x}^T \equiv (x_1, x_2, \cdots, x_n) \in \mathbb{F}^n,$$

there exists $\mathbf{x} \neq \mathbf{0}$ such that the components satisfy each of the equations of (A.14). Here $\mathbb{F}$ is a field of scalars. Think $\mathbb{R}$ or $\mathbb{C}$ for example.

   Proof: The above system is of the form

$$A\mathbf{x} = \mathbf{0}$$

where A is an $m \times n$ matrix with $m < n$. Therefore, if you form the matrix $\begin{pmatrix} A \\ 0 \end{pmatrix}$, an $n \times n$ matrix having $n - m$ rows of zeros on the bottom, it follows this matrix has determinant equal to 0. Therefore, from Theorem A.5, there exists $\mathbf{x} \neq \mathbf{0}$ such that $A\mathbf{x} = \mathbf{0}$. ∎

Definition A.8. A set of vectors in $\mathbb{F}^n, \mathbb{F} = \mathbb{R}$ or $\mathbb{C}$, $\{\mathbf{x}_1, \cdots, \mathbf{x}_k\}$ is called an **orthonormal** set of vectors if

$$\mathbf{x}_i \cdot \overline{\mathbf{x}_j} = \delta_{ij} \equiv \begin{cases} 1 \text{ if } i = j \\ 0 \text{ if } i \neq j \end{cases}$$

Theorem A.7. *Let $\mathbf{v}_1$ be a unit vector ($|\mathbf{v}_1| = 1$) in $\mathbb{F}^n$, $n > 1$. Then there exist vectors $\{\mathbf{v}_2, \cdots, \mathbf{v}_n\}$ such that*

$$\{\mathbf{v}_1, \cdots, \mathbf{v}_n\}$$

is an orthonormal set of vectors.

   Proof: The equation for $\mathbf{x}$, $\overline{\mathbf{v}_1} \cdot \mathbf{x} = 0$ has a nonzero solution $\mathbf{x}$ by Theorem A.6. Pick such a solution and divide by its magnitude to get $\mathbf{v}_2$ a unit vector such that $\mathbf{v}_1 \cdot \overline{\mathbf{v}_2} = 0$. Now suppose $\mathbf{v}_1, \cdots, \mathbf{v}_k$ have been chosen such that $\{\mathbf{v}_1, \cdots, \mathbf{v}_k\}$ is an orthonormal set of vectors. Then consider the equations

$$\overline{\mathbf{v}_j} \cdot \mathbf{x} = 0 \ \ j = 1, 2, \cdots, k$$

This amounts to the situation of Theorem A.6 in which there are more variables than equations. Therefore, by this theorem, there exists a nonzero $\mathbf{x}$ solving all these equations. Divide by its magnitude and this gives $\mathbf{v}_{k+1}$. ∎

Definition A.9. If U is an $n \times n$ matrix whose columns form an orthonormal set of vectors, then U is called an **orthogonal matrix** if it is real and a **unitary matrix** if it is complex. Note that from the way we multiply matrices,

$$U^T U = U U^T = I$$

in case U is orthogonal. Thus $U^{-1} = U^T$. If U is only unitary, then from the dot product in $\mathbb{C}^n$, we replace the above with

$$U^*U = UU^* = I.$$

Where the $*$ indicates to take the conjugate of the transpose.

Note the product of orthogonal or unitary matrices is orthogonal or unitary because

$$
\begin{aligned}
(U_1U_2)^T (U_1U_2) &= U_2^T U_1^T U_1 U_2 = I \\
(U_1U_2)^* (U_1U_2) &= U_2^* U_1^* U_1 U_2 = I.
\end{aligned}
$$

Two matrices A and B are similar if there is some invertible matrix S such that $A = S^{-1}BS$. Note that similar matrices have the same characteristic equation because by Theorem A.1 which says the determinant of a product is the product of the determinants,

$$\det\left(\lambda I - A\right) = \det\left(\lambda I - S^{-1}BS\right) = \det\left(S^{-1}\left(\lambda I - B\right)S\right)$$

$$= \det\left(S^{-1}\right)\det\left(\lambda I - B\right)\det\left(S\right) = \det\left(S^{-1}S\right)\det\left(\lambda I - B\right) = \det\left(\lambda I - B\right)$$

With this preparation, here is Schur's theorem.

Theorem A.8. *Let A be a real or complex $n \times n$ matrix. Then there exists a unitary matrix U such that*

$$U^*AU = T, \tag{A.15}$$

where T is an upper triangular matrix having the eigenvalues of A on the main diagonal, listed according to multiplicity as zeros of the characteristic equation. If A has all real entries and eigenvalues, then U can be chosen to be orthogonal.

Proof: The theorem is clearly true if A is a 1×1 matrix. Just let $U = 1$ the 1×1 matrix which has 1 down the main diagonal and zeros elsewhere. Suppose it is true for $(n-1) \times (n-1)$ matrices and let A be an $n \times n$ matrix. Then let $\mathbf{v}_1$ be a unit eigenvector for A. There exists λ_1 such that

$$A\mathbf{v}_1 = \lambda_1\mathbf{v}_1, \ |\mathbf{v}_1| = 1.$$

By Theorem A.7 there exists $\{\mathbf{v}_1, \cdots, \mathbf{v}_n\}$, an orthonormal set in $\mathbb{F}^n$. Let U_0 be a matrix whose i^{th} column is $\mathbf{v}_i$. Then from the above, it follows U_0 is unitary. Then from the way you multiply matrices $U_0^*AU_0$ is of the form

$$
\begin{pmatrix}
\lambda_1 & * & \cdots & * \\
0 & & & \\
\vdots & & A_1 & \\
0 & & &
\end{pmatrix}
$$

where A_1 is an $(n-1) \times (n-1)$ matrix. The above matrix is similar to A so it has the same eigenvalues and indeed the same characteristic equation. Now by induction there exists an $(n-1) \times (n-1)$ unitary matrix $\tilde{U}_1$ such that

$$\tilde{U}_1^* A_1 \tilde{U}_1 = T_{n-1},$$

an upper triangular matrix. Consider

$$U_1 \equiv \begin{pmatrix} 1 & \mathbf{0} \\ \mathbf{0} & \tilde{U}_1 \end{pmatrix}$$

From the way we multiply matrices, this is a unitary matrix and

$$U_1^* U_0^* A U_0 U_1 = \begin{pmatrix} 1 & \mathbf{0} \\ \mathbf{0} & \tilde{U}_1^* \end{pmatrix} \begin{pmatrix} \lambda_1 & * \\ \mathbf{0} & A_1 \end{pmatrix} \begin{pmatrix} 1 & \mathbf{0} \\ \mathbf{0} & \tilde{U}_1 \end{pmatrix} = \begin{pmatrix} \lambda_1 & * \\ \mathbf{0} & T_{n-1} \end{pmatrix} \equiv T$$

where T is upper triangular. Then let $U = U_0 U_1$. Then $U^* A U = T$. If A is real having real eigenvalues, all of the above can be accomplished using the real dot product and using real eigenvectors. Thus U can be orthogonal. $\blacksquare$

A.2.14   *Symmetric Matrices*

Recall a real matrix A is symmetric if $A = A^T$.

Lemma A.3. *A real symmetric matrix has all real eigenvalues.*

   Proof: Recall the eigenvalues are solutions λ to

$$\det(\lambda I - A) = 0$$

and so by Theorem A.5, there exists $\mathbf{x}$ a vector such that

$$A\mathbf{x} = \lambda \mathbf{x}, \ \mathbf{x} \neq \mathbf{0}$$

Of course if A is real, it is still possible that the eigenvalue could be complex and if this is the case, then the vector $\mathbf{x}$ will also end up being complex. I wish to show that the eigenvalues are all real. Suppose then that λ is an eigenvalue and let $\mathbf{x}$ be the corresponding eigenvector described above. Then letting $\overline{\mathbf{x}}$ denote the complex conjugate of $\mathbf{x}$,

$$\lambda \mathbf{x}^T \overline{\mathbf{x}} = (A\mathbf{x})^T \overline{\mathbf{x}} = \mathbf{x}^T A^T \overline{\mathbf{x}} = \mathbf{x}^T A \overline{\mathbf{x}} = \mathbf{x}^T \overline{A\mathbf{x}} = \mathbf{x}^T \overline{\mathbf{x}} \overline{\lambda}$$

and so, canceling $\mathbf{x}^T \overline{\mathbf{x}}$, it follows $\lambda = \overline{\lambda}$ showing λ is real. $\blacksquare$

Theorem A.9. *Let A be a real symmetric matrix. Then there exists a diagonal matrix D consisting of the eigenvalues of A down the main diagonal and an orthogonal matrix U such that*

$$U^T A U = D.$$

Proof: Since A has all real eigenvalues, it follows from Theorem A.8, there exists an orthogonal matrix U such that

$$U^T A U = T$$

where T is upper triangular. Now

$$T^T = U^T A^T U = U^T A U = T,$$

so T is a diagonal matrix having the eigenvalues of A down the diagonal. ∎

Theorem A.10. *Let A be a real symmetric matrix which has all positive eigenvalues* $0 < \lambda_1 \le \lambda_2 \cdots \le \lambda_n$. *Then*

$$(A\mathbf{x} \cdot \mathbf{x}) \equiv \mathbf{x}^T A \mathbf{x} \ge \lambda_1 |\mathbf{x}|^2$$

Proof: Let U be the orthogonal matrix of Theorem A.9. Then

$$
\begin{aligned}
(A\mathbf{x} \cdot \mathbf{x}) &= \mathbf{x}^T A \mathbf{x} = \left(\mathbf{x}^T U\right) D \left(U^T \mathbf{x}\right) \\
&= \left(U^T \mathbf{x}\right) D \left(U^T \mathbf{x}\right) = \sum_i \lambda_i \left|\left(U^T \mathbf{x}\right)_i\right|^2 \\
&\ge \lambda_1 \sum_i \left|\left(U^T \mathbf{x}\right)_i\right|^2 = \lambda_1 \left(U^T \mathbf{x} \cdot U^T \mathbf{x}\right) \\
&= \lambda_1 \left(U^T \mathbf{x}\right)^T U^T \mathbf{x} = \lambda_1 \mathbf{x}^T U U^T \mathbf{x} = \lambda_1 \mathbf{x}^T I \mathbf{x} = \lambda_1 |\mathbf{x}|^2 .
\end{aligned}
$$
∎

A.3 Exercises

(1) Let $m < n$ and let A be an $m \times n$ matrix. Show that A is **not** one to one. **Hint:** Consider the $n \times n$ matrix A_1 which is of the form

$$A_1 \equiv \begin{pmatrix} A \\ 0 \end{pmatrix}$$

where the 0 denotes an $(n - m) \times n$ matrix of zeros. Thus $\det A_1 = 0$ and so A_1 is not one to one. Now observe that $A_1 \mathbf{x}$ is the vector

$$A_1 \mathbf{x} = \begin{pmatrix} A\mathbf{x} \\ \mathbf{0} \end{pmatrix}$$

which equals zero if and only if $A\mathbf{x} = \mathbf{0}$.

(2) In the proof of Theorem A.3 it was claimed that $\det(I) = 1$. Here $I = (\delta_{ij})$. Prove this assertion. Also prove Corollary A.5.

(3) Let $\mathbf{v}_1, \cdots, \mathbf{v}_n$ be vectors in $\mathbb{F}^n$ and let $M(\mathbf{v}_1, \cdots, \mathbf{v}_n)$ denote the matrix whose i^{th} column equals $\mathbf{v}_i$. Define

$$d(\mathbf{v}_1, \cdots, \mathbf{v}_n) \equiv \det (M(\mathbf{v}_1, \cdots, \mathbf{v}_n)).$$

Prove that d is linear in each variable, (multilinear), that

$$d\left(\mathbf{v}_1,\cdots,\mathbf{v}_i,\cdots,\mathbf{v}_j,\cdots,\mathbf{v}_n\right) = -d\left(\mathbf{v}_1,\cdots,\mathbf{v}_j,\cdots,\mathbf{v}_i,\cdots,\mathbf{v}_n\right), \qquad \text{(A.16)}$$

and

$$d\left(\mathbf{e}_1,\cdots,\mathbf{e}_n\right) = 1 \qquad \text{(A.17)}$$

where here $\mathbf{e}_j$ is the vector in $\mathbb{F}^n$ which has a zero in every position except the jth position in which it has a one.

(4) Suppose $f : \mathbb{F}^n \times \cdots \times \mathbb{F}^n \to \mathbb{F}$ satisfies (A.16) and (A.17) and is linear in each variable. Show that $f = d$.

(5) Two $n \times n$ matrices, A and B, are similar if $B = S^{-1}AS$ for some invertible $n \times n$ matrix S. Show that if two matrices are similar, they have the same characteristic polynomials.

(6) Let A_i for $i = 1, \cdots, n$ be $n \times n$ matrices. Suppose also that for all $|\lambda|$ sufficiently large,

$$A_0 + A_1\lambda + \cdots + A_n\lambda^n = 0$$

Show that it follows that each $A_k = 0$.

(7) ↑Let F be $\mathbb{Q}, \mathbb{R}$ or $\mathbb{C}$. Let $A_k, B_k, k = 1, \cdots, n$ be $n \times n$ matrices. Show that if

$$A_0 + A_1\lambda + \cdots + A_n\lambda^n = B_0 + B_1\lambda + \cdots + B_n\lambda^n$$

for all $|\lambda|$ sufficiently large enough, then $A_i = B_i$ for each i.

(8) ↑Let A be an $n \times n$ real, complex, or rational matrix. Explain why $\det\left(\lambda I - A\right)$ is a polynomial of degree n. Explain why there are finitely many roots of $\det\left(\lambda I - A\right) = 0$. Then explain why $\left(\lambda I - A\right)^{-1}$ exists for all $|\lambda|$ large enough.

(9) ↑ In the situation of the above problem, let $C\left(\lambda\right)$ denote the adjugate matrix of $\left(\lambda I - A\right)$ for all $|\lambda|$ sufficiently large that $\left(\lambda I - A\right)^{-1}$ exists. Then by the formula for the inverse in terms of the cofactors,

$$\begin{aligned} C\left(\lambda\right) &= \det\left(\lambda I - A\right)\left(\lambda I - A\right)^{-1} \\ &= q\left(\lambda\right)\left(\lambda I - A\right)^{-1} \end{aligned}$$

Explain why each entry in $C\left(\lambda\right)$ is a polynomial in λ having degree no more than $n - 1$ so that

$$C\left(\lambda\right) = C_0 + C_1\lambda + \cdots + C_{n-1}\lambda^{n-1}$$

where each C_i is an $n \times n$ matrix. Then for all $|\lambda|$ large enough,

$$\left(\lambda I - A\right)\left(C_0 + C_1\lambda + \cdots + C_{n-1}\lambda^{n-1}\right) = q\left(\lambda\right)I$$

Now use the result of Problem 7. Conclude that the matrix coefficients of the two sides are the same. Explain why λ can now be replaced with A to conclude

$$0 = q\left(A\right)I = q\left(A\right).$$

This is the very important **Cayley - Hamilton theorem**[1].

(10) Suppose the characteristic polynomial of an $n \times n$ matrix A is of the form

$$t^n + a_{n-1}t^{n-1} + \cdots + a_1 t + a_0$$

and that $a_0 \neq 0$. Find a formula A^{-1} in terms of powers of the matrix A. Show that A^{-1} exists if and only if $a_0 \neq 0$.

---

[1] A special case was first proved by Hamilton in 1853. The general case was announced by Cayley some time later and a proof was given by Frobenius in 1878.

Appendix B

Implicit Function Theorem*

The implicit function theorem is one of the greatest theorems in mathematics. There are many versions of this theorem which are of far greater generality than the one given here. The proof given here is like one found in one of Caratheodory's books on the calculus of variations. It is not as elegant as some of the others which are based on a contraction mapping principle but it may be more accessible. However, it is an advanced topic. Do not waste your time with it unless you have first read and understood the material on rank and determinants found in the chapter on the mathematical theory of determinants. You will also need to use the extreme value theorem for a function of n variables and the chain rule as well as everything about matrix multiplication.

Definition B.1. Suppose U is an open set in $\mathbb{R}^n \times \mathbb{R}^m$ and $(\mathbf{x}, \mathbf{y})$ will denote a typical point of $\mathbb{R}^n \times \mathbb{R}^m$ with $\mathbf{x} \in \mathbb{R}^n$ and $\mathbf{y} \in \mathbb{R}^m$. Let $\mathbf{f} : U \to \mathbb{R}^p$ be in $C^1(U)$. Then define

$$D_1\mathbf{f}(\mathbf{x}, \mathbf{y}) \equiv \begin{pmatrix} f_{1,x_1}(\mathbf{x}, \mathbf{y}) & \cdots & f_{1,x_n}(\mathbf{x}, \mathbf{y}) \\ \vdots & & \vdots \\ f_{p,x_1}(\mathbf{x}, \mathbf{y}) & \cdots & f_{p,x_n}(\mathbf{x}, \mathbf{y}) \end{pmatrix},$$

$$D_2\mathbf{f}(\mathbf{x}, \mathbf{y}) \equiv \begin{pmatrix} f_{1,y_1}(\mathbf{x}, \mathbf{y}) & \cdots & f_{1,y_m}(\mathbf{x}, \mathbf{y}) \\ \vdots & & \vdots \\ f_{p,y_1}(\mathbf{x}, \mathbf{y}) & \cdots & f_{p,y_m}(\mathbf{x}, \mathbf{y}) \end{pmatrix}.$$

Thus $D\mathbf{f}(\mathbf{x}, \mathbf{y})$ is a $p \times (n + m)$ matrix of the form

$$D\mathbf{f}(\mathbf{x}, \mathbf{y}) = \left(\; D_1\mathbf{f}(\mathbf{x}, \mathbf{y}) \; \mid \; D_2\mathbf{f}(\mathbf{x}, \mathbf{y}) \; \right).$$

Note that $D_1\mathbf{f}(\mathbf{x}, \mathbf{y})$ is a $p \times n$ matrix and $D_2\mathbf{f}(\mathbf{x}, \mathbf{y})$ is a $p \times m$ matrix.

Theorem B.1. *(implicit function theorem) Suppose U is an open set in $\mathbb{R}^n \times \mathbb{R}^m$. Let $\mathbf{f} : U \to \mathbb{R}^n$ be in $C^1(U)$ and suppose*

$$\mathbf{f}(\mathbf{x}_0, \mathbf{y}_0) = \mathbf{0}, \ D_1\mathbf{f}(\mathbf{x}_0, \mathbf{y}_0)^{-1} \ exists. \tag{B.1}$$

Then there exist positive constants δ, η, such that for every $\mathbf{y} \in B(\mathbf{y}_0, \eta)$ there exists a unique $\mathbf{x}(\mathbf{y}) \in B(\mathbf{x}_0, \delta)$ such that

$$\mathbf{f}(\mathbf{x}(\mathbf{y}), \mathbf{y}) = \mathbf{0}. \tag{B.2}$$

Furthermore, the mapping $\mathbf{y} \to \mathbf{x}(\mathbf{y})$ is in $C^1(B(\mathbf{y}_0, \eta))$.

Proof: Let

$$\mathbf{f}(\mathbf{x}, \mathbf{y}) = \begin{pmatrix} f_1(\mathbf{x}, \mathbf{y}) \\ f_2(\mathbf{x}, \mathbf{y}) \\ \vdots \\ f_n(\mathbf{x}, \mathbf{y}) \end{pmatrix}.$$

Define for $(\mathbf{x}^1, \cdots, \mathbf{x}^n) \in \overline{B(\mathbf{x}_0, \delta)}^n$ and $\mathbf{y} \in B(\mathbf{y}_0, \eta)$ the following matrix.

$$J(\mathbf{x}^1, \cdots, \mathbf{x}^n, \mathbf{y}) \equiv \begin{pmatrix} f_{1,x_1}(\mathbf{x}^1, \mathbf{y}) & \cdots & f_{1,x_n}(\mathbf{x}^1, \mathbf{y}) \\ \vdots & & \vdots \\ f_{n,x_1}(\mathbf{x}^n, \mathbf{y}) & \cdots & f_{n,x_n}(\mathbf{x}^n, \mathbf{y}) \end{pmatrix}.$$

Then by the assumption of continuity of all the partial derivatives and the extreme value theorem, there exists $r > 0$ and $\delta_0, \eta_0 > 0$ such that if $\delta \le \delta_0$ and $\eta \le \eta_0$, it follows that for all $(\mathbf{x}^1, \cdots, \mathbf{x}^n) \in \overline{B(\mathbf{x}_0, \delta)}^n$ and $\mathbf{y} \in \overline{B(\mathbf{y}_0, \eta)}$,

$$\left| \det\left(J(\mathbf{x}^1, \cdots, \mathbf{x}^n, \mathbf{y})\right) \right| > r > 0. \tag{B.3}$$

and $\overline{B(\mathbf{x}_0, \delta_0)} \times \overline{B(\mathbf{y}_0, \eta_0)} \subseteq U$. By continuity of all the partial derivatives and the extreme value theorem, it can also be assumed there exists a constant K such that for all $(\mathbf{x}, \mathbf{y}) \in \overline{B(\mathbf{x}_0, \delta_0)} \times \overline{B(\mathbf{y}_0, \eta_0)}$ and $i = 1, 2, \cdots, n$, the ith row of $D_2\mathbf{f}(\mathbf{x}, \mathbf{y})$, given by $D_2 f_i(\mathbf{x}, \mathbf{y})$ satisfies

$$|D_2 f_i(\mathbf{x}, \mathbf{y})| < K, \tag{B.4}$$

and for all $(\mathbf{x}^1, \cdots, \mathbf{x}^n) \in \overline{B(\mathbf{x}_0, \delta_0)}^n$ and $\mathbf{y} \in \overline{B(\mathbf{y}_0, \eta_0)}$ the ith row of the matrix

$$J(\mathbf{x}^1, \cdots, \mathbf{x}^n, \mathbf{y})^{-1}$$

which equals $\mathbf{e}_i^T \left(J(\mathbf{x}^1, \cdots, \mathbf{x}^n, \mathbf{y})^{-1}\right)$ satisfies

$$\left| \mathbf{e}_i^T \left(J(\mathbf{x}^1, \cdots, \mathbf{x}^n, \mathbf{y})^{-1}\right) \right| < K. \tag{B.5}$$

(Recall that $\mathbf{e}_i$ is the column vector consisting of all zeros except for a 1 in the ith position.)

To begin with, it is shown that for a given $\mathbf{y} \in B(\mathbf{y}_0, \eta)$ there is at most one $\mathbf{x} \in B(\mathbf{x}_0, \delta)$ such that $\mathbf{f}(\mathbf{x}, \mathbf{y}) = \mathbf{0}$.

Pick $\mathbf{y} \in B(\mathbf{y}_0, \eta)$ and suppose there exist $\mathbf{x}, \mathbf{z} \in \overline{B(\mathbf{x}_0, \delta)}$ such that $\mathbf{f}(\mathbf{x}, \mathbf{y}) = \mathbf{f}(\mathbf{z}, \mathbf{y}) = \mathbf{0}$. Consider f_i and let

$$h(t) \equiv f_i(\mathbf{x} + t(\mathbf{z} - \mathbf{x}), \mathbf{y}).$$

Then $h(1) = h(0)$ and so by Rolle's theorem, $h'(t_i) = 0$ for some $t_i \in (0, 1)$. Therefore, from the chain rule and for this value of t_i,

$$h'(t_i) = Df_i(\mathbf{x} + t_i(\mathbf{z} - \mathbf{x}), \mathbf{y})(\mathbf{z} - \mathbf{x}) = 0. \tag{B.6}$$

Then denote by $\mathbf{x}^i$ the vector $\mathbf{x} + t_i(\mathbf{z} - \mathbf{x})$. It follows from (B.6) that

$$J(\mathbf{x}^1, \cdots, \mathbf{x}^n, \mathbf{y})(\mathbf{z} - \mathbf{x}) = \mathbf{0}$$

and so from (B.3) $\mathbf{z} - \mathbf{x} = \mathbf{0}$. (The matrix in the above is invertible since its determinant is nonzero.) Now it will be shown that if η is chosen sufficiently small, then for all $\mathbf{y} \in B(\mathbf{y}_0, \eta)$, there exists a unique $\mathbf{x}(\mathbf{y}) \in B(\mathbf{x}_0, \delta)$ such that $\mathbf{f}(\mathbf{x}(\mathbf{y}), \mathbf{y}) = \mathbf{0}$.

Claim: If η is small enough, then the function $h_{\mathbf{y}}(\mathbf{x}) \equiv |\mathbf{f}(\mathbf{x}, \mathbf{y})|^2$ achieves its minimum value on $\overline{B(\mathbf{x}_0, \delta)}$ at a point of $B(\mathbf{x}_0, \delta)$. (The existence of a point in $\overline{B(\mathbf{x}_0, \delta)}$ at which $h_{\mathbf{y}}$ achieves its minimum follows from the extreme value theorem.)

Proof of claim: Suppose this is not the case. Then there exists a sequence $\eta_k \to 0$ and for some $\mathbf{y}_k$ having $|\mathbf{y}_k - \mathbf{y}_0| < \eta_k$, the minimum of $h_{\mathbf{y}_k}$ on $\overline{B(\mathbf{x}_0, \delta)}$ occurs on a point of $\overline{B(\mathbf{x}_0, \delta)}$, $\mathbf{x}_k$ such that $|\mathbf{x}_0 - \mathbf{x}_k| = \delta$. Now taking a subsequence, still denoted by k, it can be assumed that $\mathbf{x}_k \to \mathbf{x}$ with $|\mathbf{x} - \mathbf{x}_0| = \delta$ and $\mathbf{y}_k \to \mathbf{y}_0$. This follows from the fact that $\left\{ \mathbf{x} \in \overline{B(\mathbf{x}_0, \delta)} : |\mathbf{x} - \mathbf{x}_0| = \delta \right\}$ is a closed and bounded set and is therefore sequentially compact. Then by continuity, $|\mathbf{f}(\mathbf{x}_k, \mathbf{y}_k)|^2 \to |\mathbf{f}(\mathbf{x}, \mathbf{y}_0)|^2$. However, $|\mathbf{f}(\mathbf{x}, \mathbf{y}_0)|^2$ must be equal to 0 because for large k, $|\mathbf{f}(\mathbf{x}_k, \mathbf{y}_k)|^2 \leq |\mathbf{f}(\mathbf{x}_0, \mathbf{y}_k)|^2$ which converges to 0 as $k \to \infty$. This contradicts the first part of the argument in which it was shown that for $\mathbf{y} \in B(\mathbf{y}_0, \eta)$ there is at most one point $\mathbf{x}$ of $\overline{B(\mathbf{x}_0, \delta)}$ where $\mathbf{f}(\mathbf{x}, \mathbf{y}) = \mathbf{0}$. Here two have been obtained, $\mathbf{x}_0$ and $\mathbf{x}$. This proves the claim.

Choose $\eta < \eta_0$ and also small enough that the above claim holds and let $\mathbf{x}(\mathbf{y})$ denote a point of $B(\mathbf{x}_0, \delta)$ at which the minimum of $h_{\mathbf{y}}$ on $\overline{B(\mathbf{x}_0, \delta)}$ is achieved. Since $\mathbf{x}(\mathbf{y})$ is an interior point, you can consider $h_{\mathbf{y}}(\mathbf{x}(\mathbf{y}) + t\mathbf{v})$ for $|t|$ small and conclude this function of t has a zero derivative at $t = 0$. Now

$$h_{\mathbf{y}}(\mathbf{x}(\mathbf{y}) + t\mathbf{v}) = \sum_{i=1}^{n} f_i^2(\mathbf{x}(\mathbf{y}) + t\mathbf{v}, \mathbf{y})$$

and so from the chain rule,

$$\frac{d}{dt} h_{\mathbf{y}}(\mathbf{x}(\mathbf{y}) + t\mathbf{v}) = \sum_{i=1}^{n} 2f_i(\mathbf{x}(\mathbf{y}) + t\mathbf{v}, \mathbf{y}) \frac{\partial f_i(\mathbf{x}(\mathbf{y}) + t\mathbf{v}, \mathbf{y})}{\partial x_j} v_j.$$

Therefore, letting $t = 0$, it is required that for every $\mathbf{v}$,

$$\sum_{i=1}^{n} 2f_i(\mathbf{x}(\mathbf{y}), \mathbf{y}) \frac{\partial f_i(\mathbf{x}(\mathbf{y}), \mathbf{y})}{\partial x_j} v_j = 0.$$

In terms of matrices this reduces to

$$0 = 2\mathbf{f}\left(\mathbf{x}\left(\mathbf{y}\right),\mathbf{y}\right)^{T} D_{1}\mathbf{f}\left(\mathbf{x}\left(\mathbf{y}\right),\mathbf{y}\right)\mathbf{v}$$

for every vector $\mathbf{v}$. Therefore,

$$\mathbf{0} = \mathbf{f}\left(\mathbf{x}\left(\mathbf{y}\right),\mathbf{y}\right)^{T} D_{1}\mathbf{f}\left(\mathbf{x}\left(\mathbf{y}\right),\mathbf{y}\right)$$

From (B.3), it follows $\mathbf{f}\left(\mathbf{x}\left(\mathbf{y}\right),\mathbf{y}\right) = \mathbf{0}$. This proves the existence of the function $\mathbf{y} \to \mathbf{x}\left(\mathbf{y}\right)$ such that $\mathbf{f}\left(\mathbf{x}\left(\mathbf{y}\right),\mathbf{y}\right) = \mathbf{0}$ for all $\mathbf{y} \in B\left(\mathbf{y}_{0},\eta\right)$.

It remains to verify this function is a C^{1} function. To do this, let $\mathbf{y}_{1}$ and $\mathbf{y}_{2}$ be points of $B\left(\mathbf{y}_{0},\eta\right)$. Then as before, consider the ith component of $\mathbf{f}$ and consider the same argument using the mean value theorem to write

$$
\begin{aligned}
0 &= f_{i}\left(\mathbf{x}\left(\mathbf{y}_{1}\right),\mathbf{y}_{1}\right) - f_{i}\left(\mathbf{x}\left(\mathbf{y}_{2}\right),\mathbf{y}_{2}\right) \\
&= f_{i}\left(\mathbf{x}\left(\mathbf{y}_{1}\right),\mathbf{y}_{1}\right) - f_{i}\left(\mathbf{x}\left(\mathbf{y}_{2}\right),\mathbf{y}_{1}\right) + f_{i}\left(\mathbf{x}\left(\mathbf{y}_{2}\right),\mathbf{y}_{1}\right) - f_{i}\left(\mathbf{x}\left(\mathbf{y}_{2}\right),\mathbf{y}_{2}\right) \quad (\text{B.7}) \\
&= D_{1}f_{i}\left(\mathbf{x}^{i},\mathbf{y}_{1}\right)\left(\mathbf{x}\left(\mathbf{y}_{1}\right) - \mathbf{x}\left(\mathbf{y}_{2}\right)\right) + D_{2}f_{i}\left(\mathbf{x}\left(\mathbf{y}_{2}\right),\mathbf{y}^{i}\right)\left(\mathbf{y}_{1} - \mathbf{y}_{2}\right).
\end{aligned}
$$

where $\mathbf{y}^{i}$ is a point on the line segment joining $\mathbf{y}_{1}$ and $\mathbf{y}_{2}$. Thus from (B.4) and the Cauchy Schwarz inequality,

$$\left|D_{2}f_{i}\left(\mathbf{x}\left(\mathbf{y}_{2}\right),\mathbf{y}^{i}\right)\left(\mathbf{y}_{1} - \mathbf{y}_{2}\right)\right| \leq K\left|\mathbf{y}_{1} - \mathbf{y}_{2}\right|.$$

Therefore, letting $M\left(\mathbf{y}^{1},\cdots,\mathbf{y}^{n}\right) \equiv M$ denote the matrix having the ith row equal to $D_{2}f_{i}\left(\mathbf{x}\left(\mathbf{y}_{2}\right),\mathbf{y}^{i}\right)$, it follows

$$\left|M\left(\mathbf{y}_{1} - \mathbf{y}_{2}\right)\right| \leq \left(\sum_{i} K^{2}\left|\mathbf{y}_{1} - \mathbf{y}_{2}\right|^{2}\right)^{1/2} = \sqrt{m}K\left|\mathbf{y}_{1} - \mathbf{y}_{2}\right|. \quad (\text{B.8})$$

Also, from (B.7),

$$J\left(\mathbf{x}^{1},\cdots,\mathbf{x}^{n},\mathbf{y}_{1}\right)\left(\mathbf{x}\left(\mathbf{y}_{1}\right) - \mathbf{x}\left(\mathbf{y}_{2}\right)\right) = -M\left(\mathbf{y}_{1} - \mathbf{y}_{2}\right) \quad (\text{B.9})$$

and so from (B.8) and (B.10),

$$\left|\mathbf{x}\left(\mathbf{y}_{1}\right) - \mathbf{x}\left(\mathbf{y}_{2}\right)\right| =$$

$$\left|J\left(\mathbf{x}^{1},\cdots,\mathbf{x}^{n},\mathbf{y}_{1}\right)^{-1} M\left(\mathbf{y}_{1} - \mathbf{y}_{2}\right)\right| \quad (\text{B.10})$$

$$= \left(\sum_{i=1}^{n}\left|\mathbf{e}_{i}^{T} J\left(\mathbf{x}^{1},\cdots,\mathbf{x}^{n},\mathbf{y}_{1}\right)^{-1} M\left(\mathbf{y}_{1} - \mathbf{y}_{2}\right)\right|^{2}\right)^{1/2}$$

$$\leq \left(\sum_{i=1}^{n} K^{2}\left|M\left(\mathbf{y}_{1} - \mathbf{y}_{2}\right)\right|^{2}\right)^{1/2} \leq \left(\sum_{i=1}^{n} K^{2}\left(\sqrt{m}K\left|\mathbf{y}_{1} - \mathbf{y}_{2}\right|\right)^{2}\right)^{1/2}$$

$$= K^{2}\sqrt{mn}\left|\mathbf{y}_{1} - \mathbf{y}_{2}\right| \quad (\text{B.11})$$

It follows as in the proof of the chain rule that

$$\mathbf{o}\left(\mathbf{x}\left(\mathbf{y}+\mathbf{v}\right) - \mathbf{x}\left(\mathbf{y}\right)\right) = \mathbf{o}\left(\mathbf{v}\right). \quad (\text{B.12})$$

Now let $\mathbf{y} \in B(\mathbf{y}_0, \eta)$ and let $|\mathbf{v}|$ be sufficiently small that $\mathbf{y} + \mathbf{v} \in B(\mathbf{y}_0, \eta)$. Then

$$
\begin{aligned}
\mathbf{0} &= \mathbf{f}(\mathbf{x}(\mathbf{y}+\mathbf{v}), \mathbf{y}+\mathbf{v}) - \mathbf{f}(\mathbf{x}(\mathbf{y}), \mathbf{y}) \\
&= \mathbf{f}(\mathbf{x}(\mathbf{y}+\mathbf{v}), \mathbf{y}+\mathbf{v}) - \mathbf{f}(\mathbf{x}(\mathbf{y}+\mathbf{v}), \mathbf{y}) + \mathbf{f}(\mathbf{x}(\mathbf{y}+\mathbf{v}), \mathbf{y}) - \mathbf{f}(\mathbf{x}(\mathbf{y}), \mathbf{y})
\end{aligned}
$$

$$
= D_2\mathbf{f}(\mathbf{x}(\mathbf{y}+\mathbf{v}), \mathbf{y})\mathbf{v} + D_1\mathbf{f}(\mathbf{x}(\mathbf{y}), \mathbf{y})(\mathbf{x}(\mathbf{y}+\mathbf{v}) - \mathbf{x}(\mathbf{y})) + o(|\mathbf{x}(\mathbf{y}+\mathbf{v}) - \mathbf{x}(\mathbf{y})|)
$$

$$
\begin{aligned}
&= D_2\mathbf{f}(\mathbf{x}(\mathbf{y}), \mathbf{y})\mathbf{v} + D_1\mathbf{f}(\mathbf{x}(\mathbf{y}), \mathbf{y})(\mathbf{x}(\mathbf{y}+\mathbf{v}) - \mathbf{x}(\mathbf{y})) + \\
&\quad o(|\mathbf{x}(\mathbf{y}+\mathbf{v}) - \mathbf{x}(\mathbf{y})|) + (D_2\mathbf{f}(\mathbf{x}(\mathbf{y}+\mathbf{v}), \mathbf{y}) \mathbf{v} - D_2\mathbf{f}(\mathbf{x}(\mathbf{y}), \mathbf{y})\mathbf{v}) \\
&= D_2\mathbf{f}(\mathbf{x}(\mathbf{y}), \mathbf{y})\mathbf{v} + D_1\mathbf{f}(\mathbf{x}(\mathbf{y}), \mathbf{y})(\mathbf{x}(\mathbf{y}+\mathbf{v}) - \mathbf{x}(\mathbf{y})) + o(\mathbf{v}).
\end{aligned}
$$

Therefore,

$$
\mathbf{x}(\mathbf{y}+\mathbf{v}) - \mathbf{x}(\mathbf{y}) = -D_1\mathbf{f}(\mathbf{x}(\mathbf{y}), \mathbf{y})^{-1} D_2\mathbf{f}(\mathbf{x}(\mathbf{y}), \mathbf{y})\mathbf{v} + o(\mathbf{v})
$$

which shows that $D\mathbf{x}(\mathbf{y}) = -D_1\mathbf{f}(\mathbf{x}(\mathbf{y}), \mathbf{y})^{-1} D_2\mathbf{f}(\mathbf{x}(\mathbf{y}), \mathbf{y})$ and $\mathbf{y} \to D\mathbf{x}(\mathbf{y})$ is continuous. ∎

B.2   More Continuous Partial Derivatives

The implicit function theorem will now be improved slightly. If $\mathbf{f}$ is C^k, it follows that the function which is implicitly defined is also C^k, not just C^1, meaning all mixed partial derivatives of $\mathbf{f}$ up to order k are continuous. Since the inverse function theorem comes as a case of the implicit function theorem, this shows that the inverse function also inherits the property of being C^k. First some notation is convenient. Let $\alpha = (\alpha_1, \cdots, \alpha_n)$ where each α_i is a nonnegative integer. Then letting $|\alpha| = \sum_i \alpha_i$,

$$
D^\alpha \mathbf{f}(\mathbf{x}) \equiv \frac{\partial^{|\alpha|} \mathbf{f}}{\partial^{\alpha_1} \partial^{\alpha_2} \cdots \partial^{\alpha_n}}(\mathbf{x}), \quad D^0 \mathbf{f}(\mathbf{x}) \equiv \mathbf{f}(\mathbf{x})
$$

Theorem B.2. *(implicit function theorem) Suppose U is an open set in $\mathbb{F}^n \times \mathbb{F}^m$. Let $\mathbf{f} : U \to \mathbb{F}^n$ be in $C^k(U)$ and suppose*

$$
\mathbf{f}(\mathbf{x}_0, \mathbf{y}_0) = \mathbf{0}, \quad D_1\mathbf{f}(\mathbf{x}_0, \mathbf{y}_0)^{-1} \in \mathcal{L}(\mathbb{F}^n, \mathbb{F}^n). \tag{B.13}
$$

Then there exist positive constants δ, η, such that for every $\mathbf{y} \in B(\mathbf{y}_0, \eta)$ there exists a unique $\mathbf{x}(\mathbf{y}) \in B(\mathbf{x}_0, \delta)$ such that

$$
\mathbf{f}(\mathbf{x}(\mathbf{y}), \mathbf{y}) = \mathbf{0}. \tag{B.14}
$$

Furthermore, the mapping $\mathbf{y} \to \mathbf{x}(\mathbf{y})$ is in $C^k(B(\mathbf{y}_0, \eta))$.

Proof: From the implicit function theorem $\mathbf{y} \to \mathbf{x}(\mathbf{y})$ is C^1. It remains to show that it is C^k for $k > 1$ assuming that $\mathbf{f}$ is C^k. From (B.14)

$$
\frac{\partial \mathbf{x}}{\partial y^l} = -D_1\mathbf{f}(\mathbf{x}, \mathbf{y})^{-1} \frac{\partial \mathbf{f}}{\partial y^l}.
$$

Thus the following formula holds for $q = 1$ and $|\alpha| = q$.

$$D^{\alpha}\mathbf{x}(\mathbf{y}) = \sum_{|\beta| \le q} M_{\beta}(\mathbf{x}, \mathbf{y}) D^{\beta}\mathbf{f}(\mathbf{x}, \mathbf{y}) \qquad (B.15)$$

where M_{β} is a matrix whose entries are differentiable functions of $D^{\gamma}\mathbf{x}$ for $|\gamma| < q$ and $D^{\tau}\mathbf{f}(\mathbf{x}, \mathbf{y})$ for $|\tau| \le q$. This follows easily from the description of $D_1\mathbf{f}(\mathbf{x}, \mathbf{y})^{-1}$ in terms of the cofactor matrix and the determinant of $D_1\mathbf{f}(\mathbf{x}, \mathbf{y})$. Suppose (B.15) holds for $|\alpha| = q < k$. Then by induction, this yields $\mathbf{x}$ is C^q. Then

$$\frac{\partial D^{\alpha}\mathbf{x}(\mathbf{y})}{\partial y^p} = \sum_{|\beta| \le |\alpha|} \frac{\partial M_{\beta}(\mathbf{x}, \mathbf{y})}{\partial y^p} D^{\beta}\mathbf{f}(\mathbf{x}, \mathbf{y}) + M_{\beta}(\mathbf{x}, \mathbf{y}) \frac{\partial D^{\beta}\mathbf{f}(\mathbf{x}, \mathbf{y})}{\partial y^p}.$$

By the chain rule $\frac{\partial M_{\beta}(\mathbf{x}, \mathbf{y})}{\partial y^p}$ is a matrix whose entries are differentiable functions of $D^{\tau}\mathbf{f}(\mathbf{x}, \mathbf{y})$ for $|\tau| \le q + 1$ and $D^{\gamma}\mathbf{x}$ for $|\gamma| < q + 1$. It follows, since y^p was arbitrary, that for any $|\alpha| = q + 1$, a formula like (B.15) holds with q being replaced by $q + 1$. By induction, $\mathbf{x}$ is C^k. ∎

As a simple corollary, this yields the inverse function theorem. You just let $\mathbf{F}(\mathbf{x}, \mathbf{y}) = \mathbf{y} - \mathbf{f}(\mathbf{x})$ and apply the implicit function theorem.

Theorem B.3. *(inverse function theorem) Let $\mathbf{x}_0 \in U \subseteq \mathbb{F}^n$ and let $\mathbf{f} : U \to \mathbb{F}^n$. Suppose for k a positive integer,*

$$\mathbf{f} \text{ is } C^k(U), \quad \text{and} \quad D\mathbf{f}(\mathbf{x}_0)^{-1} \in \mathcal{L}(\mathbb{F}^n, \mathbb{F}^n). \qquad (B.16)$$

Then there exist open sets W, and V such that

$$\mathbf{x}_0 \in W \subseteq U, \qquad (B.17)$$

$$\mathbf{f} : W \to V \text{ is one to one and onto}, \qquad (B.18)$$

$$\mathbf{f}^{-1} \text{ is } C^k. \qquad (B.19)$$

B.3 The Method Of Lagrange Multipliers

As an application of the implicit function theorem, consider the method of Lagrange multipliers. Recall the problem is to maximize or minimize a function subject to equality constraints. Let $f : U \to \mathbb{R}$ be a C^1 function where $U \subseteq \mathbb{R}^n$ and let

$$g_i(\mathbf{x}) = 0, \ i = 1, \cdots, m \qquad (B.20)$$

be a collection of equality constraints with $m < n$. Now consider the system of nonlinear equations

$$\begin{aligned} f(\mathbf{x}) &= a \\ g_i(\mathbf{x}) &= 0, \ i = 1, \cdots, m. \end{aligned}$$

Recall $\mathbf{x}_0$ is a local maximum if $f(\mathbf{x}_0) \geq f(\mathbf{x})$ for all $\mathbf{x}$ near $\mathbf{x}_0$ which also satisfies the constraints (B.20). A local minimum is defined similarly. Let $\mathbf{F} : U \times \mathbb{R} \to \mathbb{R}^{m+1}$ be defined by

$$\mathbf{F}(\mathbf{x},a) \equiv \begin{pmatrix} f(\mathbf{x}) - a \\ g_1(\mathbf{x}) \\ \vdots \\ g_m(\mathbf{x}) \end{pmatrix}. \tag{B.21}$$

Now consider the $m+1 \times n$ matrix

$$\begin{pmatrix} f_{x_1}(\mathbf{x}_0) & \cdots & f_{x_n}(\mathbf{x}_0) \\ g_{1x_1}(\mathbf{x}_0) & \cdots & g_{1x_n}(\mathbf{x}_0) \\ \vdots & & \vdots \\ g_{mx_1}(\mathbf{x}_0) & \cdots & g_{mx_n}(\mathbf{x}_0) \end{pmatrix}.$$

If this matrix has rank $m+1$ then some $m+1 \times m+1$ submatrix has nonzero determinant. It follows from the implicit function theorem, there exists $m+1$ variables $x_{i_1}, \cdots, x_{i_{m+1}}$ such that the system

$$\mathbf{F}(\mathbf{x},a) = \mathbf{0} \tag{B.22}$$

specifies these $m+1$ variables as a function of the remaining $n - (m+1)$ variables and a in an open set of $\mathbb{R}^{n-m}$. Thus there is a solution $(\mathbf{x},a)$ to (B.22) for some $\mathbf{x}$ close to $\mathbf{x}_0$ whenever a is in some open interval. Therefore, $\mathbf{x}_0$ cannot be either a local minimum or a local maximum. It follows that if $\mathbf{x}_0$ is either a local maximum or a local minimum, then the above matrix must have rank less than $m+1$. By Theorem A.4, this implies that some row is a linear combination of the others. Thus there exist m scalars,

$$\lambda_1, \cdots, \lambda_m,$$

and a scalar μ, not all zero such that

$$\mu \begin{pmatrix} f_{x_1}(\mathbf{x}_0) \\ \vdots \\ f_{x_n}(\mathbf{x}_0) \end{pmatrix} = \lambda_1 \begin{pmatrix} g_{1x_1}(\mathbf{x}_0) \\ \vdots \\ g_{1x_n}(\mathbf{x}_0) \end{pmatrix} + \cdots + \lambda_m \begin{pmatrix} g_{mx_1}(\mathbf{x}_0) \\ \vdots \\ g_{mx_n}(\mathbf{x}_0) \end{pmatrix}. \tag{B.23}$$

If the rank of the matrix

$$\begin{pmatrix} g_{1x_1}(\mathbf{x}_0) & \cdots & g_{mx_1}(\mathbf{x}_0) \\ \vdots & & \vdots \\ g_{1x_n}(\mathbf{x}_0) & \cdots & g_{mx_n}(\mathbf{x}_0) \end{pmatrix} \tag{B.24}$$

is n, so that its determinant is nonzero, then $\mu \neq 0$, since this would require the above matrix to not be one to one, contrary to Corollary A.6. Dividing by μ yields an expression of the form

$$\begin{pmatrix} f_{x_1}(\mathbf{x}_0) \\ \vdots \\ f_{x_n}(\mathbf{x}_0) \end{pmatrix} = \lambda_1 \begin{pmatrix} g_{1x_1}(\mathbf{x}_0) \\ \vdots \\ g_{1x_n}(\mathbf{x}_0) \end{pmatrix} + \cdots + \lambda_m \begin{pmatrix} g_{mx_1}(\mathbf{x}_0) \\ \vdots \\ g_{mx_n}(\mathbf{x}_0) \end{pmatrix} \tag{B.25}$$

at every point $\mathbf{x}_0$ which is either a local maximum or a local minimum. This proves the following theorem.

Theorem B.4. *Let U be an open subset of $\mathbb{R}^n$ and let $f : U \to \mathbb{R}$ be a C^1 function. Then if $\mathbf{x}_0 \in U$ is either a local maximum or local minimum of f subject to the constraints (B.20), then (B.23) must hold for some scalars $\mu, \lambda_1, \cdots, \lambda_m$ not all equal to zero. If the determinant of the matrix in (B.24) equals 0, it follows (B.25) holds for some choice of the λ_i.*

B.4 The Local Structure Of C^1 Mappings

In linear algebra it is shown that every invertible matrix can be written as a product of elementary matrices, those matrices which are obtained from doing a row operation to the identity matrix. Two of the row operations produce a matrix which will change exactly one entry of a vector when it is multiplied by the elementary matrix. The other row operation involves switching two rows and this has the effect of switching two entries in a vector when multiplied on the left by the elementary matrix. Thus, in terms of the effect on a vector, the mapping determined by the given matrix can be considered as a composition of mappings which either flip two entries of the vector or change exactly one. A similar local result is available for nonlinear mappings. I found this interesting result in the advanced calculus book by Rudin.

Definition B.2. Let U be an open set in $\mathbb{R}^n$ and let $\mathbf{G} : U \to \mathbb{R}^n$. Then $\mathbf{G}$ is called primitive if it is of the form

$$\mathbf{G}(\mathbf{x}) = \left(\begin{array}{ccccc} x_1 & \cdots & \alpha(\mathbf{x}) & \cdots & x_n \end{array} \right)^T.$$

Thus, $\mathbf{G}$ is primitive if it only changes one of the variables. A function $\mathbf{F} : \mathbb{R}^n \to \mathbb{R}^n$ is called a flip if

$$\mathbf{F}(x_1, \cdots, x_k, \cdots, x_l, \cdots, x_n) = (x_1, \cdots, x_l, \cdots, x_k, \cdots, x_n)^T.$$

Thus a function is a flip if it interchanges two coordinates. Also, for $m = 1, 2, \cdots, n$, define

$$P_m(\mathbf{x}) \equiv \left(\begin{array}{cccccc} x_1 & x_2 & \cdots & x_m & 0 & \cdots & 0 \end{array} \right)^T$$

It turns out that if $\mathbf{h}(\mathbf{0}) = \mathbf{0}$, $D\mathbf{h}(\mathbf{0})^{-1}$ exists, and $\mathbf{h}$ is C^1 on U, then $\mathbf{h}$ can be written as a composition of primitive functions and flips. This is a very interesting application of the inverse function theorem.

Theorem B.5. *Let $\mathbf{h} : U \to \mathbb{R}^n$ be a C^1 function with $\mathbf{h}(\mathbf{0}) = \mathbf{0}$, $D\mathbf{h}(\mathbf{0})^{-1}$ exists. Then there is an open set $V \subseteq U$ containing $\mathbf{0}$, flips $\mathbf{F}_1, \cdots, \mathbf{F}_{n-1}$, and primitive functions $\mathbf{G}_n, \mathbf{G}_{n-1}, \cdots, \mathbf{G}_1$ such that for $\mathbf{x} \in V$,*

$$\mathbf{h}(\mathbf{x}) = \mathbf{F}_1 \circ \cdots \circ \mathbf{F}_{n-1} \circ \mathbf{G}_n \circ \mathbf{G}_{n-1} \circ \cdots \circ \mathbf{G}_1(\mathbf{x}).$$

The primitive function $\mathbf{G}_j$ leaves x_i unchanged for $i \neq j$.

Proof: Let

$$\mathbf{h}_1(\mathbf{x}) \equiv \mathbf{h}(\mathbf{x}) = \left(\begin{array}{ccc} \alpha_1(\mathbf{x}) & \cdots & \alpha_n(\mathbf{x}) \end{array}\right)^T$$

$$D\mathbf{h}(\mathbf{0})\,\mathbf{e}_1 = \left(\begin{array}{ccc} \alpha_{1,1}(\mathbf{0}) & \cdots & \alpha_{n,1}(\mathbf{0}) \end{array}\right)^T$$

where $\alpha_{k,1}$ denotes $\frac{\partial \alpha_k}{\partial x_1}$. Since $D\mathbf{h}(\mathbf{0})$ is one to one, the right side of this expression cannot be zero. Hence there exists some k such that $\alpha_{k,1}(\mathbf{0}) \neq 0$. Now define

$$\mathbf{G}_1(\mathbf{x}) \equiv \left(\begin{array}{cccc} \alpha_k(\mathbf{x}) & x_2 & \cdots & x_n \end{array}\right)^T$$

Then the matrix of $D\mathbf{G}_1(\mathbf{0})$ is of the form

$$\begin{pmatrix} \alpha_{k,1}(\mathbf{0}) & \cdots & \cdots & \alpha_{k,n}(\mathbf{0}) \\ 0 & 1 & & 0 \\ \vdots & & \ddots & \vdots \\ 0 & 0 & \cdots & 1 \end{pmatrix}$$

and its determinant equals $\alpha_{k,1}(\mathbf{0}) \neq 0$. Therefore, by the inverse function theorem, there exists an open set U_1, containing $\mathbf{0}$ and an open set V_2 containing $\mathbf{0}$ such that $\mathbf{G}_1(U_1) = V_2$ and $\mathbf{G}_1$ is one to one and onto, such that it and its inverse are both C^1. Let $\mathbf{F}_1$ denote the flip which interchanges x_k with x_1. Now define

$$\mathbf{h}_2(\mathbf{y}) \equiv \mathbf{F}_1 \circ \mathbf{h}_1 \circ \mathbf{G}_1^{-1}(\mathbf{y})$$

Thus

$$\begin{aligned} \mathbf{h}_2(\mathbf{G}_1(\mathbf{x})) &\equiv \mathbf{F}_1 \circ \mathbf{h}_1(\mathbf{x}) & \text{(B.26)} \\ &= \left(\begin{array}{ccccc} \alpha_k(\mathbf{x}) & \cdots & \alpha_1(\mathbf{x}) & \cdots & \alpha_n(\mathbf{x}) \end{array}\right)^T \end{aligned}$$

Therefore,

$$P_1\mathbf{h}_2(\mathbf{G}_1(\mathbf{x})) = \left(\begin{array}{cccc} \alpha_k(\mathbf{x}) & 0 & \cdots & 0 \end{array}\right)^T.$$

Also

$$P_1(\mathbf{G}_1(\mathbf{x})) = \left(\begin{array}{cccc} \alpha_k(\mathbf{x}) & 0 & \cdots & 0 \end{array}\right)^T$$

so $P_1\mathbf{h}_2(\mathbf{y}) = P_1(\mathbf{y})$ for all $\mathbf{y} \in V_2$. Also, $\mathbf{h}_2(\mathbf{0}) = \mathbf{0}$ and $D\mathbf{h}_2(\mathbf{0})^{-1}$ exists because of the definition of $\mathbf{h}_2$ above and the chain rule. Since $\mathbf{F}_1^2 = I$, the identity map, it follows from (B.26) that

$$\mathbf{h}(\mathbf{x}) = \mathbf{h}_1(\mathbf{x}) = \mathbf{F}_1 \circ \mathbf{h}_2 \circ \mathbf{G}_1(\mathbf{x}). \qquad \text{(B.27)}$$

Note that on an open set $V_2 \equiv \mathbf{G}_1(U_1)$ containing the origin, $\mathbf{h}_2$ leaves the first entry unchanged. This is what $P_1\mathbf{h}_2(\mathbf{G}_1(\mathbf{x})) = P_1(\mathbf{G}_1(\mathbf{x}))$ says. In contrast, $\mathbf{h}_1 = \mathbf{h}$ left possibly no entries unchanged.

Suppose then, that for $m \geq 2$, $\mathbf{h}_m$ leaves the first $m-1$ entries unchanged,

$$P_{m-1}\mathbf{h}_m(\mathbf{x}) = P_{m-1}(\mathbf{x}) \qquad \text{(B.28)}$$

for all $\mathbf{x} \in U_m$, an open subset of U containing $\mathbf{0}$, and $\mathbf{h}_m(\mathbf{0}) = \mathbf{0}$, $D\mathbf{h}_m(\mathbf{0})^{-1}$ exists. From (B.28), $\mathbf{h}_m(\mathbf{x})$ must be of the form

$$\mathbf{h}_m(\mathbf{x}) = \left(\begin{array}{ccccc} x_1 & \cdots & x_{m-1} & \alpha_1(\mathbf{x}) & \cdots & \alpha_n(\mathbf{x}) \end{array}\right)^T$$

where these α_k are different than the ones used earlier. Then

$$D\mathbf{h}_m(\mathbf{0})\,\mathbf{e}_m = \left(\begin{array}{ccccc} 0 & \cdots & 0 & \alpha_{1,m}(\mathbf{0}) & \cdots & \alpha_{n,m}(\mathbf{0}) \end{array}\right)^T \neq \mathbf{0}$$

because $D\mathbf{h}_m(\mathbf{0})^{-1}$ exists. Therefore, there exists a $k \geq m$ such that $\alpha_{k,m}(\mathbf{0}) \neq 0$, not the same k as before. Define

$$\mathbf{G}_m(\mathbf{x}) \equiv \left(\begin{array}{ccccc} x_1 & \cdots & x_{m-1} & \alpha_k(\mathbf{x}) & \cdots & x_n \end{array}\right)^T \qquad (\text{B.29})$$

so a change in $\mathbf{G}_m$ occurs only in the mth slot. Then $\mathbf{G}_m(\mathbf{0}) = \mathbf{0}$ and $D\mathbf{G}_m(\mathbf{0})^{-1}$ exists similar to the above. In fact

$$\det\left(D\mathbf{G}_m(\mathbf{0})\right) = \alpha_{k,m}(\mathbf{0}).$$

Therefore, by the inverse function theorem, there exists an open set V_{m+1} containing $\mathbf{0}$ such that $V_{m+1} = \mathbf{G}_m(U_m)$ with $\mathbf{G}_m$ and its inverse being one to one, continuous and onto. Let $\mathbf{F}_m$ be the flip which flips x_m and x_k. Then define $\mathbf{h}_{m+1}$ on V_{m+1} by

$$\mathbf{h}_{m+1}(\mathbf{y}) = \mathbf{F}_m \circ \mathbf{h}_m \circ \mathbf{G}_m^{-1}(\mathbf{y}).$$

Thus for $\mathbf{x} \in U_m$,

$$\mathbf{h}_{m+1}(\mathbf{G}_m(\mathbf{x})) = (\mathbf{F}_m \circ \mathbf{h}_m)(\mathbf{x}). \qquad (\text{B.30})$$

and consequently, since $\mathbf{F}_m^2 = I$,

$$\mathbf{F}_m \circ \mathbf{h}_{m+1} \circ \mathbf{G}_m(\mathbf{x}) = \mathbf{h}_m(\mathbf{x}) \qquad (\text{B.31})$$

It follows

$$\begin{aligned} P_m \mathbf{h}_{m+1}(\mathbf{G}_m(\mathbf{x})) &= P_m(\mathbf{F}_m \circ \mathbf{h}_m)(\mathbf{x}) \\ &= \left(\begin{array}{ccccc} x_1 & \cdots & x_{m-1} & \alpha_k(\mathbf{x}) & 0 & \cdots & 0 \end{array}\right)^T \end{aligned}$$

and

$$P_m(\mathbf{G}_m(\mathbf{x})) = \left(\begin{array}{ccccc} x_1 & \cdots & x_{m-1} & \alpha_k(\mathbf{x}) & 0 & \cdots & 0 \end{array}\right)^T.$$

Therefore, for $\mathbf{y} \in V_{m+1}$,

$$P_m \mathbf{h}_{m+1}(\mathbf{y}) = P_m(\mathbf{y}).$$

As before, $\mathbf{h}_{m+1}(\mathbf{0}) = \mathbf{0}$ and $D\mathbf{h}_{m+1}(\mathbf{0})^{-1}$ exists. Therefore, we can apply (B.31) repeatedly, obtaining the following:

$$\begin{aligned} \mathbf{h}(\mathbf{x}) &= \mathbf{F}_1 \circ \mathbf{h}_2 \circ \mathbf{G}_1(\mathbf{x}) \\ &= \mathbf{F}_1 \circ \mathbf{F}_2 \circ \mathbf{h}_3 \circ \mathbf{G}_2 \circ \mathbf{G}_1(\mathbf{x}) \\ &\;\;\vdots \\ &= \mathbf{F}_1 \circ \cdots \circ \mathbf{F}_{n-1} \circ \mathbf{h}_n \circ \mathbf{G}_{n-1} \circ \cdots \circ \mathbf{G}_1(\mathbf{x}) \end{aligned}$$

where $\mathbf{h}_n$ fixes the first $n-1$ entries,

$$P_{n-1}\mathbf{h}_n(\mathbf{x}) = P_{n-1}(\mathbf{x}) = \left(\begin{array}{ccc} x_1 & \cdots & x_{n-1} & 0 \end{array}\right)^T,$$

and so $\mathbf{h}_n(\mathbf{x})$ is a primitive mapping of the form

$$\mathbf{h}_n(\mathbf{x}) = \left(\begin{array}{ccc} x_1 & \cdots & x_{n-1} & \alpha(\mathbf{x}) \end{array}\right)^T.$$

Therefore, define the primitive function $\mathbf{G}_n(\mathbf{x})$ to equal $\mathbf{h}_n(\mathbf{x})$. ∎

Appendix C

The Theory Of The Riemann Integral*

Jordan the contented dragon

The definition of the Riemann integral of a function of n variables uses the following definition.

Definition C.1. For $i = 1, \cdots, n$, let $\{\alpha_k^i\}_{k=-\infty}^{\infty}$ be points on $\mathbb{R}$ which satisfy

$$\lim_{k \to \infty} \alpha_k^i = \infty, \quad \lim_{k \to -\infty} \alpha_k^i = -\infty, \quad \alpha_k^i < \alpha_{k+1}^i. \tag{C.1}$$

For such sequences, define a grid on $\mathbb{R}^n$ denoted by $\mathcal{G}$ or $\mathcal{F}$ as the collection of boxes of the form

$$Q = \prod_{i=1}^{n} \left[\alpha_{j_i}^i, \alpha_{j_i+1}^i \right]. \tag{C.2}$$

If $\mathcal{G}$ is a grid, $\mathcal{F}$ is called a refinement of $\mathcal{G}$ if every box of $\mathcal{G}$ is the union of boxes of $\mathcal{F}$.

Lemma C.1. If $\mathcal{G}$ and $\mathcal{F}$ are two grids, they have a common refinement, denoted here by $\mathcal{G} \vee \mathcal{F}$.

Proof: Let $\{\alpha_k^i\}_{k=-\infty}^{\infty}$ be the sequences used to construct $\mathcal{G}$ and let $\{\beta_k^i\}_{k=-\infty}^{\infty}$ be the sequence used to construct $\mathcal{F}$. Now let $\{\gamma_k^i\}_{k=-\infty}^{\infty}$ denote the union of $\{\alpha_k^i\}_{k=-\infty}^{\infty}$ and $\{\beta_k^i\}_{k=-\infty}^{\infty}$. It is necessary to show that for each i these points can be arranged in order. To do so, let $\gamma_0^i \equiv \alpha_0^i$. Now if

$$\gamma_{-j}^i, \cdots, \gamma_0^i, \cdots, \gamma_j^i$$

315

have been chosen such that they are in order and all distinct, let γ^i_{j+1} be the first element of

$$\{\alpha^i_k\}^\infty_{k=-\infty} \cup \{\beta^i_k\}^\infty_{k=-\infty} \tag{C.3}$$

which is larger than γ^i_j and let $\gamma^i_{-(j+1)}$ be the last element of (C.3) which is strictly smaller than γ^i_{-j}. The assumption (C.1) insures such a first and last element exists. Now let the grid $\mathcal{G} \vee \mathcal{F}$ consist of boxes of the form

$$Q \equiv \prod^n_{i=1} \left[\gamma^i_{j_i}, \gamma^i_{j_i+1}\right]. \qquad\blacksquare$$

The Riemann integral is only defined for functions f which are bounded and are equal to zero off some bounded set D. In what follows f will always be such a function.

Definition C.2. Let f be a bounded function which equals zero off a bounded set D, and let $\mathcal{G}$ be a grid. For $Q \in \mathcal{G}$, define

$$M_Q(f) \equiv \sup\{f(\mathbf{x}) : \mathbf{x} \in Q\}, \; m_Q(f) \equiv \inf\{f(\mathbf{x}) : \mathbf{x} \in Q\}. \tag{C.4}$$

Also define for Q a box, the volume of Q, denoted by $v(Q)$ by

$$v(Q) \equiv \prod^n_{i=1}(b_i - a_i), \; Q \equiv \prod^n_{i=1}[a_i, b_i].$$

Now define upper sums, $\mathcal{U}_\mathcal{G}(f)$ and lower sums, $\mathcal{L}_\mathcal{G}(f)$ with respect to the indicated grid, by the formulas

$$\mathcal{U}_\mathcal{G}(f) \equiv \sum_{Q \in \mathcal{G}} M_Q(f) v(Q), \; \mathcal{L}_\mathcal{G}(f) \equiv \sum_{Q \in \mathcal{G}} m_Q(f) v(Q).$$

A function of n variables is Riemann integrable when there is a unique number between all the upper and lower sums. This number is the value of the integral.

Note that in this definition, $M_Q(f) = m_Q(f) = 0$ for all but finitely many $Q \in \mathcal{G}$ so there are no convergence questions to be considered here.

Lemma C.2. *If $\mathcal{F}$ is a refinement of $\mathcal{G}$ then*

$$\mathcal{U}_\mathcal{G}(f) \geq \mathcal{U}_\mathcal{F}(f), \; \mathcal{L}_\mathcal{G}(f) \leq \mathcal{L}_\mathcal{F}(f).$$

Also if $\mathcal{F}$ and $\mathcal{G}$ are two grids,

$$\mathcal{L}_\mathcal{G}(f) \leq \mathcal{U}_\mathcal{F}(f).$$

Proof: For $P \in \mathcal{G}$ let $\widehat{P}$ denote the set

$$\{Q \in \mathcal{F} : Q \subseteq P\}.$$

Then $P = \cup \widehat{P}$ and

$$\mathcal{L}_{\mathcal{F}}(f) \equiv \sum_{Q \in \mathcal{F}} m_Q(f) v(Q) = \sum_{P \in \mathcal{G}} \sum_{Q \in \widehat{P}} m_Q(f) v(Q)$$

$$\geq \sum_{P \in \mathcal{G}} m_P(f) \sum_{Q \in \widehat{P}} v(Q) = \sum_{P \in \mathcal{G}} m_P(f) v(P) \equiv \mathcal{L}_{\mathcal{G}}(f).$$

Similarly, the other inequality for the upper sums is valid.

To verify the last assertion of the lemma, use Lemma C.1 to write

$$\mathcal{L}_{\mathcal{G}}(f) \leq \mathcal{L}_{\mathcal{G} \vee \mathcal{F}}(f) \leq \mathcal{U}_{\mathcal{G} \vee \mathcal{F}}(f) \leq \mathcal{U}_{\mathcal{F}}(f). \qquad \blacksquare$$

This lemma makes it possible to define the Riemann integral.

Definition C.3. Define an upper and a lower integral as follows.

$$\overline{I}(f) \equiv \inf \{\mathcal{U}_{\mathcal{G}}(f) : \mathcal{G} \text{ is a grid}\},$$

$$\underline{I}(f) \equiv \sup \{\mathcal{L}_{\mathcal{G}}(f) : \mathcal{G} \text{ is a grid}\}.$$

Lemma C.3. $\overline{I}(f) \geq \underline{I}(f)$.

Proof: From Lemma C.2 it follows for any two grids $\mathcal{G}$ and $\mathcal{F}$,

$$\mathcal{L}_{\mathcal{G}}(f) \leq \mathcal{U}_{\mathcal{F}}(f).$$

Therefore, taking the supremum for all grids on the left in this inequality,

$$\underline{I}(f) \leq \mathcal{U}_{\mathcal{F}}(f)$$

for all grids $\mathcal{F}$. Taking the infimum in this inequality, yields the conclusion of the lemma. $\blacksquare$

Definition C.4. A bounded function f which equals zero off a bounded set D, is said to be Riemann integrable, written as $f \in \mathcal{R}(\mathbb{R}^n)$ exactly when $\underline{I}(f) = \overline{I}(f)$. In this case define

$$\int f \, dV \equiv \int f \, dx = \overline{I}(f) = \underline{I}(f).$$

As in the case of integration of functions of one variable, one obtains the Riemann criterion which is stated as the following theorem.

Theorem C.1. *(Riemann criterion)* $f \in \mathcal{R}(\mathbb{R}^n)$ *if and only if for all $\varepsilon > 0$ there exists a grid $\mathcal{G}$ such that*

$$\mathcal{U}_{\mathcal{G}}(f) - \mathcal{L}_{\mathcal{G}}(f) < \varepsilon.$$

Proof: If $f \in \mathcal{R}(\mathbb{R}^n)$, then $\overline{I}(f) = \underline{I}(f)$ and so there exist grids $\mathcal{G}$ and $\mathcal{F}$ such that

$$\mathcal{U}_{\mathcal{G}}(f) - \mathcal{L}_{\mathcal{F}}(f) \leq \overline{I}(f) + \frac{\varepsilon}{2} - \left(\underline{I}(f) - \frac{\varepsilon}{2}\right) = \varepsilon.$$

Then letting $\mathcal{H} = \mathcal{G} \vee \mathcal{F}$, Lemma C.2 implies

$$\mathcal{U}_{\mathcal{H}}(f) - \mathcal{L}_{\mathcal{H}}(f) \leq \mathcal{U}_{\mathcal{G}}(f) - \mathcal{L}_{\mathcal{F}}(f) < \varepsilon.$$

Conversely, if for all $\varepsilon > 0$ there exists $\mathcal{G}$ such that

$$\mathcal{U}_{\mathcal{G}}(f) - \mathcal{L}_{\mathcal{G}}(f) < \varepsilon,$$

then

$$\overline{I}(f) - \underline{I}(f) \leq \mathcal{U}_{\mathcal{G}}(f) - \mathcal{L}_{\mathcal{G}}(f) < \varepsilon.$$

Since $\varepsilon > 0$ is arbitrary, this proves the theorem. ∎

C.1 Basic Properties

It is important to know that certain combinations of Riemann integrable functions are Riemann integrable. The following theorem will include all the important cases.

Theorem C.2. *Let $f, g \in \mathcal{R}(\mathbb{R}^n)$ and let $\phi : K \to \mathbb{R}$ be continuous where K is a compact set in $\mathbb{R}^2$ containing $f(\mathbb{R}^n) \times g(\mathbb{R}^n)$. Also suppose that $\phi(0, 0) = 0$. Then defining*

$$h(\mathbf{x}) \equiv \phi(f(\mathbf{x}), g(\mathbf{x})),$$

it follows that h is also in $\mathcal{R}(\mathbb{R}^n)$.

Proof: Let $\varepsilon > 0$ and let $\delta_1 > 0$ be such that if $(y_i, z_i), i = 1, 2$ are points in K, such that $|z_1 - z_2| \leq \delta_1$ and $|y_1 - y_2| \leq \delta_1$, then

$$|\phi(y_1, z_1) - \phi(y_2, z_2)| < \varepsilon.$$

Let $0 < \delta < \min(\delta_1, \varepsilon, 1)$. Let $\mathcal{G}$ be a grid with the property that for $Q \in \mathcal{G}$, the diameter of Q is less than δ and also for $k = f, g$,

$$\mathcal{U}_{\mathcal{G}}(k) - \mathcal{L}_{\mathcal{G}}(k) < \delta^2. \tag{C.5}$$

Then defining for $k = f, g$,

$$\mathcal{P}_k \equiv \{Q \in \mathcal{G} : M_Q(k) - m_Q(k) > \delta\},$$

it follows

$$\delta^2 > \sum_{Q \in \mathcal{G}} (M_Q(k) - m_Q(k)) v(Q) \geq$$

$$\sum_{\mathcal{P}_k} (M_Q(k) - m_Q(k)) v(Q) \geq \delta \sum_{\mathcal{P}_k} v(Q)$$

and so for $k = f, g$,

$$\varepsilon > \delta > \sum_{\mathcal{P}_k} v(Q). \tag{C.6}$$

Suppose for $k = f, g$,

$$M_Q(k) - m_Q(k) \le \delta.$$

Then if $\mathbf{x}_1, \mathbf{x}_2 \in Q$,

$$|f(\mathbf{x}_1) - f(\mathbf{x}_2)| < \delta, \text{ and } |g(\mathbf{x}_1) - g(\mathbf{x}_2)| < \delta.$$

Therefore,

$$|h(\mathbf{x}_1) - h(\mathbf{x}_2)| \equiv |\phi(f(\mathbf{x}_1), g(\mathbf{x}_1)) - \phi(f(\mathbf{x}_2), g(\mathbf{x}_2))| < \varepsilon$$

and it follows that

$$|M_Q(h) - m_Q(h)| \le \varepsilon.$$

Now let

$$\mathcal{S} \equiv \{Q \in \mathcal{G} : 0 < M_Q(k) - m_Q(k) \le \delta, \ k = f, g\}.$$

Thus the union of the boxes in $\mathcal{S}$ is contained in some large box, R, which depends only on f and g and also, from the assumption that $\phi(0, 0) = 0$, $M_Q(h) - m_Q(h) = 0$, unless $Q \subseteq R$. Then

$$\mathcal{U}_{\mathcal{G}}(h) - \mathcal{L}_{\mathcal{G}}(h) \le \sum_{Q \in \mathcal{P}_f} (M_Q(h) - m_Q(h)) v(Q) +$$

$$\sum_{Q \in \mathcal{P}_g} (M_Q(h) - m_Q(h)) v(Q) + \sum_{Q \in \mathcal{S}} \delta v(Q).$$

Now since K is compact, it follows $\phi(K)$ is bounded and so there exists a constant C, depending only on h and ϕ such that $M_Q(h) - m_Q(h) < C$. Therefore, the above inequality implies

$$\mathcal{U}_{\mathcal{G}}(h) - \mathcal{L}_{\mathcal{G}}(h) \le C \sum_{Q \in \mathcal{P}_f} v(Q) + C \sum_{Q \in \mathcal{P}_g} v(Q) + \sum_{Q \in \mathcal{S}} \delta v(Q),$$

which by (C.6) implies

$$\mathcal{U}_{\mathcal{G}}(h) - \mathcal{L}_{\mathcal{G}}(h) \le 2C\varepsilon + \delta v(R) \le 2C\varepsilon + \varepsilon v(R).$$

Since ε is arbitrary, the Riemann criterion is satisfied and so $h \in \mathcal{R}(\mathbb{R}^n)$. ∎

Corollary C.1. *Let $f, g \in \mathcal{R}(\mathbb{R}^n)$ and let $a, b \in \mathbb{R}$. Then $af + bg$, fg, and $|f|$ are all in $\mathcal{R}(\mathbb{R}^n)$. Also,*

$$\int_{\mathbb{R}^n} (af + bg) \, dx = a \int_{\mathbb{R}^n} f \, dx + b \int_{\mathbb{R}^n} g \, dx, \tag{C.7}$$

and

$$\int |f| \, dx \ge \left| \int f \, dx \right|. \tag{C.8}$$

Proof: Each of the combinations of functions described above is Riemann integrable by Theorem C.2. For example, to see $af + bg \in \mathcal{R}(\mathbb{R}^n)$ consider $\phi(y, z) \equiv ay + bz$. This is clearly a continuous function of (y, z) such that $\phi(0, 0) = 0$. To obtain $|f| \in \mathcal{R}(\mathbb{R}^n)$, let $\phi(y, z) \equiv |y|$. It remains to verify the formulas. To do so, let $\mathcal{G}$ be a grid with the property that for $k = f, g, |f|$ and $af + bg$,

$$\mathcal{U}_{\mathcal{G}}(k) - \mathcal{L}_{\mathcal{G}}(k) < \varepsilon. \tag{C.9}$$

Consider (C.7). For each $Q \in \mathcal{G}$ pick a point in Q, $\mathbf{x}_Q$. Then

$$\sum_{Q \in \mathcal{G}} k(\mathbf{x}_Q) v(Q) \in [\mathcal{L}_{\mathcal{G}}(k), \mathcal{U}_{\mathcal{G}}(k)]$$

and so

$$\left| \int k \, dx - \sum_{Q \in \mathcal{G}} k(\mathbf{x}_Q) v(Q) \right| < \varepsilon.$$

Consequently, since

$$\sum_{Q \in \mathcal{G}} (af + bg)(\mathbf{x}_Q) v(Q)$$

$$= a \sum_{Q \in \mathcal{G}} f(\mathbf{x}_Q) v(Q) + b \sum_{Q \in \mathcal{G}} g(\mathbf{x}_Q) v(Q),$$

it follows

$$\left| \int (af + bg) \, dx - a \int f \, dx - b \int g \, dx \right| \leq$$

$$\left| \int (af + bg) \, dx - \sum_{Q \in \mathcal{G}} (af + bg)(\mathbf{x}_Q) v(Q) \right| +$$

$$\left| a \sum_{Q \in \mathcal{G}} f(\mathbf{x}_Q) v(Q) - a \int f \, dx \right| + \left| b \sum_{Q \in \mathcal{G}} g(\mathbf{x}_Q) v(Q) - b \int g \, dx \right|$$

$$\leq \varepsilon + |a| \varepsilon + |b| \varepsilon.$$

Since ε is arbitrary, this establishes (C.7) and shows the integral is linear.

It remains to establish the inequality (C.8). By (C.9), and the triangle inequality for sums,

$$\int |f| \, dx + \varepsilon \geq \sum_{Q \in \mathcal{G}} |f(\mathbf{x}_Q)| v(Q)$$

$$\geq \left| \sum_{Q \in \mathcal{G}} f(\mathbf{x}_Q) v(Q) \right| \geq \left| \int f \, dx \right| - \varepsilon.$$

Then since ε is arbitrary, this establishes the desired inequality. ∎

C.2   Which Functions Are Integrable?

Which functions are in $\mathcal{R}(\mathbb{R}^n)$? As in the case of integrals of functions of one variable, this is an important question. It turns out the Riemann integrable functions are characterized by being continuous except on a very small set. This has to do with Jordan content.

Definition C.5. A bounded set E, has Jordan content 0 or content 0 if for every $\varepsilon > 0$ there exists a grid $\mathcal{G}$ such that

$$\sum_{Q \cap E \neq \emptyset} v(Q) < \varepsilon.$$

This symbol says to sum the volumes of all boxes from $\mathcal{G}$ which have nonempty intersection with E.

Next it is necessary to define the oscillation of a function.

Definition C.6. Let f be a function defined on $\mathbb{R}^n$ and let

$$\omega_{f,r}(\mathbf{x}) \equiv \sup \{|f(\mathbf{z}) - f(\mathbf{y})| : \mathbf{z}, \mathbf{y} \in B(\mathbf{x},r)\}.$$

This is called the oscillation of f on $B(\mathbf{x},r)$. Note that this function of r is decreasing in r. Define the oscillation of f as

$$\omega_f(\mathbf{x}) \equiv \lim_{r \to 0+} \omega_{f,r}(\mathbf{x}).$$

Note that as r decreases, the function $\omega_{f,r}(\mathbf{x})$ decreases. It is also bounded below by 0, so the limit must exist and equals $\inf \{\omega_{f,r}(\mathbf{x}) : r > 0\}$. (Why?) Then the following simple lemma whose proof follows directly from the definition of continuity gives the reason for this definition.

Lemma C.4. *A function f is continuous at $\mathbf{x}$ if and only if $\omega_f(\mathbf{x}) = 0$.*

This concept of oscillation gives a way to define how discontinuous a function is at a point. The discussion will depend on the following fundamental lemma which gives the existence of something called the Lebesgue number.

Definition C.7. Let $\mathfrak{C}$ be a set whose elements are sets of $\mathbb{R}^n$ and let $K \subseteq \mathbb{R}^n$. The set $\mathfrak{C}$ is called a cover of K if every point of K is contained in some set of $\mathfrak{C}$. If the elements of $\mathfrak{C}$ are open sets, it is called an open cover.

Lemma C.5. *Let K be sequentially compact and let $\mathfrak{C}$ be an open cover of K. Then there exists $r > 0$ such that whenever $\mathbf{x} \in K$, $B(\mathbf{x}, r)$ is contained in some set of $\mathfrak{C}$.*

Proof: Suppose this is not so. Then letting $r_n = 1/n$, there exists $\mathbf{x}_n \in K$ such that $B(\mathbf{x}_n, r_n)$ is not contained in any set of $\mathfrak{C}$. Since K is sequentially compact, there is a subsequence, $\mathbf{x}_{n_k}$ which converges to a point $\mathbf{x} \in K$. But there exists

$\delta > 0$ such that $B(\mathbf{x}, \delta) \subseteq U$ for some $U \in \mathfrak{C}$. Let k be so large that $1/k < \delta/2$ and $|\mathbf{x}_{n_k} - \mathbf{x}| < \delta/2$ also. Then if $\mathbf{z} \in B(\mathbf{x}_{n_k}, r_{n_k})$, it follows

$$|\mathbf{z} - \mathbf{x}| \le |\mathbf{z} - \mathbf{x}_{n_k}| + |\mathbf{x}_{n_k} - \mathbf{x}| < \frac{\delta}{2} + \frac{\delta}{2} = \delta$$

and so $B(\mathbf{x}_{n_k}, r_{n_k}) \subseteq U$ contrary to supposition. Therefore, the desired number exists after all. ∎

Theorem C.3. *Let f be a bounded function which equals zero off a bounded set and let W denote the set of points where f fails to be continuous. Then $f \in \mathcal{R}(\mathbb{R}^n)$ if W has content zero. That is, for all $\varepsilon > 0$ there exists a grid $\mathcal{G}$ such that*

$$\sum_{Q \in \mathcal{G}_W} v(Q) < \varepsilon \qquad (C.10)$$

where

$$\mathcal{G}_W \equiv \{Q \in \mathcal{G} : Q \cap W \ne \emptyset\}.$$

Proof: Let W have content zero. Also let $|f(\mathbf{x})| < C/2$ for all $\mathbf{x} \in \mathbb{R}^n$, let $\varepsilon > 0$ be given, and let $\mathcal{G}$ be a grid which satisfies (C.10). Since f equals zero off some bounded set, there exists R such that f equals zero off of $B\left(\mathbf{0}, \frac{R}{2}\right)$. Thus $W \subseteq B\left(\mathbf{0}, \frac{R}{2}\right)$. Also note that if $\mathcal{G}$ is a grid for which (C.10) holds, then this inequality continues to hold if $\mathcal{G}$ is replaced with a refined grid. Therefore, you may assume the diameter of every box in $\mathcal{G}$ which intersects $B(\mathbf{0}, R)$ is less than $\frac{R}{3}$ and so all boxes of $\mathcal{G}$ which intersect the set where f is nonzero are contained in $B(\mathbf{0}, R)$. Since W is bounded, $\mathcal{G}_W$ contains only finitely many boxes. Letting

$$Q \equiv \prod_{i=1}^{n} [a_i, b_i]$$

be one of these boxes, enlarge the box slightly as indicated in the following picture.

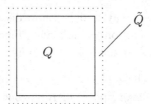

The enlarged box is an open set of the form,

$$\tilde{Q} \equiv \prod_{i=1}^{n} (a_i - \eta_i, b_i + \eta_i)$$

where η_i is chosen small enough that if

$$\prod_{i=1}^{n} (b_i + \eta_i - (a_i - \eta_i)) \equiv v\left(\tilde{Q}\right),$$

and $\widetilde{\mathcal{G}_W}$ denotes those $\tilde{Q}$ for $Q \in \mathcal{G}$ which have nonempty intersection with W, then

$$\sum_{\tilde{Q} \in \widetilde{\mathcal{G}_W}} v\left(\tilde{\tilde{Q}}\right) < \varepsilon \tag{C.11}$$

where $\tilde{\tilde{Q}}$ is the box,

$$\prod_{i=1}^{n} \left((a_i - 2\eta_i)\,,\ b_i + 2\eta_i\right).$$

For each $\mathbf{x} \in \mathbb{R}^n$, let $r_\mathbf{x} < \min\left(\eta_1/2, \cdots, \eta_n/2\right)$ be such that

$$\omega_{f,r_\mathbf{x}}\left(\mathbf{x}\right) < \varepsilon + \omega_f\left(\mathbf{x}\right). \tag{C.12}$$

Now let $\mathfrak{C}$ denote all intersections of the form $\tilde{Q} \cap B\left(\mathbf{x}, r_\mathbf{x}\right)$ such that $\mathbf{x} \in \overline{B\left(\mathbf{0}, R\right)}$ so that $\mathfrak{C}$ is an open cover of the compact set $\overline{B\left(\mathbf{0}, R\right)}$. Let δ be a Lebesgue number for this open cover of $\overline{B\left(\mathbf{0}, R\right)}$ and let $\mathcal{F}$ be a refinement of $\mathcal{G}$ such that every box in $\mathcal{F}$ has diameter less than δ. Now let $\mathcal{F}_1$ consist of those boxes of $\mathcal{F}$ which have nonempty intersection with $B\left(\mathbf{0}, R/2\right)$. Thus all boxes of $\mathcal{F}_1$ are contained in $B\left(\mathbf{0}, R\right)$ and each one is contained in some set of $\mathfrak{C}$. Let $\mathfrak{C}_W$ be those open sets of $\mathfrak{C}$, $\tilde{Q} \cap B\left(\mathbf{x}, r_\mathbf{x}\right)$, for which $\mathbf{x} \in W$. Thus each of these sets is contained in some $\tilde{Q}$ where $Q \in \mathcal{G}_W$. Let $\mathcal{F}_W$ be those sets of $\mathcal{F}_1$ which are subsets of some set of $\mathfrak{C}_W$. Thus

$$\sum_{Q \in \mathcal{F}_W} v\left(Q\right) < \varepsilon. \tag{C.13}$$

because each Q in $\mathcal{F}_W$ is contained in a set $\tilde{\tilde{Q}}$ described above and the sum of the volumes of these is less than ε by (C.11). Then

$$\mathcal{U}_\mathcal{F}\left(f\right) - \mathcal{L}_\mathcal{F}\left(f\right) = \sum_{Q \in \mathcal{F}_W} \left(M_Q\left(f\right) - m_Q\left(f\right)\right) v\left(Q\right)$$

$$+ \sum_{Q \in \mathcal{F}_1 \setminus \mathcal{F}_W} \left(M_Q\left(f\right) - m_Q\left(f\right)\right) v\left(Q\right).$$

If $Q \in \mathcal{F}_1 \setminus \mathcal{F}_W$, then Q must be a subset of some set of $\mathfrak{C} \setminus \mathfrak{C}_W$ since it is not in any set of $\mathfrak{C}_W$. Say $Q \subseteq \tilde{Q_1} \cap B\left(\mathbf{x}, r_\mathbf{x}\right)$ where $\mathbf{x} \notin W$. Therefore, from (C.12) and the observation that $\mathbf{x} \notin W$, it follows $\omega_f\left(\mathbf{x}\right) = 0$ and so

$$M_Q\left(f\right) - m_Q\left(f\right) \leq \varepsilon.$$

Therefore, from (C.13) and the estimate on f,

$$\mathcal{U}_\mathcal{F}\left(f\right) - \mathcal{L}_\mathcal{F}\left(f\right) \leq \sum_{Q \in \mathcal{F}_W} C v\left(Q\right) + \sum_{Q \in \mathcal{F}_1 \setminus \mathcal{F}_W} \varepsilon v\left(Q\right)$$

$$\leq C\varepsilon + \varepsilon \left(2R\right)^n,$$

the estimate of the second sum coming from the fact that

$$B\left(\mathbf{0}, R\right) \subseteq \prod_{i=1}^{n} [-R, R].$$

Since ε is arbitrary, this proves the theorem.[2] ∎

Definition C.8. A bounded set E is a Jordan set in $\mathbb{R}^n$, also called a contented set in $\mathbb{R}^n$ if $\mathcal{X}_E \in \mathcal{R}\left(\mathbb{R}^n\right)$. The symbol $\mathcal{X}_E$ means

$$\mathcal{X}_E\left(\mathbf{x}\right) = \left\{ \begin{array}{l} 1 \text{ if } \mathbf{x} \in E \\ 0 \text{ if } \mathbf{x} \notin E \end{array} \right.$$

It is called the indicator function because it indicates whether $\mathbf{x}$ is in E according to whether it equals 1. For a function $f \in \mathcal{R}\left(\mathbb{R}^n\right)$ and E a contented set, $f\mathcal{X}_E \in \mathcal{R}\left(\mathbb{R}^n\right)$ by Corollary C.1. Then

$$\int_E f dV \equiv \int f \mathcal{X}_E dV.$$

So what are examples of contented sets?

Theorem C.4. *Suppose E is a bounded contented set in $\mathbb{R}^n$ and $f, g : E \to \mathbb{R}$ are two functions satisfying $f\left(\mathbf{x}\right) \geq g\left(\mathbf{x}\right)$ for all $\mathbf{x} \in E$ and $f\mathcal{X}_E$ and $g\mathcal{X}_E$ are both in $\mathcal{R}\left(\mathbb{R}^n\right)$. Now define*

$$P \equiv \{(\mathbf{x}, x_{n+1}) : \mathbf{x} \in E \text{ and } g\left(\mathbf{x}\right) \leq x_{n+1} \leq f\left(\mathbf{x}\right)\}.$$

Then P is a contented set in $\mathbb{R}^{n+1}$.

Proof: Let $\mathcal{G}$ be a grid such that for $k = f\mathcal{X}_E, g\mathcal{X}_E$,

$$\mathcal{U}_{\mathcal{G}}\left(k\right) - \mathcal{L}_{\mathcal{G}}\left(k\right) < \varepsilon/4. \tag{C.14}$$

Also let $K \geq \sum_{j=1}^{m} v_n\left(Q_j\right)$ where the Q_j are the boxes which intersect E. Let $\{a_i\}_{i=-\infty}^{\infty}$ be a sequence on $\mathbb{R}$, $a_i < a_{i+1}$ for all i, which includes

$$M_{Q_j}\left(f\mathcal{X}_E\right) + \frac{\varepsilon}{4mK}, M_{Q_j}\left(f\mathcal{X}_E\right), M_{Q_j}\left(g\mathcal{X}_E\right),$$

$$m_{Q_j}\left(f\mathcal{X}_E\right), m_{Q_j}\left(g\mathcal{X}_E\right), m_{Q_j}\left(g\mathcal{X}_E\right) - \frac{\varepsilon}{4mK}$$

for all $j = 1, \cdots, m$. Now define a grid on $\mathbb{R}^{n+1}$ as follows.

$$\mathcal{G}' \equiv \{Q \times [a_i, a_{i+1}] : Q \in \mathcal{G}, i \in \mathbb{Z}\}$$

In words, this grid consists of all possible boxes of the form $Q \times [a_i, a_{i+1}]$ where $Q \in \mathcal{G}$ and a_i is a term of the sequence just described. It is necessary to verify that for $P \in \mathcal{G}'$, $\mathcal{X}_P \in \mathcal{R}\left(\mathbb{R}^{n+1}\right)$. This is done by showing that $\mathcal{U}_{\mathcal{G}'}\left(\mathcal{X}_P\right) - \mathcal{L}_{\mathcal{G}'}\left(\mathcal{X}_P\right) < \varepsilon$

---

[2]In fact one cannot do any better. It can be shown that if a function is Riemann integrable, then it must be the case that for all $\varepsilon > 0$, (C.10) is satisfied for some grid $\mathcal{G}$. This along with what was just shown is known as Lebesgue's theorem after Lebesgue who discovered it in the early years of the twentieth century. Actually, he also invented a far superior integral which made the Riemann integral which is the topic of this appendix obsolete.

and then noting that $\varepsilon > 0$ was arbitrary. For $\mathcal{G}'$ just described, denote by Q' a box in $\mathcal{G}'$. Thus $Q' = Q \times [a_i, a_{i+1}]$ for some i.

$$\mathcal{U}_{\mathcal{G}'}(\mathcal{X}_P) - \mathcal{L}_{\mathcal{G}'}(\mathcal{X}_P) \equiv \sum_{Q' \in \mathcal{G}'} (M_{Q'}(\mathcal{X}_P) - m_{Q'}(\mathcal{X}_P)) v_{n+1}(Q')$$

$$= \sum_{i=-\infty}^{\infty} \sum_{j=1}^{m} \left(M_{Q'_j}(\mathcal{X}_P) - m_{Q'_j}(\mathcal{X}_P) \right) v_n(Q_j)(a_{i+1} - a_i)$$

and all sums are bounded because the functions f and g are given to be bounded. Therefore, there are no limit considerations needed here. Thus

$$\mathcal{U}_{\mathcal{G}'}(\mathcal{X}_P) - \mathcal{L}_{\mathcal{G}'}(\mathcal{X}_P) =$$

$$\sum_{j=1}^{m} v_n(Q_j) \sum_{i=-\infty}^{\infty} \left(M_{Q_j \times [a_i, a_{i+1}]}(\mathcal{X}_P) - m_{Q_j \times [a_i, a_{i+1}]}(\mathcal{X}_P) \right) (a_{i+1} - a_i).$$

Consider the inside sum with the aid of the following picture.

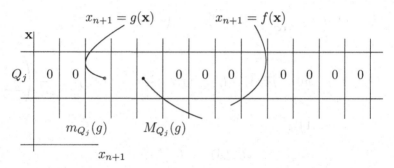

In this picture, the little rectangles represent the boxes $Q_j \times [a_i, a_{i+1}]$ for fixed j. The part of P having $\mathbf{x}$ contained in Q_j is between the two surfaces, $x_{n+1} = g(\mathbf{x})$ and $x_{n+1} = f(\mathbf{x})$ and there is a zero placed in those boxes for which

$$M_{Q_j \times [a_i, a_{i+1}]}(\mathcal{X}_P) - m_{Q_j \times [a_i, a_{i+1}]}(\mathcal{X}_P) = 0.$$

You see, $\mathcal{X}_P$ has either the value of 1 or the value of 0 depending on whether $(\mathbf{x}, y)$ is contained in P. For the boxes shown with 0 in them, either all of the box is contained in P or none of the box is contained in P. Either way,

$$M_{Q_j \times [a_i, a_{i+1}]}(\mathcal{X}_P) - m_{Q_j \times [a_i, a_{i+1}]}(\mathcal{X}_P) = 0$$

on these boxes. However, on the boxes intersected by the surfaces, the value of

$$M_{Q_j \times [a_i, a_{i+1}]}(\mathcal{X}_P) - m_{Q_j \times [a_i, a_{i+1}]}(\mathcal{X}_P)$$

is 1 because there are points in this box which are not in P as well as points which are in P. Because of the construction of $\mathcal{G}'$ which included all values of

$$M_{Q_j}(f\mathcal{X}_E) + \frac{\varepsilon}{4mK}, M_{Q_j}(f\mathcal{X}_E),$$

$$M_{Q_j}(g\mathcal{X}_E), m_{Q_j}(f\mathcal{X}_E), m_{Q_j}(g\mathcal{X}_E)$$

for all $j = 1, \cdots, m$,

$$\sum_{i=-\infty}^{\infty} \left(M_{Q_j \times [a_i, a_{i+1}]} \left(\mathcal{X}_P \right) - m_{Q_j \times [a_i, a_{i+1}]} \left(\mathcal{X}_P \right) \right) \left(a_{i+1} - a_i \right) \leq$$

$$\sum_{\{i : m_{Q_j}(g\mathcal{X}_E) \leq a_i < M_{Q_j}(g\mathcal{X}_E)\}} 1 \left(a_{i+1} - a_i \right) + \sum_{\{i : m_{Q_j}(f\mathcal{X}_E) \leq a_i < M_{Q_j}(f\mathcal{X}_E)\}} 1 \left(a_{i+1} - a_i \right)$$

$$\text{(C.15)}$$

The first of the sums in (C.15) contains all possible terms for which

$$M_{Q_j \times [a_i, a_{i+1}]} \left(\mathcal{X}_P \right) - m_{Q_j \times [a_i, a_{i+1}]} \left(\mathcal{X}_P \right)$$

might be 1 due to the graph of the bottom surface $g\mathcal{X}_E$ while the second sum contains all possible terms for which the expression might be 1 due to the graph of the top surface $f\mathcal{X}_E$.

$$\leq \left(M_{Q_j} \left(g\mathcal{X}_E \right) + \frac{\varepsilon}{4mK} - m_{Q_j} \left(g\mathcal{X}_E \right) \right) + \left(M_{Q_j} \left(f\mathcal{X}_E \right) + \frac{\varepsilon}{4mK} - m_{Q_j} \left(f\mathcal{X}_E \right) \right)$$

$$= \left(M_{Q_j} \left(g\mathcal{X}_E \right) - m_{Q_j} \left(g\mathcal{X}_E \right) \right) + \left(M_{Q_j} \left(f\mathcal{X}_E \right) - m_{Q_j} \left(f\mathcal{X}_E \right) \right) + \frac{\varepsilon}{2m} \left(\sum_{j=1}^{m} v \left(Q_j \right) \right)^{-1}.$$

Therefore, by (C.14),

$$\mathcal{U}_{g'} \left(\mathcal{X}_P \right) - \mathcal{L}_{g'} \left(\mathcal{X}_P \right) \leq$$

$$\sum_{j=1}^{m} v_n \left(Q_j \right) \left[\left(M_{Q_j} \left(g\mathcal{X}_E \right) - m_{Q_j} \left(g\mathcal{X}_E \right) \right) + \left(M_{Q_j} \left(f\mathcal{X}_E \right) - m_{Q_j} \left(f\mathcal{X}_E \right) \right) \right]$$

$$+ \sum_{j=1}^{m} v \left(Q_j \right) \frac{\varepsilon}{2m} \left(\sum_{j=1}^{m} v \left(Q_j \right) \right)^{-1}$$

$$= \mathcal{U}_{\mathcal{G}} \left(f \right) - \mathcal{L}_{\mathcal{G}} \left(f \right) + \mathcal{U}_{\mathcal{G}} \left(g \right) - \mathcal{L}_{\mathcal{G}} \left(g \right) + \frac{\varepsilon}{2}$$

$$< \frac{\varepsilon}{4} + \frac{\varepsilon}{4} + \frac{\varepsilon}{2} = \varepsilon.$$

Since $\varepsilon > 0$ is arbitrary, this proves the theorem. ∎

Corollary C.2. *Suppose f and g are continuous functions defined on E, a contented set in $\mathbb{R}^n$ and that $g(\mathbf{x}) \leq f(\mathbf{x})$ for all $\mathbf{x} \in E$. Then*

$$P \equiv \{(\mathbf{x}, x_{n+1}) : \mathbf{x} \in E \text{ and } g(\mathbf{x}) \leq x_{n+1} \leq f(\mathbf{x})\}$$

is a contented set in $\mathbb{R}^n$.

Proof: Since E is contented, meaning $\mathcal{X}_E$ is integrable, it follows from Theorem C.3 the set of discontinuities of $\mathcal{X}_E$ has Jordan content 0. But the set of discontinuities of $\mathcal{X}_E$ is ∂E defined as those points $\mathbf{x}$ such that $B(\mathbf{x}, r)$ contains points of E and points of E^C for every $r > 0$. Extend f and g to equal 0 off E. Then the set of discontinuities of these extended functions still denoted as f, g is ∂E which has Jordan content 0. This reduces to the situation of Theorem C.4. ∎

As an example of how this can be applied, it is obvious a closed interval is a contented set in $\mathbb{R}$. Therefore, if f, g are two continuous functions with $f(x) \geq g(x)$ for $x \in [a, b]$, it follows from the above theorem or its corollary that the set

$$P_1 \equiv \{(x, y) : g(x) \leq y \leq f(x)\}$$

is a contented set in $\mathbb{R}^2$. Now using the theorem and corollary again, suppose $f_1(x, y) \geq g_1(x, y)$ for $(x, y) \in P_1$ and f, g are continuous. Then the set

$$P_2 \equiv \{(x, y, z) : g_1(x, y) \leq z \leq f_1(x, y)\}$$

is a contented set in $\mathbb{R}^3$. Clearly you can continue this way obtaining examples of contented sets. ∎

Note that as a special case, it follows that every box is a contented set. Therefore, if B_i is a box, functions of the form

$$\sum_{i=1}^{m} a_i \mathcal{X}_{B_i}$$

are integrable. These functions are called step functions.

The following theorem is analogous to the fact that in one dimension, when you integrate over a point, the answer is 0.

Theorem C.5. *If a bounded set E, has Jordan content 0, then E is a Jordan (contented) set and if f is any bounded function defined on E, then $f\mathcal{X}_E \in \mathcal{R}(\mathbb{R}^n)$ and*

$$\int_E f \, dV = 0.$$

Proof: Let m be a lower bound for f and let M be an upper bound. Let $\mathcal{G}$ be a grid with

$$\sum_{Q \cap E \neq \emptyset} v(Q) < \frac{\varepsilon}{1 + (M - m)}.$$

Then

$$\mathcal{U}_{\mathcal{G}}(f\mathcal{X}_E) \leq \sum_{Q \cap E \neq \emptyset} Mv(Q) \leq \frac{\varepsilon M}{1 + (M - m)}$$

and

$$\mathcal{L}_{\mathcal{G}}(f\mathcal{X}_E) \geq \sum_{Q \cap E \neq \emptyset} mv(Q) \geq \frac{\varepsilon m}{1 + (M - m)}$$

and so

$$\mathcal{U}_{\mathcal{G}}\left(f\mathcal{X}_E\right) - \mathcal{L}_{\mathcal{G}}\left(f\mathcal{X}_E\right) \; \leq \; \sum_{Q\cap E\neq\emptyset} Mv\left(Q\right) - \sum_{Q\cap E\neq\emptyset} mv\left(Q\right)$$

$$= \; (M-m)\sum_{Q\cap E\neq\emptyset} v\left(Q\right) < \frac{\varepsilon\left(M-m\right)}{1+\left(M-m\right)} < \varepsilon.$$

This shows $f\mathcal{X}_E \in \mathcal{R}\left(\mathbb{R}^n\right)$. Now also,

$$m\varepsilon \leq \int f\mathcal{X}_E \, dV \leq M\varepsilon$$

and since ε is arbitrary, this shows

$$\int_E f \, dV \equiv \int f\mathcal{X}_E \, dV = 0$$

Why is E contented? Let $\mathcal{G}$ be a grid for which

$$\sum_{Q\cap E\neq\emptyset} v\left(Q\right) < \varepsilon$$

Then for this grid,

$$\mathcal{U}_{\mathcal{G}}\left(\mathcal{X}_E\right) - \mathcal{L}_{\mathcal{G}}\left(\mathcal{X}_E\right) \leq \sum_{Q\cap E\neq\emptyset} v\left(Q\right) < \varepsilon$$

and this proves the theorem. ∎

Corollary C.3. *If $f\mathcal{X}_{E_i} \in \mathcal{R}\left(\mathbb{R}^n\right)$ for $i = 1,2,\cdots,r$ and for all $i \neq j, E_i \cap E_j$ is either the empty set or a set of Jordan content 0, then letting $F \equiv \cup_{i=1}^r E_i$, it follows $f\mathcal{X}_F \in \mathcal{R}\left(\mathbb{R}^n\right)$ and*

$$\int f\mathcal{X}_F \, dV \equiv \int_F f \, dV = \sum_{i=1}^r \int_{E_i} f \, dV.$$

Proof: This is true if $r = 1$. Suppose it is true for r. It will be shown that it is true for $r+1$. Let $F_r = \cup_{i=1}^r E_i$ and let F_{r+1} be defined similarly. By the induction hypothesis, $f\mathcal{X}_{F_r} \in \mathcal{R}\left(\mathbb{R}^n\right)$. Also, since F_r is a finite union of the E_i, it follows that $F_r \cap E_{r+1}$ is either empty or a set of Jordan content 0.

$$-f\mathcal{X}_{F_r\cap E_{r+1}} + f\mathcal{X}_{F_r} + f\mathcal{X}_{E_{r+1}} = f\mathcal{X}_{F_{r+1}}$$

and by Theorem C.5 each function on the left is in $\mathcal{R}\left(\mathbb{R}^n\right)$ and the first one on the left has integral equal to zero. Therefore,

$$\int f\mathcal{X}_{F_{r+1}} \, dV = \int f\mathcal{X}_{F_r} \, dV + \int f\mathcal{X}_{E_{r+1}} \, dV$$

which by induction equals

$$\sum_{i=1}^r \int_{E_i} f \, dV + \int_{E_{r+1}} f \, dV = \sum_{i=1}^{r+1} \int_{E_i} f \, dV$$

and this proves the corollary. ∎

In particular, for

$$Q = \prod_{i=1}^{n} [a_i, b_i], \ Q' = \prod_{i=1}^{n} (a_i, b_i]$$

both are contented sets and

$$\int \mathcal{X}_Q dV = \int_{Q'} \mathcal{X}_{Q'} dV = v(Q). \tag{C.16}$$

This is because

$$Q \setminus Q' = \cup_{i=1}^{n} a_i \times \prod_{\substack{j \neq i}} (a_j, b_j]$$

a finite union of sets of content 0. It is obvious $\int \mathcal{X}_Q dV = v(Q)$ because you can use a grid which has Q as one of the boxes and then the upper and lower sums are the same and equal to $v(Q)$. Therefore, the claim about the equality of the two integrals in (C.16) follows right away from Corollary C.3. That $\mathcal{X}_{Q'}$ is integrable follows from

$$\mathcal{X}_{Q'} = \mathcal{X}_Q - \mathcal{X}_{Q \setminus Q'}$$

and each of the two functions on the right is integrable thanks to Theorem C.5.

In fact, here is an interesting version of the Riemann criterion which depends on these half open boxes.

Lemma C.6. *Suppose f is a bounded function which equals zero off some bounded set. Then $f \in \mathcal{R}(\mathbb{R}^n)$ if and only if for all $\varepsilon > 0$ there exists a grid $\mathcal{G}$ such that*

$$\sum_{Q \in \mathcal{G}} (M_{Q'}(f) - m_{Q'}(f)) v(Q) < \varepsilon. \tag{C.17}$$

Proof: Since $Q' \subseteq Q$,

$$M_{Q'}(f) - m_{Q'}(f) \leq M_Q(f) - m_Q(f)$$

and therefore, the only if part of the equivalence is obvious.

Conversely, let $\mathcal{G}$ be a grid such that (C.17) holds with ε replaced with $\frac{\varepsilon}{2}$. It is necessary to show that there is a grid such that (C.17) holds with no primes on the Q. Let $\mathcal{F}$ be a refinement of $\mathcal{G}$ obtained by adding the points $\alpha_k^i + \eta_k$ where $\eta_k \leq \eta$ and is also chosen so small that for each $i = 1, \cdots, n$,

$$\alpha_k^i + \eta_k < \alpha_{k+1}^i.$$

You only need to have $\eta_k > 0$ for the finitely many boxes of $\mathcal{G}$ which intersect the bounded set where f is not zero. Then for

$$Q \equiv \prod_{i=1}^{n} [\alpha_{k_i}^i, \alpha_{k_i+1}^i] \in \mathcal{G},$$

Let

$$\widehat{Q} \equiv \prod_{i=1}^{n} [\alpha_{k_i}^i + \eta_{k_i}, \alpha_{k_i+1}^i]$$

and denote by $\widehat{\mathcal{G}}$ the collection of these smaller boxes. For each set Q in $\mathcal{G}$ there is the smaller set $\widehat{Q}$ along with n boxes, $B_k, k = 1, \cdots, n$, one of whose sides is of length η_k and the remainder of whose sides are shorter than the diameter of Q such that the set Q is the union of $\widehat{Q}$ and these sets B_k. Now suppose f equals zero off the ball $B\left(\mathbf{0},\frac{R}{2}\right)$. Then without loss of generality, you may assume the diameter of every box in $\mathcal{G}$ which has nonempty intersection with $B\left(\mathbf{0},R\right)$ is smaller than $\frac{R}{3}$. (If this is not so, simply refine $\mathcal{G}$ to make it so, such a refinement leaving (C.17) valid because refinements do not increase the difference between upper and lower sums in this context either.) Suppose there are P sets of $\mathcal{G}$ contained in $B\left(\mathbf{0},R\right)$ (So these are the only sets of $\mathcal{G}$ which could have nonempty intersection with the set where f is nonzero.) and suppose that for all $\mathbf{x}$, $|f\left(\mathbf{x}\right)| < C/2$. Then

$$\sum_{Q \in \mathcal{F}} \left(M_Q\left(f\right) - m_Q\left(f\right)\right) v\left(Q\right) \leq \sum_{\widehat{Q} \in \widehat{\mathcal{G}}} \left(M_{\widehat{Q}}\left(f\right) - m_{\widehat{Q}}\left(f\right)\right) v\left(Q\right)$$

$$+ \sum_{Q \in \mathcal{F} \setminus \widehat{\mathcal{G}}} \left(M_Q\left(f\right) - m_Q\left(f\right)\right) v\left(Q\right)$$

The first term on the right of the inequality in the above is no larger than $\varepsilon/2$ because $M_{\widehat{Q}}\left(f\right) - m_{\widehat{Q}}\left(f\right) \leq M_{Q'}\left(f\right) - m_{Q'}\left(f\right)$ for each Q. Therefore, the above is dominated by

$$\leq \varepsilon/2 + CPnR^{n-1}\eta < \varepsilon$$

whenever η is small enough. Since ε is arbitrary, $f \in \mathcal{R}\left(\mathbb{R}^n\right)$ as claimed. ∎

C.3   Iterated Integrals

To evaluate an n dimensional Riemann integral, one uses iterated integrals. Formally, an iterated integral is defined as follows. For f a function defined on $\mathbb{R}^{n+m}$,

$$\mathbf{y} \rightarrow f\left(\mathbf{x}, \mathbf{y}\right)$$

is a function of $\mathbf{y}$ for each $\mathbf{x} \in \mathbb{R}^n$. Therefore, it might be possible to integrate this function of $\mathbf{y}$ and write

$$\int_{\mathbb{R}^m} f\left(\mathbf{x}, \mathbf{y}\right) dV_y.$$

Now the result is clearly a function of $\mathbf{x}$ and so, it might be possible to integrate this and write

$$\int_{\mathbb{R}^n} \int_{\mathbb{R}^m} f\left(\mathbf{x}, \mathbf{y}\right) dV_y \, dV_x.$$

This symbol is called an iterated integral, because it involves the iteration of two lower dimensional integrations. Under what conditions are the two iterated integrals equal to the integral

$$\int_{\mathbb{R}^{n+m}} f\left(\mathbf{z}\right) dV?$$

Definition C.9. Let $\mathcal{G}$ be a grid on $\mathbb{R}^{n+m}$ defined by the $n+m$ sequences,

$$\{\alpha_k^i\}_{k=-\infty}^{\infty} \quad i = 1, \cdots, n+m.$$

Let $\mathcal{G}_n$ be the grid on $\mathbb{R}^n$ obtained by considering only the first n of these sequences and let $\mathcal{G}_m$ be the grid on $\mathbb{R}^m$ obtained by considering only the last m of the sequences. Thus a typical box in $\mathcal{G}_m$ would be

$$\prod_{i=n+1}^{n+m} \left[\alpha_{k_i}^i, \alpha_{k_i+1}^i\right], \; k_i \geq n+1$$

and a box in $\mathcal{G}_n$ would be of the form

$$\prod_{i=1}^{n} \left[\alpha_{k_i}^i, \alpha_{k_i+1}^i\right], \; k_i \leq n.$$

Lemma C.7. *Let $\mathcal{G}$, $\mathcal{G}_n$, and $\mathcal{G}_m$ be the grids defined above. Then*

$$\mathcal{G} = \{R \times P : R \in \mathcal{G}_n \text{ and } P \in \mathcal{G}_m\}.$$

Proof: If $Q \in \mathcal{G}$, then Q is clearly of this form. On the other hand, if $R \times P$ is one of the sets described above, then from the above description of R and P, it follows $R \times P$ is one of the sets of $\mathcal{G}$. ∎

Now let $\mathcal{G}$ be a grid on $\mathbb{R}^{n+m}$ and suppose

$$\phi(\mathbf{z}) = \sum_{Q \in \mathcal{G}} \phi_Q \mathcal{X}_{Q'}(\mathbf{z}) \tag{C.18}$$

where ϕ_Q equals zero for all but finitely many Q. Thus ϕ is a step function. Recall that for

$$Q = \prod_{i=1}^{n+m} [a_i, b_i], \; Q' \equiv \prod_{i=1}^{n+m} (a_i, b_i]$$

The function

$$\phi = \sum_{Q \in \mathcal{G}} \phi_Q \mathcal{X}_{Q'}$$

is integrable because it is a finite sum of integrable functions, each function in the sum being integrable because the set of discontinuities has Jordan content 0. (why?) Letting $(\mathbf{x}, \mathbf{y}) = \mathbf{z}$,

$$\phi(\mathbf{z}) = \phi(\mathbf{x}, \mathbf{y}) = \sum_{R \in \mathcal{G}_n} \sum_{P \in \mathcal{G}_m} \phi_{R \times P} \mathcal{X}_{R' \times P'}(\mathbf{x}, \mathbf{y})$$

$$= \sum_{R \in \mathcal{G}_n} \sum_{P \in \mathcal{G}_m} \phi_{R \times P} \mathcal{X}_{R'}(\mathbf{x}) \mathcal{X}_{P'}(\mathbf{y}). \tag{C.19}$$

For a function of two variables h, denote by $h(\cdot, \mathbf{y})$ the function $\mathbf{x} \to h(\mathbf{x}, \mathbf{y})$ and $h(\mathbf{x}, \cdot)$ the function $\mathbf{y} \to h(\mathbf{x}, \mathbf{y})$. The following lemma is a preliminary version of Fubini's theorem.

Lemma C.8. *Let ϕ be a step function as described in (C.18). Then*

$$\phi(\mathbf{x}, \cdot) \in \mathcal{R}(\mathbb{R}^m), \tag{C.20}$$

$$\int_{\mathbb{R}^m} \phi(\cdot, \mathbf{y})\, dV_y \in \mathcal{R}(\mathbb{R}^n),$$ (C.21)

and

$$\int_{\mathbb{R}^n}\int_{\mathbb{R}^m} \phi(\mathbf{x}, \mathbf{y})\, dV_y\, dV_x = \int_{\mathbb{R}^{n+m}} \phi(\mathbf{z})\, dV.$$ (C.22)

Proof: To verify (C.20), note that $\phi(\mathbf{x}, \cdot)$ is the step function

$$\phi(\mathbf{x}, \mathbf{y}) = \sum_{P \in \mathcal{G}_m} \phi_{R \times P} \mathcal{X}_{P'}(\mathbf{y}).$$

Where $\mathbf{x} \in R'$ and this is a finite sum of integrable functions because each has set of discontinuities with Jordan content 0. From the description in (C.19),

$$\int_{\mathbb{R}^m} \phi(\mathbf{x}, \mathbf{y})\, dV_y = \sum_{R \in \mathcal{G}_n} \sum_{P \in \mathcal{G}_m} \phi_{R \times P} \mathcal{X}_{R'}(\mathbf{x})\, v(P)$$

$$= \sum_{R \in \mathcal{G}_n} \left(\sum_{P \in \mathcal{G}_m} \phi_{R \times P} v(P) \right) \mathcal{X}_{R'}(\mathbf{x}),$$ (C.23)

another step function. Therefore,

$$\int_{\mathbb{R}^n}\int_{\mathbb{R}^m} \phi(\mathbf{x}, \mathbf{y})\, dV_y\, dV_x = \sum_{R \in \mathcal{G}_n} \sum_{P \in \mathcal{G}_m} \phi_{R \times P} v(P)\, v(R)$$

$$= \sum_{Q \in \mathcal{G}} \phi_Q v(Q) = \int_{\mathbb{R}^{n+m}} \phi(\mathbf{z})\, dV. \qquad \blacksquare$$

From (C.23),

$$M_{R_1'}\left(\int_{\mathbb{R}^m} \phi(\cdot, \mathbf{y})\, dV_y \right) \equiv \sup\left\{ \sum_{R \in \mathcal{G}_n} \left(\sum_{P \in \mathcal{G}_m} \phi_{R \times P} v(P) \right) \mathcal{X}_{R'}(\mathbf{x}) : \mathbf{x} \in R_1' \right\}$$

$$= \sum_{P \in \mathcal{G}_m} \phi_{R_1 \times P} v(P)$$ (C.24)

because $\int_{\mathbb{R}^m} \phi(\cdot, \mathbf{y})\, dV_y$ has the constant value given in (C.24) for $\mathbf{x} \in R_1'$. Similarly,

$$m_{R_1'}\left(\int_{\mathbb{R}^m} \phi(\cdot, \mathbf{y})\, dV_y \right) \equiv \inf\left\{ \sum_{R \in \mathcal{G}_n} \left(\sum_{P \in \mathcal{G}_m} \phi_{R \times P} v(P) \right) \mathcal{X}_{R'}(\mathbf{x}) : \mathbf{x} \in R_1' \right\}$$

$$= \sum_{P \in \mathcal{G}_m} \phi_{R_1 \times P} v(P).$$ (C.25)

Theorem C.6. *(Fubini) Let $f \in \mathcal{R}(\mathbb{R}^{n+m})$ and suppose also that $f(\mathbf{x}, \cdot) \in \mathcal{R}(\mathbb{R}^m)$ for each $\mathbf{x}$. Then*

$$\int_{\mathbb{R}^m} f(\cdot, \mathbf{y})\, dV_y \in \mathcal{R}(\mathbb{R}^n)$$ (C.26)

and

$$\int_{\mathbb{R}^{n+m}} f(\mathbf{z})\, dV = \int_{\mathbb{R}^n}\int_{\mathbb{R}^m} f(\mathbf{x}, \mathbf{y})\, dV_y\, dV_x.$$ (C.27)

Proof: Let $\mathcal{G}$ be a grid such that $\mathcal{U}_\mathcal{G}(f) - \mathcal{L}_\mathcal{G}(f) < \varepsilon$ and let $\mathcal{G}_n$ and $\mathcal{G}_m$ be as defined above. Let

$$\phi(\mathbf{z}) \equiv \sum_{Q \in \mathcal{G}} M_{Q'}(f) \mathcal{X}_{Q'}(\mathbf{z}), \ \psi(\mathbf{z}) \equiv \sum_{Q \in \mathcal{G}} m_{Q'}(f) \mathcal{X}_{Q'}(\mathbf{z}).$$

Observe that $M_{Q'}(f) \leq M_Q(f)$ and $m_{Q'}(f) \geq m_Q(f)$. Then

$$\mathcal{U}_\mathcal{G}(f) \geq \int \phi \, dV, \ \mathcal{L}_\mathcal{G}(f) \leq \int \psi \, dV.$$

Also $f(\mathbf{z}) \in (\psi(\mathbf{z}), \phi(\mathbf{z}))$ for all $\mathbf{z}$. Thus from (C.24),

$$M_{R'}\left(\int_{\mathbb{R}^m} f(\cdot, \mathbf{y}) \, dV_y\right) \leq M_{R'}\left(\int_{\mathbb{R}^m} \phi(\cdot, \mathbf{y}) \, dV_y\right) = \sum_{P \in \mathcal{G}_m} M_{R' \times P'}(f) v(P)$$

and from (C.25),

$$m_{R'}\left(\int_{\mathbb{R}^m} f(\cdot, \mathbf{y}) \, dV_y\right) \geq m_{R'}\left(\int_{\mathbb{R}^m} \psi(\cdot, \mathbf{y}) \, dV_y\right) = \sum_{P \in \mathcal{G}_m} m_{R' \times P'}(f) v(P).$$

Therefore,

$$\sum_{R \in \mathcal{G}_n}\left[M_{R'}\left(\int_{\mathbb{R}^m} f(\cdot, \mathbf{y}) \, dV_y\right) - m_{R'}\left(\int_{\mathbb{R}^m} f(\cdot, \mathbf{y}) \, dV_y\right)\right] v(R) \leq$$

$$\sum_{R \in \mathcal{G}_n} \sum_{P \in \mathcal{G}_m} [M_{R' \times P'}(f) - m_{R' \times P'}(f)] v(P) v(R) \leq \mathcal{U}_\mathcal{G}(f) - \mathcal{L}_\mathcal{G}(f) < \varepsilon.$$

This shows, from Lemma C.6 and the Riemann criterion, that $\int_{\mathbb{R}^m} f(\cdot, \mathbf{y}) \, dV_y \in \mathcal{R}(\mathbb{R}^n)$. It remains to verify (C.27). First note

$$\int_{\mathbb{R}^{n+m}} f(\mathbf{z}) \, dV \in [\mathcal{L}_\mathcal{G}(f), \mathcal{U}_\mathcal{G}(f)].$$

Next,

$$\mathcal{L}_\mathcal{G}(f) \leq \int_{\mathbb{R}^{n+m}} \psi \, dV = \int_{\mathbb{R}^n} \int_{\mathbb{R}^m} \psi \, dV_y \, dV_x \leq \int_{\mathbb{R}^n} \int_{\mathbb{R}^m} f(\mathbf{x}, \mathbf{y}) \, dV_y \, dV_x$$

$$\leq \int_{\mathbb{R}^n} \int_{\mathbb{R}^m} \phi(\mathbf{x}, \mathbf{y}) \, dV_y \, dV_x = \int_{\mathbb{R}^{n+m}} \phi \, dV \leq \mathcal{U}_\mathcal{G}(f).$$

Therefore,

$$\left|\int_{\mathbb{R}^n} \int_{\mathbb{R}^m} f(\mathbf{x}, \mathbf{y}) \, dV_y \, dV_x - \int_{\mathbb{R}^{n+m}} f(\mathbf{z}) \, dV\right| \leq \varepsilon$$

and since $\varepsilon > 0$ is arbitrary, this proves Fubini's theorem[3]. ∎

Corollary C.4. *Suppose E is a bounded contented set in $\mathbb{R}^n$ and let ϕ, ψ be continuous functions defined on E such that $\phi(\mathbf{x}) \geq \psi(\mathbf{x})$. Also suppose f is a continuous bounded function defined on the set*

$$P \equiv \{(\mathbf{x}, y) : \psi(\mathbf{x}) \leq y \leq \phi(\mathbf{x})\},$$

It follows $f\mathcal{X}_P \in \mathcal{R}(\mathbb{R}^{n+1})$ and

$$\int_P f \, dV = \int_E \int_{\psi(\mathbf{x})}^{\phi(\mathbf{x})} f(\mathbf{x}, y) \, dy \, dV_x.$$

---

[3]Actually, Fubini's theorem usually refers to a much more profound result in the theory of Lebesgue integration.

Proof: Since f is continuous, there is no problem in writing $f(\mathbf{x}, \cdot)\, \mathcal{X}_{[\psi(\mathbf{x}), \phi(\mathbf{x})]}(\cdot) \in \mathcal{R}(\mathbb{R}^1)$. Also, $f\mathcal{X}_P \in \mathcal{R}(\mathbb{R}^{n+1})$ because P is contented thanks to Corollary C.2. Therefore, by Fubini's theorem

$$\int_P f\, dV = \int_{\mathbb{R}^n} \int_{\mathbb{R}} f\mathcal{X}_P \, dy \, dV_x$$

$$= \int_E \int_{\psi(\mathbf{x})}^{\phi(\mathbf{x})} f(\mathbf{x}, y) \, dy \, dV_x$$

proving the corollary. ∎

Other versions of this corollary are immediate and should be obvious whenever encountered.

C.4   The Change Of Variables Formula

First recall Theorem B.5 on Page 312 which is listed here for convenience.

Theorem C.7. *Let* $\mathbf{h} : U \to \mathbb{R}^n$ *be a* C^1 *function with* $\mathbf{h}(\mathbf{0}) = \mathbf{0}, D\mathbf{h}(\mathbf{0})^{-1}$ *exists. Then there exists an open set* $V \subseteq U$ *containing* $\mathbf{0}$ *flips,* $\mathbf{F}_1, \cdots, \mathbf{F}_{n-1}$, *and primitive functions* $\mathbf{G}_n, \mathbf{G}_{n-1}, \cdots, \mathbf{G}_1$ *such that for* $\mathbf{x} \in V$,

$$\mathbf{h}(\mathbf{x}) = \mathbf{F}_1 \circ \cdots \circ \mathbf{F}_{n-1} \circ \mathbf{G}_n \circ \mathbf{G}_{n-1} \circ \cdots \circ \mathbf{G}_1(\mathbf{x}).$$

Also recall Theorem 5.6 on Page 105.

Theorem C.8. *Let* $\phi : [a, b] \to [c, d]$ *be one to one and suppose* ϕ' *exists and is continuous on* $[a, b]$. *Then if* f *is a continuous function defined on* $[a, b]$,

$$\int_c^d f(s) \, ds = \int_a^b f(\phi(t)) \, |\phi'(t)| \, dt$$

The following is a simple corollary to this theorem.

Corollary C.5. *Let* $\phi : [a, b] \to [c, d]$ *be one to one and suppose* ϕ' *exists and is continuous on* $[a, b]$. *Then if* f *is a continuous function defined on* $[a, b]$,

$$\int_{\mathbb{R}} \mathcal{X}_{[a,b]} \left(\phi^{-1}(x)\right) f(x) \, dx = \int_{\mathbb{R}} \mathcal{X}_{[a,b]}(t) f(\phi(t)) \, |\phi'(t)| \, dt$$

Lemma C.9. *Let* $\mathbf{h} : V \to \mathbb{R}^n$ *be a* C^1 *function and suppose* H *is a compact subset of* V. *Then there exists a constant* C *independent of* $\mathbf{x} \in H$ *such that*

$$|D\mathbf{h}(\mathbf{x})\, \mathbf{v}| \le C\, |\mathbf{v}|.$$

Proof: Consider the compact set $H \times \partial B(\mathbf{0}, 1) \subseteq \mathbb{R}^{2n}$. Let $f : H \times \partial B(\mathbf{0}, 1) \to \mathbb{R}$ be given by $f(\mathbf{x}, \mathbf{v}) = |D\mathbf{h}(\mathbf{x})\, \mathbf{v}|$. Then let C denote the maximum value of f. It follows that for $\mathbf{v} \in \mathbb{R}^n$,

$$\left| D\mathbf{h}(\mathbf{x}) \frac{\mathbf{v}}{|\mathbf{v}|} \right| \le C$$

and so the desired formula follows when you multiply both sides by $|\mathbf{v}|$. ∎

Definition C.10. Let A be an open set. Write $C^k\left(A;\mathbb{R}^n\right)$ to denote a C^k function whose domain is A and whose range is in $\mathbb{R}^n$. Let U be an open set in $\mathbb{R}^n$. Then $\mathbf{h} \in C^k\left(\overline{U};\mathbb{R}^n\right)$ if there exists an open set $V \supseteq \overline{U}$ and a function $\mathbf{g} \in C^1\left(V;\mathbb{R}^n\right)$ such that $\mathbf{g} = \mathbf{h}$ on $\overline{U}$. $f \in C^k\left(\overline{U}\right)$ means the same thing except that f has values in $\mathbb{R}$. Also recall that $\mathbf{x} \in \partial U$ means that every open set which contains $\mathbf{x}$ contains points of U and points of U^C.

Theorem C.9. *Let U be a bounded open set such that ∂U has zero content and let $\mathbf{h} \in C\left(\overline{U};\mathbb{R}^n\right)$ be one to one and $D\mathbf{h}\left(\mathbf{x}\right)^{-1}$ exists for all $\mathbf{x} \in U$. Then $\mathbf{h}\left(\partial U\right) = \partial\left(\mathbf{h}\left(U\right)\right)$ and $\partial\left(\mathbf{h}\left(U\right)\right)$ has zero content.*

Proof: Let $\mathbf{x} \in \partial U$ and let $\mathbf{g} = \mathbf{h}$ where $\mathbf{g}$ is a C^1 function defined on an open set containing $\overline{U}$. By the inverse function theorem, $\mathbf{g}$ is locally one to one and an open mapping near $\mathbf{x}$. Thus $\mathbf{g}\left(\mathbf{x}\right) = \mathbf{h}\left(\mathbf{x}\right)$ and is in an open set containing points of $\mathbf{g}\left(U\right)$ and points of $\mathbf{g}\left(U^C\right)$. These points of $\mathbf{g}\left(U^C\right)$ cannot equal any points of $\mathbf{h}\left(U\right)$ because $\mathbf{g}$ is one to one locally. Thus $\mathbf{h}\left(\mathbf{x}\right) \in \partial\left(\mathbf{h}\left(U\right)\right)$ and so $\mathbf{h}\left(\partial U\right) \subseteq \partial\left(\mathbf{h}\left(U\right)\right)$. Now suppose $\mathbf{y} \in \partial\left(\mathbf{h}\left(U\right)\right)$. By the inverse function theorem $\mathbf{y}$ cannot be in the open set $\mathbf{h}\left(U\right)$. Since $\mathbf{y} \in \partial\left(\mathbf{h}\left(U\right)\right)$, every ball centered at $\mathbf{y}$ contains points of $\mathbf{h}\left(U\right)$ and so $\mathbf{y} \in \overline{\mathbf{h}\left(U\right)} \setminus \mathbf{h}\left(U\right)$. Thus there exists a sequence, $\{\mathbf{x}_n\} \subseteq U$ such that $\mathbf{h}\left(\mathbf{x}_n\right) \to \mathbf{y}$. But then, by the continuity of $\mathbf{h}^{-1}$ which comes from the inverse function theorem, $\mathbf{x}_n \to \mathbf{h}^{-1}\left(\mathbf{y}\right)$ and so $\mathbf{h}^{-1}\left(\mathbf{y}\right) \notin U$ but is in $\overline{U}$. Thus $\mathbf{h}^{-1}\left(\mathbf{y}\right) \in \partial U$. (Why?) Therefore, $\mathbf{y} \in \mathbf{h}\left(\partial U\right)$, and this proves the two sets are equal. It remains to verify the claim about content.

First let H denote a compact set whose interior contains $\overline{U}$ which is also in the interior of the domain of $\mathbf{g}$. Now since ∂U has content zero, it follows that for $\varepsilon > 0$ given, there exists a grid $\mathcal{G}$ such that if $\mathcal{G}'$ are those boxes of $\mathcal{G}$ which have nonempty intersection with ∂U, then

$$\sum_{Q \in \mathcal{G}'} v\left(Q\right) < \varepsilon$$

and by refining the grid if necessary, no box of $\mathcal{G}$ has nonempty intersection with both $\overline{U}$ and H^C. Refining this grid still more, you can also assume that for all boxes in $\mathcal{G}'$,

$$\frac{l_i}{l_j} < 2$$

where l_i is the length of the ith side. (Thus the boxes are not too far from being cubes.)

Let C be the constant of Lemma C.9 applied to $\mathbf{g}$ on H.

Now consider one of these boxes, $Q \in \mathcal{G}'$. If $\mathbf{x}, \mathbf{y} \in Q$, it follows from the chain rule that

$$\mathbf{g}\left(\mathbf{y}\right) - \mathbf{g}\left(\mathbf{x}\right) = \int_0^1 D\mathbf{g}\left(\mathbf{x} + t\left(\mathbf{y} - \mathbf{x}\right)\right)\left(\mathbf{y} - \mathbf{x}\right)dt$$

By Lemma C.9 applied to H

$$|\mathbf{g}(\mathbf{y}) - \mathbf{g}(\mathbf{x})| \leq \int_0^1 |D\mathbf{g}(\mathbf{x} + t(\mathbf{y} - \mathbf{x}))(\mathbf{y} - \mathbf{x})| \, dt$$

$$\leq C \int_0^1 |\mathbf{x} - \mathbf{y}| \, dt \leq C \operatorname{diam}(Q)$$

$$= C \left(\sum_{i=1}^n l_i^2 \right)^{1/2} \leq C\sqrt{n}L$$

where L is the length of the longest side of Q. Thus $\operatorname{diam}(\mathbf{g}(Q)) \leq C\sqrt{n}L$ and so $\mathbf{g}(Q)$ is contained in a cube having sides equal to $C\sqrt{n}L$ and volume equal to

$$C^n n^{n/2} L^n \leq C^n n^{n/2} 2^n l_1 l_2 \cdots l_n = C^n n^{n/2} 2^n v(Q).$$

Denoting by P_Q this cube, it follows

$$\mathbf{h}(\partial U) \subseteq \cup_{Q \in \mathcal{G}'} v(P_Q)$$

and

$$\sum_{Q \in \mathcal{G}'} v(P_Q) \leq C^n n^{n/2} 2^n \sum_{Q \in \mathcal{G}'} v(Q) < \varepsilon C^n n^{n/2} 2^n.$$

Since $\varepsilon > 0$ is arbitrary, this shows $\mathbf{h}(\partial U)$ has content zero as claimed. ∎

Theorem C.10. *Suppose* $f \in C(\overline{U})$ *where* U *is a bounded open set with* ∂U *having content 0. Then* $f\mathcal{X}_U \in \mathcal{R}(\mathbb{R}^n)$.

Proof: Let H be a compact set whose interior contains $\overline{U}$ which is also contained in the domain of g where g is a continuous functions whose restriction to U equals f. Consider $g\mathcal{X}_U$, a function whose set of discontinuities has content 0. Then $g\mathcal{X}_U = f\mathcal{X}_U \in \mathcal{R}(\mathbb{R}^n)$ as claimed. This is by the big theorem which tells which functions are Riemann integrable. ∎

The symbol $U - \mathbf{p}$ is defined as $\{\mathbf{x} - \mathbf{p} : \mathbf{x} \in U\}$. It merely slides U by the vector $\mathbf{p}$. The following lemma is obvious from the definition of the integral.

Lemma C.10. *Let* U *be a bounded open set and let* $f\mathcal{X}_U \in \mathcal{R}(\mathbb{R}^n)$. *Then*

$$\int f(\mathbf{x} + \mathbf{p}) \mathcal{X}_{U-\mathbf{p}}(\mathbf{x}) \, dx = \int f(\mathbf{x}) \mathcal{X}_U(\mathbf{x}) \, dx$$

A few more lemmas are needed.

Lemma C.11. *Let* S *be a nonempty subset of* $\mathbb{R}^n$. *Define*

$$f(\mathbf{x}) \equiv \operatorname{dist}(\mathbf{x}, S) \equiv \inf\{|\mathbf{x} - \mathbf{y}| : \mathbf{y} \in S\}.$$

Then f *is continuous.*

Proof: Consider $|f(\mathbf{x}) - f(\mathbf{x}_1)|$ and suppose without loss of generality that $f(\mathbf{x}_1) \geq f(\mathbf{x})$. Then choose $\mathbf{y} \in S$ such that $f(\mathbf{x}) + \varepsilon > |\mathbf{x} - \mathbf{y}|$. Then

$$
\begin{aligned}
|f(\mathbf{x}_1) - f(\mathbf{x})| &= f(\mathbf{x}_1) - f(\mathbf{x}) \leq f(\mathbf{x}_1) - |\mathbf{x} - \mathbf{y}| + \varepsilon \\
&\leq |\mathbf{x}_1 - \mathbf{y}| - |\mathbf{x} - \mathbf{y}| + \varepsilon \\
&\leq |\mathbf{x} - \mathbf{x}_1| + |\mathbf{x} - \mathbf{y}| - |\mathbf{x} - \mathbf{y}| + \varepsilon \\
&= |\mathbf{x} - \mathbf{x}_1| + \varepsilon.
\end{aligned}
$$

Since ε is arbitrary, it follows that $|f(\mathbf{x}_1) - f(\mathbf{x})| \leq |\mathbf{x} - \mathbf{x}_1|$ and this proves the lemma. ∎

Theorem C.11. *(Urysohn's lemma for $\mathbb{R}^n$) Let H be a closed subset of an open set U. Then there exists a continuous function $g : \mathbb{R}^n \to [0,1]$ such that $g(\mathbf{x}) = 1$ for all $\mathbf{x} \in H$ and $g(\mathbf{x}) = 0$ for all $\mathbf{x} \notin U$.*

Proof: If $\mathbf{x} \notin C$, a closed set, then dist $(\mathbf{x}, C) > 0$ because there exists $\delta > 0$ such that $B(\mathbf{x}, \delta) \cap C = \emptyset$. This is because, since C is closed, its complement is open. Therefore, dist $(\mathbf{x}, H) + $ dist $(\mathbf{x}, U^C) > 0$ for all $\mathbf{x} \in \mathbb{R}^n$. Now define a continuous function g as

$$
g(\mathbf{x}) \equiv \frac{\text{dist}\left(\mathbf{x}, U^C\right)}{\text{dist}(\mathbf{x}, H) + \text{dist}(\mathbf{x}, U^C)}.
$$

It is easy to see this verifies the conclusions of the theorem and this proves the theorem. ∎

Definition C.11. Define spt(f) (support of f) to be the closure of the set $\{x : f(x) \neq 0\}$. If V is an open set, $C_c(V)$ will be the set of continuous functions f, defined on $\mathbb{R}^n$ having spt$(f) \subseteq V$.

Definition C.12. If K is a compact subset of an open set V, then $K \prec \phi \prec V$ if

$$
\phi \in C_c(V), \ \phi(K) = \{1\}, \ \phi(\mathbb{R}^n) \subseteq [0,1].
$$

Also for $\phi \in C_c(\mathbb{R}^n)$, $K \prec \phi$ if

$$
\phi(\mathbb{R}^n) \subseteq [0,1] \text{ and } \phi(K) = 1.
$$

and $\phi \prec V$ if

$$
\phi(\mathbb{R}^n) \subseteq [0,1] \text{ and } \text{spt}(\phi) \subseteq V.
$$

Theorem C.12. *(Partition of unity) Let K be a compact subset of $\mathbb{R}^n$ and suppose*

$$
K \subseteq V = \cup_{i=1}^n V_i, \ V_i \ \text{open and bounded.}
$$

Then there exist $\psi_i \prec V_i$ with

$$
\sum_{i=1}^n \psi_i(\mathbf{x}) = 1
$$

for all $\mathbf{x} \in K$.

Proof: Let $K_1 = K \setminus \cup_{i=2}^n V_i$. Thus K_1 is compact because it is the intersection of a closed set with a compact set and $K_1 \subseteq V_1$. Let $K_1 \subseteq W_1 \subseteq \overline{W}_1 \subseteq V_1$ with $\overline{W}_1$ compact. To obtain W_1, use Theorem C.11 to get f such that $K_1 \prec f \prec V_1$ and let $W_1 \equiv \{\mathbf{x} : f(\mathbf{x}) \neq 0\}$. Thus $W_1, V_2, \cdots V_n$ covers K and $\overline{W}_1 \subseteq V_1$. Let $K_2 = K \setminus (\cup_{i=3}^n V_i \cup W_1)$. Then K_2 is compact and $K_2 \subseteq V_2$. Let $K_2 \subseteq W_2 \subseteq \overline{W}_2 \subseteq V_2$ $\overline{W}_2$ compact. Continue this way finally obtaining $W_1, \cdots, W_n$, $K \subseteq W_1 \cup \cdots \cup W_n$, and $\overline{W}_i \subseteq V_i; \overline{W}_i$ compact. Now let $\overline{W}_i \subseteq U_i \subseteq \overline{U}_i \subseteq V_i$, $\overline{U}_i$ compact.

By Theorem C.11, there exist functions ϕ_i, γ such that $\overline{U}_i \prec \phi_i \prec V_i$, $\cup_{i=1}^n \overline{W}_i \prec \gamma \prec \cup_{i=1}^n U_i$. Define

$$\psi_i(\mathbf{x}) = \begin{cases} \gamma(\mathbf{x})\phi_i(\mathbf{x}) / \sum_{j=1}^n \phi_j(\mathbf{x}) \text{ if } \sum_{j=1}^n \phi_j(\mathbf{x}) \neq 0, \\ 0 \text{ if } \sum_{j=1}^n \phi_j(\mathbf{x}) = 0. \end{cases}$$

If $\mathbf{x}$ is such that $\sum_{j=1}^n \phi_j(\mathbf{x}) = 0$, then $\mathbf{x} \notin \cup_{i=1}^n \overline{U}_i$. Consequently $\gamma(\mathbf{y}) = 0$ for all $\mathbf{y}$ near $\mathbf{x}$ and so $\psi_i(\mathbf{y}) = 0$ for all $\mathbf{y}$ near $\mathbf{x}$. Hence ψ_i is continuous at such $\mathbf{x}$. If $\sum_{j=1}^n \phi_j(\mathbf{x}) \neq 0$, this situation persists near $\mathbf{x}$ and so ψ_i is continuous at such points. Therefore ψ_i is continuous. If $\mathbf{x} \in K$, then $\gamma(\mathbf{x}) = 1$ and so $\sum_{j=1}^n \psi_j(\mathbf{x}) = 1$. Clearly $0 \leq \psi_i(\mathbf{x}) \leq 1$ and $\text{spt}(\psi_j) \subseteq V_j$. ∎

The next lemma contains the main ideas. See [27] and [23] for similar proofs.

Lemma C.12. *Let U be a bounded open set with ∂U having content 0. Also let $\mathbf{h} \in C^1(\overline{U}; \mathbb{R}^n)$ be one to one on U with $D\mathbf{h}(\mathbf{x})^{-1}$ exists for all $\mathbf{x} \in U$. Let $f \in C(\overline{U})$ be nonnegative. Then*

$$\int \mathcal{X}_{\mathbf{h}(U)}(\mathbf{z}) f(\mathbf{z}) \, dV_n = \int \mathcal{X}_U(\mathbf{x}) f(\mathbf{h}(\mathbf{x})) |\det D\mathbf{h}(\mathbf{x})| \, dV_n$$

Proof: Let $\varepsilon > 0$ be given. Then by Theorem C.10,

$$\mathbf{x} \to \mathcal{X}_U(\mathbf{x}) f(\mathbf{h}(\mathbf{x})) |\det D\mathbf{h}(\mathbf{x})|$$

is Riemann integrable. Therefore, there exists a grid $\mathcal{G}$ such that, letting

$$g(\mathbf{x}) = \mathcal{X}_U(\mathbf{x}) f(\mathbf{h}(\mathbf{x})) |\det D\mathbf{h}(\mathbf{x})|,$$

$$\mathcal{L}_{\mathcal{G}}(g) + \varepsilon > \mathcal{U}_{\mathcal{G}}(g).$$

Let K denote the union of the boxes Q of $\mathcal{G}$ which intersect $\overline{U}$. Thus K is a compact subset of V where V is a bounded open set containing $\overline{U}$, and it is only the terms from these boxes which contribute anything nonzero to the lower sum. By Theorem B.5 on Page 312 which is stated above and the inverse function theorem, it follows that for $\mathbf{p} \in K$, there exists an open set contained in U which contains $\mathbf{p}$, denoted as $O_{\mathbf{p}}$ such that for $\mathbf{x} \in O_{\mathbf{p}} - \mathbf{p}$,

$$\mathbf{h}(\mathbf{x} + \mathbf{p}) - \mathbf{h}(\mathbf{p}) = \mathbf{F}_1 \circ \cdots \circ \mathbf{F}_{n-1} \circ \mathbf{G}_n \circ \cdots \circ \mathbf{G}_1(\mathbf{x})$$

where the $\mathbf{G}_i$ are primitive functions, and the $\mathbf{F}_j$ are flips. Also $\mathbf{h}(O_j)$ is an open set.

Finitely many of these open sets $\{O_j\}_{j=1}^q$ cover K. Let the distinguished point for O_j be denoted by $\mathbf{p}_j$. Now refine $\mathcal{G}$ if necessary, such that the diameter of every cell of the new $\mathcal{G}$ which intersects $\overline{U}$ is smaller than a Lebesgue number for this open cover. Denote by $\mathcal{G}'$ those boxes of the new $\mathcal{G}$ which intersect $\overline{U}$. Thus the union of these boxes of $\mathcal{G}'$ equals the set K and every box of $\mathcal{G}'$ is contained in one of these O_j. By Theorem C.12, there exists a partition of unity $\{\psi_j\}$ on $\mathbf{h}(K)$ such that $\psi_j \prec \mathbf{h}(O_j)$. Then

$$\mathcal{L}_{\mathcal{G}}(g) \leq \sum_{Q \in \mathcal{G}'} \int \mathcal{X}_Q(\mathbf{x}) f(\mathbf{h}(\mathbf{x})) \left| \det D\mathbf{h}(\mathbf{x}) \right| d\mathbf{x}$$

$$= \sum_{Q \in \mathcal{G}'} \sum_{j=1}^q \int \mathcal{X}_Q(\mathbf{x}) (\psi_j f)(\mathbf{h}(\mathbf{x})) \left| \det D\mathbf{h}(\mathbf{x}) \right| d\mathbf{x}. \qquad (C.28)$$

Consider the term $\int \mathcal{X}_Q(\mathbf{x}) (\psi_j f)(\mathbf{h}(\mathbf{x})) \left| \det D\mathbf{h}(\mathbf{x}) \right| d\mathbf{x}$. By Lemma C.10 and Fubini's theorem this equals

$$\int_{\mathbb{R}^{n-1}} \int_{\mathbb{R}} \mathcal{X}_{Q-\mathbf{p}_j}(\mathbf{x}) (\psi_j f)(\mathbf{h}(\mathbf{p}_i) + \mathbf{F}_1 \circ \cdots \circ \mathbf{F}_{n-1} \circ \mathbf{G}_n \circ \cdots \circ \mathbf{G}_1(\mathbf{x})) \cdot$$

$$\left| D\mathbf{F}(\mathbf{G}_n \circ \cdots \circ \mathbf{G}_1(\mathbf{x})) \right| \left| D\mathbf{G}_n(\mathbf{G}_{n-1} \circ \cdots \circ \mathbf{G}_1(\mathbf{x})) \right| \cdot$$

$$\left| D\mathbf{G}_{n-1}(\mathbf{G}_{n-2} \circ \cdots \circ \mathbf{G}_1(\mathbf{x})) \right| \qquad (C.29)$$

$$\cdots \left| D\mathbf{G}_2(\mathbf{G}_1(\mathbf{x})) \right| \left| D\mathbf{G}_1(\mathbf{x}) \right| dx_1 dV_{n-1}. \qquad (C.30)$$

The vertical lines in the above signify the absolute value of the determinant of the matrix on the inside. Here dV_{n-1} is with respect to the variables $x_2, \cdots, x_n$. Also $\mathbf{F}$ denotes $\mathbf{F}_1 \circ \cdots \circ \mathbf{F}_{n-1}$. Now

$$\mathbf{G}_1(\mathbf{x}) = (\alpha(\mathbf{x}), x_2, \cdots, x_n)^T$$

and is one to one. Therefore, fixing $x_2, \cdots, x_n$, $x_1 \rightarrow \alpha(\mathbf{x})$ is one to one. Also

$$\left| D\mathbf{G}_1(\mathbf{x}) \right| = \left| \alpha_{x_1}(\mathbf{x}) \right|$$

Fixing $x_2, \cdots, x_n$, change the variable,

$$y_1 = \alpha(x_1, x_2, \cdots, x_n), \quad dy_1 = \alpha_{x_1}(x_1, x_2, \cdots, x_n) dx_1$$

Thus

$$\mathbf{x} = (x_1, x_2, \cdots, x_n)^T = \mathbf{G}_1^{-1}(y_1, x_2, \cdots, x_n) \equiv \mathbf{G}_1^{-1}(\mathbf{x}')$$

Then in (C.30) you can use Corollary C.5 to write (C.30) as

$$\int_{\mathbb{R}^{n-1}} \int_{\mathbb{R}} \mathcal{X}_{Q-\mathbf{p}_j}(\mathbf{G}_1^{-1}(\mathbf{x}')) (\psi_j f)$$

$$(\mathbf{h}(\mathbf{p}_i) + \mathbf{F}_1 \circ \cdots \circ \mathbf{F}_{n-1} \circ \mathbf{G}_n \circ \cdots \circ \mathbf{G}_1(\mathbf{G}_1^{-1}(\mathbf{x}')))$$

$$\cdot \left| DF \left(\mathbf{G}_n \circ \cdots \circ \mathbf{G}_1 \left(\mathbf{G}_1^{-1} \left(\mathbf{x'} \right) \right) \right) \right| \left| DG_n \left(\mathbf{G}_{n-1} \circ \cdots \circ \mathbf{G}_1 \left(\mathbf{G}_1^{-1} \left(\mathbf{x'} \right) \right) \right) \right| \cdot$$

$$\left| DG_{n-1} \left(\mathbf{G}_{n-2} \circ \cdots \circ \mathbf{G}_1 \left(\mathbf{G}_1^{-1} \left(\mathbf{x'} \right) \right) \right) \right| \cdots \left| DG_2 \left(\mathbf{G}_1 \left(\mathbf{G}_1^{-1} \left(\mathbf{x'} \right) \right) \right) \right| dy_1 dV_{n-1}$$

which reduces to

$$\int_{\mathbb{R}^n} \mathcal{X}_{Q-\mathbf{p}_j} \left(\mathbf{G}_1^{-1} \left(\mathbf{x'} \right) \right) \left(\psi_j f \right) \left(\mathbf{h} \left(\mathbf{p}_i \right) + \mathbf{F}_1 \circ \cdots \circ \mathbf{F}_{n-1} \circ \mathbf{G}_n \circ \cdots \circ \mathbf{G}_2 \left(\mathbf{x'} \right) \right)$$

$$\cdot \left| DF \left(\mathbf{G}_n \circ \cdots \circ \mathbf{G}_2 \left(\mathbf{x'} \right) \right) \right| \left| DG_n \left(\mathbf{G}_{n-1} \circ \cdots \circ \mathbf{G}_2 \left(\mathbf{x'} \right) \right) \right| \cdot$$
$$\left| DG_{n-1} \left(\mathbf{G}_{n-2} \circ \cdots \circ \mathbf{G}_2 \left(\mathbf{x'} \right) \right) \right| \cdots \left| DG_2 \left(\mathbf{x'} \right) \right| dV_n.$$

Now use Fubini's theorem again to make the inside integral taken with respect to x_2. Note that the term $\left| D\mathbf{G}_1 \left(\mathbf{x} \right) \right|$ disappeared. Exactly the same process yields

$$\int_{\mathbb{R}^{n-1}} \int_{\mathbb{R}} \mathcal{X}_{Q-\mathbf{p}_j} \left(\mathbf{G}_1^{-1} \circ \mathbf{G}_2^{-1} \left(\mathbf{x''} \right) \right) \left(\psi_j f \right)$$
$$\left(\mathbf{h} \left(\mathbf{p}_i \right) + \mathbf{F}_1 \circ \cdots \circ \mathbf{F}_{n-1} \circ \mathbf{G}_n \circ \cdots \circ \mathbf{G}_3 \left(\mathbf{x''} \right) \right)$$

$$\cdot \left| DF \left(\mathbf{G}_n \circ \cdots \circ \mathbf{G}_3 \left(\mathbf{x''} \right) \right) \right| \left| DG_n \left(\mathbf{G}_{n-1} \circ \cdots \circ \mathbf{G}_3 \left(\mathbf{x''} \right) \right) \right| \cdot$$
$$\left| DG_{n-1} \left(\mathbf{G}_{n-2} \circ \cdots \circ \mathbf{G}_3 \left(\mathbf{x''} \right) \right) \right| \cdots dy_2 dV_{n-1}.$$

Now $\mathbf{F}$ is just a composition of flips, so $\left| DF \left(\mathbf{G}_n \circ \cdots \circ \mathbf{G}_3 \left(\mathbf{x''} \right) \right) \right| = 1$, and so this term can be replaced with 1. Continuing this process, eventually yields an expression of the form

$$\int_{\mathbb{R}^n} \mathcal{X}_{Q-\mathbf{p}_j} \left(\mathbf{G}_1^{-1} \circ \cdots \circ \mathbf{G}_{n-2}^{-1} \circ \mathbf{G}_{n-1}^{-1} \circ \mathbf{G}_n^{-1} \circ \mathbf{F}^{-1} \left(\mathbf{y} \right) \right) \left(\psi_j f \right) \left(\mathbf{h} \left(\mathbf{p}_i \right) + \mathbf{y} \right) dV_n.$$
$$\text{(C.31)}$$

Denoting by $\mathbf{G}^{-1}$ the expression, $\mathbf{G}_1^{-1} \circ \cdots \circ \mathbf{G}_{n-2}^{-1} \circ \mathbf{G}_{n-1}^{-1} \circ \mathbf{G}_n^{-1}$,

$$\mathcal{X}_{Q-\mathbf{p}_j} \left(\mathbf{G}_1^{-1} \circ \cdots \circ \mathbf{G}_{n-2}^{-1} \circ \mathbf{G}_{n-1}^{-1} \circ \mathbf{G}_n^{-1} \circ \mathbf{F}^{-1} \left(\mathbf{y} \right) \right) = 1$$

exactly when $\mathbf{G}^{-1} \circ \mathbf{F}^{-1} \left(\mathbf{y} \right) \in Q - \mathbf{p}_j$. Now recall that

$$\mathbf{h} \left(\mathbf{p}_j + \mathbf{x} \right) - \mathbf{h} \left(\mathbf{p}_j \right) = \mathbf{F} \circ \mathbf{G} \left(\mathbf{x} \right)$$

and so the above holds exactly when

$$\begin{aligned} \mathbf{y} &= \mathbf{h} \left(\mathbf{p}_j + \mathbf{G}^{-1} \circ \mathbf{F}^{-1} \left(\mathbf{y} \right) \right) - \mathbf{h} \left(\mathbf{p}_j \right) \in \mathbf{h} \left(\mathbf{p}_j + Q - \mathbf{p}_j \right) - \mathbf{h} \left(\mathbf{p}_j \right) \\ &= \mathbf{h} \left(Q \right) - \mathbf{h} \left(\mathbf{p}_j \right). \end{aligned}$$

Thus (C.31) reduces to

$$\int_{\mathbb{R}^n} \mathcal{X}_{\mathbf{h}(Q)-\mathbf{h}(\mathbf{p}_j)} \left(\mathbf{y} \right) \left(\psi_j f \right) \left(\mathbf{h} \left(\mathbf{p}_i \right) + \mathbf{y} \right) dV_n$$
$$= \int_{\mathbb{R}^n} \mathcal{X}_{\mathbf{h}(Q)} \left(\mathbf{z} \right) \left(\psi_j f \right) \left(\mathbf{z} \right) dV_n.$$

It follows from (C.28),

$$\mathcal{U}_{\mathcal{G}}\left(g\right) - \varepsilon \;\leq\; \mathcal{L}_{\mathcal{G}}\left(g\right) \leq \int \mathcal{X}_U\left(\mathbf{x}\right) f\left(\mathbf{h}\left(\mathbf{x}\right)\right) \left|\det D\mathbf{h}\left(\mathbf{x}\right)\right| dV_n$$

$$\leq \sum_{Q \in \mathcal{G}'} \int \mathcal{X}_Q\left(\mathbf{x}\right) f\left(\mathbf{h}\left(\mathbf{x}\right)\right) \left|\det D\mathbf{h}\left(\mathbf{x}\right)\right| dx$$

$$= \sum_{Q \in \mathcal{G}'} \sum_{j=1}^{q} \int \mathcal{X}_Q\left(\mathbf{x}\right) \left(\psi_j f\right)\left(\mathbf{h}\left(\mathbf{x}\right)\right) \left|\det D\mathbf{h}\left(\mathbf{x}\right)\right| dx$$

$$= \sum_{Q \in \mathcal{G}'} \sum_{j=1}^{q} \int_{\mathbb{R}^n} \mathcal{X}_{\mathbf{h}(Q)}\left(\mathbf{z}\right) \left(\psi_j f\right)\left(\mathbf{z}\right) dV_n$$

$$= \sum_{Q \in \mathcal{G}'} \int_{\mathbb{R}^n} \mathcal{X}_{\mathbf{h}(Q)}\left(\mathbf{z}\right) f\left(\mathbf{z}\right) dV_n = \int \mathcal{X}_{\mathbf{h}(U)}\left(\mathbf{z}\right) f\left(\mathbf{z}\right) dV_n$$

which implies the inequality,

$$\int \mathcal{X}_U\left(\mathbf{x}\right) f\left(\mathbf{h}\left(\mathbf{x}\right)\right) \left|\det D\mathbf{h}\left(\mathbf{x}\right)\right| dV_n \leq \int \mathcal{X}_{\mathbf{h}(U)}\left(\mathbf{z}\right) f\left(\mathbf{z}\right) dV_n$$

But now you can use the same information just derived to obtain equality.

$$\mathbf{x} = \mathbf{h}^{-1}\left(\mathbf{z}\right)$$

and so from what was just done,

$$\int \mathcal{X}_U\left(\mathbf{x}\right) f\left(\mathbf{h}\left(\mathbf{x}\right)\right) \left|\det D\mathbf{h}\left(\mathbf{x}\right)\right| dV_n$$

$$= \int \mathcal{X}_{\mathbf{h}^{-1}(\mathbf{h}(U))}\left(\mathbf{x}\right) f\left(\mathbf{h}\left(\mathbf{x}\right)\right) \left|\det D\mathbf{h}\left(\mathbf{x}\right)\right| dV_n$$

$$\geq \int \mathcal{X}_{\mathbf{h}(U)}\left(\mathbf{z}\right) f\left(\mathbf{z}\right) \left|\det D\mathbf{h}\left(\mathbf{h}^{-1}\left(\mathbf{z}\right)\right)\right| \left|\det D\mathbf{h}^{-1}\left(\mathbf{z}\right)\right| dV_n$$

$$= \int \mathcal{X}_{\mathbf{h}(U)}\left(\mathbf{z}\right) f\left(\mathbf{z}\right) dV_n$$

from the chain rule. In fact,

$$I = D\mathbf{h}\left(\mathbf{h}^{-1}\left(\mathbf{z}\right)\right) D\mathbf{h}^{-1}\left(\mathbf{z}\right),$$

so

$$1 = \left|\det D\mathbf{h}\left(\mathbf{h}^{-1}\left(\mathbf{z}\right)\right)\right| \left|\det D\mathbf{h}^{-1}\left(\mathbf{z}\right)\right|. \qquad \blacksquare$$

The change of variables theorem follows.

Theorem C.13. *Let U be a bounded open set with ∂U having content 0. Also let $\mathbf{h} \in C^1\left(\overline{U}; \mathbb{R}^n\right)$ be one to one on U and $D\mathbf{h}\left(\mathbf{x}\right)^{-1}$ exists for all $\mathbf{x} \in U$. Let $f \in C\left(\overline{U}\right)$. Then*

$$\int \mathcal{X}_{\mathbf{h}(U)}\left(\mathbf{z}\right) f\left(\mathbf{z}\right) dz = \int \mathcal{X}_U\left(\mathbf{x}\right) f\left(\mathbf{h}\left(\mathbf{x}\right)\right) \left|\det D\mathbf{h}\left(\mathbf{x}\right)\right| dx$$

Proof: You note that the formula holds for $f^+ \equiv \frac{|f|+f}{2}$ and $f^- \equiv \frac{|f|-f}{2}$. Now $f = f^+ - f^-$ and so

$$\int \mathcal{X}_{\mathbf{h}(U)}(\mathbf{z}) f(\mathbf{z}) \, dz$$

$$= \int \mathcal{X}_{\mathbf{h}(U)}(\mathbf{z}) f^+(\mathbf{z}) \, dz - \int \mathcal{X}_{\mathbf{h}(U)}(\mathbf{z}) f^-(\mathbf{z}) \, dz$$

$$= \int \mathcal{X}_U(\mathbf{x}) f^+(\mathbf{h}(\mathbf{x})) |\det Dh(\mathbf{x})| \, dx - \int \mathcal{X}_U(\mathbf{x}) f^-(\mathbf{h}(\mathbf{x})) |\det Dh(\mathbf{x})| \, dx$$

$$= \int \mathcal{X}_U(\mathbf{x}) f(\mathbf{h}(\mathbf{x})) |\det Dh(\mathbf{x})| \, dx. \qquad \blacksquare$$

C.5 Some Observations

Some of the above material is very technical. This is because it gives complete answers to the fundamental questions on existence of the integral and related theoretical considerations. However, most of the difficulties are artifacts. They should not even be considered! It was realized early in the twentieth century that these difficulties occur because, from the point of view of mathematics, this is not the right way to define an integral! Better results are obtained much more easily using the Lebesgue integral. Many of the technicalities related to Jordan content disappear almost magically when the right integral is used. However, the Lebesgue integral is more abstract than the Riemann integral and it is not traditional to consider it in a beginning calculus course. If you are interested in the fundamental properties of the integral and the theory behind it, you should abandon the Riemann integral which is an antiquated relic and begin to study the integral of the last century. An introduction to it is in [23]. Another very good source is [12]. This advanced calculus text does everything in terms of the Lebesgue integral and never bothers to struggle with the inferior Riemann integral. A more general treatment is found in [18], [19], [24], and [20]. There is also a still more general integral called the generalized Riemann integral. A recent book on this subject is [5]. It is far easier to define than the Lebesgue integral but the convergence theorems are much harder to prove. An introduction is also in [19].

Appendix D

Volumes In Higher Dimensions

D.1 p Dimensional Surfaces

Everything which was done for 2 dimensional surfaces in $\mathbb{R}^3$ can be done for more general situations. The main change is in the level of generality. You need to first define carefully what is meant by the area of such a surface. This involves Definition 3.2. You should look at this definition first.

Suppose then that $\mathbf{x} = \mathbf{f}(\mathbf{u})$ where $\mathbf{u} \in U$, a subset of $\mathbb{R}^p$, and $\mathbf{x}$ is a point in V, a subset of n dimensional space where $n \geq p$. Thus, letting the Cartesian coordinates of $\mathbf{x}$ be given by $\mathbf{x} = (x_1, \cdots, x_n)^T$, each x_i being a function of $\mathbf{u}$, an infinitesimal box located at $\mathbf{u}_0$ corresponds to an infinitesimal parallelepiped located at $\mathbf{f}(\mathbf{u}_0)$ which is determined by the p vectors $\left\{ \frac{\partial \mathbf{x}(\mathbf{u}_0)}{\partial u_i} du_i \right\}_{i=1}^{p}$, each of which is tangent to the surface defined by $\mathbf{x} = \mathbf{f}(\mathbf{u})$. (No sum on the repeated index.) From Definition 3.2, the volume of this infinitesimal parallelepiped located at $\mathbf{f}(\mathbf{u}_0)$ is given by

$$\det \left(\frac{\partial \mathbf{x}(\mathbf{u}_0)}{\partial u_i} du_i \cdot \frac{\partial \mathbf{x}(\mathbf{u}_0)}{\partial u_j} du_j \right)^{1/2}. \tag{D.1}$$

The matrix in the above formula is a $p \times p$ matrix. Denoting

$$\frac{\partial \mathbf{x}(\mathbf{u}_0)}{\partial u_i} = \mathbf{x}_{,i}$$

to save space, this matrix is of the form

$$
\overbrace{\begin{pmatrix} du_1 & 0 & \cdots & 0 \\ 0 & du_2 & \cdots & 0 \\ \vdots & \vdots & \ddots & \vdots \\ 0 & 0 & \cdots & du_p \end{pmatrix}}^{p \times p}
\overbrace{\begin{pmatrix} \mathbf{x}_{,1}^T \\ \mathbf{x}_{,2}^T \\ \vdots \\ \mathbf{x}_{,p}^T \end{pmatrix}}^{p \times n}
\overbrace{\left(\mathbf{x}_{,1} \ \mathbf{x}_{,2} \ \cdots \ \mathbf{x}_{,p} \right)}^{n \times p}
\overbrace{\begin{pmatrix} du_1 & 0 & \cdots & 0 \\ 0 & du_2 & \cdots & 0 \\ \vdots & \vdots & \ddots & \vdots \\ 0 & 0 & \cdots & du_p \end{pmatrix}}^{p \times p}
$$

Therefore, by the theorem which says the determinant of a product equals the product of the determinants, the determinant of the above product equals

$$
\left(\det \begin{pmatrix} du_1 & 0 & \cdots & 0 \\ 0 & du_2 & \cdots & 0 \\ \vdots & \vdots & \ddots & \vdots \\ 0 & 0 & \cdots & du_p \end{pmatrix} \right)^2 \det \left(\begin{pmatrix} \mathbf{x}_{,1}^T \\ \mathbf{x}_{,2}^T \\ \vdots \\ \mathbf{x}_{,p}^T \end{pmatrix} \left(\mathbf{x}_{,1} \ \mathbf{x}_{,2} \ \cdots \ \mathbf{x}_{,p} \right) \right) =
$$

$$\det\left(\left(\begin{array}{c} \mathbf{x}_{,1}^T \\ \mathbf{x}_{,2}^T \\ \vdots \\ \mathbf{x}_{,p}^T \end{array}\right)\left(\begin{array}{cccc} \mathbf{x}_{,1} & \mathbf{x}_{,2} & \cdots & \mathbf{x}_{,p} \end{array}\right)\right)(du_1 du_2 \cdots du_p)^2$$

and so, taking the square root implies the volume of the infinitesimal parallelepiped at $\mathbf{x} = \mathbf{f}(\mathbf{u}_0)$ is

$$\left(\det\left(\frac{\partial \mathbf{x}(\mathbf{u}_0)}{\partial u_i} \cdot \frac{\partial \mathbf{x}(\mathbf{u}_0)}{\partial u_j}\right)\right)^{1/2} du_1\, du_2 \cdots du_p =$$

$$\det\left(D\mathbf{f}(\mathbf{u})^T D\mathbf{f}(\mathbf{u})\right)^{1/2} du_1\, du_2 \cdots du_p$$

Definition D.1. Let $\mathbf{x} = \mathbf{f}(\mathbf{u})$ be as described above. Then the symbol $\frac{\partial(x_1,\cdots x_n)}{\partial(u_1,\cdots,u_p)}$, is defined by

$$\det\left(\frac{\partial \mathbf{x}(\mathbf{u}_0)}{\partial u_i} \cdot \frac{\partial \mathbf{x}(\mathbf{u}_0)}{\partial u_j}\right)^{1/2} \equiv \left|\frac{\partial(x_1,\cdots x_n)}{\partial(u_1,\cdots,u_p)}\right|.$$

Also, the symbol $dV_p \equiv \left|\frac{\partial(x_1,\cdots x_n)}{\partial(u_1,\cdots,u_p)}\right| du_1 \cdots du_p$ is called the volume element or area element. Note the use of the subscript p. This indicates the p dimensional volume element. When $p = 2$ it is customary to write dA. Also, continue referring to $\frac{\partial(x_1,\cdots x_n)}{\partial(u_1,\cdots,u_p)}$ as the Jacobian.

This motivates the following fundamental procedure.

Procedure D.1. Suppose U is a subset of $\mathbb{R}^p$ and suppose $\mathbf{f} : U \to \mathbf{f}(U) \subseteq \mathbb{R}^n$ is a one to one and C^1 function. Then if $h : \mathbf{f}(U) \to \mathbb{R}$, define the p dimensional surface integral $\int_{\mathbf{f}(U)} h(\mathbf{x})\, dV_p$ according to the following formula.

$$\int_{\mathbf{f}(U)} h(\mathbf{x})\, dV_p \equiv \int_U h(\mathbf{f}(\mathbf{u})) \frac{\partial(x_1,\cdots x_n)}{\partial(u_1,\cdots,u_p)}\, dV.$$

As discussed earlier in the context of Lemma 3.1 this new definition of p dimensional volume does not contradict the earlier definition of 2 dimensional area.

Now here is an example of a three dimensional surface in $\mathbb{R}^4$.

Example D.1. Let $U = \{(x,y,z) : x^2 + y^2 + z^2 \le 4\}$ and for $(x,y,z) \in U$ let $\mathbf{f}(x,y,z) = (x,y,x+y,z)$. Find the three dimensional volume of $\mathbf{f}(U)$.

Note there is no picture here because I am unable to draw one in four dimensions. Nevertheless it is a three dimensional volume which is being computed. Everything is done the same as before.

$$D\mathbf{f}(x,y,z) = \begin{pmatrix} 1 & 0 & 0 \\ 0 & 1 & 0 \\ 1 & 1 & 0 \\ 0 & 0 & 1 \end{pmatrix}$$

and so

$$Df(x, y, z)^T Df(x, y, z) = \begin{pmatrix} 2 & 1 & 0 \\ 1 & 2 & 0 \\ 0 & 0 & 1 \end{pmatrix}$$

and so the volume element is $3\,dx\,dy\,dz$. Therefore, the volume of $\mathbf{f}(U)$ is

$$\int_U 3\,dx\,dy\,dz = 3\left(\frac{4}{3}\pi\,(8)\right) = 32\pi.$$

One other issue is worth mentioning. Suppose $\mathbf{f}_i : U_i \to \mathbb{R}^n$ where U_i are sets in $\mathbb{R}^p$ and suppose $\mathbf{f}_1(U_1)$ intersects $\mathbf{f}_2(U_2)$ along C where $C = \mathbf{h}(V)$ where the points of V have more than one variable held constant. Then define integrals and areas over $\mathbf{f}_1(U_1) \cup \mathbf{f}_2(U_2)$ as follows.

$$\int_{\mathbf{f}_1(U_1)\cup\mathbf{f}_2(U_2)} g\,dV_p \equiv \int_{\mathbf{f}_1(U_1)} g\,dV_p + \int_{\mathbf{f}_2(U_2)} g\,dV_p.$$

Admittedly, the set C gets added in twice but this does not matter because its p dimensional volume equals zero, and therefore, the integrals over this set will also be zero. Why is this? It is because the rank of $Dh(\mathbf{u})$ is at most $k < p$ which implies that $Dh(\mathbf{u})\mathbf{y} = \mathbf{0}$ for some $\mathbf{y} \neq \mathbf{0}$. (Why?) Thus it is also the case that $\det\left(Dh(\mathbf{u})^T Dh(\mathbf{u})\right) = 0$. (Why?) This shows the p dimensional volume element is zero and so this makes no contribution to the integral as claimed. Clearly something similar holds in the case of many surfaces joined in this way.

I have been purposely vague about precise mathematical conditions necessary for the above procedures. This is because the precise mathematical conditions which are usually cited are very technical and at the same time far too restrictive. The most general conditions under which these sorts of procedures are valid include things like Lipschitz functions defined on very general sets. These are functions satisfying a Lipschitz condition of the form $|\mathbf{f}(\mathbf{x}) - \mathbf{f}(\mathbf{y})| \leq K|\mathbf{x} - \mathbf{y}|$. For example, $y = |x|$ is Lipschitz continuous. However, this function does not have a derivative at every point. So it is with Lipschitz functions. However, it turns out that these functions have derivatives at enough points to push everything through but this requires considerations involving the Lebesgue integral. Lipschitz functions are also not the most general kind of function for which the above is valid. Much more can be said than this short introduction.

D.2 Spherical Coordinates In Many Dimensions*

Sometimes there is a need to deal with spherical coordinates in more than three dimensions. In this section, this concept is defined and formulas are derived for these coordinate systems. Recall polar coordinates are of the form

$$x_1 = \rho\cos\theta$$
$$x_2 = \rho\sin\theta$$

where $\rho > 0$ and $\theta \in [0, 2\pi)$. Here I am writing ρ in place of r to emphasize a pattern which is about to emerge. I will consider polar coordinates as spherical coordinates in two dimensions. This is also the reason I am writing x_1 and x_2 instead of the more usual x and y. Now consider what happens when you go to three dimensions. The situation is depicted in the following picture.

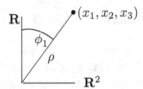

From this picture, you see that $x_3 = \rho \cos \phi_1$. Also the distance between (x_1, x_2) and $(0, 0)$ is $\rho \sin (\phi_1)$. Therefore, using polar coordinates to write (x_1, x_2) in terms of θ and this distance,

$$x_1 = \rho \sin \phi_1 \cos \theta,$$
$$x_2 = \rho \sin \phi_1 \sin \theta,$$
$$x_3 = \rho \cos \phi_1.$$

where $\phi_1 \in [0, \pi]$. What was done is to replace ρ with $\rho \sin \phi_1$ and then to add in $x_3 = \rho \cos \phi_1$. Having done this, there is no reason to stop with three dimensions. Consider the following picture:

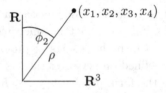

From this picture, you see that $x_4 = \rho \cos \phi_2$. Also the distance between (x_1, x_2, x_3) and $(0, 0, 0)$ is $\rho \sin (\phi_2)$. Therefore, using polar coordinates to write (x_1, x_2, x_3) in terms of θ, ϕ_1, and this distance,

$$x_1 = \rho \sin \phi_2 \sin \phi_1 \cos \theta,$$
$$x_2 = \rho \sin \phi_2 \sin \phi_1 \sin \theta,$$
$$x_3 = \rho \sin \phi_2 \cos \phi_1,$$
$$x_4 = \rho \cos \phi_2$$

where $\phi_2 \in [0, \pi]$.

Continuing this way, given spherical coordinates in $\mathbb{R}^n$, to get the spherical coordinates in $\mathbb{R}^{n+1}$, you let $x_{n+1} = \rho \cos \phi_{n-1}$ and then replace every occurrence of ρ with $\rho \sin \phi_{n-1}$ to obtain $x_1 \cdots x_n$ in terms of $\phi_1, \phi_2, \cdots, \phi_{n-1}, \theta$, and ρ.

For spherical coordinates, it is always the case that ρ measures the distance from the point in $\mathbb{R}^n$ to the origin in $\mathbb{R}^n$, $\mathbf{0}$. Each $\phi_i \in [0, \pi]$, and $\theta \in [0, 2\pi)$. I leave it as an exercise using math induction to prove that these coordinates map $\prod_{i=1}^{n-2}[0,\pi]\times[0,2\pi)\times(0,\infty)$ one to one onto $\mathbb{R}^n\setminus\{\mathbf{0}\}$.

It is customary to write S^{n-1} for the set $\{\mathbf{x}\in\mathbb{R}^n:|\mathbf{x}|=1\}$. Thus a parametrization for this level surface is given by letting $\rho=1$ in spherical coordinates. I will denote by $S^{n-1}(a)$ the sphere having radius $a>0$. What would the $n-1$ dimensional volume element on $S^{n-1}(a)$ be? For $S^1(a)$, there is only one parameter θ. Therefore, the one dimensional volume element is

$$\det\left((-a\sin\theta,a\cos\theta)\begin{pmatrix}-a\sin\theta\\a\cos\theta\end{pmatrix}\right)^{1/2}d\theta=a\,d\theta$$

where $\theta\in[0,2\pi)$.

Next consider S^2. In this case the two dimensional volume element is

$$\det\left(\begin{pmatrix}-a\sin\phi_1\sin\theta & a\sin\phi_1\cos\theta & 0\\a\cos\phi_1\cos\theta & a\cos\phi_1\sin\theta & -a\sin\phi_1\end{pmatrix}\cdot\begin{pmatrix}-a\sin\phi_1\sin\theta & a\cos\phi_1\cos\theta\\a\sin\phi_1\cos\theta & a\cos\phi_1\sin\theta\\0 & -a\sin\phi_1\end{pmatrix}\right)^{1/2}d\phi_1 d\theta$$

$$=\det\begin{pmatrix}a^2\sin^2\phi_1 & 0\\0 & a^2\end{pmatrix}^{1/2}d\phi_1 d\theta=a^2(\sin\phi_1)\,d\phi_1 d\theta.$$

What of $S^3(a)\equiv\{\mathbf{x}\in\mathbb{R}^4:|\mathbf{x}|=a\}$ and $S^n(a)=\{\mathbf{x}\in\mathbb{R}^{n+1}:|\mathbf{x}|=a\}$. Let $\mathbf{x}_n$ denote the vector $(x_1,\cdots,x_n)$. That is, $\mathbf{x}_n$ consists of the first n components of $\mathbf{x}$. Let $D_{\theta\cdots\phi_{n-2}}$ denote the derivative with respect to the vector $(\theta,\phi_1\cdots\phi_{n-2})$. Then the volume element on $S^n(a)$ is of the form

$$\det\left(\begin{pmatrix}\left(\sin(\phi_{n-1})\left(D_{\theta\cdots\phi_{n-2}}\mathbf{x}_n\right)_{n\times n-1}\right)^T_{(n-1)\times n} & (0)_{(n-1)\times 1}\\(*)_{1\times n} & -a\sin\phi_{n-1}\end{pmatrix}\cdot\begin{pmatrix}\left(\sin(\phi_{n-1})\left(D_{\theta\cdots\phi_{n-2}}\mathbf{x}_n\right)\right)_{n\times(n-1)} & (*)_{n\times 1}\\(0)_{1\times n-1} & -a\sin\phi_{n-1}\end{pmatrix}\right)^{1/2}$$

$$\cdot d\phi_{n-1}\cdots d\phi_1 d\theta$$

Now from the way we multiply matrices, this reduces to

$$a\sin(\phi_{n-1})\det\left(\left(\left(D_{\theta\cdots\phi_{n-2}}\mathbf{x}_n\right)_{n\times n-1}\right)^T_{(n-1)\times n}\left(D_{\theta\cdots\phi_{n-2}}\mathbf{x}_n\right)_{n\times(n-1)}\right)^{1/2}$$

$$\cdot d\phi_{n-1}\cdots d\phi_1 d\theta.$$

That is, to get the volume element in $S^n(a)$, you multiply the volume element on $S^{n-1}(a)$ by $a\sin(\phi_{n-1})\,d\phi_{n-1}$. Consequently, beginning with the volume element on $S^1(a)$, you obtain the succession of volume elements for $S^1(a)$, $S^2(a)$, $S^3(a)$, $S^4(a)$.

$$a\,d\theta,\ a^2\sin\phi_1 d\phi_1 d\theta,\ a^3\sin\phi_2\sin\phi_1 d\phi_2 d\phi_1 d\theta,$$

$$a^4\sin\phi_3\sin\phi_2\sin\phi_1 d\phi_3 d\phi_2 d\phi_1 d\theta,\ \text{etc.}$$

In general, the n dimensional volume element on $S^n(a)$ is

$$a^n \left(\prod_{i=1}^{n-1} \sin \phi_i \right) \left(\prod_{i=1}^{n-1} d\phi_i \right) d\theta. \tag{D.2}$$

Using similar reasoning, the n dimensional volume element in terms of the spherical coordinates is

$$\rho^{n-1} \left(\prod_{i=1}^{n-2} \sin \phi_i \right) \left(\prod_{i=1}^{n-2} d\phi_i \right) d\theta d\rho. \tag{D.3}$$

Formulas (D.2) and (D.3) are very useful in estimating integrals.

Example D.2. For what values of s is the integral $\int_{B(0,R)} \left(1 + |\mathbf{x}|^2 \right)^s dV$ bounded independent of R? Here $B(0, R)$ is the ball, $\{\mathbf{x} \in \mathbb{R}^n : |\mathbf{x}| \leq R\}$.

I think you can see immediately that s must be negative but exactly how negative? It turns out it depends on n and using spherical coordinates, you can find just exactly what is needed. Write the above integral in n dimensional spherical coordinates.

$$\int_0^R \int_0^\pi \cdots \int_0^\pi \int_0^{2\pi} \rho^{n-1} \left(1 + \rho^2 \right)^s \left(\prod_{i=1}^{n-2} \sin \phi_i \right) \left(\prod_{i=1}^{n-2} d\phi_i \right) d\theta d\rho$$

$$= \int_0^R \rho^{n-1} \left(1 + \rho^2 \right)^s \int_{S^{n-1}} dS^{n-1} d\rho = \omega_n \int_0^R \rho^{n-1} \left(1 + \rho^2 \right)^s d\rho$$

where $dS^{n-1} = \left(\prod_{i=1}^{n-2} \sin \phi_i \right) \left(\prod_{i=1}^{n-2} d\phi_i \right) d\theta$.

$$\omega_n \equiv \int_0^\pi \cdots \int_0^\pi \int_0^{2\pi} \left(\prod_{i=1}^{n-2} \sin \phi_i \right) \left(\prod_{i=1}^{n-2} d\phi_i \right) d\theta$$

and from the above explanation this equals the area of S^{n-1}, the unit sphere,

$$\{\mathbf{x} \in \mathbb{R}^n : |\mathbf{x}| = 1\}.$$

Now the very hard problem has been reduced to considering an easy one variable problem of finding when

$$\int_0^R \rho^{n-1} \left(1 + \rho^2 \right)^s d\rho$$

is bounded independent of R. I leave it to you to verify using standard one variable calculus that you need $2s + (n - 1) < -1$ so you need $s < -n/2$.

Appendix E

Some Physical Applications

E.1 Product Rule For Matrices

Here is the concept of the product rule extended to matrix multiplication.

Definition E.1. Let $A(t)$ be an $m \times n$ matrix. Say $A(t) = (A_{ij}(t))$. Suppose also that $A_{ij}(t)$ is a differentiable function for all i, j. Then define $A'(t) \equiv (A'_{ij}(t))$. That is, $A'(t)$ is the matrix which consists of replacing each entry by its derivative. Such an $m \times n$ matrix in which the entries are differentiable functions is called a differentiable matrix.

The next lemma is just a version of the product rule.

Lemma E.1. *Let $A(t)$ be an $m \times n$ matrix and let $B(t)$ be an $n \times p$ matrix with the property that all the entries of these matrices are differentiable functions. Then*

$$(A(t) B(t))' = A'(t) B(t) + A(t) B'(t).$$

Proof: $(A(t) B(t))' = (C'_{ij}(t))$ where $C_{ij}(t) = A_{ik}(t) B_{kj}(t)$ and the repeated index summation convention is being used. Therefore,

$$
\begin{aligned}
C'_{ij}(t) &= A'_{ik}(t) B_{kj}(t) + A_{ik}(t) B'_{kj}(t) \\
&= (A'(t) B(t))_{ij} + (A(t) B'(t))_{ij} \\
&= (A'(t) B(t) + A(t) B'(t))_{ij}
\end{aligned}
$$

Therefore, the ij^{th} entry of $A(t) B(t)$ equals the ij^{th} entry of $A'(t) B(t) + A(t) B'(t)$ and this proves the lemma.

E.2 Moving Coordinate Systems

Let $\mathbf{i}(t), \mathbf{j}(t), \mathbf{k}(t)$ be a right handed[4] orthonormal basis of vectors for each t. It is assumed these vectors are C^1 functions of t. Letting the positive x axis extend in the direction of $\mathbf{i}(t)$, the positive y axis extend in the direction of $\mathbf{j}(t)$, and the

---

[4] Recall that right handed implies $\mathbf{i} \times \mathbf{j} = \mathbf{k}$.

positive z axis extend in the direction of $\mathbf{k}(t)$, yields a moving coordinate system. Now let $\mathbf{u} = (u_1, u_2, u_3) \in \mathbb{R}^3$ and let t_0 be some reference time. For example you could let $t_0 = 0$. Then define the components of $\mathbf{u}$ with respect to these vectors $\mathbf{i}, \mathbf{j}, \mathbf{k}$ at time t_0 as

$$\mathbf{u} \equiv u_1 \mathbf{i}(t_0) + u_2 \mathbf{j}(t_0) + u_3 \mathbf{k}(t_0).$$

Let $\mathbf{u}(t)$ be defined as the vector which has the same components with respect to $\mathbf{i}, \mathbf{j}, \mathbf{k}$ but at time t. Thus

$$\mathbf{u}(t) \equiv u_1 \mathbf{i}(t) + u_2 \mathbf{j}(t) + u_3 \mathbf{k}(t)$$

and the vector has changed although the components have not.

For example, this is exactly the situation in the case of apparently fixed basis vectors on the earth if $\mathbf{u}$ is a position vector from the given spot on the earth's surface to a point regarded as fixed with the earth due to its keeping the same coordinates relative to coordinate axes which are fixed with the earth.

Now define a linear transformation $Q(t)$ mapping $\mathbb{R}^3$ to $\mathbb{R}^3$ by

$$Q(t) \mathbf{u} \equiv u_1 \mathbf{i}(t) + u_2 \mathbf{j}(t) + u_3 \mathbf{k}(t)$$

where

$$\mathbf{u} \equiv u_1 \mathbf{i}(t_0) + u_2 \mathbf{j}(t_0) + u_3 \mathbf{k}(t_0)$$

Thus letting $\mathbf{v}, \mathbf{u} \in \mathbb{R}^3$ be vectors and α, β, scalars,

$$
\begin{aligned}
Q(t)(\alpha \mathbf{u} + \beta \mathbf{v}) &\equiv (\alpha u_1 + \beta v_1) \mathbf{i}(t) + (\alpha u_2 + \beta v_2) \mathbf{j}(t) + (\alpha u_3 + \beta v_3) \mathbf{k}(t) \\
&= (\alpha u_1 \mathbf{i}(t) + \alpha u_2 \mathbf{j}(t) + \alpha u_3 \mathbf{k}(t)) + (\beta v_1 \mathbf{i}(t) + \beta v_2 \mathbf{j}(t) + \beta v_3 \mathbf{k}(t)) \\
&= \alpha(u_1 \mathbf{i}(t) + u_2 \mathbf{j}(t) + u_3 \mathbf{k}(t)) + \beta(v_1 \mathbf{i}(t) + v_2 \mathbf{j}(t) + v_3 \mathbf{k}(t)) \\
&\equiv \alpha Q(t) \mathbf{u} + \beta Q(t) \mathbf{v}
\end{aligned}
$$

showing that $Q(t)$ is a linear transformation. Also, $Q(t)$ preserves all distances because, since the vectors $\mathbf{i}(t), \mathbf{j}(t), \mathbf{k}(t)$ form an orthonormal set,

$$|Q(t) \mathbf{u}| = \left(\sum_{i=1}^{3} (u^i)^2\right)^{1/2} = |\mathbf{u}|.$$

For simplicity, let

$$\mathbf{i}(t) = \mathbf{e}_1(t), \mathbf{j}(t) = \mathbf{e}_2(t), \mathbf{k}(t) = \mathbf{e}_3(t)$$

and

$$\mathbf{i}(t_0) = \mathbf{e}_1(t_0), \mathbf{j}(t_0) = \mathbf{e}_2(t_0), \mathbf{k}(t_0) = \mathbf{e}_3(t_0).$$

Then using the repeated index summation convention,

$$\mathbf{u}(t) = u_j \mathbf{e}_j(t) = u_j \mathbf{e}_j(t) \cdot \mathbf{e}_i(t_0) \mathbf{e}_i(t_0).$$

Thus, from the definition of matrix multiplication, the components of $\mathbf{u}(t)$ with respect to $\mathbf{i}(t_0), \mathbf{j}(t_0), \mathbf{k}(t_0)$ are obtained in the form $Q(t) \mathbf{u}$ where $\mathbf{u} = u_1 \mathbf{i}(t_0) + u_2 \mathbf{j}(t_0) + u_3 \mathbf{k}(t_0)$, where $Q_{ij}(t) = \mathbf{e}_j(t) \cdot \mathbf{e}_i(t_0)$.

Lemma E.2. *Suppose $Q(t)$ is a real, differentiable $n \times n$ matrix which preserves distances. Then $Q(t) Q(t)^T = Q(t)^T Q(t) = I$. Also, if $\mathbf{u}(t) \equiv Q(t) \mathbf{u}$, then there exists a vector $\boldsymbol{\Omega}(t)$ such that*

$$\mathbf{u}'(t) = \boldsymbol{\Omega}(t) \times \mathbf{u}(t).$$

Proof: Recall that $(\mathbf{z} \cdot \mathbf{w}) = \frac{1}{4}\left(|\mathbf{z} + \mathbf{w}|^2 - |\mathbf{z} - \mathbf{w}|^2\right)$. Therefore,

$$
\begin{aligned}
(Q(t)\,\mathbf{u}\cdot Q(t)\,\mathbf{w}) &= \frac{1}{4}\left(|Q(t)(\mathbf{u} + \mathbf{w})|^2 - |Q(t)(\mathbf{u} - \mathbf{w})|^2\right) \\
&= \frac{1}{4}\left(|\mathbf{u} + \mathbf{w}|^2 - |\mathbf{u} - \mathbf{w}|^2\right) \\
&= (\mathbf{u} \cdot \mathbf{w}).
\end{aligned}
$$

This implies

$$
\left(Q(t)^T Q(t)\,\mathbf{u} \cdot \mathbf{w}\right) = (\mathbf{u} \cdot \mathbf{w})
$$

for all $\mathbf{u}, \mathbf{w}$. Therefore, $Q(t)^T Q(t)\,\mathbf{u} = \mathbf{u}$, so $Q(t)^T Q(t) = Q(t) Q(t)^T = I$. This proves the first part of the lemma.

It follows from the product rule, Lemma E.1 that

$$
Q'(t) Q(t)^T + Q(t) Q'(t)^T = 0,
$$

so

$$
Q'(t) Q(t)^T = -\left(Q'(t) Q(t)^T\right)^T. \tag{E.1}
$$

From the definition, $Q(t)\,\mathbf{u} = \mathbf{u}(t)$,

$$
\mathbf{u}'(t) = Q'(t)\,\mathbf{u} = Q'(t) \overbrace{Q(t)^T \mathbf{u}(t)}^{=\mathbf{u}}.
$$

Then it follows from (E.1) that the matrix of $Q'(t) Q(t)^T$ is of the form

$$
\begin{pmatrix}
0 & -\omega_3(t) & \omega_2(t) \\
\omega_3(t) & 0 & -\omega_1(t) \\
-\omega_2(t) & \omega_1(t) & 0
\end{pmatrix}
$$

for some time dependent scalars, ω_i. (This matrix $Q'Q^T$ is skew symmetric is the message of E.1.) Therefore,

$$
\begin{aligned}
\begin{pmatrix} u_1 \\ u_2 \\ u_3 \end{pmatrix}'(t) &= \begin{pmatrix}
0 & -\omega_3(t) & \omega_2(t) \\
\omega_3(t) & 0 & -\omega_1(t) \\
-\omega_2(t) & \omega_1(t) & 0
\end{pmatrix}\begin{pmatrix} u_1 \\ u_2 \\ u_3 \end{pmatrix}(t) \\
&= \begin{pmatrix}
\omega_2(t)\,u_3(t) - \omega_3(t)\,u_2(t) \\
\omega_3(t)\,u_1(t) - \omega_1(t)\,u_3(t) \\
\omega_1(t)\,u_2(t) - \omega_2(t)\,u_1(t)
\end{pmatrix}
\end{aligned}
$$

where the u_i are the components of the vector $\mathbf{u}(t)$ in terms of the fixed vectors

$$
\mathbf{i}(t_0), \mathbf{j}(t_0), \mathbf{k}(t_0).
$$

Therefore,

$$
\mathbf{u}'(t) = \mathbf{\Omega}(t) \times \mathbf{u}(t) = Q'(t) Q(t)^T \mathbf{u}(t) \tag{E.2}
$$

where

$$\Omega\left(t\right) = \omega_1\left(t\right)\mathbf{i}\left(t_0\right) + \omega_2\left(t\right)\mathbf{j}\left(t_0\right) + \omega_3\left(t\right)\mathbf{k}\left(t_0\right).$$

because

$$\Omega\left(t\right) \times \mathbf{u}\left(t\right) \equiv \begin{vmatrix} \mathbf{i}\left(t_0\right) & \mathbf{j}\left(t_0\right) & \mathbf{k}\left(t_0\right) \\ w_1 & w_2 & w_3 \\ u_1 & u_2 & u_3 \end{vmatrix} \equiv$$

$$\mathbf{i}\left(t_0\right)\left(w_2 u_3 - w_3 u_2\right) + \mathbf{j}\left(t_0\right)\left(w_3 u_1 - w_1 u_3\right) + \mathbf{k}\left(t_0\right)\left(w_1 u_2 - w_2 u_1\right). \qquad \blacksquare$$

The above lemma yields the existence part of the following theorem.

Theorem E.1. *Let* $\mathbf{i}\left(t\right), \mathbf{j}\left(t\right), \mathbf{k}\left(t\right)$ *be as described. Then there exists a unique vector* $\Omega\left(t\right)$ *such that if* $\mathbf{u}\left(t\right)$ *is a vector whose components are constant with respect to* $\mathbf{i}\left(t\right), \mathbf{j}\left(t\right), \mathbf{k}\left(t\right),$ *then*

$$\mathbf{u}'\left(t\right) = \Omega\left(t\right) \times \mathbf{u}\left(t\right).$$

Proof: It only remains to prove uniqueness. Suppose Ω_1 also works. Then $\mathbf{u}\left(t\right) = Q\left(t\right)\mathbf{u}$ and so $\mathbf{u}'\left(t\right) = Q'\left(t\right)\mathbf{u}$ and

$$Q'\left(t\right)\mathbf{u} = \Omega \times Q\left(t\right)\mathbf{u} = \Omega_1 \times Q\left(t\right)\mathbf{u}$$

for all $\mathbf{u}$. Therefore,

$$\left(\Omega - \Omega_1\right) \times Q\left(t\right)\mathbf{u} = \mathbf{0}$$

for all $\mathbf{u}$, and since $Q\left(t\right)$ is one to one and onto, this implies $\left(\Omega - \Omega_1\right) \times \mathbf{w} = \mathbf{0}$ for all $\mathbf{w}$, and thus $\Omega - \Omega_1 = \mathbf{0}$. $\blacksquare$

Definition E.2. A **rigid body** in $\mathbb{R}^3$ has a moving coordinate system with the property that for an observer on the rigid body, the vectors $\mathbf{i}\left(t\right), \mathbf{j}\left(t\right), \mathbf{k}\left(t\right)$ are constant. More generally, a vector $\mathbf{u}\left(t\right)$ is said to be fixed with the body if to a person on the body, the vector appears to have the same magnitude and same direction independent of t. Thus $\mathbf{u}\left(t\right)$ is fixed with the body if $\mathbf{u}\left(t\right) = u_1\mathbf{i}\left(t\right) + u_2\mathbf{j}\left(t\right) + u_3\mathbf{k}\left(t\right)$.

The following comes from the above discussion.

Theorem E.2. *Let* $B\left(t\right)$ *be the set of points in three dimensions occupied by a rigid body. Then there exists a vector* $\Omega\left(t\right)$ *such that whenever* $\mathbf{u}\left(t\right)$ *is fixed with the rigid body,*

$$\mathbf{u}'\left(t\right) = \Omega\left(t\right) \times \mathbf{u}\left(t\right).$$

E.2.1   *The Spinning Top*

To begin with, consider a spinning top as illustrated in the following picture.

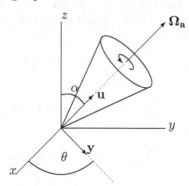

For the purpose of this discussion, consider the top as a large number of point masses m_i, located at the positions, $\mathbf{r}_i(t)$ for $i = 1, 2, \cdots, N$, and these masses are symmetrically arranged relative to the axis of the top. As the top spins, the axis of symmetry is observed to move around the z axis. This is called precession and you will see it occur whenever you spin a top. What is the speed of this precession? In other words, what is θ'? The following discussion follows one given in Sears and Zemansky [26].

Imagine a coordinate system which is fixed relative to the moving top. Thus, in this coordinate system, the points of the top are fixed. Let the standard unit vectors of the coordinate system moving with the top be denoted by $\mathbf{i}(t), \mathbf{j}(t), \mathbf{k}(t)$. From Theorem E.1, there exists an angular velocity vector $\boldsymbol{\Omega}(t)$ such that if $\mathbf{u}(t)$ is the position vector of a point fixed in the top, $(\mathbf{u}(t) = u_1\mathbf{i}(t) + u_2\mathbf{j}(t) + u_3\mathbf{k}(t))$,
$$\mathbf{u}'(t) = \boldsymbol{\Omega}(t) \times \mathbf{u}(t).$$
The vector $\boldsymbol{\Omega}_a$ shown in the picture is the vector for which
$$\mathbf{r}'_i(t) \equiv \boldsymbol{\Omega}_a \times \mathbf{r}_i(t)$$
is the velocity of the ith point mass due to rotation about the axis of the top. Thus $\boldsymbol{\Omega}(t) = \boldsymbol{\Omega}_a(t) + \boldsymbol{\Omega}_p(t)$ and it is assumed $\boldsymbol{\Omega}_p(t)$ is very small relative to $\boldsymbol{\Omega}_a$. In other words, it is assumed the axis of the top moves very slowly relative to the speed of the points in the top which are spinning very fast around the axis of the top. The angular momentum, $\mathbf{L}$ is defined by
$$\mathbf{L} \equiv \sum_{i=1}^{N} \mathbf{r}_i \times m_i \mathbf{v}_i \tag{E.3}$$
where $\mathbf{v}_i$ equals the velocity of the ith point mass. Thus $\mathbf{v}_i = \boldsymbol{\Omega}(t) \times \mathbf{r}_i$, and from the above assumption, $\mathbf{v}_i$ may be taken equal to $\boldsymbol{\Omega}_a \times \mathbf{r}_i$. Therefore, $\mathbf{L}$ is essentially given by
$$\mathbf{L} \equiv \sum_{i=1}^{N} m_i \mathbf{r}_i \times (\boldsymbol{\Omega}_a \times \mathbf{r}_i) = \sum_{i=1}^{N} m_i \left(|\mathbf{r}_i|^2 \boldsymbol{\Omega}_a - (\mathbf{r}_i \cdot \boldsymbol{\Omega}_a) \mathbf{r}_i \right).$$

This follows from an identity involving the cross product. By symmetry of the top, this last expression equals a multiple of $\mathbf{\Omega}_a$. Thus $\mathbf{L}$ is parallel to $\mathbf{\Omega}_a$. Also, beginning with the first equality above and using the identities involving the cross product from Volume 1, (Switch the dot and the cross.)

$$
\begin{aligned}
\mathbf{L} \cdot \mathbf{\Omega}_a &= \sum_{i-1}^{N} m_i \mathbf{\Omega}_a \cdot \mathbf{r}_i \times (\mathbf{\Omega}_a \times \mathbf{r}_i) \\
&= \sum_{i=1}^{N} m_i (\mathbf{\Omega}_a \times \mathbf{r}_i) \cdot (\mathbf{\Omega}_a \times \mathbf{r}_i) \\
&= \sum_{i=1}^{N} m_i |\mathbf{\Omega}_a \times \mathbf{r}_i|^2 = \sum_{i=1}^{N} m_i |\mathbf{\Omega}_a|^2 |\mathbf{r}_i|^2 \sin^2 (\beta_i)
\end{aligned}
$$

where β_i denotes the angle between the position vector of the ith point mass and the axis of the top. Since this expression is positive, this also shows $\mathbf{L}$ has the same direction as $\mathbf{\Omega}_a$. Let $\omega \equiv |\mathbf{\Omega}_a|$. Then the above expression is of the form

$$\mathbf{L} \cdot \mathbf{\Omega}_a = I \omega^2,$$

where

$$I \equiv \sum_{i=1}^{N} m_i |\mathbf{r}_i|^2 \sin^2 (\beta_i).$$

Thus, to get I you take the mass of the ith point mass, multiply it by the square of its distance to the axis of the top and add all these up. This is defined as the moment of inertia of the top about the axis of the top.

Letting $\mathbf{u}$ denote a unit vector in the direction of the axis of the top, this implies

$$\mathbf{L} = I \omega \mathbf{u}. \tag{E.4}$$

Note the simple description of the angular momentum in terms of the moment of inertia. Referring to the above picture, define the vector $\mathbf{y}$ to be the projection of the vector $\mathbf{u}$ on the xy plane. Thus

$$\mathbf{y} = \mathbf{u} - (\mathbf{u} \cdot \mathbf{k}) \mathbf{k}$$

and

$$(\mathbf{u} \cdot \mathbf{i}) = (\mathbf{y} \cdot \mathbf{i}) = \sin \alpha \cos \theta. \tag{E.5}$$

Now also from (E.3),

$$
\frac{d\mathbf{L}}{dt} = \sum_{i=1}^{N} m_i \overbrace{\mathbf{r}_i' \times \mathbf{v}_i}^{=0} + \mathbf{r}_i \times m_i \mathbf{v}_i' = \sum_{i=1}^{N} \mathbf{r}_i \times m_i \mathbf{v}_i' = - \sum_{i=1}^{N} \mathbf{r}_i \times m_i g \mathbf{k}
$$

where g is the acceleration of gravity. From (E.4), (E.5), and the above,

$$\frac{d\mathbf{L}}{dt} \cdot \mathbf{i} = I\omega\left(\frac{d\mathbf{u}}{dt}\cdot\mathbf{i}\right) = I\omega\left(\frac{d\mathbf{y}}{dt}\cdot\mathbf{i}\right)$$

$$= (-I\omega\sin\alpha\sin\theta)\,\theta' = -\sum_{i=1}^{N}\mathbf{r}_i\times m_i g\mathbf{k}\cdot\mathbf{i}$$

$$= -\sum_{i=1}^{N}m_i g\mathbf{r}_i\cdot\mathbf{k}\times\mathbf{i} = -\sum_{i=1}^{N}m_i g\mathbf{r}_i\cdot\mathbf{j}. \tag{E.6}$$

To simplify this further, recall the following definition of the center of mass.

Definition E.3. Define the total mass M by

$$M = \sum_{i=1}^{N}m_i$$

and the center of mass $\mathbf{r}_0$ by

$$\mathbf{r}_0 \equiv \frac{\sum_{i=1}^{N}\mathbf{r}_i m_i}{M}. \tag{E.7}$$

In terms of the center of mass, the last expression equals

$$\begin{aligned}-Mg\mathbf{r}_0\cdot\mathbf{j} &= -Mg\left(\mathbf{r}_0-(\mathbf{r}_0\cdot\mathbf{k})\,\mathbf{k}+(\mathbf{r}_0\cdot\mathbf{k})\,\mathbf{k}\right)\cdot\mathbf{j}\\ &= -Mg\left(\mathbf{r}_0-(\mathbf{r}_0\cdot\mathbf{k})\,\mathbf{k}\right)\cdot\mathbf{j}\\ &= -Mg\left|\mathbf{r}_0-(\mathbf{r}_0\cdot\mathbf{k})\,\mathbf{k}\right|\cos\theta\\ &= -Mg\left|\mathbf{r}_0\right|\sin\alpha\cos\left(\frac{\pi}{2}-\theta\right).\end{aligned}$$

Note that by symmetry, $\mathbf{r}_0(t)$ is on the axis of the top, is in the same direction as $\mathbf{L}$, $\mathbf{u}$, and $\boldsymbol{\Omega}_a$, and also $|\mathbf{r}_0|$ is independent of t. Therefore, from the second line of (E.6),

$$(-I\omega\sin\alpha\sin\theta)\,\theta' = -Mg\left|\mathbf{r}_0\right|\sin\alpha\sin\theta.$$

which shows

$$\theta' = \frac{Mg\left|\mathbf{r}_0\right|}{I\omega}. \tag{E.8}$$

From (E.8), the angular velocity of precession does not depend on α in the picture. It also is slower when ω is large and I is large.

The above discussion is a considerable simplification of the problem of a spinning top obtained from an assumption that $\boldsymbol{\Omega}_a$ is approximately equal to $\boldsymbol{\Omega}$. It also leaves out all considerations of friction and the observation that the axis of symmetry wobbles. This is wobbling is called **nutation**. The full mathematical treatment of this problem involves the Euler angles and some fairly complicated differential equations obtained using techniques discussed in advanced physics classes. Lagrange studied these types of problems back in the 1700's.

E.2.2   *Kinetic Energy*

The next problem is that of understanding the total kinetic energy of a collection of moving point masses. Consider a possibly large number of point masses m_i located at the positions $\mathbf{r}_i$ for $i = 1, 2, \cdots, N$. Thus the velocity of the ith point mass is $\mathbf{r}_i' = \mathbf{v}_i$. The kinetic energy of the mass m_i is defined by

$$\frac{1}{2} m_i \left| \mathbf{r}_i' \right|^2 .$$

(This is a very good time to review the concept of kinetic energy.) The total kinetic energy of the collection of masses is then

$$E = \sum_{i=1}^{N} \frac{1}{2} m_i \left| \mathbf{r}_i' \right|^2 . \tag{E.9}$$

As these masses move about, so does the center of mass $\mathbf{r}_0$. Thus $\mathbf{r}_0$ is a function of t just as the other $\mathbf{r}_i$. From (E.9) the total kinetic energy is

$$
\begin{aligned}
E &= \sum_{i=1}^{N} \frac{1}{2} m_i \left| \mathbf{r}_i' - \mathbf{r}_0' + \mathbf{r}_0' \right|^2 \\
&= \sum_{i=1}^{N} \frac{1}{2} m_i \left[\left| \mathbf{r}_i' - \mathbf{r}_0' \right|^2 + \left| \mathbf{r}_0' \right|^2 + 2 \left(\mathbf{r}_i' - \mathbf{r}_0' \cdot \mathbf{r}_0' \right) \right] .
\end{aligned} \tag{E.10}
$$

Now

$$
\sum_{i=1}^{N} m_i \left(\mathbf{r}_i' - \mathbf{r}_0' \cdot \mathbf{r}_0' \right) = \left(\sum_{i=1}^{N} m_i \left(\mathbf{r}_i - \mathbf{r}_0 \right) \right)' \cdot \mathbf{r}_0'
$$
$$= 0$$

because from (E.7)

$$
\begin{aligned}
\sum_{i=1}^{N} m_i \left(\mathbf{r}_i - \mathbf{r}_0 \right) &= \sum_{i=1}^{N} m_i \mathbf{r}_i - \sum_{i=1}^{N} m_i \mathbf{r}_0 \\
&= \sum_{i=1}^{N} m_i \mathbf{r}_i - \sum_{i=1}^{N} m_i \left(\frac{\sum_{i=1}^{N} \mathbf{r}_i m_i}{\sum_{i=1}^{N} m_i} \right) = \mathbf{0}.
\end{aligned}
$$

Let $M \equiv \sum_{i=1}^{N} m_i$ be the total mass. Then (E.10) reduces to

$$
\begin{aligned}
E &= \sum_{i=1}^{N} \frac{1}{2} m_i \left[\left| \mathbf{r}_i' - \mathbf{r}_0' \right|^2 + \left| \mathbf{r}_0' \right|^2 \right] \\
&= \frac{1}{2} M \left| \mathbf{r}_0' \right|^2 + \sum_{i=1}^{N} \frac{1}{2} m_i \left| \mathbf{r}_i' - \mathbf{r}_0' \right|^2 .
\end{aligned} \tag{E.11}
$$

The first term is just the kinetic energy of a point mass equal to the sum of all the masses involved, located at the center of mass of the system of masses while the second term represents kinetic energy which comes from the relative velocities of the masses taken with respect to the center of mass. It is this term which is considered more carefully in the case where the system of masses maintain distance between each other.

To illustrate the contrast between the case where the masses maintain a constant distance and one in which they do not, take a hard boiled egg and spin it and then take a raw egg and give it a spin. You will certainly feel a big difference in the way the two eggs respond. Incidentally, this is a good way to tell whether the egg has been hard boiled or is raw and can be used to prevent messiness which could occur if you think it is hard boiled and it really is not.

Now let $e_1(t)$, $e_2(t)$, and $e_3(t)$ be an orthonormal set of vectors which is fixed in the body undergoing rigid body motion. This means that $\mathbf{r}_i(t) - \mathbf{r}_0(t)$ has components which are constant in t with respect to the vectors $e_i(t)$. By Theorem E.1 on Page 352 there exists a vector $\mathbf{\Omega}(t)$ which does not depend on i such that

$$\mathbf{r}_i'(t) - \mathbf{r}_0'(t) = \mathbf{\Omega}(t) \times (\mathbf{r}_i(t) - \mathbf{r}_0(t)).$$

Now using this in (E.11),

$$
\begin{aligned}
E &= \frac{1}{2}M\,|\mathbf{r}_0'|^2 + \sum_{i=1}^{N}\frac{1}{2}m_i\,|\mathbf{\Omega}(t)\times(\mathbf{r}_i(t)-\mathbf{r}_0(t))|^2 \\
&= \frac{1}{2}M\,|\mathbf{r}_0'|^2 + \frac{1}{2}\left(\sum_{i=1}^{N}m_i\,|\mathbf{r}_i(t)-\mathbf{r}_0(t)|^2\sin^2\theta_i\right)|\mathbf{\Omega}(t)|^2 \\
&= \frac{1}{2}M\,|\mathbf{r}_0'|^2 + \frac{1}{2}\left(\sum_{i=1}^{N}m_i\,|\mathbf{r}_i(0)-\mathbf{r}_0(0)|^2\sin^2\theta_i\right)|\mathbf{\Omega}(t)|^2
\end{aligned}
$$

where θ_i is the angle between $\mathbf{\Omega}(t)$ and the vector $\mathbf{r}_i(t) - \mathbf{r}_0(t)$. Therefore, $|\mathbf{r}_i(t) - \mathbf{r}_0(t)|\sin\theta_i$ is the distance between the point mass m_i located at $\mathbf{r}_i$ and a line through the center of mass $\mathbf{r}_0$ with direction $\mathbf{\Omega}$ as indicated in the following picture.

Thus the expression, $\sum_{i=1}^{N}m_i\,|\mathbf{r}_i(0)-\mathbf{r}_0(0)|^2\sin^2\theta_i$ plays the role of a mass in the definition of kinetic energy except, instead of the speed, substitute the angular speed $|\mathbf{\Omega}(t)|$. It is this expression which is called the moment of inertia about the line whose direction is $\mathbf{\Omega}(t)$.

In both of these examples, the center of mass and the moment of inertia occurred in a natural way.

E.3   Acceleration With Respect To Moving Coordinate Systems[*]

You have a coordinate system which is moving, and this results in strange forces experienced relative to these moving coordinates systems. A good example is what we experience every day living on a rotating ball. Relative to our supposedly fixed coordinate system, we experience forces which account for many phenomena which are observed.

E.3.1   *The Coriolis Acceleration*

Imagine a point on the surface of the earth. Now consider unit vectors, one pointing South, one pointing East and one pointing directly away from the center of the earth.

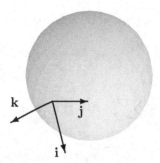

Denote the first as $\mathbf{i}$, the second as $\mathbf{j}$ and the third as $\mathbf{k}$. If you are standing on the earth, you will consider these vectors as fixed, but of course they are not. As the earth turns, they change direction and so each is in reality a function of t. Nevertheless, it is with respect to these apparently fixed vectors that you wish to understand acceleration, velocities, and displacements.

In general, let $\mathbf{i}(t), \mathbf{j}(t), \mathbf{k}(t)$ be an orthonormal basis of vectors for each t, like the vectors described in the first paragraph. It is assumed these vectors are C^1 functions of t. Letting the positive x axis extend in the direction of $\mathbf{i}(t)$, the positive y axis extend in the direction of $\mathbf{j}(t)$, and the positive z axis extend in the direction of $\mathbf{k}(t)$, yields a moving coordinate system. By Theorem E.1 on Page 352, there exists an angular velocity vector $\boldsymbol{\Omega}(t)$ such that if $\mathbf{u}(t)$ is any vector which has constant components with respect to $\mathbf{i}(t), \mathbf{j}(t)$, and $\mathbf{k}(t)$, then

$$\boldsymbol{\Omega} \times \mathbf{u} = \mathbf{u}'. \tag{E.12}$$

Now let $\mathbf{R}(t)$ be a position vector of the moving coordinate system and let

$$\mathbf{r}(t) = \mathbf{R}(t) + \mathbf{r}_B(t)$$

where

$$\mathbf{r}_B(t) \equiv x(t)\,\mathbf{i}(t) + y(t)\,\mathbf{j}(t) + z(t)\,\mathbf{k}(t).$$

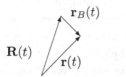

In the example of the earth, $\mathbf{R}(t)$ is the position vector of a point $\mathbf{p}(t)$ on the earth's surface and $\mathbf{r}_B(t)$ is the position vector of another point from $\mathbf{p}(t)$, thus regarding $\mathbf{p}(t)$ as the origin. $\mathbf{r}_B(t)$ is the position vector of a point as perceived by the observer on the earth with respect to the vectors he thinks of as fixed. Similarly, $\mathbf{v}_B(t)$ and $\mathbf{a}_B(t)$ will be the velocity and acceleration relative to $\mathbf{i}(t), \mathbf{j}(t), \mathbf{k}(t)$, and so $\mathbf{v}_B = x'\mathbf{i} + y'\mathbf{j} + z'\mathbf{k}$ and $\mathbf{a}_B = x''\mathbf{i} + y''\mathbf{j} + z''\mathbf{k}$. Then

$$\mathbf{v} \equiv \mathbf{r}' = \mathbf{R}' + x'\mathbf{i} + y'\mathbf{j} + z'\mathbf{k} + x\mathbf{i}' + y\mathbf{j}' + z\mathbf{k}'.$$

By (E.12), if $\mathbf{e} \in \{\mathbf{i}, \mathbf{j}, \mathbf{k}\}$, $\mathbf{e}' = \mathbf{\Omega} \times \mathbf{e}$ because the components of these vectors with respect to $\mathbf{i}, \mathbf{j}, \mathbf{k}$ are constant. Therefore,

$$\begin{aligned} x\mathbf{i}' + y\mathbf{j}' + z\mathbf{k}' &= x\mathbf{\Omega} \times \mathbf{i} + y\mathbf{\Omega} \times \mathbf{j} + z\mathbf{\Omega} \times \mathbf{k} \\ &= \mathbf{\Omega} \times (x\mathbf{i} + y\mathbf{j} + z\mathbf{k}) \end{aligned}$$

and consequently,

$$\mathbf{v} = \mathbf{R}' + x'\mathbf{i} + y'\mathbf{j} + z'\mathbf{k} + \mathbf{\Omega} \times \mathbf{r}_B = \mathbf{R}' + x'\mathbf{i} + y'\mathbf{j} + z'\mathbf{k} + \mathbf{\Omega} \times (x\mathbf{i} + y\mathbf{j} + z\mathbf{k}).$$

Now consider the acceleration. Quantities which are relative to the moving coordinate system are distinguished by using the subscript B.

$$\mathbf{a} = \mathbf{v}' = \mathbf{R}'' + x''\mathbf{i} + y''\mathbf{j} + z''\mathbf{k} + \overbrace{x'\mathbf{i}' + y'\mathbf{j}' + z'\mathbf{k}'}^{\mathbf{\Omega} \times \mathbf{v}_B} + \mathbf{\Omega}' \times \mathbf{r}_B$$

$$+ \mathbf{\Omega} \times \left(\underbrace{x'\mathbf{i} + y'\mathbf{j} + z'\mathbf{k}}_{\mathbf{v}_B} + \overbrace{x\mathbf{i}' + y\mathbf{j}' + z\mathbf{k}'}^{\mathbf{\Omega} \times \mathbf{r}_B(t)} \right)$$

$$= \mathbf{R}'' + \mathbf{a}_B + \mathbf{\Omega}' \times \mathbf{r}_B + 2\mathbf{\Omega} \times \mathbf{v}_B + \mathbf{\Omega} \times (\mathbf{\Omega} \times \mathbf{r}_B).$$

The acceleration $\mathbf{a}_B$ is that perceived by an observer for whom the moving coordinate system is fixed. The term $\mathbf{\Omega} \times (\mathbf{\Omega} \times \mathbf{r}_B)$ is called the centripetal acceleration. Solving for $\mathbf{a}_B$,

$$\mathbf{a}_B = \mathbf{a} - \mathbf{R}'' - \mathbf{\Omega}' \times \mathbf{r}_B - 2\mathbf{\Omega} \times \mathbf{v}_B - \mathbf{\Omega} \times (\mathbf{\Omega} \times \mathbf{r}_B). \tag{E.13}$$

Here the term $-(\mathbf{\Omega} \times (\mathbf{\Omega} \times \mathbf{r}_B))$ is called the centrifugal acceleration, it being an acceleration felt by the observer relative to the moving coordinate system which he regards as fixed, and the term $-2\mathbf{\Omega} \times \mathbf{v}_B$ is called the Coriolis acceleration, an acceleration experienced by the observer as he moves relative to the moving coordinate system. The mass multiplied by the Coriolis acceleration defines the Coriolis force.

There is a ride found in some amusement parks in which the victims stand next to a circular wall covered with a carpet or some rough material. Then the whole circular room begins to revolve faster and faster. At some point, the bottom drops out and the victims are held in place by friction. The force they feel which keeps them stuck to the wall is called centrifugal force and it causes centrifugal acceleration. It is not necessary to move relative to coordinates fixed with the revolving wall in order to feel this force and it is pretty predictable. However, if the nauseated victim moves relative to the rotating wall, he will feel the effects of the Coriolis force and this force is really strange. The difference between these forces is that the Coriolis force is caused by movement relative to the moving coordinate system and the centrifugal force is not.

E.3.2 *The Coriolis Acceleration On The Rotating Earth*

Now consider the earth. Let $\mathbf{i}^*, \mathbf{j}^*, \mathbf{k}^*$, be the usual basis vectors attached to the rotating earth. Thus $\mathbf{k}^*$ is fixed in space with $\mathbf{k}^*$ pointing in the direction of the north pole from the center of the earth while $\mathbf{i}^*$ and $\mathbf{j}^*$ point to fixed points on the surface of the earth. Thus $\mathbf{i}^*$ and $\mathbf{j}^*$ depend on t while $\mathbf{k}^*$ does not. Let $\mathbf{i}, \mathbf{j}, \mathbf{k}$ be the unit vectors described earlier with $\mathbf{i}$ pointing South, $\mathbf{j}$ pointing East, and $\mathbf{k}$ pointing away from the center of the earth at some point of the rotating earth's surface $\mathbf{p}$. Letting $\mathbf{R}\,(t)$ be the position vector of the point $\mathbf{p}$, from the center of the earth, observe the coordinates of $\mathbf{R}\,(t)$ are constant with respect to $\mathbf{i}\,(t)\,, \mathbf{j}\,(t)\,, \mathbf{k}\,(t)$. Also, since the earth rotates from West to East and the speed of a point on the surface of the earth relative to an observer fixed in space is $\omega\,|\mathbf{R}|\sin\phi$ where ω is the angular speed of the earth about an axis through the poles, it follows from the geometric definition of the cross product that

$$\mathbf{R}' = \omega\mathbf{k}^* \times \mathbf{R}$$

Therefore, $\boldsymbol{\Omega} = \omega\mathbf{k}^*$ and so

$$\mathbf{R}'' = \overbrace{\boldsymbol{\Omega}' \times \mathbf{R}}^{=0} + \boldsymbol{\Omega} \times \mathbf{R}' = \boldsymbol{\Omega} \times (\boldsymbol{\Omega} \times \mathbf{R})$$

since $\boldsymbol{\Omega}$ does not depend on t. Formula (E.13) implies

$$\mathbf{a}_B = \mathbf{a} - \boldsymbol{\Omega} \times (\boldsymbol{\Omega} \times \mathbf{R}) - 2\boldsymbol{\Omega} \times \mathbf{v}_B - \boldsymbol{\Omega} \times (\boldsymbol{\Omega} \times \mathbf{r}_B). \tag{E.14}$$

In this formula, you can totally ignore the term $\boldsymbol{\Omega} \times (\boldsymbol{\Omega} \times \mathbf{r}_B)$ because it is so small whenever you are considering motion near some point on the earth's surface. To see this, note $\omega \overbrace{(24)\,(3600)}^{\text{seconds in a day}} = 2\pi$, and so $\omega = 7.2722 \times 10^{-5}$ in radians per second. If you are using seconds to measure time and feet to measure distance, this term is therefore, no larger than

$$\left(7.2722 \times 10^{-5}\right)^2 |\mathbf{r}_B|.$$

Clearly this is not worth considering in the presence of the acceleration due to gravity which is approximately 32 feet per second squared near the surface of the earth.

If the acceleration $\mathbf{a}$, is due to gravity, then

$$\mathbf{a}_B = \mathbf{a} - \mathbf{\Omega} \times (\mathbf{\Omega} \times \mathbf{R}) - 2\mathbf{\Omega} \times \mathbf{v}_B =$$

$$\overbrace{-\frac{GM(\mathbf{R} + \mathbf{r}_B)}{|\mathbf{R} + \mathbf{r}_B|^3}}^{\equiv \mathbf{g}} - \mathbf{\Omega} \times (\mathbf{\Omega} \times \mathbf{R}) - 2\mathbf{\Omega} \times \mathbf{v}_B \equiv \mathbf{g} - 2\mathbf{\Omega} \times \mathbf{v_B}.$$

Note that from cross product identities,

$$\mathbf{\Omega} \times (\mathbf{\Omega} \times \mathbf{R}) = (\mathbf{\Omega} \cdot \mathbf{R})\mathbf{\Omega} - |\mathbf{\Omega}|^2 \mathbf{R}$$

and so $\mathbf{g}$, the acceleration relative to the moving coordinate system on the earth is not directed exactly toward the center of the earth except at the poles and at the equator, although the components of acceleration which are in other directions are very small when compared with the acceleration due to the force of gravity and are often neglected. Therefore, if the only force acting on an object is due to gravity, the following formula describes the acceleration relative to a coordinate system moving with the earth's surface.

$$\mathbf{a}_B = \mathbf{g} - 2(\mathbf{\Omega} \times \mathbf{v}_B)$$

While the vector $\mathbf{\Omega}$ is quite small, if the relative velocity, $\mathbf{v}_B$ is large, the Coriolis acceleration could be significant. This is described in terms of the vectors $\mathbf{i}(t), \mathbf{j}(t), \mathbf{k}(t)$ next.

Letting (ρ, θ, ϕ) be the usual spherical coordinates of the point $\mathbf{p}(t)$ on the surface taken with respect to $\mathbf{i}^*, \mathbf{j}^*, \mathbf{k}^*$ the usual way with ϕ the polar angle, it follows the $\mathbf{i}^*, \mathbf{j}^*, \mathbf{k}^*$ coordinates of this point are

$$\begin{pmatrix} \rho \sin(\phi) \cos(\theta) \\ \rho \sin(\phi) \sin(\theta) \\ \rho \cos(\phi) \end{pmatrix}.$$

It follows,

$$\mathbf{i} = \cos(\phi) \cos(\theta) \mathbf{i}^* + \cos(\phi) \sin(\theta) \mathbf{j}^* - \sin(\phi) \mathbf{k}^*$$

$$\mathbf{j} = -\sin(\theta) \mathbf{i}^* + \cos(\theta) \mathbf{j}^* + 0\mathbf{k}^*$$

and

$$\mathbf{k} = \sin(\phi) \cos(\theta) \mathbf{i}^* + \sin(\phi) \sin(\theta) \mathbf{j}^* + \cos(\phi) \mathbf{k}^*.$$

It is necessary to obtain $\mathbf{k}^*$ in terms of the vectors $\mathbf{i}, \mathbf{j}, \mathbf{k}$. Thus the following equation needs to be solved for a, b, c to find $\mathbf{k}^* = a\mathbf{i} + b\mathbf{j} + c\mathbf{k}$

$$\overbrace{\begin{pmatrix} 0 \\ 0 \\ 1 \end{pmatrix}}^{\mathbf{k}^*} = \begin{pmatrix} \cos(\phi)\cos(\theta) & -\sin(\theta) & \sin(\phi)\cos(\theta) \\ \cos(\phi)\sin(\theta) & \cos(\theta) & \sin(\phi)\sin(\theta) \\ -\sin(\phi) & 0 & \cos(\phi) \end{pmatrix} \begin{pmatrix} a \\ b \\ c \end{pmatrix} \qquad \text{(E.15)}$$

The first column is $\mathbf{i}$, the second is $\mathbf{j}$ and the third is $\mathbf{k}$ in the above matrix. The solution is $a = -\sin(\phi), b = 0$, and $c = \cos(\phi)$.

Now the Coriolis acceleration on the earth equals

$$2\left(\boldsymbol{\Omega} \times \mathbf{v}_B\right) = 2\omega \left(\overbrace{-\sin(\phi)\,\mathbf{i} + 0\mathbf{j} + \cos(\phi)\,\mathbf{k}}^{\mathbf{k}^*} \right) \times (x'\mathbf{i} + y'\mathbf{j} + z'\mathbf{k}).$$

This equals

$$2\omega\left[(-y'\cos\phi)\,\mathbf{i} + (x'\cos\phi + z'\sin\phi)\,\mathbf{j} - (y'\sin\phi)\,\mathbf{k}\right]. \tag{E.16}$$

Remember ϕ is fixed and pertains to the fixed point $\mathbf{p}(t)$ on the earth's surface. Therefore, if the acceleration $\mathbf{a}$ is due to gravity,

$$\mathbf{a}_B = \mathbf{g} - 2\omega\left[(-y'\cos\phi)\,\mathbf{i} + (x'\cos\phi + z'\sin\phi)\,\mathbf{j} - (y'\sin\phi)\,\mathbf{k}\right]$$

where $\mathbf{g} = -\frac{GM(\mathbf{R}+\mathbf{r}_B)}{|\mathbf{R}+\mathbf{r}_B|^3} - \boldsymbol{\Omega} \times (\boldsymbol{\Omega} \times \mathbf{R})$ as explained above. The term $\boldsymbol{\Omega} \times (\boldsymbol{\Omega} \times \mathbf{R})$ is pretty small and so it will be neglected. However, the Coriolis force will not be neglected.

Example E.3. Suppose a rock is dropped from a tall building. Where will it strike?

Assume $\mathbf{g} = -g\mathbf{k}$. Thus the $\mathbf{j}$ component of $\mathbf{a}_B$ is approximately

$$-2\omega(x'\cos\phi + z'\sin\phi).$$

The dominant term in this expression is clearly the second one because x' will be small. Also, the $\mathbf{i}$ and $\mathbf{k}$ contributions will be very small. Therefore, the following equation is descriptive of the situation.

$$\mathbf{a}_B = -g\mathbf{k} - 2z'\omega\sin\phi\mathbf{j}.$$

$z' = -gt$ approximately. Therefore, considering the $\mathbf{j}$ component, this is

$$2gt\omega\sin\phi.$$

Two integrations give $(\omega gt^3/3)\sin\phi$ for the $\mathbf{j}$ component of the relative displacement at time t.

This shows the rock does not fall directly towards the center of the earth as expected but slightly to the east.

Example E.4. In 1851 Foucault set a pendulum vibrating and observed the earth rotate out from under it. It was a very long pendulum with a heavy weight at the end so that it would vibrate for a long time without stopping.[5] This is what allowed him to observe the earth rotate out from under it. Clearly such a pendulum will take 24 hours for the plane of vibration to appear to make one complete revolution at the north pole. It is also reasonable to expect that no such observed rotation would take place on the equator. Is it possible to predict what will take place at various latitudes?

---

[5]There is such a pendulum in the Eyring building at BYU and to keep people from touching it, there is a little sign which says Warning! 10,000 ohms.

Using (E.16), in (E.14),

$$\mathbf{a}_B = \mathbf{g} - \mathbf{\Omega} \times (\mathbf{\Omega} \times \mathbf{R})$$

$$-2\omega\left[(-y'\cos\phi)\,\mathbf{i} + (x'\cos\phi + z'\sin\phi)\,\mathbf{j} - (y'\sin\phi)\,\mathbf{k}\right].$$

Neglecting the small term $\mathbf{\Omega} \times (\mathbf{\Omega} \times \mathbf{R})$, this becomes

$$= -g\mathbf{k} + \mathbf{T}/m - 2\omega\left[(-y'\cos\phi)\,\mathbf{i} + (x'\cos\phi + z'\sin\phi)\,\mathbf{j} - (y'\sin\phi)\,\mathbf{k}\right]$$

where $\mathbf{T}$, the tension in the string of the pendulum, is directed towards the point at which the pendulum is supported, and m is the mass of the pendulum bob. The pendulum can be thought of as the position vector from $(0,0,l)$ to the surface of the sphere $x^2 + y^2 + (z - l)^2 = l^2$. Therefore,

$$\mathbf{T} = -T\frac{x}{l}\mathbf{i} - T\frac{y}{l}\mathbf{j} + T\frac{l-z}{l}\mathbf{k}$$

and consequently, the differential equations of relative motion are

$$x'' = -T\frac{x}{ml} + 2\omega y'\cos\phi$$

$$y'' = -T\frac{y}{ml} - 2\omega\left(x'\cos\phi + z'\sin\phi\right)$$

and

$$z'' = T\frac{l-z}{ml} - g + 2\omega y'\sin\phi.$$

If the vibrations of the pendulum are small so that for practical purposes, $z'' = z = 0$, the last equation may be solved for T to get

$$gm - 2\omega y'\sin\left(\phi\right)m = T.$$

Therefore, the first two equations become

$$x'' = -\left(gm - 2\omega m y'\sin\phi\right)\frac{x}{ml} + 2\omega y'\cos\phi$$

and

$$y'' = -\left(gm - 2\omega m y'\sin\phi\right)\frac{y}{ml} - 2\omega\left(x'\cos\phi + z'\sin\phi\right).$$

All terms of the form xy' or $y'y$ can be neglected because it is assumed x and y remain small. Also, the pendulum is assumed to be long with a heavy weight so that x' and y' are also small. With these simplifying assumptions, the equations of motion become

$$x'' + g\frac{x}{l} = 2\omega y'\cos\phi, \quad y'' + g\frac{y}{l} = -2\omega x'\cos\phi.$$

These equations are of the form

$$x'' + a^2 x = by', \quad y'' + a^2 y = -bx' \tag{E.17}$$

where $a^2 = \frac{g}{l}$ and $b = 2\omega\cos\phi$. Then it is fairly tedious but routine to verify that for each constant c,

$$x = c\sin\left(\frac{bt}{2}\right)\sin\left(\frac{\sqrt{b^2 + 4a^2}}{2}t\right), \quad y = c\cos\left(\frac{bt}{2}\right)\sin\left(\frac{\sqrt{b^2 + 4a^2}}{2}t\right) \qquad \text{(E.18)}$$

yields a solution to (E.17) along with the initial conditions,

$$x(0) = 0, y(0) = 0, x'(0) = 0, y'(0) = \frac{c\sqrt{b^2 + 4a^2}}{2}. \qquad \text{(E.19)}$$

It is clear from experiments with the pendulum that the earth does indeed rotate out from under it causing the plane of vibration of the pendulum to appear to rotate. The purpose of this discussion is not to establish these self evident facts, but to predict how long it takes for the plane of vibration to make one revolution. Therefore, there will be some instant in time at which the pendulum will be vibrating in a plane determined by $\mathbf{k}$ and $\mathbf{j}$. (Recall $\mathbf{k}$ points away from the center of the earth and $\mathbf{j}$ points East.) At this instant in time, defined as $t = 0$, the conditions of (E.19) will hold for some value of c and so the solution to (E.17) having these initial conditions will be those of (E.18) by uniqueness of the initial value problem. Writing these solutions differently,

$$\begin{pmatrix} x(t) \\ y(t) \end{pmatrix} = c\begin{pmatrix} \sin\left(\frac{bt}{2}\right) \\ \cos\left(\frac{bt}{2}\right) \end{pmatrix}\sin\left(\frac{\sqrt{b^2 + 4a^2}}{2}t\right)$$

This is very interesting! The vector $c\begin{pmatrix} \sin\left(\frac{bt}{2}\right) \\ \cos\left(\frac{bt}{2}\right) \end{pmatrix}$ always has magnitude equal to $|c|$ but its direction changes very slowly because b is very small. The plane of vibration is determined by this vector and the vector $\mathbf{k}$. The term $\sin\left(\frac{\sqrt{b^2+4a^2}}{2}t\right)$ changes relatively fast and takes values between -1 and 1. This is what describes the actual observed vibrations of the pendulum. Thus the plane of vibration will have made one complete revolution when $t = P$ for $\frac{bP}{2} \equiv 2\pi$. Therefore, the time it takes for the earth to turn out from under the pendulum is $P = \frac{4\pi}{2\omega\cos\phi} = \frac{2\pi}{\omega}\sec\phi$. Since ω is the angular speed of the rotating earth, it follows $\omega = \frac{2\pi}{24} = \frac{\pi}{12}$ in radians per hour. Therefore, the above formula implies $P = 24\sec\phi$.

I think this is really amazing. You could actually determine latitude, not by taking readings with instruments using the North Star but by doing an experiment with a big pendulum. You would set it vibrating, observe P in hours, and then solve the above equation for ϕ. Also note the pendulum would not appear to change its plane of vibration at the equator because $\lim_{\phi\to\pi/2}\sec\phi = \infty$.

The Coriolis acceleration is also responsible for the phenomenon of the next example.

Example E.5. It is known that low pressure areas rotate counterclockwise as seen from above in the Northern hemisphere but clockwise in the Southern hemisphere. Why?

Neglect accelerations other than the Coriolis acceleration and the following acceleration which comes from an assumption that the point $\mathbf{p}(t)$ is the location of the lowest pressure. $\mathbf{a} = -a(r_B)\mathbf{r}_B$ where $r_B = r$ will denote the distance from the fixed point $\mathbf{p}(t)$ on the earth's surface which is also the lowest pressure point. Of course the situation could be more complicated but this will suffice to explain the above question. Then the acceleration observed by a person on the earth relative to the apparently fixed vectors $\mathbf{i}, \mathbf{k}, \mathbf{j}$, is $\mathbf{a}_B =$

$$-a(r_B)(x\mathbf{i}+y\mathbf{j}+z\mathbf{k}) - 2\omega\left[-y'\cos(\phi)\mathbf{i}+(x'\cos(\phi)+z'\sin(\phi))\mathbf{j}-(y'\sin(\phi)\mathbf{k})\right]$$

Therefore, one obtains some differential equations from $\mathbf{a}_B = x''\mathbf{i} + y''\mathbf{j} + z''\mathbf{k}$ by matching the components. These are

$$x'' + a(r_B)x = 2\omega y'\cos\phi$$
$$y'' + a(r_B)y = -2\omega x'\cos\phi - 2\omega z'\sin(\phi)$$
$$z'' + a(r_B)z = 2\omega y'\sin\phi$$

Now remember, the vectors $\mathbf{i}, \mathbf{j}, \mathbf{k}$ are fixed relative to the earth and so are constant vectors. Therefore, from the properties of the determinant and the above differential equations,

$$(\mathbf{r}'_B \times \mathbf{r}_B)' = \begin{vmatrix} \mathbf{i} & \mathbf{j} & \mathbf{k} \\ x' & y' & z' \\ x & y & z \end{vmatrix}' = \begin{vmatrix} \mathbf{i} & \mathbf{j} & \mathbf{k} \\ x'' & y'' & z'' \\ x & y & z \end{vmatrix} = \begin{vmatrix} \mathbf{i} & \mathbf{j} & \mathbf{k} \\ A & B & C \\ x & y & z \end{vmatrix}$$

where

$$A = -a(r_B)x + 2\omega y'\cos\phi,$$
$$B = -a(r_B)y - 2\omega x'\cos\phi - 2\omega z'\sin(\phi),$$
$$C = -a(r_B)z + 2\omega y'\sin\phi$$

Then the $\mathbf{k}^{th}$ component of this cross product equals

$$\omega\cos(\phi)\left(y^2 + x^2\right)' + 2\omega xz'\sin(\phi).$$

The first term will be negative because it is assumed $\mathbf{p}(t)$ is the location of low pressure causing $y^2 + x^2$ to be a decreasing function. If it is assumed there is not a substantial motion in the $\mathbf{k}$ direction, so that z is fairly constant and the last term can be neglected, then the $\mathbf{k}^{th}$ component of $(\mathbf{r}'_B \times \mathbf{r}_B)'$ is negative provided $\phi \in \left(0, \frac{\pi}{2}\right)$ and positive if $\phi \in \left(\frac{\pi}{2}, \pi\right)$. Beginning with a point at rest, this implies $\mathbf{r}'_B \times \mathbf{r}_B = \mathbf{0}$ initially and then the above implies its $\mathbf{k}^{th}$ component is negative in the upper hemisphere when $\phi < \pi/2$ and positive in the lower hemisphere when $\phi > \pi/2$. Using the right hand and the geometric definition of the cross product, this shows clockwise rotation in the lower hemisphere and counter clockwise rotation in the upper hemisphere.

Note also that as ϕ gets close to $\pi/2$ near the equator, the above reasoning tends to break down because $\cos(\phi)$ becomes close to zero. Therefore, the motion towards the low pressure has to be more pronounced in comparison with the motion in the $\mathbf{k}$ direction in order to draw this conclusion.

E.4   Exercises

(1) An illustration used in many beginning physics books is that of firing a rifle horizontally and dropping an identical bullet from the same height above the perfectly flat ground followed by an assertion that the two bullets will hit the ground at exactly the same time. Is this true on the rotating earth assuming the experiment takes place over a large perfectly flat field so the curvature of the earth is not an issue? Explain. What other irregularities will occur? Recall the Coriolis force is $2\omega\left[(-y'\cos\phi)\mathbf{i} + (x'\cos\phi + z'\sin\phi)\mathbf{j} - (y'\sin\phi)\mathbf{k}\right]$ where $\mathbf{k}$ points away from the center of the earth, $\mathbf{j}$ points East, and $\mathbf{i}$ points South.

(2) Suppose you have n masses $m_1, \cdots, m_n$. Let the position vector of the ith mass be $\mathbf{r}_i(t)$. The center of mass of these is defined to be

$$\mathbf{R}(t) \equiv \frac{\sum_{i=1}^n \mathbf{r}_i m_i}{\sum_{i=1}^n m_i} \equiv \frac{\sum_{i=1}^n \mathbf{r}_i(t) m_i}{M}.$$

Let $\mathbf{r}_{Bi}(t) = \mathbf{r}_i(t) - \mathbf{R}(t)$. Show that $\sum_{i=1}^n m_i \mathbf{r}_i(t) - \sum_i m_i \mathbf{R}(t) = \mathbf{0}$.

(3) Suppose you have n masses $m_1, \cdots, m_n$ which make up a moving rigid body. Let $\mathbf{R}(t)$ denote the position vector of the center of mass of these n masses. Find a formula for the total kinetic energy in terms of this position vector, the angular velocity vector, and the position vector of each mass from the center of mass. **Hint:** Use Problem 2.

Appendix F

Curvilinear Coordinates

F.1 Basis vectors

This section will make routine use of the repeated summation convention and the identities involving the permutation symbol. The usual basis vectors are denoted by $\mathbf{i}, \mathbf{j}, \mathbf{k}$ and are as the following picture describes.

The vectors $\mathbf{i}, \mathbf{j}, \mathbf{k}$, are fixed. If $\mathbf{v}$ is a vector, there are unique scalars called components such that $\mathbf{v} = v^1 \mathbf{i} + v^2 \mathbf{j} + v^3 \mathbf{k}$. This is what it means to say $\mathbf{i}, \mathbf{j}, \mathbf{k}$ is a basis.

Now suppose $\mathbf{e}_1, \mathbf{e}_2, \mathbf{e}_3$ are three vectors which satisfy

$$\mathbf{e}_1 \times \mathbf{e}_2 \cdot \mathbf{e}_3 \neq 0.$$

Recall this means the volume of the box spanned by the three vectors is not zero.

Suppose $\mathbf{e}_1, \mathbf{e}_2, \mathbf{e}_3$ are as just described. Does it follow that they form a basis? That is, do there exist unique scalars, v^1, v^2, and v^3 such that

$$\mathbf{v} = v^i \mathbf{e}_i.$$

This is the content of the following theorem.

Theorem F.1. *If $\mathbf{e}_1, \mathbf{e}_2, \mathbf{e}_3$ are three vectors, then they form a basis if and only if*

$$\mathbf{e}_1 \times \mathbf{e}_2 \cdot \mathbf{e}_3 \neq 0.$$

Proof: Suppose first the above condition holds. Let $\mathbf{i}_1 \equiv \mathbf{i}, \mathbf{i}_2 \equiv \mathbf{j}, \mathbf{i}_3 \equiv \mathbf{k}$ and suppose $\mathbf{v} = u^j \mathbf{i}_j$. Therefore, u^1, u^2, u^3 are the components of the vector $\mathbf{v}$ with respect to the usual basis vectors. Also let

$$\mathbf{e}_i = a_i^j \mathbf{i}_j$$

thus writing each $\mathbf{e}_i$ in terms of the vectors $\mathbf{i}_j$. Then from the definition of the box product in terms of the usual basis vectors,

$$0 \neq \mathbf{e}_1 \times \mathbf{e}_2 \cdot \mathbf{e}_3 = \det\left(a_i^j\right)$$

$$\equiv \det\begin{pmatrix} a_1^1 & a_1^2 & a_1^3 \\ a_2^1 & a_2^2 & a_2^3 \\ a_3^1 & a_3^2 & a_3^3 \end{pmatrix} = \det\begin{pmatrix} a_1^1 & a_2^1 & a_3^1 \\ a_1^2 & a_2^2 & a_3^2 \\ a_1^3 & a_2^3 & a_3^3 \end{pmatrix} \tag{F.1}$$

Does there exists a unique solution to

$$\mathbf{v} \equiv u^j \mathbf{i}_j = v^i \mathbf{e}_i = a_i^j v^i \mathbf{i}_j? \tag{F.2}$$

In other words, is there a unique solution to the system of equations

$$u^j = a_i^j v^i, \quad j = 1, 2, 3.$$

This is in the form

$$\mathbf{u} = A\mathbf{v} \tag{F.3}$$

where A is a matrix which, by (F.1), has non zero determinant. Therefore, there exists a unique solution.

If the box product is equal to zero, then the system which needs to be solved is of the form (F.3) where $\det(A) = 0$. Therefore, A has an eigenvector $\mathbf{v} = (v_1, v_2, v_3)^T$ which corresponds to the eigenvalue $\lambda = 0$. Therefore, (F.2) shows $\mathbf{0}$ has more than one set of components with respect to the vectors $\mathbf{e}_1, \mathbf{e}_2, \mathbf{e}_3$ and this set of vectors is therefore not a basis. ∎

This gives a simple geometric condition which determines whether a list of three vectors forms a basis in $\mathbb{R}^3$. One simply takes the box product. If the box product is not equal to zero, then the vectors form a basis. If not, the list of three vectors does not form a basis. This condition generalizes to $\mathbb{R}^n$ as follows. If $\mathbf{e}_i = a_i^j \mathbf{i}_j$, then $\{\mathbf{e}_i\}_{i=1}^n$ forms a basis if and only if $\det\left(a_i^j\right) \neq 0$.

These vectors $\{\mathbf{e}_i\}$ may or may not be orthonormal. In any case, it is convenient to define something called the dual basis.

Definition F.1. Let $\{\mathbf{e}_i\}_{i=1}^n$ form a basis for $\mathbb{R}^n$. Then $\{\mathbf{e}^i\}_{i=1}^n$ is called the dual basis if

$$\mathbf{e}^i \cdot \mathbf{e}_j = \delta_j^i \equiv \begin{cases} 1 \text{ if } i = j \\ 0 \text{ if } i \neq j \end{cases}. \tag{F.4}$$

Theorem F.2. *If $\{\mathbf{e}_i\}_{i=1}^n$ is a basis then $\{\mathbf{e}^i\}_{i=1}^n$ is also a basis provided (F.4) holds.*

Proof: Suppose

$$\mathbf{v} = v_i \mathbf{e}^i. \tag{F.5}$$

Then taking the dot product of both sides of (F.5) with $\mathbf{e}_j$, yields

$$v_j = \mathbf{v} \cdot \mathbf{e}_j. \tag{F.6}$$

Thus there is at most one choice of scalars v_j such that $\mathbf{v} = v_j \mathbf{e}^j$ and it is given by (F.6).

$$\left(\mathbf{v} - \mathbf{v} \cdot \mathbf{e}_j \mathbf{e}^j \right) \cdot \mathbf{e}_k = 0$$

and so, since $\{\mathbf{e}_i\}_{i=1}^n$ is a basis,

$$\left(\mathbf{v} - \mathbf{v} \cdot \mathbf{e}_j \mathbf{e}^j \right) \cdot \mathbf{w} = 0$$

for all vectors $\mathbf{w}$. It follows $\mathbf{v} - \mathbf{v} \cdot \mathbf{e}_j \mathbf{e}^j = \mathbf{0}$ and this shows $\{\mathbf{e}^i\}_{i=1}^n$ is a basis. ∎

The above gives formulas for the components of a vector $\mathbf{v}$, v_i, with respect to the dual basis, found to be $v_j = \mathbf{v} \cdot \mathbf{e}_j$. In the same way, the components of a vector with respect to the basis $\{\mathbf{e}_i\}_{i=1}^n$ can be obtained. Let $\mathbf{v}$ be any vector and let

$$\mathbf{v} = v^j \mathbf{e}_j. \tag{F.7}$$

Then using (F.4) and taking the dot product of both sides of (F.7) with $\mathbf{e}^i$, $v^i = \mathbf{e}^i \cdot \mathbf{v}$.

Does there exist a dual basis and is it uniquely determined?

Theorem F.3. *If $\{\mathbf{e}_i\}_{i=1}^n$ is a basis for $\mathbb{R}^n$, then there exists a unique dual basis, $\{\mathbf{e}^j\}_{j=1}^n$ satisfying*

$$\mathbf{e}^j \cdot \mathbf{e}_i = \delta_i^j.$$

Proof: First the dual basis is unique. Suppose $\{\mathbf{f}^j\}_{j=1}^n$ is another set of vectors which satisfies $\mathbf{f}^j \cdot \mathbf{e}_i = \delta_i^j$. Then

$$\mathbf{f}^j = \mathbf{f}^j \cdot \mathbf{e}_i \mathbf{e}^i = \delta_i^j \mathbf{e}^i = \mathbf{e}^j.$$

This shows uniqueness.

What about existence? Note that from the definition, the dual basis to $\{\mathbf{i}_j\}_{j=1}^n$ is just $\mathbf{i}^j = \mathbf{i}_j$. Letting

$$\mathbf{e}_i = a_i^j \mathbf{i}_j$$

where the vectors $\{\mathbf{i}_j\}_{j=1}^n$ are the standard basis vectors, it follows, since the $\mathbf{e}_i$ form a basis, that the matrix whose ij^{th} entry is a_i^j is an invertible matrix. Letting

$$\mathbf{e}^k = b_r^k \mathbf{i}^r,$$

b_r^k must be chosen such that

$$\overbrace{b_r^k \mathbf{i}^r}^{\mathbf{e}^k} \cdot a_i^j \mathbf{i}_j = b_r^k a_i^j \delta_j^r = b_j^k a_i^j = \delta_i^k.$$

But this is nothing more than the matrix equation for $B \equiv (b_r^k)$ which is of the form

$$AB = I$$

where $A = (a_i^j)$ and has an inverse. There exists a unique solution to this equation given by $B = A^{-1}$ and this proves the existence of the dual basis. ∎

Summarizing what has been shown so far, $\{\mathbf{e}_i\}_{i=1}^n$ is a basis for $\mathbb{R}^n$ if and only if when $\mathbf{e}_i = a_i^j \mathbf{i}_j$,

$$\det\left(a_i^j\right) \neq 0. \tag{F.8}$$

If $\{\mathbf{e}_i\}_{i=1}^n$ is a basis, then there exists a unique dual basis, $\{\mathbf{e}^j\}_{j=1}^n$ satisfying

$$\mathbf{e}^j \cdot \mathbf{e}_i = \delta_i^j, \tag{F.9}$$

and that if $\mathbf{v}$ is any vector

$$\mathbf{v} = v_j \mathbf{e}^j, \ \mathbf{v} = v^j \mathbf{e}_j. \tag{F.10}$$

The components of $\mathbf{v}$ which have the index on the top are called the contravariant components of the vector while the components which have the index on the bottom are called the covariant components. In general $v_i \neq v^j$! The formulas for these components in terms of the dot product are

$$v_j = \mathbf{v} \cdot \mathbf{e}_j, \ v^j = \mathbf{v} \cdot \mathbf{e}^j. \tag{F.11}$$

Define $g_{ij} \equiv \mathbf{e}_i \cdot \mathbf{e}_j$ and $g^{ij} \equiv \mathbf{e}^i \cdot \mathbf{e}^j$. The next theorem describes the process of raising or lowering an index.

Theorem F.4. *The following hold.*

$$g^{ij}\mathbf{e}_j = \mathbf{e}^i, \ g_{ij}\mathbf{e}^j = \mathbf{e}_i, \tag{F.12}$$

$$g^{ij}v_j = v^i, \ g_{ij}v^j = v_i, \tag{F.13}$$

$$g^{ij}g_{jk} = \delta_k^i, \tag{F.14}$$

$$\det(g_{ij}) > 0, \ \det(g^{ij}) > 0. \tag{F.15}$$

Proof: First,

$$\mathbf{e}^i = \mathbf{e}^i \cdot \mathbf{e}^j \mathbf{e}_j = g^{ij}\mathbf{e}_j$$

by (F.10) and (F.11). Similarly, by (F.10) and (F.11),

$$\mathbf{e}_i = \mathbf{e}_i \cdot \mathbf{e}_j \mathbf{e}^j = g_{ij}\mathbf{e}^j.$$

This verifies (F.12). To verify (F.13),

$$v^i = \mathbf{e}^i \cdot \mathbf{v} = g^{ij}\mathbf{e}_j \cdot \mathbf{v} = g^{ij}v_j.$$

The proof of the remaining formula in (F.13) is similar.

To verify (F.14),

$$g^{ij}g_{jk} = \mathbf{e}^i \cdot \mathbf{e}^j \mathbf{e}_j \cdot \mathbf{e}_k = \left((\mathbf{e}^i \cdot \mathbf{e}^j)\,\mathbf{e}_j\right) \cdot \mathbf{e}_k = \mathbf{e}^i \cdot \mathbf{e}_k = \delta_k^i.$$

This shows the two determinants in (F.15) are non zero because the two matrices are inverses of each other. It only remains to verify that one of these is greater than zero. Letting $\mathbf{e}_i = a_i^j \mathbf{i}_j = b_j^i \mathbf{i}^j$, it follows since $\mathbf{i}_j = \mathbf{i}^j, a_i^j = b_j^i$. Therefore,

$$\mathbf{e}_i \cdot \mathbf{e}_j = a_i^r \mathbf{i}_r \cdot b_k^j \mathbf{i}^k = a_i^r b_k^j \delta_r^k = a_i^k b_k^j = a_i^k a_j^k.$$

It follows that for G the matrix whose ij^{th} entry is $\mathbf{e}_i \cdot \mathbf{e}_j$, $G = AA^T$ where the ik^{th} entry of A is a_i^k. Therefore, $\det(G) = \det(A)\det(A^T) = \det(A)^2 > 0$. It follows from (F.14) that if H is the matrix whose ij^{th} entry is g^{ij}, then $GH = I$ and so $H = G^{-1}$ and

$$\det(G)\det(G^{-1}) = \det(g^{ij})\det(G) = 1.$$

Therefore, $\det(G^{-1}) > 0$ also. ∎

Definition F.2. The matrix $(g_{ij}) = G$ is called the metric tensor.

F.2 Exercises

(1) Let $\mathbf{e}_1 = \mathbf{i} + \mathbf{j}, \mathbf{e}_2 = \mathbf{i} - \mathbf{j}, \mathbf{e}_3 = \mathbf{j} + \mathbf{k}$. Find $\mathbf{e}^1, \mathbf{e}^2, \mathbf{e}^3$, $(g_{ij}), (g^{ij})$. If $\mathbf{v} = \mathbf{i} + 2\mathbf{j} + \mathbf{k}$, find v^i and v_j, the contravariant and covariant components of the vector.

(2) Let $\mathbf{e}^1 = 2\mathbf{i} + \mathbf{j}, \mathbf{e}^2 = \mathbf{i} - 2\mathbf{j}, \mathbf{e}^3 = \mathbf{k}$. Find $\mathbf{e}_1, \mathbf{e}_2, \mathbf{e}_3$, $(g_{ij}), (g^{ij})$. If $\mathbf{v} = 2\mathbf{i} - 2\mathbf{j} + \mathbf{k}$, find v^i and v_j, the contravariant and covariant components of the vector.

(3) Suppose $\mathbf{e}_1, \mathbf{e}_2, \mathbf{e}_3$ have the property that $\mathbf{e}_i \cdot \mathbf{e}_j = 0$ whenever $i \neq j$. Show that the same is true of the dual basis and that in fact, $\mathbf{e}^i$ is a multiple of $\mathbf{e}_i$.

(4) Let $\mathbf{e}_1, \cdots, \mathbf{e}_3$ be a basis for $\mathbb{R}^n$ and let $\mathbf{v} = v^i \mathbf{e}_i = v_i \mathbf{e}^i, \mathbf{w} = w^j \mathbf{e}_j = w_j \mathbf{e}^j$ be two vectors. Show that

$$\mathbf{v} \cdot \mathbf{w} = g_{ij} v^i w^j = g^{ij} v_i w_j.$$

(5) Show that if $\{\mathbf{e}_i\}_{i=1}^3$ is a basis in $\mathbb{R}^3$

$$\mathbf{e}^1 = \frac{\mathbf{e}_2 \times \mathbf{e}_3}{\mathbf{e}_2 \times \mathbf{e}_3 \cdot \mathbf{e}_1}, \quad \mathbf{e}^2 = \frac{\mathbf{e}_1 \times \mathbf{e}_2}{\mathbf{e}_1 \times \mathbf{e}_3 \cdot \mathbf{e}_2}, \quad \mathbf{e}^3 = \frac{\mathbf{e}_1 \times \mathbf{e}_2}{\mathbf{e}_1 \times \mathbf{e}_2 \cdot \mathbf{e}_3}.$$

(6) Let $\{\mathbf{e}_i\}_{i=1}^n$ be a basis and define

$$\mathbf{e}_i^* \equiv \frac{\mathbf{e}_i}{|\mathbf{e}_i|}, \quad \mathbf{e}^{*i} \equiv \mathbf{e}^i |\mathbf{e}_i|.$$

Show that $\mathbf{e}^{*i} \cdot \mathbf{e}_j^* = \delta_j^i$.

(7) If $\mathbf{v}$ is a vector, v_i^* and v^{*i}, are defined by

$$\mathbf{v} \equiv v_i^* \mathbf{e}^{*i} \equiv v^{*i} \mathbf{e}_i^*.$$

These are called the physical components of $\mathbf{v}$. Show that

$$v_i^* = \frac{v_i}{|\mathbf{e}_i|}, \quad v^{*i} = v^i |\mathbf{e}_i| \quad (\text{ No summation on } i).$$

F.3   Curvilinear Coordinates

With the algebraic preparation of the last section, consider curvilinear coordinates. Let $D \subseteq \mathbb{R}^n$ be an open set and let $\mathbf{M} : D \to \mathbb{R}^n$ satisfy

$$\mathbf{M} \text{ is } C^2, \tag{F.16}$$

$$\mathbf{M} \text{ is one to one.} \tag{F.17}$$

Letting $\mathbf{x} \in D$,

$$\mathbf{M}(\mathbf{x}) = M^k(\mathbf{x}) \mathbf{i}_k$$

where, as usual, $\mathbf{i}_k$ are the standard basis vectors for $\mathbb{R}^n$, $\mathbf{i}_k$ being the vector in $\mathbb{R}^n$ which has a one in the kth coordinate and a 0 in every other spot. For a fixed $\mathbf{x} \in D$, consider the curves,

$$t \to \mathbf{M}(\mathbf{x} + t\mathbf{i}_k)$$

for $t \in I$, some open interval containing 0. Then for the point $\mathbf{x}$, let

$$\mathbf{e}_k \equiv \frac{\partial \mathbf{M}}{\partial x^k}(\mathbf{x}) \equiv \frac{d}{dt}(\mathbf{M}(\mathbf{x} + t\mathbf{i}_k))|_{t=0}.$$

Denote this vector as $\mathbf{e}_k(\mathbf{x})$ to emphasize its dependence on $\mathbf{x}$. The following picture illustrates the situation in $\mathbb{R}^3$.

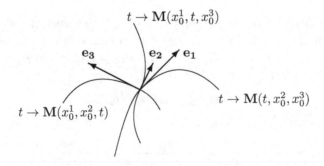

Is $\{\mathbf{e}_k\}_{k=1}^n$ a basis? This will be so if

$$\det\left(\frac{\partial M^i}{\partial x^k}\right) \neq 0. \tag{F.18}$$

Let

$$y^i = M^i(\mathbf{x}) \quad i = 1, \cdots, n \tag{F.19}$$

so that the y^i are the usual coordinates with respect to the usual basis vectors $\{\mathbf{i}_k\}_{k=1}^n$ of the point $\mathbf{M}(\mathbf{x})$. Letting $\mathbf{x} \equiv (x^1, \cdots, x^n)$, it follows from the inverse function theorem of advanced calculus that $\mathbf{M}(D)$ is open, and that (F.18), (F.16), and (F.17) imply the equations (F.19) define each x^i as a C^2 function of $\mathbf{y} \equiv \left(y^1, \cdots, y^n\right)^T$. Thus, abusing notation slightly, the equations (F.19) are equivalent to

$$x^i = x^i \left(\mathbf{y}\right), \ i = 1, \cdots, n$$

where x^i is a C^2 function. Thus

$$\nabla x^k \left(\mathbf{y}\right) = \frac{\partial x^k \left(\mathbf{y}\right)}{\partial y^j} \mathbf{i}^j.$$

Then

$$\nabla x^k \left(\mathbf{y}\right) \cdot \mathbf{e}_j = \frac{\partial x^k}{\partial y^s} \mathbf{i}^s \cdot \frac{\partial y^r}{\partial x^j} \mathbf{i}_r = \frac{\partial x^k}{\partial y^s} \frac{\partial y^s}{\partial x^j} = \delta_j^k$$

by the chain rule. Therefore, the dual basis is given by

$$\mathbf{e}^k \left(\mathbf{x}\right) = \nabla x^k \left(\mathbf{y}\right). \tag{F.20}$$

Notice that it might be hard or even impossible to solve algebraically for x^i in terms of the y^j. Thus the straight forward approach to finding $\mathbf{e}^k$ by (F.20) might be impossible. Also, this approach leads to an expression in terms of the $\mathbf{y}$ coordinates rather than the desired $\mathbf{x}$ coordinates. Therefore, it is expedient to use another method to obtain these vectors. The vectors $\mathbf{e}^k \left(\mathbf{x}\right)$ may always be found by using formula (F.12) and the result is in terms of the curvilinear coordinates $\mathbf{x}$. Consider the following example.

Example F.6. $D \equiv (0, \infty) \times (0, \pi) \times (0, 2\pi)$ and

$$\begin{pmatrix} y^1 \\ y^2 \\ y^3 \end{pmatrix} = \begin{pmatrix} x^1 \sin\left(x^2\right) \cos\left(x^3\right) \\ x^1 \sin\left(x^2\right) \sin\left(x^3\right) \\ x^1 \cos\left(x^2\right) \end{pmatrix}$$

(These are spherical coordinates, usually written as

$$\begin{pmatrix} x \\ y \\ z \end{pmatrix} = \begin{pmatrix} \rho \sin\left(\phi\right) \cos\left(\theta\right) \\ \rho \sin\left(\phi\right) \sin\left(\theta\right) \\ \rho \cos\left(\phi\right) \end{pmatrix}$$

where (ρ, ϕ, θ) are the spherical coordinates. Thus

$$\mathbf{e}_1 \left(\mathbf{x}\right) = \sin\left(x^2\right) \cos\left(x^3\right) \mathbf{i}_1 + \sin\left(x^2\right) \sin\left(x^3\right) \mathbf{i}_2 + \cos\left(x^2\right) \mathbf{i}_3,$$

$$\mathbf{e}_2 \left(\mathbf{x}\right) = x^1 \cos\left(x^2\right) \cos\left(x^3\right) \mathbf{i}_1$$

$$+ x^1 \cos\left(x^2\right) \sin\left(x^3\right) \mathbf{i}_2 - x^1 \sin\left(x^2\right) \mathbf{i}_3,$$

$$\mathbf{e}_3 \left(\mathbf{x}\right) = -x^1 \sin\left(x^2\right) \sin\left(x^3\right) \mathbf{i}_1 + x^1 \sin\left(x^2\right) \cos\left(x^3\right) \mathbf{i}_2 + 0 \mathbf{i}_3.$$

It follows the metric tensor is

$$G = \begin{pmatrix} 1 & 0 & 0 \\ 0 & \left(x^1\right)^2 & 0 \\ 0 & 0 & \left(x^1\right)^2 \sin^2\left(x^2\right) \end{pmatrix} = \left(g_{ij}\right) = \left(\mathbf{e}_i \cdot \mathbf{e}_j\right). \tag{F.21}$$

Therefore, by Theorem F.4

$$G^{-1} = (g^{ij})$$

$$= (e^i, e^j) = \begin{pmatrix} 1 & 0 & 0 \\ 0 & (x^1)^{-2} & 0 \\ 0 & 0 & (x^1)^{-2} \sin^{-2}(x^2) \end{pmatrix}.$$

To obtain the dual basis, use Theorem F.4 to write

$$e^1 = g^{1j}e_j = e_1$$

$$e^2 = g^{2j}e_j = (x^1)^{-2} e_2$$

$$e^3 = g^{3j}e_j = (x^1)^{-2} \sin^{-2}(x^2) e_3.$$

It is natural to ask if there exists a transformation $\mathbf{M}$ such that

$$\frac{\partial \mathbf{M}}{\partial x^1} = \mathbf{i} = \mathbf{i}_1, \frac{\partial \mathbf{M}}{\partial x^2} = \mathbf{j} = \mathbf{i}_2, \frac{\partial \mathbf{M}}{\partial x^3} = \mathbf{k} = \mathbf{i}_3. \tag{F.22}$$

Let

$$\mathbf{M}\left(x^1, x^2, x^3\right) \equiv x^1\mathbf{i} + x^2\mathbf{j} + x^3\mathbf{k}.$$

Then (F.22) holds for this transformation.

F.4   Exercises

(1) Let

$$\begin{pmatrix} y^1 \\ y^2 \\ y^3 \end{pmatrix} = \begin{pmatrix} x^1 + 2x^2 \\ x^2 + x^3 \\ x^1 - 2x^2 \end{pmatrix}$$

where the y^i are the rectangular coordinates of the point. Find $e^i, e_i, i = 1, 2, 3$, and find $(g_{ij})(\mathbf{x})$ and $(g^{ij}(\mathbf{x}))$.

(2) Let $\mathbf{y} = \mathbf{y}(\mathbf{x}, t)$ where t signifies time and $\mathbf{x} \in U \subseteq \mathbb{R}^m$ for U an open set, while $\mathbf{y} \in \mathbb{R}^n$ and suppose $\mathbf{x}$ is a function of t. Physically, this corresponds to an object moving over a surface in $\mathbb{R}^n$ which may be changing as a function of t. The point $\mathbf{y} = \mathbf{y}(\mathbf{x}(t), t)$ is the point in $\mathbb{R}^n$ corresponding to t. For example, consider the pendulum

in which $n = 2, l$ is fixed and $y^1 = l \sin\theta, y^2 = l - l \cos\theta$. Thus, in this simple example, $m = 1$. If l were changing in a known way with respect to t, then this would be of the form $\mathbf{y} = \mathbf{y}\,(\mathbf{x}, t)$. The kinetic energy is defined as

$$T \equiv \frac{1}{2}m\dot{\mathbf{y}} \cdot \dot{\mathbf{y}} \tag{*}$$

where the dot on the top signifies differentiation with respect to t. Show that

$$\frac{\partial T}{\partial \dot{x}^k} = m\dot{\mathbf{y}} \cdot \frac{\partial \mathbf{y}}{\partial x^k}.$$

Hint: First show that

$$\dot{\mathbf{y}} = \frac{\partial \mathbf{y}}{\partial x^j}\dot{x}^j + \frac{\partial \mathbf{y}}{\partial t} \tag{**}$$

and so

$$\frac{\partial \dot{\mathbf{y}}}{\partial \dot{x}^j} = \frac{\partial \mathbf{y}}{\partial x^j}.$$

(3) ↑ Show that

$$\frac{d}{dt}\left(\frac{\partial T}{\partial \dot{x}^k}\right) = m\ddot{\mathbf{y}} \cdot \frac{\partial \mathbf{y}}{\partial x^k} + m\dot{\mathbf{y}} \cdot \frac{\partial^2 \mathbf{y}}{\partial x^k \partial x^r}\dot{x}^r + m\dot{\mathbf{y}} \cdot \frac{\partial^2 \mathbf{y}}{\partial t \partial x^k}.$$

(4) ↑ Show that

$$\frac{\partial T}{\partial x^k} = m\dot{\mathbf{y}} \cdot \left(\frac{\partial^2 \mathbf{y}}{\partial x^r \partial x^k}\dot{x}^r + \frac{\partial^2 \mathbf{y}}{\partial t \partial x^k}\right).$$

Hint: Use $*$ and $**$.

(5) ↑ Now show from Newton's second law (mass times acceleration equals force) that for $\mathbf{F}$ the force,

$$\frac{d}{dt}\left(\frac{\partial T}{\partial \dot{x}^k}\right) - \frac{\partial T}{\partial x^k} = m\ddot{\mathbf{y}} \cdot \frac{\partial \mathbf{y}}{\partial x^k} = \mathbf{F} \cdot \frac{\partial \mathbf{y}}{\partial x^k}. \tag{***}$$

(6) ↑ In the example of the simple pendulum above,

$$\mathbf{y} = \begin{pmatrix} l\sin\theta \\ l - l\cos\theta \end{pmatrix} = l\sin\theta\mathbf{i} + (l - l\cos\theta)\mathbf{j}.$$

Use $***$ to find a differential equation which describes the vibrations of the pendulum in terms of θ. First write the kinetic energy and then consider the force acting on the mass which is

$$-mg\mathbf{j}.$$

(7) The above problem is fairly easy to do without the formalism developed. Now consider the case where $\mathbf{x} = (\rho, \theta, \phi)$, spherical coordinates, and write differential equations for $\rho, \theta,$ and ϕ to describe the motion of an object in terms of these coordinates given a force $\mathbf{F}$.

(8) Suppose the pendulum is not assumed to vibrate in a plane. Let it be suspended at the origin and consider spherical coordinates. Find differential equations for θ and ϕ.

(9) If there are many masses $m_\alpha, \alpha = 1, \cdots, R$, the kinetic energy is the sum of the kinetic energies of the individual masses. Thus,

$$T \equiv \frac{1}{2} \sum_{\alpha=1}^{R} m_\alpha \left| \dot{\mathbf{y}}_\alpha \right|^2 .$$

Generalize the above problems to show that, assuming

$$\mathbf{y}_\alpha = \mathbf{y}_\alpha \left(\mathbf{x}, t \right),$$

$$\frac{d}{dt} \left(\frac{\partial T}{\partial \dot{x}^k} \right) - \frac{\partial T}{\partial x^k} = \sum_{\alpha=1}^{R} \mathbf{F}_\alpha \cdot \frac{\partial \mathbf{y}_\alpha}{\partial x^k}$$

where $\mathbf{F}_\alpha$ is the force acting on m_α.

(10) Discuss the equivalence of these formulae with Newton's second law, force equals mass times acceleration. What is gained from the above so called Lagrangian formalism?

(11) The double pendulum has two masses instead of only one.

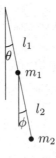

Write differential equations for θ and ϕ to describe the motion of the double pendulum.

F.5 Transformation Of Coordinates

The scalars $\left\{ x^i \right\}$ are called cuvilinear coordinates. Note they can be used to identify a point in $\mathbb{R}^n$ and $\mathbf{x} = \left(x^1, \cdots, x^n \right)$ is a point in $\mathbb{R}^n$. The basis vectors associated with this particular set of curvilinear coordinates at a point identified by $\mathbf{x}$ are denoted by $\mathbf{e}_i \left(\mathbf{x} \right)$ and the dual basis vectors at this point are denoted by $\mathbf{e}^j \left(\mathbf{x} \right)$. What if other curvilinear coordinates are used? How do you write $\mathbf{e}^k \left(\mathbf{x} \right)$ in terms of the vectors $\mathbf{e}^j \left(\mathbf{z} \right)$ where $\mathbf{z}$ is some other type of curvilinear coordinates? This is next.

Consider the following picture in which U is an open set in $\mathbb{R}^n$, D, and $\widehat{D}$ are open sets in $\mathbb{R}^n$, and $\mathbf{M}, \mathbf{N}$ are C^2 mappings which are one to one from D and $\widehat{D}$

respectively. Suppose that a point in U is identified by the curvilinear coordinates $\mathbf{x}$ in D and $\mathbf{z}$ in $\widehat{D}$.

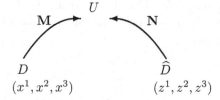

$$U$$
$$\mathbf{M} \xrightarrow{\quad} \xleftarrow{\quad} \mathbf{N}$$

$$D \qquad\qquad \widehat{D}$$
$$(x^1, x^2, x^3) \qquad (z^1, z^2, z^3)$$

Thus $\mathbf{M}(\mathbf{x}) = \mathbf{N}(\mathbf{z})$. Now by the chain rule,

$$\mathbf{e}_i(\mathbf{z}) \equiv \frac{\partial \mathbf{N}}{\partial z^i} = \frac{\partial \mathbf{M}}{\partial x^j}\frac{\partial x^j}{\partial z^i} = \frac{\partial x^j}{\partial z^i}\mathbf{e}_j(\mathbf{x}). \tag{F.23}$$

Define the covariant and contravariant coordinates for the various curvilinear coordinates in the obvious way. Thus,

$$\mathbf{v} = v_i(\mathbf{x})\,\mathbf{e}^i(\mathbf{x}) = v^i(\mathbf{x})\,\mathbf{e}_i(\mathbf{x}) = v_j(\mathbf{z})\,\mathbf{e}^j(\mathbf{z}) = v^j(\mathbf{z})\,\mathbf{e}_j(\mathbf{z}).$$

Then here is a fundamental theorem about transforming the vectors and coordinates.

Theorem F.5. *The following transformation rules hold for pairs of curvilinear coordinates.*

$$v_i(\mathbf{z}) = \frac{\partial x^j}{\partial z_i}v_j(\mathbf{x}), \quad v^i(\mathbf{z}) = \frac{\partial z^i}{\partial x^j}v^j(\mathbf{x}), \tag{F.24}$$

$$\mathbf{e}_i(\mathbf{z}) = \frac{\partial x^j}{\partial z_i}\mathbf{e}_j(\mathbf{x}), \quad \mathbf{e}^i(\mathbf{z}) = \frac{\partial z^i}{\partial x^j}\mathbf{e}^j(\mathbf{x}), \tag{F.25}$$

$$g_{ij}(\mathbf{z}) = \frac{\partial x^r}{\partial z^i}\frac{\partial x^s}{\partial z^j}g_{rs}(\mathbf{x}), \quad g^{ij}(\mathbf{z}) = \frac{\partial z^i}{\partial x^r}\frac{\partial z^j}{\partial x^s}g^{rs}(\mathbf{x}). \tag{F.26}$$

Proof: The first part of (F.25) is shown in (F.23). Then, from (F.23),

$$\mathbf{e}^i(\mathbf{z}) = \mathbf{e}^i(\mathbf{z}) \cdot \mathbf{e}_j(\mathbf{x})\,\mathbf{e}^j(\mathbf{x}) = \mathbf{e}^i(\mathbf{z}) \cdot \frac{\partial z^k}{\partial x^j}\mathbf{e}_k(\mathbf{z})\,\mathbf{e}^j(\mathbf{x})$$

$$= \delta^i_k \frac{\partial z^k}{\partial x^j}\mathbf{e}^j(\mathbf{x}) = \frac{\partial z^i}{\partial x^j}\mathbf{e}^j(\mathbf{x})$$

and this proves the second part of (F.25). Now to show (F.24),

$$v_i(\mathbf{z}) = \mathbf{v} \cdot \mathbf{e}_i(\mathbf{z}) = \mathbf{v} \cdot \frac{\partial x^j}{\partial z_i}\mathbf{e}_j(\mathbf{x}) = \frac{\partial x^j}{\partial z_i}v_j(\mathbf{x})$$

and

$$v^i(\mathbf{z}) = \mathbf{v} \cdot \mathbf{e}^i(\mathbf{z}) = \mathbf{v} \cdot \frac{\partial z^i}{\partial x^j}\mathbf{e}^j(\mathbf{x}) = \frac{\partial z^i}{\partial x^j}v^j(\mathbf{x}).$$

To verify (F.26),

$$g_{ij}(\mathbf{z}) = \mathbf{e}_i(\mathbf{z}) \cdot \mathbf{e}_j(\mathbf{z}) = \mathbf{e}_r(\mathbf{x})\frac{\partial x^r}{\partial z^i} \cdot \mathbf{e}_s(\mathbf{x})\frac{\partial x^s}{\partial z^j} = g_{rs}(\mathbf{x})\frac{\partial x^r}{\partial z^i}\frac{\partial x^s}{\partial z^j}. \quad\blacksquare$$

Denote by $\mathbf{y}$ the curvilinear coordinates with the property that

$$\mathbf{e}^k(\mathbf{y}) = \mathbf{i}_k = \mathbf{e}_k(\mathbf{y}).$$

F.6   Differentiation and Christoffel Symbols

Let $\mathbf{F} : U \to \mathbb{R}^n$ be differentiable. Thus $\mathbf{F}$ is a vector field and it is used to model force, velocity, acceleration, or any other vector quantity which may change from point to point in U. Then

$$\frac{\partial \mathbf{F}(\mathbf{x})}{\partial x^j}$$

is a vector and so there exist scalars, $F^i_{,j}(\mathbf{x})$ and $F_{i,j}(\mathbf{x})$ such that

$$\frac{\partial \mathbf{F}(\mathbf{x})}{\partial x^j} = F^i_{,j}(\mathbf{x})\,\mathbf{e}_i(\mathbf{x}) = F_{i,j}(\mathbf{x})\,\mathbf{e}^j(\mathbf{x}). \tag{F.27}$$

Consider how these scalars, $F^i_{,j}(\mathbf{x})$ and $F_{i,j}(\mathbf{x})$ transform when the coordinates are changed.

Theorem F.6. *If $\mathbf{x}$ and $\mathbf{z}$ are curvilinear coordinates,*

$$F^r_{,s}(\mathbf{x}) = F^i_{,j}(\mathbf{z})\,\frac{\partial x^r}{\partial z^i}\frac{\partial z^j}{\partial x^s}, \quad F_{r,s}(\mathbf{x})\,\frac{\partial x^r}{\partial z^i}\frac{\partial x^s}{\partial z^j} = F_{i,j}(\mathbf{z}). \tag{F.28}$$

Proof:

$$F^r_{,s}(\mathbf{x})\,\mathbf{e}_r(\mathbf{x}) \equiv \frac{\partial \mathbf{F}(\mathbf{x})}{\partial x^s} = \frac{\partial \mathbf{F}(\mathbf{z})}{\partial z^j}\frac{\partial z^j}{\partial x^s} \equiv$$

$$F^i_{,j}(\mathbf{z})\,\mathbf{e}_i(\mathbf{z})\,\frac{\partial z^j}{\partial x^s} = F^i_{,j}(\mathbf{z})\,\frac{\partial x^r}{\partial z^i}\frac{\partial z^j}{\partial x^s}\mathbf{e}_r(\mathbf{x})$$

which shows the first formula of Theorem F.6. To show the other formula,

$$F_{i,j}(\mathbf{z})\,\mathbf{e}^i(\mathbf{z}) \equiv \frac{\partial \mathbf{F}(\mathbf{z})}{\partial z^j} = \frac{\partial \mathbf{F}(\mathbf{x})}{\partial x^s}\frac{\partial x^s}{\partial z^j} \equiv$$

$$F_{r,s}(\mathbf{x})\,\mathbf{e}^r(\mathbf{x})\,\frac{\partial x^s}{\partial z^j} = F_{r,s}(\mathbf{x})\,\frac{\partial x^r}{\partial z^i}\frac{\partial x^s}{\partial z^j}\mathbf{e}^i(\mathbf{z}),$$

and this shows the second formula for transforming these scalars. ∎

Now $\mathbf{F}(\mathbf{x}) = F^i(\mathbf{x})\,\mathbf{e}_i(\mathbf{x})$ and so by the product rule,

$$\frac{\partial \mathbf{F}}{\partial x^j} = \frac{\partial F^i}{\partial x^j}\mathbf{e}_i(\mathbf{x}) + F^i(\mathbf{x})\frac{\partial \mathbf{e}_i(\mathbf{x})}{\partial x^j}.$$

Now $\frac{\partial \mathbf{e}_i(\mathbf{x})}{\partial x^j}$ is a vector and so there exist scalars, $\left\{ \begin{matrix} k \\ ij \end{matrix} \right\}$ such that

$$\frac{\partial \mathbf{e}_i(\mathbf{x})}{\partial x^j} =$$

Therefore,

$$\frac{\partial \mathbf{F}}{\partial x^j} = \frac{\partial F^k}{\partial x^j}\mathbf{e}_k(\mathbf{x}) + F^i(\mathbf{x})\left\{ \begin{matrix} k \\ ij \end{matrix} \right\}\mathbf{e}_k(\mathbf{x})$$

which shows

$$F^k_{,j}(\mathbf{x}) = \frac{\partial F^k}{\partial x^j} + F^i(\mathbf{x}) \left\{ \begin{matrix} k \\ ij \end{matrix} \right\}.$$

This is sometimes called the covariant derivative.

These scalars are called the Christoffel symbols of the second kind. The next theorem is devoted to properties of these Christoffel symbols. Before stating the theorem, recall that the mapping $\mathbf{M}$ which defines the curvilinear coordinates is C^2. The reason for this is to be able to assert the mixed partial derivatives are equal.

Theorem F.7. *The Christoffel symbols of the second kind satisfy the following*

$$\frac{\partial \mathbf{e}_i(\mathbf{x})}{\partial x^j} = \left\{ \begin{matrix} k \\ ij \end{matrix} \right\} \mathbf{e}_k(\mathbf{x}), \tag{F.29}$$

$$\frac{\partial \mathbf{e}^i(\mathbf{x})}{\partial x^j} = - \left\{ \begin{matrix} i \\ kj \end{matrix} \right\} \mathbf{e}^k(\mathbf{x}), \tag{F.30}$$

$$\left\{ \begin{matrix} k \\ ij \end{matrix} \right\} = \left\{ \begin{matrix} k \\ ji \end{matrix} \right\}, \tag{F.31}$$

$$\left\{ \begin{matrix} m \\ ik \end{matrix} \right\} = \frac{g^{jm}}{2} \left[\frac{\partial g_{ij}}{\partial x^k} + \frac{\partial g_{kj}}{\partial x^i} - \frac{\partial g_{ik}}{\partial x^j} \right]. \tag{F.32}$$

Proof: Formula (F.29) is the definition of the Christoffel symbols. Consider (F.30) next. To do so, note

$$\mathbf{e}^i(\mathbf{x}) \cdot \mathbf{e}_k(\mathbf{x}) = \delta^i_k.$$

Then from the product rule,

$$\frac{\partial \mathbf{e}^i(\mathbf{x})}{\partial x^j} \cdot \mathbf{e}_k(\mathbf{x}) + \mathbf{e}^i(\mathbf{x}) \cdot \frac{\partial \mathbf{e}_k(\mathbf{x})}{\partial x^j} = 0.$$

Now from the definition,

$$\frac{\partial \mathbf{e}^i(\mathbf{x})}{\partial x^j} \cdot \mathbf{e}_k(\mathbf{x}) = -\mathbf{e}^i(\mathbf{x}) \cdot \left\{ \begin{matrix} r \\ kj \end{matrix} \right\} \mathbf{e}_r(\mathbf{x}) = - \left\{ \begin{matrix} i \\ kj \end{matrix} \right\}.$$

But also,

$$\frac{\partial \mathbf{e}^i(\mathbf{x})}{\partial x^j} = \frac{\partial \mathbf{e}^i(\mathbf{x})}{\partial x^j} \cdot \mathbf{e}_k(\mathbf{x}) \mathbf{e}^k(\mathbf{x}) = - \left\{ \begin{matrix} i \\ kj \end{matrix} \right\} \mathbf{e}^k(\mathbf{x}).$$

This verifies (F.30).

Letting $\frac{\partial \mathbf{M}(\mathbf{x})}{\partial x^j} = \mathbf{e}_j(\mathbf{x})$, it follows from equality of mixed partial derivatives,

$$\left\{ \begin{matrix} k \\ ij \end{matrix} \right\} \mathbf{e}_k(\mathbf{x}) = \frac{\partial \mathbf{e}_i}{\partial x^j} \equiv \frac{\partial^2 \mathbf{M}}{\partial x^j \partial x^i} = \frac{\partial^2 \mathbf{M}}{\partial x^i \partial x^j} = \frac{\partial \mathbf{e}_j}{\partial x^i} = \left\{ \begin{matrix} k \\ ji \end{matrix} \right\} \mathbf{e}_k(\mathbf{x}),$$

which shows (F.31). It remains to show (F.32).

$$\frac{\partial g_{ij}}{\partial x^k} = \frac{\partial \mathbf{e}_i}{\partial x^k} \cdot \mathbf{e}_j + \mathbf{e}_i \cdot \frac{\partial \mathbf{e}_j}{\partial x^k} = \left\{ \begin{matrix} r \\ ik \end{matrix} \right\} \mathbf{e}_r \cdot \mathbf{e}_j + \mathbf{e}_i \cdot \mathbf{e}_r \left\{ \begin{matrix} r \\ jk \end{matrix} \right\}.$$

Therefore,

$$\frac{\partial g_{ij}}{\partial x^k} = \left\{ \begin{matrix} r \\ ik \end{matrix} \right\} g_{rj} + \left\{ \begin{matrix} r \\ jk \end{matrix} \right\} g_{ri}. \tag{F.33}$$

Switching i and k while remembering (F.31) yields

$$\frac{\partial g_{kj}}{\partial x^i} = \left\{ \begin{matrix} r \\ ik \end{matrix} \right\} g_{rj} + \left\{ \begin{matrix} r \\ ji \end{matrix} \right\} g_{rk}. \tag{F.34}$$

Now switching j and k in (F.33),

$$\frac{\partial g_{ik}}{\partial x^j} = \left\{ \begin{matrix} r \\ ij \end{matrix} \right\} g_{rk} + \left\{ \begin{matrix} r \\ jk \end{matrix} \right\} g_{ri}. \tag{F.35}$$

Adding (F.33) to (F.34) and subtracting (F.35) yields

$$\frac{\partial g_{ij}}{\partial x^k} + \frac{\partial g_{kj}}{\partial x^i} - \frac{\partial g_{ik}}{\partial x^j} = 2 \left\{ \begin{matrix} r \\ ik \end{matrix} \right\} g_{rj}.$$

Now multiplying both sides by g^{jm} and using the fact shown earlier in Theorem F.4 that

$$g_{rj} g^{jm} = \delta_r^m,$$

$$2 \left\{ \begin{matrix} m \\ ik \end{matrix} \right\} = g^{jm} \left(\frac{\partial g_{ij}}{\partial x^k} + \frac{\partial g_{kj}}{\partial x^i} - \frac{\partial g_{ik}}{\partial x^j} \right)$$

which proves (F.32). ■

This is a very interesting formula because it shows the Christoffel symbols are completely determined by the metric tensor and its derivatives.

F.7   Gradients And Divergence

What is the gradient and divergence of a vector field in general curvilinear coordinates? As before, $\mathbf{y}$ will denote the standard coordinates with respect to the usual basis vectors. Thus

$$\mathbf{N}(\mathbf{y}) \equiv y^k \mathbf{i}_k, \quad \mathbf{e}_k(\mathbf{y}) = \mathbf{i}_k = \mathbf{e}^k(\mathbf{y}).$$

Let $\phi : U \to \mathbb{R}$ be a differentiable scalar function, sometimes called a "scalar field" in this subject. Write $\phi(\mathbf{x})$ to denote the value of ϕ at the point whose coordinates are $\mathbf{x}$. In general, this convention is followed for any field, vector or scalar. Thus $\mathbf{F}(\mathbf{x})$ is the value of a vector field at the point of U determined by the coordinates $\mathbf{x}$. In the standard coordinates, the formula for the gradient of ϕ is known. It is given by the following formula.

$$\nabla\phi\left(\mathbf{y}\right) = \frac{\partial\phi\left(\mathbf{y}\right)}{\partial y^k}\mathbf{e}^k\left(\mathbf{y}\right).$$

(Recall $\mathbf{e}^k\left(\mathbf{y}\right) = \mathbf{e}_k\left(\mathbf{y}\right) = \mathbf{i}_k$.) Therefore, using the chain rule, if the coordinates of the point of U are given in terms of the curvilinear coordinates $\mathbf{x}$,

$$\nabla\phi\left(\mathbf{x}\right) = \nabla\phi\left(\mathbf{y}\right)$$

$$= \frac{\partial\phi\left(\mathbf{x}\right)}{\partial x^r}\frac{\partial x^r}{\partial y^k}\frac{\partial y^k}{\partial x^s}\mathbf{e}^s\left(\mathbf{x}\right) = \frac{\partial\phi\left(\mathbf{x}\right)}{\partial x^r}\delta^r_s\mathbf{e}^s\left(\mathbf{x}\right) = \frac{\partial\phi\left(\mathbf{x}\right)}{\partial x^r}\mathbf{e}^r\left(\mathbf{x}\right).$$

This shows the covariant components of $\nabla\phi\left(\mathbf{x}\right)$ are

$$\left(\nabla\phi\left(\mathbf{x}\right)\right)_r = \frac{\partial\phi\left(\mathbf{x}\right)}{\partial x^r}. \tag{F.36}$$

To find the contravariant components, "raise the index" in the usual way. Thus

$$\left(\nabla\phi\left(\mathbf{x}\right)\right)^r = g^{rk}\left(\mathbf{x}\right)\left(\nabla\phi\left(\mathbf{x}\right)\right)_k = g^{rk}\left(\mathbf{x}\right)\frac{\partial\phi\left(\mathbf{x}\right)}{\partial x^k}. \tag{F.37}$$

What about the divergence of a vector field? The divergence of a vector field, $\mathbf{F}$ defined on U is a scalar field, $\mathrm{div}\left(\mathbf{F}\right)$ which in terms of the usual coordinates $\mathbf{y}$ is

$$\frac{\partial F^k}{\partial y^k}\left(\mathbf{y}\right) = F^k_{,k}\left(\mathbf{y}\right)$$

in terms of the usual coordinates $\mathbf{y}$. The reason the above equation holds in this case is that $\mathbf{e}_k\left(\mathbf{y}\right)$ is a constant and so the Christoffel symbols are zero. What is the correct expression for the divergence in an arbitrary coordinate system? From Theorem F.6,

$$F^i_{,j}\left(\mathbf{y}\right) = F^r_{,s}\left(\mathbf{x}\right)\frac{\partial x^s}{\partial y^j}\frac{\partial y^i}{\partial x^r}$$

$$= \left(\frac{\partial F^r\left(\mathbf{x}\right)}{\partial x^s} + F^k\left(\mathbf{x}\right)\left\{\begin{matrix}r\\ks\end{matrix}\right\}\left(\mathbf{x}\right)\right)\frac{\partial x^s}{\partial y^j}\frac{\partial y^i}{\partial x^r}.$$

Letting $j = i$ yields

$$\mathrm{div}\left(\mathbf{F}\right) = \left(\frac{\partial F^r\left(\mathbf{x}\right)}{\partial x^s} + F^k\left(\mathbf{x}\right)\left\{\begin{matrix}r\\ks\end{matrix}\right\}\left(\mathbf{x}\right)\right)\frac{\partial x^s}{\partial y^i}\frac{\partial y^i}{\partial x^r}$$

$$= \left(\frac{\partial F^r\left(\mathbf{x}\right)}{\partial x^s} + F^k\left(\mathbf{x}\right)\left\{\begin{matrix}r\\ks\end{matrix}\right\}\left(\mathbf{x}\right)\right)\delta^s_r$$

$$= \left(\frac{\partial F^r\left(\mathbf{x}\right)}{\partial x^r} + F^k\left(\mathbf{x}\right)\left\{\begin{matrix}r\\kr\end{matrix}\right\}\left(\mathbf{x}\right)\right). \tag{F.38}$$

The Christoffel symbols $\left\{\begin{matrix}r\\kr\end{matrix}\right\}$ are simplified using the description of it in Theorem F.7. Thus, from this theorem,

$$\left\{\begin{matrix}r\\rk\end{matrix}\right\} = \frac{g^{jr}}{2}\left[\frac{\partial g_{rj}}{\partial x^k} + \frac{\partial g_{kj}}{\partial x^r} - \frac{\partial g_{rk}}{\partial x^j}\right]$$

Now consider $\frac{g^{jr}}{2}$ times the last two terms in $[\cdot]$. Relabeling the indices r and j in the second term implies

$$\frac{g^{jr}}{2}\frac{\partial g_{kj}}{\partial x^r} - \frac{g^{jr}}{2}\frac{\partial g_{rk}}{\partial x^j} = \frac{g^{jr}}{2}\frac{\partial g_{kj}}{\partial x^r} - \frac{g^{rj}}{2}\frac{\partial g_{jk}}{\partial x^r} = 0.$$

Therefore,

$$\left\{ \begin{array}{c} r \\ rk \end{array} \right\} = \frac{g^{jr}}{2}\frac{\partial g_{rj}}{\partial x^k}. \tag{F.39}$$

Now recall $g \equiv \det(g_{ij}) = \det(G) > 0$ from Theorem F.4. Also from the formula for the inverse of a matrix and this theorem,

$$g^{jr} = A^{rj}(\det G)^{-1} = A^{jr}(\det G)^{-1}$$

where A^{rj} is the rj^{th} cofactor of the matrix (g_{ij}). Also recall that

$$g = \sum_{r=1}^{n} g_{rj}A^{rj} \text{ no sum on } j.$$

Therefore, g is a function of the variables $\{g_{rj}\}$ and

$$\frac{\partial g}{\partial g_{rj}} = A^{rj}.$$

From (F.39),

$$\left\{ \begin{array}{c} r \\ rk \end{array} \right\} = \frac{g^{jr}}{2}\frac{\partial g_{rj}}{\partial x^k} = \frac{1}{2g}\frac{\partial g_{rj}}{\partial x^k}A^{jr} = \frac{1}{2g}\frac{\partial g}{\partial g_{rj}}\frac{\partial g_{rj}}{\partial x^k} = \frac{1}{2g}\frac{\partial g}{\partial x^k}$$

and so from (F.38),

$$\text{div}\,(\mathbf{F}) = \frac{\partial F^k(\mathbf{x})}{\partial x^k} +$$

$$+ F^k(\mathbf{x})\frac{1}{2g(\mathbf{x})}\frac{\partial g(\mathbf{x})}{\partial x^k} = \frac{1}{\sqrt{g(\mathbf{x})}}\frac{\partial}{\partial x^i}\left(F^i(\mathbf{x})\sqrt{g(\mathbf{x})}\right). \tag{F.40}$$

This is the formula for the divergence of a vector field in general curvilinear coordinates.

The Laplacian of a scalar field is nothing more than the divergence of the gradient. In symbols,

$$\Delta\phi \equiv \nabla \cdot \nabla\phi$$

From (F.40) and (F.37) it follows

$$\Delta\phi(\mathbf{x}) = \frac{1}{\sqrt{g(\mathbf{x})}}\frac{\partial}{\partial x^i}\left(g^{ik}(\mathbf{x})\frac{\partial\phi(\mathbf{x})}{\partial x^k}\sqrt{g(\mathbf{x})}\right). \tag{F.41}$$

Summarizing all this yields the following theorem.

Theorem F.8. *The following formulas hold for the gradient, divergence and Laplacian in general curvilinear coordinates.*

$$(\nabla\phi(\mathbf{x}))_r = \frac{\partial\phi(\mathbf{x})}{\partial x^r}, \tag{F.42}$$

$$(\nabla\phi(\mathbf{x}))^r = g^{rk}(\mathbf{x})\frac{\partial\phi(\mathbf{x})}{\partial x^k}, \tag{F.43}$$

$$div(\mathbf{F}) = \frac{1}{\sqrt{g(\mathbf{x})}}\frac{\partial}{\partial x^i}\left(F^i(\mathbf{x})\sqrt{g(\mathbf{x})}\right), \tag{F.44}$$

$$\Delta\phi(\mathbf{x}) = \frac{1}{\sqrt{g(\mathbf{x})}}\frac{\partial}{\partial x^i}\left(g^{ik}(\mathbf{x})\frac{\partial\phi(\mathbf{x})}{\partial x^k}\sqrt{g(\mathbf{x})}\right). \tag{F.45}$$

F.8   Exercises

(1) Let $y^1 = x^1 + 2x^2, y^2 = x^2 + 3x^3, y^3 = x^1 + x^3$. Let
$$\mathbf{F}(\mathbf{x}) = x^1\mathbf{e}_1(\mathbf{x}) + x^2\mathbf{e}_2(\mathbf{x}) + \left(x^3\right)^2\mathbf{e}(\mathbf{x}).$$
Find div $(\mathbf{F})(\mathbf{x})$.

(2) For the coordinates of the preceding problem, and ϕ a scalar field, find
$$(\nabla\phi(\mathbf{x}))^3$$
in terms of the partial derivatives of ϕ taken with respect to the variables x^i.

(3) Let $y^1 = 7x^1 + 2x^2, y^2 = x^2 + 3x^3, y^3 = x^1 + x^3$. Let ϕ be a scalar field. Find $\nabla^2\phi(\mathbf{x})$.

(4) Derive $\nabla^2 u$ in cylindrical coordinates r, θ, z, where u is a scalar field on $\mathbb{R}^3$.
$$x = r\cos\theta,\ y = r\sin\theta,\ z = z.$$

(5) ↑ Find all solutions to $\nabla^2 u = 0$ which depend only on r where $r \equiv \sqrt{x^2 + y^2}$.

(6) Let u be a scalar field on $\mathbb{R}^3$. Find all solutions to $\nabla^2 u = 0$ which depend only on
$$\rho \equiv \sqrt{x^2 + y^2 + z^2}.$$

(7) The temperature, u, in a solid satisfies $\nabla^2 u = 0$ after a long time. Suppose in a long pipe of inner radius 9 and outer radius 10 the exterior surface is held at $100°$ while the inner surface is held at $200°$ find the temperature in the solid part of the pipe.

(8) Show
$$\left\{\begin{matrix} l \\ ij \end{matrix}\right\} = \frac{\partial\mathbf{e}_i}{\partial x^j}\cdot\mathbf{e}^l.$$
Find the Christoffel symbols of the second kind for spherical coordinates in which $x^1 = \phi$, $x^2 = \theta$, and $x^3 = \rho$. Do the same for cylindrical coordinates letting $x^1 = r$, $x^2 = \theta$, $x^3 = z$.

(9) Show that velocity can be expressed as $\mathbf{v} = v_i(\mathbf{x})\mathbf{e}^i(\mathbf{x})$, where
$$v_i(\mathbf{x}) = \frac{\partial r_i}{\partial x^j}\frac{\partial x^j}{\partial t} - \Gamma_p(\mathbf{x})\left\{\begin{matrix} p \\ ik \end{matrix}\right\}\frac{dx^k}{dt}$$
and $r_i(\mathbf{x})$ are the covariant components of the displacement vector,
$$\mathbf{r} = r_i(\mathbf{x})\mathbf{e}^i(\mathbf{x})$$

(10) ↑ Using Problem 8 and (9), show that the covariant components of velocity in spherical coordinates are

$$v_1 = \rho^2 \frac{d\phi}{dt}, \; v_2 = \rho^2 \sin^2(\phi) \frac{d\theta}{dt}, \; v_3 = \frac{d\rho}{dt}.$$

Hint: First observe that if $\mathbf{r}$ is the position vector from the origin, then $\mathbf{r} = \rho \mathbf{e}_3$ so $r_1 = 0 = r_2$, and $r_3 = \rho$. Now use (9).

F.9   Curl And Cross Products

What about the curl and cross product in general curvilinear coordinates in $\mathbb{R}^3$. In all of this, assume that for $\mathbf{x}$ a set of curvilinear coordinates,

$$\det \left(\frac{\partial y^i}{\partial x^j} \right) > 0 \tag{F.46}$$

Where the y_i are the usual coordinates in which $\mathbf{e}_k(\mathbf{y}) = \mathbf{i}_k$. This is a statement about orientation.

Theorem F.9. *Let (F.46) hold. Then*

$$\det \left(\frac{\partial y^i}{\partial x^j} \right) = \sqrt{g(\mathbf{x})} \tag{F.47}$$

and

$$\det \left(\frac{\partial x^i}{\partial y^j} \right) = \frac{1}{\sqrt{g(\mathbf{x})}}. \tag{F.48}$$

   Proof:

$$\mathbf{e}_i(\mathbf{x}) = \frac{\partial y^k}{\partial x^i} \mathbf{i}_k$$

and so

$$g_{ij}(\mathbf{x}) = \frac{\partial y^k}{\partial x^i} \mathbf{i}_k \cdot \frac{\partial y^l}{\partial x^j} \mathbf{i}_l = \frac{\partial y^k}{\partial x^i} \frac{\partial y^k}{\partial x^j}.$$

Therefore, $g = \det(g_{ij}(\mathbf{x})) = \left(\det \left(\frac{\partial y^k}{\partial x^i} \right) \right)^2$. By (F.46), $\sqrt{g} = \det \left(\frac{\partial y^k}{\partial x^i} \right)$ as claimed. Now

$$\frac{\partial y^k}{\partial x^i} \frac{\partial x^i}{\partial y^r} = \delta^k_r$$

and so

$$\det \left(\frac{\partial x^i}{\partial y^r} \right) = \frac{1}{\sqrt{g(\mathbf{x})}}. \qquad \blacksquare$$

To get the curl and cross product in curvilinear coordinates, let ϵ^{ijk} be the usual permutation symbol. Thus,

$$\epsilon^{123} = 1$$

and when any two indices in ϵ^{ijk} are switched, the sign changes. Thus

$$\epsilon^{132} = -1, \epsilon^{312} = 1, \text{ etc.}$$

Now define

$$\varepsilon^{ijk}(\mathbf{x}) \equiv \epsilon^{ijk} \frac{1}{\sqrt{g(\mathbf{x})}}.$$

Then for $\mathbf{x}$ and $\mathbf{z}$ satisfying (F.46),

$$\varepsilon^{ijk}(\mathbf{x}) \frac{\partial z^r}{\partial x^i} \frac{\partial z^s}{\partial x^j} \frac{\partial z^t}{\partial x^k} = \epsilon^{ijk} \det\left(\frac{\partial x^p}{\partial y^q}\right) \frac{\partial z^r}{\partial x^i} \frac{\partial z^s}{\partial x^j} \frac{\partial z^t}{\partial x^k}$$

$$= \epsilon^{rst} \det\left(\frac{\partial x^p}{\partial y^q}\right) \det\left(\frac{\partial z^i}{\partial x^k}\right) = \epsilon^{rst} \det(MN)$$

where N is the matrix whose pq^{th} entry is $\frac{\partial x^p}{\partial y^q}$ and M is the matrix whose ik^{th} entry is $\frac{\partial z^i}{\partial x^k}$. Therefore, from the definition of matrix multiplication and the chain rule, this equals

$$= \epsilon^{rst} \det\left(\frac{\partial z^i}{\partial y^p}\right) \equiv \varepsilon^{rst}(\mathbf{z})$$

from the above discussion.

Now $\varepsilon^{ijk}(\mathbf{y}) = \epsilon^{ijk}$ and for a vector field $\mathbf{F}$

$$\text{curl}(\mathbf{F}) \equiv \varepsilon^{ijk}(\mathbf{y}) F_{k,j}(\mathbf{y}) \mathbf{e}_i(\mathbf{y}).$$

Therefore, since it is known how these things transform, assuming (F.46), it is routine to write this in terms of $\mathbf{x}$.

$$\text{curl}(\mathbf{F}) = \varepsilon^{rst}(\mathbf{x}) \frac{\partial y^i}{\partial x^r} \frac{\partial y^j}{\partial x^s} \frac{\partial y^k}{\partial x^t} F_{p,q}(\mathbf{x}) \frac{\partial x^p}{\partial y^k} \frac{\partial x^q}{\partial y^j} \mathbf{e}_m(\mathbf{x}) \frac{\partial x^m}{\partial y^i}$$

$$= \varepsilon^{rst}(\mathbf{x}) \delta_r^m \delta_s^q \delta_t^p F_{p,q}(\mathbf{x}) \mathbf{e}_m(\mathbf{x}) = \varepsilon^{mqp}(\mathbf{x}) F_{p,q}(\mathbf{x}) \mathbf{e}_m(\mathbf{x}). \tag{F.49}$$

More simplification is possible. Recalling the definition of $F_{p,q}(\mathbf{x})$,

$$\frac{\partial \mathbf{F}}{\partial x^q} \equiv F_{p,q}(\mathbf{x}) \mathbf{e}^p(\mathbf{x}) = \frac{\partial}{\partial x^q} [F_p(\mathbf{x}) \mathbf{e}^p(\mathbf{x})]$$

$$= \frac{\partial F_p(\mathbf{x})}{\partial x^q} \mathbf{e}^p(\mathbf{x}) + F_p(\mathbf{x}) \frac{\partial \mathbf{e}^p}{\partial x^q} = \frac{\partial F_p(\mathbf{x})}{\partial x^q} \mathbf{e}^p(\mathbf{x}) - F_r(\mathbf{x}) \left\{ \begin{array}{c} r \\ pq \end{array} \right\} \mathbf{e}^p(\mathbf{x})$$

by Theorem F.7. Therefore,

$$F_{p,q}(\mathbf{x}) = \frac{\partial F_p(\mathbf{x})}{\partial x^q} - F_r(\mathbf{x}) \left\{ \begin{array}{c} r \\ pq \end{array} \right\}$$

and so

$$\text{curl}(\mathbf{F}) = \varepsilon^{mqp}(\mathbf{x}) \frac{\partial F_p(\mathbf{x})}{\partial x^q} \mathbf{e}_m(\mathbf{x}) - \varepsilon^{mqp}(\mathbf{x}) F_r(\mathbf{x}) \left\{ \begin{array}{c} r \\ pq \end{array} \right\} \mathbf{e}_m(\mathbf{x}).$$

However, because $\left\{ \begin{array}{c} r \\ pq \end{array} \right\} = \left\{ \begin{array}{c} r \\ qp \end{array} \right\}$, the second term in this expression equals 0. To see this,

$$\varepsilon^{mqp}\left(\mathbf{x}\right)\left\{ \begin{array}{c} r \\ pq \end{array} \right\} = \varepsilon^{mpq}\left(\mathbf{x}\right)\left\{ \begin{array}{c} r \\ qp \end{array} \right\} = -\varepsilon^{mqp}\left(\mathbf{x}\right)\left\{ \begin{array}{c} r \\ pq \end{array} \right\}.$$

Therefore, by (F.49),

$$\operatorname{curl}\left(\mathbf{F}\right) = \varepsilon^{mqp}\left(\mathbf{x}\right)\frac{\partial F_p\left(\mathbf{x}\right)}{\partial x^q}\mathbf{e}_m\left(\mathbf{x}\right). \tag{F.50}$$

What about the cross product of two vector fields? Let $\mathbf{F}$ and $\mathbf{G}$ be two vector fields. Then in terms of standard coordinates $\mathbf{y}$,

$$\mathbf{F} \times \mathbf{G} = \varepsilon^{ijk}\left(\mathbf{y}\right) F_j\left(\mathbf{y}\right) G_k\left(\mathbf{y}\right) \mathbf{e}_i\left(\mathbf{y}\right)$$

$$= \varepsilon^{rst}\left(\mathbf{x}\right)\frac{\partial y^i}{\partial x^r}\frac{\partial y^j}{\partial x^s}\frac{\partial y^k}{\partial x^t}F_p\left(\mathbf{x}\right)\frac{\partial x^p}{\partial y^j}G_q\left(\mathbf{x}\right)\frac{\partial x^q}{\partial y^k}\mathbf{e}_l\left(\mathbf{x}\right)\frac{\partial x^l}{\partial y^i}$$

$$= \varepsilon^{rst}\left(\mathbf{x}\right)\delta_s^p\delta_t^q\delta_r^l F_p\left(\mathbf{x}\right) G_q\left(\mathbf{x}\right)\mathbf{e}_l\left(\mathbf{x}\right) = \varepsilon^{lpq}\left(\mathbf{x}\right)F_p\left(\mathbf{x}\right) G_q\left(\mathbf{x}\right)\mathbf{e}_l\left(\mathbf{x}\right). \tag{F.51}$$

This implies

Theorem F.10. *Suppose $\mathbf{x}$ is a system of curvilinear coordinates in $\mathbb{R}^3$ such that*

$$\det\left(\frac{\partial y^i}{\partial x^j}\right) > 0.$$

Let

$$\varepsilon^{ijk}\left(\mathbf{x}\right) \equiv \epsilon^{ijk}\frac{1}{\sqrt{g\left(\mathbf{x}\right)}}.$$

Then the following formulas for curl and cross product hold in this system of coordinates.

$$\operatorname{curl}\left(\mathbf{F}\right) = \varepsilon^{mqp}\left(\mathbf{x}\right)\frac{\partial F_p\left(\mathbf{x}\right)}{\partial x^q}\mathbf{e}_m\left(\mathbf{x}\right),$$

and

$$\mathbf{F} \times \mathbf{G} = \varepsilon^{lpq}\left(\mathbf{x}\right)F_p\left(\mathbf{x}\right) G_q\left(\mathbf{x}\right)\mathbf{e}_l\left(\mathbf{x}\right).$$

Bibliography

Apostol, T. M., *Calculus.*, second edition, Wiley, 1967.

Apostol T., *Calculus Volume II*, second edition, Wiley, 1969.

Apostol, T. M., *Mathematical Analysis*, Addison Wesley Publishing Co., 1974.

Baker, Roger, *Linear Algebra*, Rinton Press, 2001.

Bartle R.G., *A Modern Theory of Integration*, Grad. Studies in Math., Amer. Math. Society, Providence, RI, 2000.

Chahal J. S., *Historical Perspective of Mathematics* 2000 B.C. - 2000 A.D.

Davis H. and Snider A., *Vector Analysis*, Wm. C. Brown, 1995.

D'Angelo, J. and West D., *Mathematical Thinking Problem Solving and Proofs*, Prentice Hall, 1997.

Edwards C.H., *Advanced Calculus of Several Variables*, Dover, 1994.

Euclid, *The Thirteen Books of the Elements*, Dover, 1956.

Fitzpatrick P. M., *Advanced Calculus a Course in Mathematical Analysis*, PWS Publishing Company, 1996.

Fleming W., *Functions of Several Variables*, Springer Verlag, 1976.

Greenberg, M., *Advanced Engineering Mathematics, second edition*, Prentice Hall, 1998.

Gurtin M., *An Introduction to Continuum Mechanics*, Academic press, 1981.

Hardy G., *A Course Of Pure Mathematics*, tenth edition, Cambridge University Press, 1992.

Horn R. and Johnson C., *Matrix Analysis*, Cambridge University Press, 1985.

Karlin S. and Taylor H., *A First Course in Stochastic Processes*, Academic Press, 1975.

Kuttler K. L., *Basic Analysis*, Rinton.

Kuttler K.L., *Modern Analysis*, CRC Press, 1998.

Lang S., *Real and Functional Analysis*, third edition, Springer Verlag 1993. Press, 2001.

Nobel B. and Daniel J., *Applied Linear Algebra*, Prentice Hall, 1977.

Rose, David, A., The College Math Journal, vol. 22, No.2 March 1991.

Rudin, W., *Principles of Mathematical Analysis*, third edition, McGraw Hill, 1976.

Rudin W., *Real and Complex Analysis*, third edition, McGraw-Hill, 1987.

Salas S. and Hille E., *Calculus One and Several Variables*, Wiley, 1990.

Sears and Zemansky, *University Physics*, third edition, Addison Wesley, 1963.

Spivak M., *Calculus on Manifolds*, Benjamin, 1965.

Tierney John, *Calculus and Analytic Geometry*, fourth edition, Allyn and Bacon, Boston, 1969.

Yosida K., *Functional Analysis*, Springer Verlag, 1978.

Appendix G

Answers To Selected Exercises

Exercises 1.4

(1) $\begin{pmatrix} 3 & 6 & 9 \\ 6 & 3 & 21 \end{pmatrix}$

$\begin{pmatrix} 8 & -5 & 3 \\ -11 & 5 & -4 \end{pmatrix}$

AC makes no sense.

$\begin{pmatrix} -3 & 3 & 4 \\ 6 & -1 & 7 \end{pmatrix}$

AE makes no sense

EA makes no sense.

(2) $\begin{pmatrix} -3 & -6 \\ -9 & -6 \\ -3 & 3 \end{pmatrix}$

$3B - A$ makes no sense.

$\begin{pmatrix} 11 & 2 \\ 13 & 6 \\ -4 & 2 \end{pmatrix}$

CA makes no sense.

$\begin{pmatrix} 7 \\ 9 \\ -2 \end{pmatrix}$

EA makes no sense.

(3) $\begin{pmatrix} -3 & -9 & -3 \\ -6 & -6 & 3 \end{pmatrix}$

$\begin{pmatrix} 5 & -18 & 5 \\ -11 & 4 & 4 \end{pmatrix}$

$\begin{pmatrix} 11 & 2 \\ 13 & 6 \\ -4 & 2 \end{pmatrix}$

CA makes no sense.

$\begin{pmatrix} 7 \\ 9 \\ -2 \end{pmatrix}$

$\begin{pmatrix} -7 & 1 & 5 \end{pmatrix}$

BE makes no sense.

$\begin{pmatrix} 2 \\ -5 \end{pmatrix}$

$\begin{pmatrix} 1 & 3 \\ 3 & 9 \end{pmatrix}$

10

(4) Find

(a) $\begin{pmatrix} 3 & 0 & -4 \\ -4 & 1 & 6 \\ 5 & 1 & -6 \end{pmatrix}$

(b) $\begin{pmatrix} 1 & -2 \\ -2 & -3 \end{pmatrix}$

(c) AC makes no sense.

(d) $\begin{pmatrix} -4 & -6 \\ -5 & -3 \\ -1 & -2 \end{pmatrix}$

(e) CB makes no sense.

(f) $\begin{pmatrix} 8 & 1 & -3 \\ 7 & 6 & -6 \end{pmatrix}$

(5) $k = 4$.

(6) There is no way to make these equal because $3 \neq 7$.

(7) $\begin{pmatrix} 0 & -1 & -2 \\ 0 & -1 & -2 \\ 0 & 1 & 2 \end{pmatrix}$

$$\left(\begin{array}{ccc} 0 & 1 & 2 \end{array} \right) \left(\begin{array}{c} -1 \\ -1 \\ 1 \end{array} \right) = 1$$

(8) $\left(\begin{array}{cc} \frac{1}{2}\sqrt{2} & -\frac{1}{2}\sqrt{2} \\ \frac{1}{2}\sqrt{2} & \frac{1}{2}\sqrt{2} \end{array} \right)$

(9) $\left(\begin{array}{cc} \frac{1}{2} & \frac{1}{2}\sqrt{3} \\ -\frac{1}{2}\sqrt{3} & \frac{1}{2} \end{array} \right)$

(10) $\left(\begin{array}{cc} -\frac{1}{2} & -\frac{1}{2}\sqrt{3} \\ \frac{1}{2}\sqrt{3} & -\frac{1}{2} \end{array} \right)$

(11) $\left(\begin{array}{cc} \frac{1}{4}\sqrt{2}+\frac{1}{4}\sqrt{3}\sqrt{2} & \frac{1}{4}\sqrt{2}-\frac{1}{4}\sqrt{3}\sqrt{2} \\ \frac{1}{4}\sqrt{3}\sqrt{2}-\frac{1}{4}\sqrt{2} & \frac{1}{4}\sqrt{2}+\frac{1}{4}\sqrt{3}\sqrt{2} \end{array} \right)$

(12) $\left(\begin{array}{cc} 1 & 4 \\ 2 & 1 \end{array} \right) \left(\begin{array}{c} x_1 \\ x_2 \end{array} \right)$

(13) $\left(\begin{array}{cc} 1 & -1 \\ 1 & 0 \\ 1 & 3 \\ 5 & 3 \end{array} \right) \left(\begin{array}{c} x_1 \\ x_2 \end{array} \right)$

(14) $\left(\begin{array}{cccc} 1 & -1 & 2 & 0 \\ 1 & 0 & 2 & 0 \\ 0 & 0 & 3 & 0 \\ 1 & 3 & 0 & 3 \end{array} \right) \left(\begin{array}{c} x_1 \\ x_2 \\ x_3 \\ x_4 \end{array} \right)$

(15) $T(1,0) = \left(\begin{array}{c} 1 \\ 2 \end{array} \right), T(2,0) =$

$\left(\begin{array}{c} 4 \\ 4 \end{array} \right) \neq 2T(1,0)$.

(16) Suppose A and B are square matrices of the same size. Which of the following are correct?

(a) Not true.

(b) Not true

(c) Not true

(d) True

(e) True

(f) Not true

(g) Not true

(17) $\left(\begin{array}{cc} x & y \\ -x & -y \end{array} \right)$ where $x, y \in \mathbb{R}$.

(18) $-A$ is just $(-1)A$. 0 is the matrix of the same size which has all zero entries.

(19) $A = \frac{1}{2}\left(A + A^T\right) + \frac{1}{2}\left(A - A^T\right) =$ symmetric + skew symmetric.

(20) $\left(A^n\right)^T = \left(A^T\right)^n = (-1)^n A^n$.

(21) You need $a_{ii} = -a_{ii}$ and so $a_{ii} = 0$.

(23) $0' = 0' + 0 = 0$.

(24) $0A = (0+0)A = 0A + 0A$ and so, adding $-0A$ to both sides, this yields $0 = 0A$.

(25) $A + (-1)A = (1 + -1)A = 0A = 0$. Since the additive inverse is unique, it follows $-A = (-1)A$.

(27) $(I_m A)_{ij} \equiv \sum_k \delta_{ik} A_{kj} = A_{ij}$ and so $I_m A = A$.

(28) $\left(\begin{array}{cc} -\frac{1}{3} & \frac{2}{3} \\ \frac{2}{3} & -\frac{1}{3} \end{array} \right)$

(29) $\left(\begin{array}{cc} 1 & 0 \\ -\frac{2}{3} & \frac{1}{3} \end{array} \right)$

(30) $\left(\begin{array}{cc} -1 & 2 \\ 2 & -3 \end{array} \right)$

(31) $\left(\begin{array}{cc} 1 & 1 \\ 1 & 1 \end{array} \right) \left(\begin{array}{cc} 1 & -1 \\ -1 & 1 \end{array} \right) = \left(\begin{array}{cc} 0 & 0 \\ 0 & 0 \end{array} \right)$

$\left(\begin{array}{cc} 1 & 1 \\ 1 & 1 \end{array} \right) \left(\begin{array}{cc} 0 & 0 \\ 0 & 0 \end{array} \right) = \left(\begin{array}{cc} 0 & 0 \\ 0 & 0 \end{array} \right)$

(32) Yes. Just multiply by the inverse of A on the left on both sides.

(34) Multiply on the right on both sides by A^{-1}.

(35) $\left(\begin{array}{cc} 1 & 0 \\ 0 & -1 \end{array} \right)^2 = \left(\begin{array}{cc} 1 & 0 \\ 0 & 1 \end{array} \right)$

(36) $A^{-1} = A^{-1}I = A^{-1}(AB) = (A^{-1}A)B = IB = B$

(37) $ABB^{-1}A^{-1} = AIA^{-1} = I$
$B^{-1}A^{-1}AB = B^{-1}IB = I$
Since the inverse is unique, it follows $(AB)^{-1}$ exists and equals $B^{-1}A^{-1}$.

(38) $\left(A^{-1}\right)^T A^T = \left(AA^{-1}\right)^T = I^T = I$
$A^T \left(A^{-1}\right)^T = \left(A^{-1}A\right)^T = I^T = I$
By uniqueness of the inverse, it follows A^T has an inverse and it equals $\left(A^{-1}\right)^T$.

(39) Multiply on both sides on the left

by A^{-1}.

(40) Multiply on both sides on the left by A^{-1}.

(41) $ABCC^{-1}B^{-1}A^{-1} = ABIB^{-1}A^{-1} = AIA^{-1} = I$

$C^{-1}B^{-1}A^{-1}ABC = C^{-1}B^{-1}IBC = C^{-1}IC = I$

By uniqueness of the inverse, it follows $(ABC)^{-1}$ exists and equals $C^{-1}B^{-1}A^{-1}$.

Exercises 2.3

(1) Find the determinants of the following matrices.

   (a) 31

   (b) 375

   (c) -2

(2) 6

(3) 2

(4) 6

(5) -4

(6) -6

(7) -32

(8) 63

(9) 211

(10) It doesn't do anything.

(11) It will multiply by -1. Two rows were switched.

(12) It is unchanged. The top row was added to the bottom.

(13) It multiplies by 2. A row was multiplied by 2.

(14) It multiplies by -1. Two columns were switched.

(15) Tell whether the statement is true or false.

   (a) False.

   (b) True.

   (c) False in general.

   (d) False if $n > 1$.

   (e) True.

   (f) True.

   (g) True.

   (h) True.

   (i) False. The rule is useless in this case because you would have to divide by 0.

   (j) True.

   (k) True.

(17) ± 1

(18) $\begin{pmatrix} \frac{-1}{\sqrt{2}} & \frac{1}{\sqrt{6}} & \frac{\sqrt{12}}{6} \\ \frac{1}{\sqrt{2}} & \sqrt{\frac{1}{6}} & \frac{1}{3}\sqrt{3} \\ 0 & \frac{\sqrt{6}}{3} & -\frac{1}{3}\sqrt{3} \end{pmatrix}$

(19) $1 = \det I = \det\left(AA^{-1}\right) = \det(A)\det\left(A^{-1}\right).$

(20) No

(29) $\det(A) = 0$

(30) 1,0

(31) 1,0,0

(32) $\det \begin{pmatrix} 1 & 2 & 3 \\ 0 & 2 & 1 \\ 3 & 1 & 0 \end{pmatrix} = -13$ It has an inverse.

(33) No. Its determinant is nonzero.

(34) $t = -\sqrt[3]{2}.$

(35) No. Its determinant is nonzero.

(36) Any value of t will work.

(37) $\begin{pmatrix} 2 & -1 \\ -1 & 1 \end{pmatrix}, \begin{pmatrix} -\frac{1}{15} & -\frac{1}{15} & \frac{4}{15} \\ -\frac{4}{15} & \frac{11}{15} & \frac{1}{15} \\ \frac{8}{15} & -\frac{7}{15} & -\frac{2}{15} \end{pmatrix}$. The third matrix has no inverse.

(38) $A^{-1} = \begin{pmatrix} e^{-t} & 0 & 0 \\ 0 & (\cos t + \sin t)\,e^{-t} & -e^{-t}\sin t \\ 0 & (-\cos t + \sin t)\,e^{-t} & e^{-t}\cos t \end{pmatrix}$

(39) $A^{-1} = \begin{pmatrix} \frac{1}{2}e^{-t} & 0 & \frac{1}{2}e^{-t} \\ \frac{1}{2}\cos t + \frac{1}{2}\sin t & -\sin t & -\frac{1}{2}\cos t + \frac{1}{2}\sin t \\ -\frac{1}{2}\cos t + \frac{1}{2}\sin t & \cos t & -\frac{1}{2}\sin t - \frac{1}{2}\cos t \end{pmatrix}$

Exercises 3.2

(1) For $\mathbf{v} \neq \mathbf{0}$, $A\mathbf{v} = \lambda\mathbf{v}$ for some λ a scalar.

(4) Yes consider $\begin{pmatrix} 0 & 1 \\ 0 & 0 \end{pmatrix}$

(6) $\begin{pmatrix} 2 \\ 1 \\ 1 \end{pmatrix} \leftrightarrow 2, \begin{pmatrix} 3 \\ 1 \\ 1 \end{pmatrix} \leftrightarrow 1.$

(7) $\begin{pmatrix} 1 \\ 2 \\ 1 \end{pmatrix} \leftrightarrow 2, \begin{pmatrix} 2 \\ 1 \\ 1 \end{pmatrix} \leftrightarrow 3, \begin{pmatrix} 3 \\ 1 \\ 1 \end{pmatrix} \leftrightarrow 1$

(8) $\begin{pmatrix} 2 \\ 1 \\ 1 \end{pmatrix} \leftrightarrow 2, \begin{pmatrix} 5 \\ 0 \\ 1 \end{pmatrix}, \begin{pmatrix} -2 \\ 1 \\ 0 \end{pmatrix} \leftrightarrow -1$

(9) $\begin{pmatrix} 1 \\ 2 \\ -2 \end{pmatrix} \leftrightarrow 3, \begin{pmatrix} 0 \\ 0 \\ 1 \end{pmatrix} \leftrightarrow 6$

(10) $\begin{pmatrix} -1 \\ 1 \\ 1 \end{pmatrix} \leftrightarrow 6, \begin{pmatrix} 1 \\ 0 \\ 0 \end{pmatrix}, \begin{pmatrix} 0 \\ 1 \\ -2 \end{pmatrix} \leftrightarrow 3$

(11) $\begin{pmatrix} 1 \\ 2 \\ 1 \end{pmatrix} \leftrightarrow -1, \begin{pmatrix} 3 \\ 1 \\ 1 \end{pmatrix} \leftrightarrow 1$

(12) $\begin{pmatrix} 0 \\ 5 \\ 2 \end{pmatrix}, \begin{pmatrix} 1 \\ -3 \\ 0 \end{pmatrix} \leftrightarrow -1$

(13) $\begin{pmatrix} 9 \\ 3 \\ 13 \end{pmatrix} \leftrightarrow -3, \begin{pmatrix} 3 \\ 1 \\ 4 \end{pmatrix} \leftrightarrow -1, \begin{pmatrix} 1 \\ 2 \\ 2 \end{pmatrix} \leftrightarrow 2$

(14) $\begin{pmatrix} -1 \\ 0 \\ 1 \end{pmatrix} \leftrightarrow 18, \begin{pmatrix} 1 \\ -2 \\ -3 \end{pmatrix} \leftrightarrow 0, \begin{pmatrix} -2 \\ 1 \\ 0 \end{pmatrix} \leftrightarrow -12$

(15) $\begin{pmatrix} 3 \\ 1 \\ 4 \end{pmatrix} \leftrightarrow 0$

(16) $\begin{pmatrix} 3 \\ 1 \\ 4 \end{pmatrix} \leftrightarrow 1$

(17) $\begin{pmatrix} 1 \\ 0 \\ 1 \end{pmatrix}, \begin{pmatrix} -1 \\ 1 \\ 0 \end{pmatrix} \leftrightarrow 1, \begin{pmatrix} 1 \\ 2 \\ 2 \end{pmatrix} \leftrightarrow 2$

(18) $\begin{pmatrix} 1 \\ -1 \\ 1 \end{pmatrix} \leftrightarrow 4, \begin{pmatrix} 1 \\ 1 \\ -i \end{pmatrix} \leftrightarrow 2 + 2i, \begin{pmatrix} 1 \\ 1 \\ i \end{pmatrix} \leftrightarrow 2 - 2i$

(19) $\begin{pmatrix} 1 \\ 0 \end{pmatrix} \leftrightarrow 1, \begin{pmatrix} 0 \\ 1 \end{pmatrix} \leftrightarrow -1$

(20) $\begin{pmatrix} i \\ 1 \end{pmatrix} \leftrightarrow i, \begin{pmatrix} -i \\ 1 \end{pmatrix} \leftrightarrow -i$

(21) $\begin{pmatrix} 0 \\ 0 \\ 1 \end{pmatrix} \leftrightarrow -1, \begin{pmatrix} 0 \\ 1 \\ 0 \end{pmatrix}, \begin{pmatrix} 1 \\ 0 \\ 0 \end{pmatrix} \leftrightarrow 1$

(23) $\sqrt{218}$

(24) $\sqrt{152}$

(25) 0. This should equal 0. Three dimensional volume in two dimensional space should be 0.

(26) It is that $n+1$ dimensional volume in $\mathbb{R}^n$ must be 0.

Exercises 4.3

(1) In order, $(-(y-1),x),(x,y),(x,-y)$

(2) $z \neq 0, x \neq 0$, and $x^2 y^2 \leq 6$.

(3) Intersection of the ball $x^2 + y^2 + z^2 \leq 4$ with the complement of the hyperbola $1 + x^2 - y^2 = 0$.

(4) Find the expressions.

   (a) $(xy, -x^2, xyz)$
   (b) $(-xy, x^2, -xyz)$
   (c) $(x,y,0) \cdot (0,yz,x) = y^2 z$
   (d) $x^3 + xyz^2$
   (e) $(0,0,-y(yz^2 + x^2))$
   (f) $xy(yz^2 + x^2)$

(5) $(y^2 + z, x, y)$

(6) $(x, z, x^2 - 1)$.

(7) Which make sense?

   (a) Nonsense
   (b) Nonsense.
   (c) Nonsense.
   (d) This makes perfect sense.
   (e) Nonsense.
   (f) This makes perfect sense.

(8) The components of the population vector cycle around 1. The vector field vanishes at $(1,1)$.

Exercises 4.7

(1) $t(t+1) + 1 + t^2 + \frac{t^2}{(t+1)(1+t^2)}$

(2) $\left(t - \frac{t}{t+1}, t - \frac{t^2}{1+t^2}, t - (1+t^2)(t+1)\right)$

(3) You need $z \neq 0, w \neq 0, x^2 y^2 \leq 6$.

(4) $4(\sin t) t^3 - 5(\sin t) t^4 + (\cos t) t^4 - (\cos t) t^5 + 3t^2 - 2t$

(7) For given $\varepsilon > 0$, $\delta = \varepsilon^{1/\alpha}/K^{1/\alpha}$.

(11) Does not exist. Look at $y = x$ and $y = 0$.

(12) Find the following limits if possible

   (a) Does not exist.

(b) $\lim_{(x,y) \to (0,0)} \frac{x(x^2 - y^2)}{(x^2 + y^2)} = 0$

(c) Does not exist.

(d) $\lim_{(x,y) \to (0,0)} x \sin\left(\frac{1}{x^2+y^2}\right) = 0$

(e) 1

(13) No

(15) Yes for first question. No for the second.

(18) Yes, the maximum exists.

(20) $\{(a,b,c) : a, b, c \text{ are integers.}\}$

(23) 1

Exercises 4.9

(1) Open

(2) Closed

(3) Open

(4) Closed

Exercises 4.11
Exercises 5.3

(1) Find the following limits if possible

   (a) $(1,1,1)$
   (b) $(1,1,1)$
   (c) $\left(8, 11, \frac{4}{5}\right)$
   (d) $(0,1,0)$

(2) $(0,8,4)$

(4) $t = 0$ at $(4,0,0)$

(5) (x,y,z) $=$ $(\sin 4, 4, 5)$ $+$ $t(2\cos 4, 4, 2)$

(6) (x,y,z) $=$ $(2, \sin 4, 3)$ $+$ $t(1, 4\cos 4, 1)$

(7) $(\sin 2, 4, \cos 4) + t(\cos 2, 4, -4\sin 4) = (x,y,z)$

(8) $(\cos 3, -2\sin 9, 1)$

(9) $(\cos 3, 6, 1)$.

(10) $\left(1, \frac{3}{5}, 1\right)$.

(15) $\mathbf{y}'(0) = -\mathbf{x}_0 n + n\mathbf{x}_1, \mathbf{y}'(1) = n\mathbf{x}_n - n\mathbf{x}_{n-1}$.

(16) $\mathbf{r}'(t) \times \mathbf{s}(t) \cdot \mathbf{p}(t) + \mathbf{r}(t) \times \mathbf{s}'(t) \cdot \mathbf{p}(t)$
$\quad + \mathbf{r}(t) \times \mathbf{s}(t) \cdot \mathbf{p}'(t)$

Exercises 5.5

(1) $\frac{3}{2} + \ln 2$

(2) $2\sqrt{3} - \frac{4}{3}\sqrt{2}$

(3) $\sqrt{10}$

(4) $4\pi^2$

(5) Find the work.

 (a) $\frac{5}{2}$

 (b) $\pi + \frac{1}{2}\pi^2 + 2$

 (c) $\frac{37}{6}$

 (d) 3

(6) The curve consists of straight line segments which go from $(0,0,0)$ to $(1,1,1)$ and finally to $(1,2,3)$. Find the work done if the force field is

 (a) 9

 (b) 21

 (c) $\sin 1 + 2 + \cos 2$

 (d) $3 + \sin 2$

(8) $\frac{1}{8}\pi^2$

(10) $\int_0^3 (t^2, t^2, t^2) \cdot (1,1,1)\, dt = 27$
$\int_0^3 (t^2, t^2, t^2) \cdot (1, 2t, 1)\, dt = \frac{117}{2}$

(11) $\int_0^3 (t, t^2, t^2) \cdot (1,1,1)\, dt = \frac{45}{2}$
$\int_0^3 (t, t^2, t^2) \cdot (1, 2t, 1)\, dt = 54$

(12) $\emptyset$

(13) $16\pi^2 - 16\pi$

(14) $\frac{4}{3}$

(15) 1

Exercises 6.3

(1) $\left(-t - \frac{8}{7}, -t + \frac{2}{7}, t\right), t \in \mathbb{R}$

(2) $z = -3 - 3\cos(t) - \sin(t),$
$x = \cos(t), y = \sin(t), t \in [0, 2\pi].$

(3) $x = 3\cos t, y = 1 - 2\cos(t) - \frac{3}{4}\sin(t),$
$z = \frac{3}{2}\sin(t), t \in [0, 2\pi].$

(4) $(1,2,1) + t(-2,2,3), t \in \mathbb{R}.$

(5) $x = t, y = \frac{5}{3} - 2t^2, z = -1 + 3t^2, t \in \mathbb{R}$

(6) $\frac{|\sin(x)|}{(1+\cos^2(x))^{3/2}}$

(7) $a_N = \frac{e^{2t}}{1+e^{2t}}$. It does not achieve a maximum.

(8) $\frac{|x'(t)y''(t) - y'(t)x''(t)|}{(x'(t)^2 + y'(t)^2)^{3/2}}$

(9) $a_T = 0, a_N = 9.$

(10) $a_N = 1, a_T = 0$

(11) $\kappa = \frac{\left(765 + 324t^2\right)^{1/2}}{(9+4t^2)^{3/2}}, a_N = $
$3\frac{\sqrt{(85+36t^2)}}{\sqrt{(4t^2+9)}}, a_T = \frac{4t}{\sqrt{(4t^2+9)}}$

(12) $\left(\left(x - \frac{\sqrt{2}}{2}\right), \left(y - \frac{\sqrt{2}}{2}\right), \left(z - \frac{\pi}{2}\right)\right)$ ·
$(2, -2, \sqrt{2}) = 0$

(13) $a_N = r\omega^2 = \frac{|v|^2}{r}.$

(16) $\sqrt{2}t$

(17) $4\sqrt{2}t$

(18) $\left(1 - e^{-10t}\right)e^{5t}$

(19) $\frac{|f''(x)|}{(1+f'(x)^2)^{3/2}}$

(20) $\left(-\frac{1}{2}\ln 2, \frac{1}{2}\sqrt{2}\right)$

(23) $\frac{9}{5}\sqrt{29}$

Exercises 7.3

(2) $z = x^2 + 3y^2, z = x^2 - y^2, z = 3x^2 + y^2$ in order from left to right.

(3) $z = x + y, z = x - y, z = x^2 - 3y^2$, in order from left to right.

(4) $z = (x+y)^2, z = x + y$, and $z = \sin(x+y)$ in order from left to right.

(5) Do the limits exist?

 (a) Does not exist.

 (b) -1

 (c) 1

    (d) 1

    (e) Does not exist.

    (f) Does not exist.

  (6) Find the limits.

    (a) 0

    (b) 0

    (c) Does not exist. Try the line $y = 0$ and the line $y = x$.

    (d) 0

  ## Exercises 7.5

  (1) $\frac{1}{11}\sqrt{11}$

  (2) $\frac{5}{6}(\cos 2)\sqrt{6} - \frac{1}{6}\sqrt{6}$

  (3) $-\frac{1}{6}\sqrt{3}$

  (5) $\frac{1}{2}\sqrt{21}$

  (6) $-\sqrt{(\sin 4 + 4(\cos 4))^2 + (8\cos(4))^2 + 4}$

  (7) $(4\sin(5) + 4\cos(5), 4\cos(5))$

  (8) $\left(2y, z^2(\cos xy)\,x, 0\right)^T$.

  (9) $\left(w, z\sin xy + zx(\cos xy)\,y, z^3\right)^T$

  (10) Find $\frac{\partial f}{\partial x}, \frac{\partial f}{\partial y}$, and $\frac{\partial f}{\partial z}$ for $f =$

    (a) $2xy^2z, 2x^2yz, x^2y^2$

    (b) $y, x, 2z$

    (c) $-(\sin yx)\,y, -(\sin yx)\,x, 2\left(\cos z^2\right)z$

    (d) $2\frac{x}{x^2+y^2+1}, 2\frac{y}{x^2+y^2+1}, e^z$

    (e) $(\cos xyz)\,yz - (\sin yx)\,y,$
        $(\cos xyz)\,xz - (\sin yx)\,x, (\cos xyz)\,yx$

  (11) Find $\frac{\partial f}{\partial x}, \frac{\partial f}{\partial y}$, and $\frac{\partial f}{\partial z}$ for $f =$

    (a) $x^2y + \cos(xy) + z^3y$
        $2xy - (\sin xy)\,y,$
        $x^2 - (\sin xy)\,x + z^3, 3z^2y$

    (b) $e^{x^2+y^2}z\sin(x+y)$
        $2xe^{x^2+y^2}z\sin(x+y) + e^{x^2+y^2}z\cos(x+y),$
        $2ye^{x^2+y^2}z\sin(x+y) + e^{x^2+y^2}z\cos(x+y), e^{x^2+y^2}\sin(x+y)$

    (c) $z^2\sin^3\left(e^{x^2+y^3}\right)$
        $6z^2\left(\cos\left(e^{x^2+y^3}\right)\right)xe^{x^2+y^3} - 6z^2\left(\cos^3\left(e^{x^2+y^3}\right)\right)xe^{x^2+y^3},$
        $9z^2\left(\cos\left(e^{x^2+y^3}\right)\right)y^2e^{x^2+y^3} - 9z^2\left(\cos^3\left(e^{x^2+y^3}\right)\right)y^2e^{x^2+y^3}$
        $2z\sin\left(e^{x^2+y^3}\right) - 2z\sin\left(e^{x^2+y^3}\right)\cos^2\left(e^{x^2+y^3}\right)$

(d) $x^2 \cos\left(\sin\left(\tan\left(z^2+y^2\right)\right)\right)$

$2x \cos\left(\sin\left(\cos\left(\tan\left(z^2+y^2\right)\right)\right)\right)$

$-2x^2 \left(\sin\left(\sin\left(\cos\left(\tan\left(z^2+y^2\right)\right)\right)\right)\right) \cdot$

$\cos\left(\tan\left(z^2+y^2\right)\right)\right) \frac{y}{\cos^2\left(z^2+y^2\right)}$

$-2x^2 \left(\sin\left(\sin\left(\cos\left(\tan\left(z^2+y^2\right)\right)\right)\right)\right) \cdot$

$\cos\left(\tan\left(z^2+y^2\right)\right)\right) \frac{z}{\cos^2\left(z^2+y^2\right)}$

(e) x^{y^2+z}

$x^{y^2+z-1}\left(y^2+z\right)$, $2x^{y^2+z}y\ln x$, $x^{y^2+z}\ln x$

(12) $\frac{7}{3}, \frac{13}{2}$

Exercises 7.8

(1) Find $f_x, f_y, f_z, f_{xy}, f_{yx}, f_{xz}, f_{zx}, f_{zy}, f_{yz}$ for the following. Verify the mixed partial derivatives are equal.

(a) $x^2y^3z^4 + \sin(xyz)$

$2xy^3z^4 + (\cos xyz)\,yz$, $3x^2y^2z^4 +$

$(\cos xyz)\,xz$,

$4x^2y^3z^3 + (\cos xyz)\,xy$,

$6xy^2z^4 - (\sin xyz)\,xz^2y + (\cos xyz)\,z$,

$6xy^2z^4 - (\sin xyz)\,xz^2y + (\cos xyz)\,z$,

$8xy^3z^3 - (\sin xyz)\,xy^2z + (\cos xyz)\,y$,

$8xy^3z^3 - (\sin xyz)\,xy^2z + (\cos xyz)\,y$,

$12x^2y^2z^3 - (\sin xyz)\,x^2zy + (\cos xyz)\,x$,

$12x^2y^2z^3 - (\sin xyz)\,x^2zy + (\cos xyz)\,x$

(b) $\sin(xyz) + x^2yz$

$(\cos xyz)\,yz + 2xyz$,

$(\cos xyz)\,xz + x^2z$,

$(\cos xyz)\,xy + x^2y$,

$-(\sin xyz)\,xz^2y + (\cos xyz)\,z + 2xz$,

$-(\sin xyz)\,xz^2y + (\cos xyz)\,z + 2xz$,

$-(\sin xyz)\,xy^2z + (\cos xyz)\,y +$

$2xy$,

$-(\sin xyz)\,xy^2z + (\cos xyz)\,y +$

$2xy$,

$-(\sin xyz)\,x^2zy + (\cos xyz)\,x +$

x^2,

$-(\sin xyz)\,x^2zy + (\cos xyz)\,x +$

x^2

(c) $z\ln\left|x^2+y^2+1\right|$

$2\frac{z}{x^2+y^2+1}x$, $2\frac{z}{x^2+y^2+1}y$,

$\ln\left|x^2+y^2+1\right|$, $-4zx\frac{y}{(x^2+y^2+1)^2}$

,

$-4zx\frac{y}{(x^2+y^2+1)^2}$,

$\frac{2}{x^2+y^2+1}x$, $\frac{2}{x^2+y^2+1}x$,

$\frac{2}{x^2+y^2+1}y$, $\frac{2}{x^2+y^2+1}y$

(d) $e^{x^2+y^2+z^2}$

$2xe^{x^2+y^2+z^2}$,

$2ye^{x^2+y^2+z^2}$, $2ze^{x^2+y^2+z^2}$,

$4xye^{x^2+y^2+z^2}$,

$4xye^{x^2+y^2+z^2}$, $4xze^{x^2+y^2+z^2}$,

$4xze^{x^2+y^2+z^2}$, $4zye^{x^2+y^2+z^2}$,

$4zye^{x^2+y^2+z^2}$

(e) $\tan(xyz)$

$\left(\sec^2(xyz)\right)zy$, $\left(\sec^2(xyz)\right)xz$,

$\left(\sec^2(xyz)\right)xy$,

$2(\tan xyz)\,xz^2y +$

$2(\tan^3 xyz)\,xz^2y + z +$

$z\tan^2 xyz$,

$2(\tan xyz)\,xz^2y +$

$2(\tan^3 xyz)\,xz^2y + z +$

$z \tan^2 xyz,$

$2 (\tan xyz) xy^2 z +$

$2 (\tan^3 xyz) xy^2 z \quad + \quad y \quad +$

$y \tan^2 xyz,$

$2 (\tan xyz) xy^2 z +$

$2 (\tan^3 xyz) xy^2 z \quad + \quad y \quad +$

$y \tan^2 xyz,$

$2 (\tan xyz) x^2 zy +$

$2 (\tan^3 xyz) x^2 zy \quad + \quad x \quad +$

$x \tan^2 xyz,$

$2 (\tan xyz) x^2 zy +$

$2 (\tan^3 xyz) x^2 zy \quad + \quad x \quad +$

$x \tan^2 xyz$

(3) $b = \dfrac{\left(\sum_{i=1}^{P} t_i^2\right) \sum_{i=1}^{P} x_i - \left(\sum_{i=1}^{P} t_i x_i\right) \sum_{i=1}^{P} t_i}{\left(\sum_{i=1}^{P} t_i^2\right) p - \left(\sum_{i=1}^{P} t_i\right)^2}$

$a = \dfrac{\left(\sum_{i=1}^{P} t_i x_i\right) p - \left(\sum_{i=1}^{P} t_i\right) \sum_{i=1}^{P} x_i}{\left(\sum_{i=1}^{P} t_i^2\right) p - \left(\sum_{i=1}^{P} t_i\right)^2}$

(7) Satisfies Laplace's equation?

(a) yes

(b) yes

(c) yes

(d) yes

(e) No

(f) yes

(g) No

Exercises 8.2

(1) $o(h)$?

(a) $o(h)$

(b) $o(h)$

(c) $o(h)$

(d) $o(h)$

(e) $o(h)$

(f) Not $o(h)$

(g) $o(h)$

(h) $o(h)$

(2) $o(\mathbf{v})$?

(a) $o(\mathbf{v})$

(b) $o(\mathbf{v})$

(c) Not $o(\mathbf{v})$

(d) $o(\mathbf{v})$

(e) $o(\mathbf{v})$

(f) $o(\mathbf{v})$

(g) $o(\mathbf{v})$

(3) $\mathbf{o}(\mathbf{v})$?

(a) $\mathbf{o}(\mathbf{v})$

(b) $\mathbf{o}(\mathbf{v})$

(c) $\mathbf{o}(\mathbf{v})$

(d) Not $\mathbf{o}(\mathbf{v})$

(e) $\mathbf{o}(\mathbf{v})$

(f) $\mathbf{o}(\mathbf{v})$

(g) $\mathbf{o}(\mathbf{v})$

Exercises 8.4

(1) Use the definition of the derivative to find the 1×1 matrix which is the derivative of the following functions.

(a) $(2t + 1)$.

(b) $3t^2$.

(c) $t \cos(t) + \sin(t)$.

(d) $\dfrac{2t}{1+t^2}$

(e) $2 |t|$.

(4) $(\sin(y), x \cos(y))$.

(5) $(2x \sin(y), x^2 \cos(y))$.

(6) $\begin{pmatrix} 2x & 1 \\ 0 & 2y \end{pmatrix}$

(7) $\begin{pmatrix} 2xy & x^2 \\ 1 & 2y \end{pmatrix}$

(9) $\begin{pmatrix} 2x \sin(y) & x^2 \cos(y) \\ 2x & 1 \end{pmatrix}$

(10) 3.333×10^{-3}

(11) $D(f\mathbf{g})(\mathbf{x})\mathbf{h} = (Df(\mathbf{x})\mathbf{h})\mathbf{g}(\mathbf{x}) \quad + \quad f(\mathbf{x}) D\mathbf{g}(\mathbf{x})\mathbf{h}$

(14) Let $f(x, y)$ be defined on $\mathbb{R}^2$ as follows. $f(x, x^2) = 1$ if $x \neq 0$. Define $f(0, 0) = 0$, and $f(x, y) = 0$ if $y \neq x^2$. Show that f is not continuous at $(0, 0)$ but that

$$\lim_{h \to 0} \frac{f(ha, hb) - f(0, 0)}{h} = 0$$

for (a, b) an arbitrary unit vector. Thus the directional derivative exists at $(0, 0)$ in every direction but f is not even continuous there. Regardless the value of the unit vector (a, b), for all h small enough, $bh \neq ah^2$. This is because ah^2 converges to 0 faster than bh. Therefore, for small enough h the difference quotient for the directional derivative equals 0 and so the directional derivative equals 0. However, this function is not continuous because it equals 1 along the curve (x, x^2) and in every open set containing $(0, 0)$, there are points of this sort. Therefore, along this curve the limit equals 0 but along the line $y = 0$, the limit equals 0.

Exercises 8.7

(1) Let $z = f(x_1, \cdots, x_n)$ be as given and let $x_i = g_i(t_1, \cdots, t_m)$ as given. Find $\frac{\partial z}{\partial t_i}$ which is indicated.

(a) $\frac{\partial z}{\partial t_1} = \frac{\partial z}{\partial x_1} \frac{\partial x_1}{\partial t_1} + \frac{\partial z}{\partial x_2} \frac{\partial x_2}{\partial t_1} = 3x_1^2 (\cos(t_1)) + 2t_1 t_2^2$

(b) $x_2^2 (t_2^2 t_3) + 4x_1 x_2 t_1 t_2$

(c) $x_2^2 (t_2^2 t_3) + 4x_1 x_2 t_1 t_2$

(d) $x_2^2 (t_1 t_2^2)$

(e) $2x_1 x_2^2 (2t_1 t_2 t_3) + 2x_1^2 x_2 (2t_1 t_2)$

(f) $(2x_1 x_2) t_1 + (x_1^2) t_1 t_4$

(g) $2x_3 (\cos(t_3))$

(h) $(2x_1 x_2) t_2 + x_1^2 t_2 t_4$

(2) $\frac{\partial z}{\partial x_1} = 2y_1 - \cos y_2 + \sec^2 y_3$,
$\frac{\partial z}{\partial x_2} = 2y_1 + (\cos y_2)(2x_2 + 1)$
$+ (\sec^2 y_3)(2x_2 + \cos x_2)$

(3) $(2y_1 + 1 + \cot^2 y_2 + \cos y_3 ,$
$(-1 - \cot^2 y_2)(2x_2 + 1) ,$
$2y_1, 2y_1 + \cos y_3 \cos x_4)$

(4) $(2y_1 - 2y_2 + \cos y_3 ,$
$1 + 2y_2 (2x_2 + 1) + 2(\cos y_3) x_2 ,$
$2y_1, 2y_1 + 1 + \cos y_3 \cos x_4)$

(5) $\begin{pmatrix} A & B \\ C & D \end{pmatrix}$ where

$A = 2y_1 - \cos y_2 + \sec^2 y_3$
$B = 2y_1 + 3(\cos y_2) x_2 +$
$(\sec^2 y_3)(2x_2 + \cos x_2)$
$C = 2y_1 y_2 - y_1^2 + 1$
$D = 2y_1 y_2 + 3y_1^2 x_2$
$+ 2x_2 + \cos x_2$

(6) $\begin{pmatrix} A & B & C & D \\ E & F & G & H \\ I & J & K & L \end{pmatrix}$ where

$\begin{pmatrix} A \\ E \\ I \end{pmatrix} = \begin{pmatrix} 2y_1 - \cos y_2 + \sec^2 y_3 \\ 2y_1 y_2 - y_1^2 + 1 \\ -2y_1 \sin y_1^2 - 3y_2^2 y_3 + y_2^2 \end{pmatrix}$

$\begin{pmatrix} B \\ F \\ J \end{pmatrix} = \begin{pmatrix} 2(\cos y_2) x_2 + \sec^2 y_3 \cos x_2 \\ 2y_1^2 x_2 + \cos x_2 \\ 6y_2^2 y_3 x_2 + y_2^2 \cos x_2 \end{pmatrix}$

$\begin{pmatrix} C \\ G \\ K \end{pmatrix} = \begin{pmatrix} \cos y_2 + 2(\sec^2 y_3) x_3 \\ y_1^2 + 2x_3 \\ 3y_2^2 y_3 + 2y_2^2 x_3 \end{pmatrix}$

$\begin{pmatrix} D \\ H \\ L \end{pmatrix} = \begin{pmatrix} 2y_1 \\ 2y_1 y_2 \\ -2y_1 \sin y_1^2 \end{pmatrix}$

(7) $D(f \circ g \circ h)(\mathbf{x}) = Df(g \circ h(\mathbf{x})) Dg(h(\mathbf{x})) Dh(\mathbf{x})$

Exercises 8.8

(2) $Df(0,0) = 0$.

$Df(x,y) = \left(y \sin \frac{1}{x} - \frac{1}{x} y \cos \frac{1}{x}, x \sin \frac{1}{x} \right)$

$Df(0,y)$ does not exist.

(4) $\begin{pmatrix} 2\sin 2 & \cos 2 & 27 \\ \cos(3) - 9\sin(1) & \cos(3) & 27\cos 1 \end{pmatrix}$

(5) $\begin{pmatrix} \tan y & x\sec^2(y) & 3z^2 \\ -\sin(x+y) - z^3 \sin x & -\sin(x+y) & 3z^2 \cos x \end{pmatrix}$

(6) $\begin{pmatrix} \sin y & x\cos y & 3z^2 \\ \cos(x+y) - z^3 \sin x & \cos(x+y) & 3z^2 \cos x \\ 5x^4 & 2y & 0 \end{pmatrix}$

(9) $21.991\,148\,6$

(10) $2\sin 1 - \cos 1 + 1$

(11) -1000

(12) -400

(13) $P' = \frac{3}{800}k$

(17) $u_{rr} + \frac{1}{r}u_r + \frac{1}{r^2}u_{\theta\theta}$

Exercises 8.11

(1) Find the gradient of $f =$

    (a) $(2,1,12)$

    (b) $(4\ln 2, 4\ln 2, \sin 1)$

    (c) $\left(2\ln 2, \frac{1}{2}, \frac{1}{2}, \frac{1}{2}, 1 \right)$

    (d) $(-\pi, -1, 3)$

    (e) $\left(\frac{1}{x+y^2}, 2\frac{y}{x+y^2}, \frac{1}{z} \right)$

    (f) $\left(\frac{y}{4}, 0, \ln(4) \right)$

(2) Find the directional derivatives of f at the indicated point in the direction, $\left(\frac{1}{2}, \frac{1}{2}, \frac{1}{\sqrt{2}} \right)$.

    (a) $\frac{3}{2} + \frac{3}{2}\sqrt{2}$

    (b) $4\ln 2 + \frac{1}{2}(\sin 1)\sqrt{2}$

    (c) $\frac{3}{2} + 3\sqrt{2}$

    (d) $\frac{1}{2} + \frac{1}{2}\sqrt{2}$

    (e) $\frac{1}{2} + \frac{1}{2}\sqrt{2}$

    (f) $\cos(\sin 1)\cos 1 + \frac{1}{2}\sqrt{2}$

(3) Find the directional derivatives of the given function at the indicated point in the indicated direction.

    (a) $\frac{2}{3}\sqrt{6}$

    (b) $\frac{2}{5}\sqrt{5}$

    (c) $\frac{1}{11}\sqrt{11}$

(4) Find the tangent plane to the indicated level surface at the indicated point.

    (a) $0 = 2x - 6 + y + 3z$

    (b) $(4\cos 1 + 4\ln 2)(x - 1)$
        $+ (2\cos 1 + 4\ln 2)(y - 1)$
        $+ (\sin 1)(z - 2) = 0$

    (c) $z = 2$

(5) $(1, 1, \sqrt{2}) + t(2, 2, 2\sqrt{2})$

(6) $\left(\frac{\sqrt{2}}{2}, \frac{\sqrt{2}}{2}, 1 \right) \times (\sqrt{2}, \sqrt{2}, 1)$
    $= \left(-\frac{1}{2}\sqrt{2}, \frac{1}{2}\sqrt{2}, 0 \right)$

(8) $(1, 1, \sqrt{2}) + t(\sqrt{2}, -2\sqrt{2}, 1)$.

(9) $(x'(t), y'(t), z'(t)) = \nabla f_1(x, y, z) \times \nabla f_2(x, y, z)$

Exercises 9.2

(1) Find the points where possible local minima or local maxima occur in the following functions.

    (a) $(1, 2)$.

    (b) $(1, 1)$.

    (c) $\left(0, \frac{1}{2} \right)$

(d) $\left(n\pi, (2m+1)\frac{\pi}{4}\right), n, m$ integers.

(e) $(1,1)$

(f) $(x,0),(1,y)$ for x,y arbitrary.

(2) $\frac{8}{9}\sqrt{3}$

(3) $\frac{8}{9}abc\sqrt{3}$

(4) $12,12$, and 12

(5) $x = \frac{1}{\sqrt{3}} = y = z$

(6) The numbers are $\left(\frac{2}{\sqrt{3}}, \frac{2}{\sqrt{3}}, \frac{2}{\sqrt{3}}\right)$

(7) $\theta = \pi/3$

(8) $x = 2\sqrt[3]{5}, y = 2\sqrt[3]{5},$
$z = \frac{40}{(2\sqrt[3]{5})^2} = 2\sqrt[3]{5}$

(9) 2

(10) $\left(0, -\frac{1}{2}, \frac{1}{2}\right)$

(11) $\left(\frac{13}{11}, \frac{26}{11}, \frac{20}{11}\right), \left(\frac{18}{11}, \frac{31}{11}, \frac{35}{11}\right)$

Exercises 9.4

(3) $\left(\frac{1}{2}, -\frac{21}{4}\right)$ local maximum,
$(0,-4),(1,-4)$ saddle point.

(4) $\left(\frac{1}{2}, -\frac{53}{12}\right)$ local maximum,
$(0,-4),(1,-4)$ saddle points.

(5) $\left(\frac{1}{2}, \frac{37}{20}\right)$ saddle,
$(0,2),(1,2)$ local minimum.

(6) $\left(\frac{1}{2}, -\frac{17}{8}\right)$ saddle, $(0,-2),$
$(1,-2)$ local minimum.

(7) $(-2,3,-5)$ saddle point.

(8) $(1,1,0)$ saddle point.

(9) $(2,1,1)$ local maximum.

(10) $(-1,-1,1)$ saddle point.

(12) Consider $y = x^4$,
$y = -x^4$, and $y = x^3$

(13) $\left(\frac{1}{2}, -\frac{9}{2}\right)$ minimum, $(0,-5)$,
$(1,-5)$ saddle.

(14) $\left(1, -\frac{11}{2}\right),(0,-5),$ saddle $(2,-5)$,
middle.

(15) $\left(\frac{3}{2}, \frac{27}{20}\right)$ saddle, $(0,0)$ minimum,
and $(3,0)$ minimum.

(16) $(2,0,2)$ a local minimum.

(17) $(-2,3,-5)$ is a saddle point.

(18) $(-3,4,-7)$ is a saddle point.

(19) $(5,-4,9)$ and it is a local minimum.

(20) All are saddle points.

(21) $\left(1, -3, -\frac{1}{3}\right)$ saddle.
$(2,-3,0),(0,-3,0)$ local max.

(22) All saddle points.

(23) $(0,-1,0),(4,-1,0)$ saddle.
$(2,-1,-12)$ local minimum.

Exercises 9.6

(1) $(1,1,1)$

(2) $(-3,3,-3)$

(3) No maximum and no minimum.

(4) $r^2/2$

(5) $\left(\frac{8}{5}, \frac{9}{5}\right)$

(6) $\left(\frac{1}{\sqrt{37}}, \frac{18}{\sqrt{37}}\right)$

(7) $\left(\frac{1}{\sqrt{10}}, \frac{9}{\sqrt{10}}, 0\right), \sqrt{10}$

(8) $\left(-\frac{1}{\sqrt{11}}, -\frac{1}{\sqrt{11}}, -\frac{1}{\sqrt{11}}\right)$ and minimum is $\frac{3}{\sqrt{11}}$

(9) $(2, 2\sqrt{2}), (2, -2\sqrt{2})$

(10) $(1/\sqrt{2}, 1), (-1/\sqrt{2}, -1)$

(11) There is no farthest point.

(12) $r = \sqrt[3]{18}, h = 2\sqrt[3]{18}$.

(13) $r = \frac{1}{12}\sqrt[3]{60}, h = \frac{432}{5}\sqrt[3]{60}$

(14) $\pm a/\sqrt{n}$,

(15) $(\pm 1/2, 0, 0)$

(16) $\left(\frac{3}{5} - \frac{2}{5}\sqrt{11}, \frac{6}{5} + \frac{1}{5}\sqrt{11}, 0\right)$
and $\left(\frac{3}{5} + \frac{2}{5}\sqrt{11}, \frac{6}{5} - \frac{1}{5}\sqrt{11}, 0\right)$

(17) Any point on the curve.

(18) $\left(0, -\frac{3}{2}, \frac{5}{2}\right)$

(20) $(3,3,4)$ and $(3,3,4)$

(21) $(0,-r), \left(\frac{\sqrt{3}r}{2}, \frac{r}{2}\right), \left(-\frac{\sqrt{3}r}{2}, \frac{r}{2}\right)$

(22) $\left(-\frac{1}{2} + \frac{1}{2}\sqrt{3}, -\frac{1}{2} + \frac{1}{2}\sqrt{3}\right.$
$\left. , \frac{1}{2}\left(\sqrt{3}-1\right)^2\right)$

(23) $-\frac{r^2}{3\sqrt{3}}$.

(24) Therefore, they all equal 8.

(25) $\left(\frac{4}{5}\left(\sqrt[4]{5}\right)^3, 4\sqrt[4]{5} - \frac{8}{5}\left(\sqrt[4]{5}\right)^3, 0\right),$
$\left(-\frac{4}{5}\left(\sqrt[4]{5}\right)^3, -4\sqrt[4]{5} + \frac{8}{5}\left(\sqrt[4]{5}\right)^3, 0\right)$

(26) $\left(\frac{1}{\sqrt{3}}, \frac{1}{\sqrt{3}}, \frac{1}{\sqrt{3}}\right)$

(27) $\left(\frac{4}{\sqrt{14}}, \frac{9}{\sqrt{14}}, \frac{1}{\sqrt{14}}\right)$

(28) $60, 30, 20, 15, 12$

(29) $x_1/x_n = 1$

(30) $2ab$

(33) $a = b = 1$.

Exercises 10.2

(1) $\int_0^4 \int_0^{3y} x \, dx \, dy = 96$

$\int_0^{12} \int_{x/3}^4 x \, dy \, dx = 96$

(2) $\int_0^3 \int_0^{3y} y \, dx \, dy = 27$

$\int_0^9 \int_{x/3}^3 y \, dy \, dx = 27$

(3) $\int_0^2 \int_0^{2y} (x + 1) \, dx \, dy = \frac{28}{3}$

$\int_0^4 \int_{x/2}^2 (x + 1) \, dy \, dx = \frac{28}{3}$

(4) $\int_0^3 \int_0^y \sin(x) \, dx \, dy = 3 - \sin 3$

$\int_0^3 \int_x^3 \sin(x) \, dy \, dx = 3 - \sin 3$

(5) $\int_0^1 \int_0^y \exp(y) \, dx \, dy = 1$

$\int_0^1 \int_x^1 \exp(y) \, dy \, dx = 1$

(6) $\int_0^1 \int_{y^2}^{3-2y} y \, dx \, dy = \frac{7}{12}$

$\int_0^1 \int_0^{\sqrt{x}} y \, dy \, dx +$

$\int_1^3 \int_0^{(3-x)/2} y \, dy \, dx = \frac{7}{12}$

(7) $\int_0^1 \int_{(1/4)x}^{\sqrt{x}} x \, dy \, dx +$

$\int_1^2 \int_{(1/4)x}^{(3-x)/2} x \, dy \, dx = \frac{49}{60}$

$\int_0^{1/2} \int_{y^2}^{4y} x \, dx \, dy + \int_{1/2}^1 \int_{y^2}^{3-2y} x \, dx \, dy = \frac{49}{60}$

(8) $\int_0^1 \int_x^{2x} (y + 1) \, dy \, dx +$

$\int_1^{3/2} \int_x^{3-x} (y + 1) \, dy \, dx = \frac{13}{8}$

$\int_0^{3/2} \int_{y/2}^y (1 + y) \, dx \, dy +$

$\int_{3/2}^2 \int_{y/2}^{3-y} (y + 1) \, dx \, dy = \frac{13}{8}$

(9) $\frac{16}{15}$

(10) $\int_0^{1/2} \int_x^{3x} (x + 1) \, dy \, dx$

$+ \int_{1/2}^1 \int_x^{2-x} (x + 1) \, dy \, dx = \frac{3}{4}$

$\int_0^1 \int_{y/3}^y (x + 1) \, dx \, dy$

$+ \int_1^{3/2} \int_{y/3}^{2-y} (x + 1) \, dx \, dy = \frac{3}{4}$

(11) $\int_0^{2/3} \int_x^{5x} 1 \, dy \, dx +$

$\int_{2/3}^{10/7} \int_x^{(10-5x)/2} 1 \, dy \, dx = \frac{40}{21}$

$\int_0^{10/7} \int_{y/5}^y 1 \, dx \, dy +$

$\int_{10/7}^{10/3} \int_{y/5}^{(10-2y)/5} 1 \, dx \, dy = \frac{40}{21}$

(12) $e^4 - 1$

(13) $\frac{4}{3}e^6 - \frac{4}{3}$

(14) $\frac{1}{2}$

(15) 1

(16) 2

(17) $\int_{-3}^3 \int_{-x}^x x^2 \, dy \, dx = 0$

$\int_3^0 \int_{-3}^{-y} x^2 \, dx \, dy + \int_0^{-3} \int_{-3}^y x^2 \, dx \, dy$

$+ \int_0^3 \int_y^3 x^2 \, dx \, dy + \int_{-3}^0 \int_{-y}^3 x^2 \, dx \, dy$

(18) $\int_{-2}^2 \int_{-x}^x x^2 \, dy \, dx = 0$

$\int_{-2}^0 \int_{-y}^2 x^2 \, dx \, dy + \int_0^2 \int_y^2 x^2 \, dx \, dy$

$+ \int_0^{-2} \int_{-2}^y x^2 \, dx \, dy + \int_2^0 \int_{-2}^{-y} x^2 \, dx \, dy$

Exercises 10.4

(1) Find the following iterated integrals.

(a) $\frac{64}{3}$

(b) $\frac{11}{120}$

(c) $-5\cos 3 + 10 \sin 3 - \frac{5}{2}$

(d) $\frac{1}{2}$

(e) $-\frac{124}{3}$

(f) -666

(g) -12

(h) -3π

(i) -8

(2) $\int_0^1 \int_y^1 \int_0^z f(x, y, z) \, dx \, dz \, dy,$

$\int_0^2 \int_{x/2}^1 \int_0^z f(x, y, z) \, dy \, dz \, dx,$

$\int_0^1 \left[\int_0^x \int_x^1 f(x, y, z) \, dz \, dy + \right.$

$\left. \int_x^1 \int_y^1 f(x, y, z) \, dz \, dy \right] dx,$

$\int_0^{1/2} \int_{y^2}^{2y} \int_0^{y+z} f(x, y, z) \, dx \, dz \, dy +$

$\int_{1/2}^1 \int_{y^2}^1 \int_0^{y+z} f(x, y, z) \, dx \, dz \, dy,$

$\int_0^3 \int_2^5 \int_5^7 f(x, y, z) \, dz \, dy \, dx$

(3) $\frac{10}{3}$

(4) $\frac{20}{3}$

(5) 10

(6) $\frac{23}{4}$

(7) 128

(8) $\frac{16}{3}$

(9) $\int_0^3 \int_0^{x^2} \int_0^{3-x} dy\, dz\, dx,$

$\int_0^9 \int_{\sqrt{z}}^3 \int_0^{3-x} dy\, dx\, dz,$

$\int_0^9 \int_0^{3-\sqrt{z}} \int_{\sqrt{z}}^{3-y} dx\, dy\, dz,$

$\int_0^3 \int_0^{3-y} \int_0^{x^2} dz\, dx\, dy,$

$\int_0^3 \int_0^{(3-y)^2} \int_{\sqrt{z}}^{3-y} dx\, dz\, dy$

(10) $\frac{621}{32}$

(11) $\frac{135}{8}$

(12) $\frac{14}{3}$

(13) $\frac{250}{3}$

(14) 54

Exercises 10.5

(1) $\frac{1024}{3}$

(2) 144

(3) 20π

(4) $\frac{1}{2}\cos 1 - \frac{1}{2}\cos 9$

(5) $-\frac{3}{4}e^{25} - \frac{33}{4} + \frac{3}{4}e^{36}$

(6) $-e^{16} - 20 + e^{36}$

(7) $-2\sin 1 \cos 5 + 2\cos 1 \sin 5$
$+2 - 2\sin 5$

(8) 2

(9) $\frac{64}{15}$

(10) 8

(11) $\frac{10}{3}$

(12) $-4\sin 1\cos 3 + 4\cos 1\sin 3 + 4 - 4\sin 3$

(13) $-5\sin 2\cos 6 + 5\cos 2\sin 6 + 10 - 5\sin 6$

(14) $\frac{4}{9}$

(15) $\frac{7}{8}$

(16) $\int_{1-\sqrt{2}}^{1+\sqrt{2}} \int_{-\sqrt{2-(x-1)^2}}^{\sqrt{2-(x-1)^2}} \int_{x^2+y^2}^{2\sqrt{x+y}} dz\, dy\, dx$

(17) $8\pi.$

(18) $\frac{1}{4}$

(19) $\frac{1}{6}$

(20) π

Exercises 11.2

(1) $\frac{9}{2}\pi$

(2) $\frac{1}{2}\pi$

(3) $\frac{7}{4}\pi$

(4) 12π

(5) $\frac{19}{4}\pi$

(6) π

(7) $\frac{11}{4}\pi$

(8) $m_x = \int_0^{2\pi} \int_0^{2+\sin(\theta)} r\cos(\theta)$
$(2 - r\cos(\theta))\, r\, dr\, d\theta = -\frac{177}{32}\pi$
$m_y = \int_0^{2\pi} \int_0^{2+\sin(\theta)} r\sin(\theta)$
$(2 - r\cos(\theta))\, r\, dr\, d\theta = \frac{17}{2}\pi$

(9) $\frac{89}{32}\pi = m_x,\ 0 = m_y$

(10) $\int_0^{2\pi} \int_0^{1+\cos(\theta)} (1 + r\cos\theta + r\sin\theta)$
$(r\cos\theta)\, r\, dr\, d\theta = \frac{89}{32}\pi = m_x$
$\int_0^{2\pi} \int_0^{1+\cos(\theta)} (1 + r\cos\theta + r\sin\theta)$
$(r\sin\theta)\, r\, dr\, d\theta = \frac{21}{32}\pi = m_y$

(11) $\frac{1}{5}\pi e^{125} - \frac{1}{5}\pi$

(12) $\frac{1}{9}\pi \sin 144$

(13) $\frac{1}{2}\int_a^b \left(\frac{f(\theta)^2}{2} - \frac{g(\theta)^2}{2} \right) d\theta$

(14) $-\frac{3}{8}\pi\cos 8 + \frac{3}{8}\pi$

(15) $\frac{1}{4}\pi e^8 - \frac{1}{4}\pi$

(16) $\frac{3}{4}\ln 2$

(17) $\frac{1}{2}\sqrt{3}$

(18) $\frac{8}{3}\pi$

(19) 4π

Exercises 11.4

(1) $\frac{256}{9}$

(2) $\frac{32}{3}\pi$

(3) $\frac{9}{2}\pi$

(4) $\frac{81}{2}\pi$

(5) $\frac{1000}{3}\pi$

(6) $\frac{7}{6}\pi$

(7) $\frac{20}{3}\pi\sqrt{5}$

(8) $\frac{64}{3}\pi$

(9) $\int_0^{2\pi} \int_0^{\sqrt{77}} \int_{-\sqrt{9-r^2}}^{2} \sqrt{r^2 + z^2}\, dz\, dr\, d\theta$
$+ \int_0^{2\pi} \int_{\sqrt{77}}^9 \int_{-\sqrt{9-r^2}}^{\sqrt{9-r^2}} \sqrt{r^2 + z^2}\, dz\, dr\, d\theta$

(10) $\frac{56}{3}\pi\sqrt{3} + \frac{56}{3}\sqrt{2}\pi$

(13) Convert and compute.

(a) $\frac{64}{15}$

(b) 0

(c) $-\frac{1}{24}(\sin 1)\pi^3 + \frac{1}{8}\pi^2$

(d) $\frac{4}{3}\pi a^3$

(e) $-\frac{4}{3}\pi\left(3\sqrt{3}-8\right)$

(14) Convert and compute.

(a) $\frac{4}{3}\pi a^3$

(b) $\frac{2}{3}\pi$

(c) $-\frac{8}{3}\pi\sqrt{2} + \frac{16}{3}\pi$

(d) $\frac{5}{3}\pi$

(e) $\frac{4}{3}\pi\left(8-3\sqrt{3}\right)$

Exercises 11.6

(1) $\rho^2 \sin(\phi)\,d\rho d\theta d\phi$

(2) $\frac{12}{7}$

(3) $\frac{16}{15}$

(4) $\frac{21}{100}$

(5) $-4\ln 2 + 4\ln 3$

(6) $\frac{7}{3}\ln 5 - \frac{14}{3}\ln 2$

(7) $\frac{1}{2}|ad-bc|$

(8) $\frac{8\pi}{3}$

(9) 42

(10) 12

(11) 160

(12) $\frac{4}{3}\pi R^3$

Exercises 11.8

(1) $z_c = 4/3$.

(2) $z_c = 2$.

(3) Mass $= 16\pi, x_c = y_c = 0, z_c = 14/3$.

(4) mass $= \frac{32}{3}\pi, z_c = 11/2, x_c = y_c = 0$.

(5) Mass $= \frac{2048}{45}, x_c = y_c = z_c = 0$.

(6) $x_c = y_c = 0, z_c = 1$

(7) $z_c = \frac{1}{20}, x_c = 0, y_c = \frac{9}{20}$.

(8) $\frac{243}{2}\pi$

(9) $\frac{1215}{2}\pi$

(10) $\frac{2}{5}MR^2$

(11) $\frac{4}{9}R^2 M$

(12) $\frac{1}{2}MR^2$

(13) The ball wins.

(14) $-\frac{256}{5}\frac{\pi}{64\pi\sqrt{3}+64\sqrt{2\pi}}$

(15) $512\pi\sqrt{3} + \frac{5120}{9}\sqrt{2\pi}$

(16) $\emptyset$

(17) $Ma^2 + \frac{2}{5}MR^2$

(19) $\frac{1}{12}l^2 M$

(20) $\frac{1}{3}l^2 M$

(21) $\frac{3}{4}h$

(22) $\frac{1}{10}\pi h^5 \frac{1-2\cos^2\alpha + \cos^4\alpha}{\cos^4\alpha}$

(23) $\frac{3}{8}R$ for all three.$=$ The others are the same.

Exercises 12.3

(1) $y = \frac{11}{10} - \frac{6}{5}z, x = \frac{1}{5} - \frac{2}{5}z$

(2) $x = \cos(t), y = \sin(t),$
$z = 1 - 3\cos(t) - \sin(t). \ t \in [0, 2\pi].$

(3) $x = 4\cos(t), z = 2\sin(t),$
$y = \frac{4-12\cos(t) - 8\sin(t)}{2}$

(4) $(1,3,1) + t(3,-2,-2), t \in \mathbb{R}.$

(5) $x = t, y = -3t + 5 - 6t^2,$
$z = 4t - 6 + 9t^2$

(6) 8π

(7) $4\pi h^2\sqrt{17}$

(9) $\frac{9}{2}\pi$

(10) π

(11) $4\sqrt{57}$

(12) $6h\pi$

(13) 8π

(14) $16\pi^2$

(15) $8\pi^2$

(16) $24\pi^2$

(17) $-2(\cos c)\pi R^2 + 2\pi R^2$

(18) $\int_{-\pi/2}^{\pi/2}\int_0^{\cos\theta} \frac{r^2}{\sqrt{1+4r^2}} r dr d\theta$

(19) $\frac{20}{3}\pi$

(20) $\int_0^{2\pi}\int_{\sqrt{5}}^{\sqrt{8}} \sqrt{9-r^2} r dr d\theta = \frac{14}{3}\pi$

Exercises 13.3

(1) Find div $\mathbf{f}$ and curl $\mathbf{f}$.

(a) $yz + \frac{1}{y} + 1$
$\left(0, xy - 2\left(\cos x^2\right)x, 2x + \frac{1}{x} - xz\right)$

(b) $\cos x + \cos y + \cos z, (0,0,0)$

(c) $f'(x) + g'(y) + h'(z), (0,0,0)$

(d) $3, (0,0,0)$

(e) $2x - \sin z, (0,0,0)$

(f) $h_z(y,z), (h_y - g_z, f_z, g_x - f_y)$

(2) $(\nabla \times \nabla\phi)_i = \varepsilon_{ijk}\partial_j (\partial_k\phi) = 0$

(3) $(\text{curl}(u\nabla v))_i = \varepsilon_{ijk}\partial_j (u\partial_k v)$
$= \varepsilon_{ijk}\partial_j u\partial_k v + u\varepsilon_{ijk}\partial_j\partial_k u$
$= \varepsilon_{ijk}\partial_j u\partial_k v = (\nabla u \times \nabla v)_i.$

(4) $2f_j\partial_i g_j\mathbf{e}_i$

(5) $i(2v_1) + j(2v_2) + k(2v_3)$

(6) $\partial_i(v\partial_i u) = \nabla v \cdot \nabla u + v\Delta u$

(8) It equals $\partial_i(\partial_i(uv))$
$= \partial_i((\partial_i u)v + u(\partial_i v))$
$= v\nabla^2 u + 2\nabla u \cdot \nabla v + u\nabla^2 v$

(14) Follows from $\nabla \times \nabla\phi = \mathbf{0}$.

Exercises 13.6

(2) $\frac{\partial c}{\partial t} = \nabla \cdot (k\nabla c) + f$

(3) $\frac{\partial c}{\partial t} = \sum_i \frac{\partial}{\partial x_i}\left(\sum_j K_{ij}\frac{\partial c}{\partial x_j}\right) + f$

(9) Both give $45/2$.

(11) 144

(12) 6π

(13) $\frac{512}{3}\pi$

(17) $\left(\int_U \nabla \times \mathbf{F}dx\right)_i = \int_U \varepsilon_{ijk}\partial_j F_k dx$
$= \int_{\partial U} \varepsilon_{ijk}n_j F_k = \left(\int_{\partial U} \mathbf{n} \times \mathbf{F}dS\right)_i$

Exercises 14.2

(1) $\int_{-\pi/2}^{\pi/2} \int_0^{2\cos\theta} r\cos(\theta)\sqrt{2}rdrd\theta = \pi\sqrt{2}$

(2) 4π

(3) 2π

(4) $8\pi - 16$

(6) $62.5\,(4\pi a^3)$

(7) $-\frac{7}{3}\pi$

(8) $\frac{81}{2}\pi$

(9) $\frac{27}{2}\pi$

(10) 40π

(11) $-\frac{5}{12}\pi\sqrt{2}a^3 + \frac{3}{8}\pi\sqrt{3}a^3$

(12) $\frac{5}{12}\pi a^3 + \frac{3}{4}a^2\pi$

(13) 0

(14) $\frac{2}{3}\pi a^3$

(15) 0

(16) -2π

(17) -1

(18) -1

(19) 2π. The reason there is no contradiction for Green's theorem is that the functions P, Q are not C^1. They are not even defined at the origin.

(20) $-\frac{1}{2}$

Exercises 14.5

(1) $f(x,y,z) = x^2y^3\sin(z^4) + y + z + C$

(2) $f(x,y,z) = xy^2 + xz + x^2y^3\sin z$

(3) Not conservative.

(4) Find scalar potentials for the following vector fields if it is possible to do so. If it is not possible to do so, explain why. These problems are just like the three above.

 (a) $xy^2 + z^2 + y\sin z = f(x,y,z)$

 (b) $f(x,y,z) = z^2 + z\sin x^2\cos y^2 + z\cos x^2\sin y^2$

 (c) $F(x) + G(y) + H(z)$ where $F' = f, G' = g, H' = h$.

 (d) Not conservative.

 (e) $xz + z^3 + \ln(x^2 + y^2 + 1) + C$

(5) -2π

(7) 4

(8) 0

(9) $6\sin 4 + 8\cos 4 - 4\cos 2$

(10) $\frac{53}{6} - \frac{1}{2}\sin 9 + \frac{1}{2}\sin 4$

(11) $\frac{9}{2}\sin 2 - \frac{3}{2}\sin 3$
$-4\cos 2 - 6\sin 1 + 2\cos 1$

(12) 6π

(14) This happens because each of the surface integrals equals a line integral over C where the two line integrals are taken over a curve with opposite orientations. Hence the two add to 0.

(15) $\text{curl}\,(\psi\nabla\phi)_i = \varepsilon_{ijk}\partial_j\,(\psi\partial_k\phi)$
$= \varepsilon_{ijk}\,(\partial_j\psi)\,(\partial_k\phi) + \psi\varepsilon_{ijk}\partial_j\,(\partial_k\phi)$
$= \varepsilon_{ijk}\,(\partial_j\psi)\,(\partial_k\phi) = (\nabla\psi\times\nabla\phi)_i$

(16) $\alpha u^{\alpha-1}\Delta u + \alpha\,(\alpha-1)\,u^{\alpha-2}\,|\nabla u|^2$

(17) $6a^2\pi + 2\pi^2$

(18) $\frac{3}{8}\pi$

(20) Not simply connected.

(21) $\text{curl}\,(\mathbf{F})\cdot\mathbf{n} = \lim_{r\to 0}\frac{1}{\pi r^2}\int_{\partial D_r}\mathbf{F}\cdot d\mathbf{R}$

(22) -8π

(23) -3π.

(24) 3π.

(25) 0.

(26) 16π

(27) 0

(31) Simply connected?

   (a) No

   (b) No

   (c) Yes

   (d) Yes

   (e) No

   (f) No.

(32) $\sum_{i=1}^{n}\frac{1}{2}\,(x_{i+1}+x_i)\,(y_{i+1}-y_i)$

Index